Beck and Braithwaite's

INVERTEBRATE ZOOLOGY

A Laboratory Manual

Beck and Braithwaite's

INVERTEBRATE ZOOLOGY

A Laboratory Manual

4th
Edition

Robert L. Wallace
Ripon College

Walter K. Taylor
University of Central Florida

James R. Litton, Jr.

Prentice Hall
Upper Saddle River, NJ 07458

A Simon & Schuster Company
Upper Saddle River, New Jersey 07458

Printed in the United States of America
15 14 13 12 11 10 9 8

ISBN 0-02-307763-8

Prentice-Hall International (UK) Limited, *London*
Prentice-Hall of Australia Pty. Limited, *Sydney*
Prentice-Hall Canada Inc., *Toronto*
Prentice-Hall Hispanoamericana, S.A., *Mexico*
Prentice-Hall of India Private Limited, *New Delhi*
Prentice-Hall of Japan, Inc., *Tokyo*
Simon & Schuster Asia Pte. Ltd., *Singapore*
Editora Prentice-Hall do Brasil, Ltda., *Rio de Janeiro*

PREFACE TO THE FOURTH EDITION

Invertebrate zoology is an enormous field. At least 96 percent of all animal species lack backbones, but this assessment may be too conservative. Recent estimates suggest that more than 10 million insect species in the Amazonian forests have yet to be described. If this prediction is true, the numerical dominance of invertebrate species will be approximately 99 percent. Nevertheless, humans seem preoccupied with organisms possessing vertebral columns, especially if the animal resembles, in any way, those cuddly toys made for young children. Invertebrates, on the other hand, are often viewed with disgust, evoking unwarranted fears and horrific screams of terror when encountered in a disused corner of a basement, in a half-eaten apple, or crawling on one's body. We do not mean to imply that invertebrates do not cause human suffering or seriously damage agricultural products. They do, and it is for these reasons and because as a group the invertebrates possess such diverse and rich biologies that they are worthy of intensive study.

We dedicate this manual to our teachers who introduced us to the invertebrate world, our families who put up with our interests, and new students of invertebrates who will discover additional wonders about animals without backbones.

This manual began as a revision of Beck and Braithwaite's *Invertebrate Zoology: Laboratory Workbook* and therefore fulfills Dr. L. F. Braithwaite's dedication appearing in the third edition that the comprehensive laboratory study guide of invertebrate zoology begun by Dr. D Elden Beck should not become extinct. This was our goal too. Beck and Braithwaite's manual has been an excellent source of instruction for nearly three generations of students. Our purpose in undertaking a revision was to retain the same pedagogical approach of the manual while updating the text to take into account advances of the discipline and to improve coverage of some important groups. We believe that a study based on morphology is an excellent way to achieve a comprehensive understanding of invertebrates. We do not suppose that this is the only way in which invertebrates may be studied (for example, one may do a study of systems across taxa, the functional approach). However, for the student who has had a course in introductory zoology or general biology, the approach of this manual will provide a solid conceptual framework for advanced work on behavior, ecology, physiology, and related subjects.

Although numerous changes are evident in this edition, we have attempted to keep the original intent of the manual: "To excite the interest of the student to get acquainted with the world of invertebrates about them wherever they may live. It isn't necessary to travel to far off forest or shore to find exciting animals to study. Wherever one lives, those animals usually listed as 'common' are often the least known and frequently make the best subjects for study and research" (Beck and Braithwaite 1968: i). Both live and preserved organisms are used throughout the manual and it is designed so that instructors may omit sections that are inappropriate to their particular interests and/or to substitute other studies.

The approach we took in writing this laboratory manual was to develop exercises using representatives of invertebrate phyla for which specimens are readily available from commercial suppliers. Therefore, some of the smaller phyla are not covered here because we have not found commercial material to support meaningful laboratory work. However, some groups have received expanded treatment (for example, insects). The use of fossils, parasitic forms, and larvae also has been expanded.

Several new features have been included in this manual that were not present in the workbook of Beck and Braithwaite. Most of the artwork has been redrawn and many new figures and some photographs have been added. We have also added simple figures of the geological time scale for phyla that have a significant fossil record. These figures do not indicate diversity throughout geologic time, but they will help students appreciate the enormous length of time some phyla have been in existence. We also have expanded greatly the number of figures of fossils, parasitic forms, and larvae. A simple phylogenetic descriptive insert has been added before each new group of invertebrates is studied (see Taxonomic Considerations and Evolution of the Investebrates on page ix). Three other new features of the laboratory manual are a simple pronuncation guide, an etymon for each phylum name, and use of **boldface type** for important terms. We believe that they will make the learning process easier for students.

Acknowledgments

Many people helped us in the production of the Laboratory Manual and the Instructor's Guide. To them we offer our most sincere thanks, as they have added much to these works by their efforts. We are especially grateful to Michele Johnson, who did all the artwork. Without her dedication, this revision would not have been completed. We also thank Jan Richardson, who offered her ideas, enthusiasm, and considerable logistic help in the beginning of the project and Paul Conant, President of Triarch Inc. of Ripon, Wisconsin, who prepared for us several special order slides. Thanks also are due to over 150 instructors from around the country who took the time to fill out a questionnaire concerning their approach to laboratory instruction of invertebrate zoology and the types of features they wished to see in the fourth edition of the laboratory manual. Those who acted as reviewers of the manuscript or who otherwise contributed to the fourth edition deserve mention: John A. Allen, University Marine Biological Station Millport, Scotland (Bivalvia); Kenneth J. Boss, Museum of Comparative Zoology, Harvard University (Polyplacophora, Monoplacophora, Aplacophora, Scaphopoda); C. Bradford Calloway, Museum of Comparative Zoology, Harvard University (Priapulida); Ralph O. Brinkhurst, Ocean Ecology Laboratory, British Columbia (Oligochaeta); William S. Brooks, Ripon College (general editing); John L. Cisne, Cornell University (Trilobita); G. Arthur Cooper, Smithsonian Institution (Brachiopoda); Edward B. Cutler, Utica College, Syracuse University (Pogonophora, Sipuncula); William C. Dewel, Appalachian State University (Cnidaria); W. T. Edmondson, University of Washington (Rotifera); Stan Edwards, South Australian Museum (Echiura); Leslie L. Ellis, University of Central Florida (Insecta); Christian C. Emig, Station Marine D' Endoume et Centre D'Oceanographie (Phoronida); K. Fauchald, Smithsonian Institution (Polychaeta); William McKay Fender, Soil Biology Associates, McMinnville, Oregon (Oligochaeta); Merrill W. Foster, Bradley University (Brachiopoda); Tom Frost, University of Wisconsin-Madison (Porifera); John J. Gilbert, Dartmouth College (Porifera, Rotifera); K. Herrmann, Universität Erlangen-Nürnberg (Phoronida); Robert P. Higgins, National Museum of Natural History (Kinorhyncha); Meg Hummon, Ohio University (Gastrotricha); William Hummon, Ohio University (Gastrotricha); Alan Kohn, University of Washington (Gastropoda): Denis H. Lynn, University of Guelph (Protozoa); A. R. Maggenti, University of California-Davis (Nematoda); Janice Moore, Colorado State University (Acanthocephala); Brian Morton, University of Hong Kong (Bivalvia); Brent Nickol, University of Nebraska (Acanthocephala); Claus Nielsen, Zoologisk Museum, Denmark (Entoprocta); David L. Pawson, National Museum of Natural History (Echinodermata); Marian H. Pettibone, National Museum of Natural History (Polychaeta); Leland W. Pollock, Drew University (Tardigrada); Mary E. Rice, National Museum of Natural History (Sipuncula); Reinhard M. Rieger, Institut für Zoologie der Universität Innsbruck

(Turbellaria); Edward E. Ruppert, Clemson University (Gnathostomulida); J. S. Ryland, University College of Swansea (Bryozoa); Steward C. Schell, University of Idaho (Trematoda); Gerald Schmidt, University of Northern Colorado (Acanthocephala, Cestoda); J. Teague Self, The University of Oklahoma (Pentastomida); Tracy Simpson, University of Hartford (Porifera); R. W. Sims, British Museum (Natural History) (Oligochaeta); Eve Southward, Marine Biological Laboratory, Plymouth, UK (Progonophora); Peter L. Starkweather, University of Nevada-Las Vegas (Rotifera); Wolgang E. Sterrer, Bermuda Biological Station (Gnathostomulida); Margaret E. Stevens, Ripon College (general editing, artwork consultant); Julia Stuart, Trentham, New Zealand (Hirudinea); Sidney L. Tamm, Boston University, Marine Biology Laboratory, Woods Hole (Ctenophora): Bryn H. Tracy, Carolina Power and Light, New Hill, North Carolina (Entoprocta); Jean C. Tryon, Ripon College (Porifera, Gnathostomulida, Acanthocephala, Gastrotricha, Tardigrada, Pentastomida); Seth Tyler, University of Maine (Turbellaria); James W. Valentine, University of California (Phoronida); Gilbert L. Voss, Rosentiel School of Marine and Atmospheric Science (Cephalopoda); and Craig Williamson, Lehigh University (Porifera). Thanks also are due to three anonymous reviewers who read and improved the manuscript. However, we take responsibility for any errors that remain and welcome constructive criticism from instructors and students of invertebrate zoology.

R. L. W.
W. K. T.
J. R. L.

TAXONOMIC CONSIDERATIONS AND EVOLUTION OF THE INVERTEBRATES

This laboratory manual outlines procedures for the study of representatives of invertebrate phyla for which specimens are readily available from commercial suppliers. The way we chose to arrange the phyla in this manual generally reflects that of most of the current textbooks of invertebrate zoology and may be used to support any one of them. In the third edition of Beck and Braithwaite's workbook the authors offered a taxonomic list for the classification of invertebrates. They intended it as a workable outline that would be reviewed by the student before each taxon was studied in the laboratory. We believe that this list was useful, but lacked the impact necessary to achieve its purpose as a teaching aid. Therefore, we have replaced the taxonomic list with a series of Phylogenetic Descriptions of the major groups. Each time organisms with a distinct body plan are considered (for example, acoelomate worms, pseudocoelomates, mollusks, lophophorates) a new phylogenetic description is presented. Instructors may have their students use these descriptions as given or they may wish to augment them, perhaps by providing a phylogenetic tree to illustrate putative relationships among the invertebrates. No phylogenetic tree has been offered here, as we believe that consideration of the various evolutionary trees that have been proposed in the literature should be left to instructors. We hope that the addition of the phylogenetic descriptions will provide a structured framework so that students do not become overwhelmed by the tremendous diversity of invertebrates.

CONTENTS

Contents

EXERCISE 1

Subkingdom Protozoa

Over 60,000 species of flagellates, amebas (or amoebas), opalinids, spore formers, and ciliates, often called collectively the protozoa, represent some of the most fascinating organisms to both novice and experienced microscopists. The tremendous number of individuals found in one drop of water can be overwhelming. Until recently, these mostly microscopic and primarily unicellular organisms were grouped in a single phylum, Protozoa (PRO-to-ZO-a; G., *proto,* first + G., *zoon,* animal). Most protozoologists now agree that this grouping constituted a heterogeneous assemblage of distantly related forms. Currently, some biologists place the protozoans within the subkingdom Protozoa (kingdom **Animalia**) containing at least six, probably polyphyletic phyla (Levine et al. 1980). Others prefer the term *protists* instead of *protozoa* and follow Whittaker (1969) in grouping the subkingdom under the kingdom **Protista.** Regardless of which grouping is followed, the protozoans play important roles in the web of life and are fascinating and challenging organisms to study.

Few places on the earth are devoid of protozoans. Free-living forms occur in aquatic habitats, moist soils, and decaying organic matter. Many protozoans are important parasites of plants, animals, and humans. Nearly one-half of the 60,000 protozoan species are fossils; some have been in existence at least since the **Precambrian** (Fig. 1.1).

The basic body plan is the single eukaryotic cell representing a functionally complete organism that performs all physiological processes found in multicellular animals, the **Metazoa** (G., *meta,* after or between). Some biologists prefer to call protozoans "acellular" (G., *a,* without) organisms instead of "unicellular," to emphasize the complete, functional organismal viewpoint.

Protozoan organelles tend to be more specialized than those found in the typical cell of metazoans. Some of the common organelles (together with their functions) are as follows: **food vacuole** or **phagosome** (digestion); **contractile vacuole** or **water expulsion vesicle** (water regulation); **myoneme** (contractile); **paraflagellar swelling** and **stigma** (sensory); **extrusome** (food getting and defense); and **pseudopodium, flagellum,** and **cilium** (food getting and locomotion).

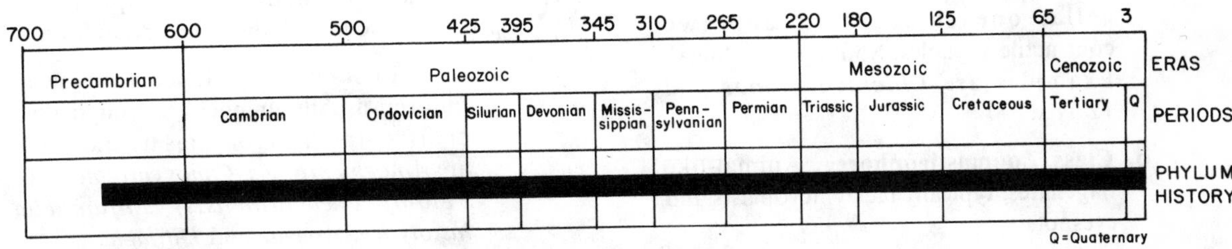

Figure 1.1. Geologic history of the subkingdom Protozoa.

Some species are colonial and can be seen with the unaided eye. However, there are no tissues, organs, or germ layers. Digestion is intracellular, occurring within food vacuoles. Respiration and excretion are accomplished by diffusion across the cell membrane. The cell covering may be an unmodified or highly modified membrane. Reproduction is by **budding, fission, conjugation,** or **syngamy**. The motile form is often called a **trophozoite** (trophont), whereas the nonmotile form is called a **cyst.**

Selected representatives of the major phyla will be studied in the following exercises. Your instructor may provide additional examples for study.

Classification

I. **Phylum Sarcomastigophora**. Monomorphic nucleus; reproduction by sexual means, when present, is by syngamy; flagellum, cilium, and pseudopodium locomotory organelles.

A. **Subphylum Mastigophora.** Flagella typically on trophozoite; reproduction usually interkinetal (symmetrogenic) binary fission.

a. **Class Phytomastigophorea.** Plantlike flagellates; typically with chloroplasts and eyespots (stigmas). Botanists call these flagellates algae.

1. **Order Dinoflagellida.** Two flagella; body with transverse and ventral grooves; body typically covered with cellulosic plates. Examples are *Glenodinium, Gymnodinium,* and *Ceratium.*
2. **Order Euglenida.** One to two flagella that arise anteriorly within the reservoir; chloroplast-containing and colorless forms present. Examples are *Euglena* and *Peranema.*
3. **Order Volvocida**. Usually with two flagella, one chloroplast, and two contractile vacuoles. Solitary or colonial. Examples are *Chlamydomonas* and *Volvox.*

b. **Class Zoomastigophorea.** Animal-like flagellates; typically lack chloroplasts and eyespots.

1. **Order Choanoflagellida**. Single anterior flagellum encircled by funnel-shaped collar; solitary or colonial. A choanoflagellate is similar to the choanocyte (flagellate cell) of sponges, This cell type is considered in exercise 2. Examples are *Monosiga, Codosiga,* and *Proterospongia.*
2. **Order Trichomonadida**. Parasitic in invertebrates and vertebrates; trophozoite typically with four to six flagella; cysts usually lacking. Examples are *Trichomonas* and *Tritrichomonas.*
3. **Order Diplomonadida.** Parasitic primarily in alimentary canal of host; trophozoite with one to two karyomastigonts (e.g., complex of flagellum, nucleus, and associated organelles), each with one to four flagella and associated organs. Examples are *Giardia* and *Hexamita.*
4. **Order Hypermastigida**. Intestinal inhabitants of termites, cockroaches, and woodroaches, many flagella. Example is *Trichonympha.*
5. **Order Kinetoplastida**. Parasitic, endocommensal, and free living; cells with one or two flagella; kinetoplast largest extranuclear repository of DNA for any cell type; life cycle includes vertebrates and invertebrate hosts. Examples are *Leishmania* and *Trypanosoma.*

B. **Subphylum Opalinata.** Numerous cilia in oblique rows over entire body; no cytostome or cytopharynx; monomorphic nuclei; reproduction usually interkinetal (symmetrogenic) binary fission. Example is *Opalina.*

C. **Subphylum Sarcodina.** Pseudopodia typically on trophozoite; body naked or with test or internal skeleton; reproduction primarily by binary fission.

1. **Superclass Rhizopodea.** Locomotion by lobopodia, filopodia, reticulopodia, or protoplasmic flow without pseudopodia.

a. **Class Lobosea.** Pseudopodia lobose to filiform; usually uninucleate; body naked or in test; free living or parasitic. Examples are *Amoeba proteus, Chaos carolinense* (= *Pelomyxa carolinensis*), *Entamoeba histolytica, Arcella,* and *Difflugia.*

b. **Class Filosea.** Filiform pseudopodia that often branch and sometimes anastomose; no flagellate stage known; body naked or in a test. Examples are *Lecythium* and *Euglypha.*

c. **Class Granuloreticulosea.** Foraminiferidans. Primarily with delicate reticulopodia; majority with a test of one to many chambers. Examples are *Globigerina, Elphidium,* and *Allogromia.*

2. **Superclass Actinopodea.** Radially arranged axopods; mostly planktonic; capsular membrane between ectoplasm and endoplasm in most forms.

a. **Class Acantharea.** Radiolarians. Skeleton of radial spines centrally joined; skeleton of strontium sulfate; mostly planktonic; all marine. Example is *Acanthometra.*

b. **Class Polycystinea.** Radiolarians. Skeleton of silica; marine and planktonic. Example is *Collosphaera.*

c. **Class Phaeodarea.** Radiolarians. Skeleton, when present, of silica and organic matter; spines hollow; marine and planktonic. Example is *Circoporus.*

d. **Class Heliozoea.** Sun animalcules. Skeleton, when present, siliceous or chitinoid; no capsular membrane; primarily freshwater. Examples are *Actinosphaerium* and *Actinophrys.*

II. **Phylum Apicomplexa.** All parasitic; no cilia or flagella (except for flagellated microgametes), nucleus vesicular; asexual and sexual phases in life cycle.

a. **Class Perkinsea.** Flagellated sporozoites. Polar capsular tubules form incomplete cone. Example is *Perkinsus marinus* of oysters (only species).

b. **Class Sporozoea.** Infective stage is a sporozoite, either naked or enclosed; both asexual and sexual phases in complex life cycles. Examples are *Monocystis, Gregarina, Eimeria,* and *Plasmodium.*

III. **Phylum Microspora.** Intracellular parasites (mainly of insects) with polar filaments. Example is *Nosema.*

IV. **Phylum Myxozoa.** Obligate, extracellular parasites of annelids and poikilothermic vertebrates (mainly fish); usually, two polar capsules. Examples are *Myxosoma* and *Triactinomyxon.*

V. **Phylum Ciliophora.** Cilia typically present; macronucleus and micronucleus present; reproduction primarily transverse binary fission (homothetogenic) and conjugation.

a. **Class Spirotrichea.** Somatic dikinetids present usually with postciliodesmata; usually, with right and left oral ciliature; paroral membranes may be present on right side of oral region. Examples are *Stentor, Spirostomum, Euplotes, Stylonychia, Kerona,* and *Blepharisma.*

b. **Class Prostomatea.** Somatic ciliature of monokinetids with radial transverse ribbon; ciliary crown of dikinetids around cytostome; somatic ciliature of monokinetids; cytostome apical or subapical. Example is *Coleps.*

c. **Class Litostomatea.** Somatic ciliature of monokinetids with tangential transverse ribbon; oral cilia simple. Example is *Didinium.*

d. **Class Phyllopharyngea.** Somatic ciliature mostly as monokinetids with reduced transverse ribbon; oral region bears radially arranged microtubular ribbons called phyllae. Examples are the suctorians, *Ephelota* and *Acineta.*

e. **Class Nassophorea.** Somatic ciliature as monokinetids, dikinetids, or polykinetids; oral region bears microtubular rods called nematodesmata; extrusomes, when present, are trichocysts. Example is *Paramecium.*

f. **Class Oligohymenophorea.** Somatic ciliature of monokinetids with radial transverse ribbon; oral apparatus, when present, of paroral dikinetid and usually three polykinetids. Examples are *Vorticella* and *Tetrahymena.*

g. **Class Colpodea.** Somatic ciliature as dikinetids with overlapping posterior transverse ribbons; oral region with left and right ciliature; resting cysts common. Example is *Bursaria.*

A. Phylum Sarcomastigophora

Protozoans commonly called flagellates, opalinids, and amebas belong to the large phylum Sarcomastigophora (SAR-ko-MAS-ti-GOF-or-a; G., *sarkos* fleshy + G., *mastix*, whip + G., *phora*, to bear). Members of the phylum move by **flagella** (flagellates), **cilia** (opalinids), and **pseudopodia** (amebas). Studies with the electron microscope show that the structural configuration of the flagellum and cilium is the same and that both arise from a **basal body (blepharoplast, kinetosome)**. Three subphyla occur in Sarcomastigophora: (1) Mastigophora (MAS-ti-GOF-or-a: G., *mastix*, whip + G., *phora*, to bear), the flagellates; (2) Opalinata (O-pa-LIN-a-ta; N.F., *opaline*, like an opal in appearance), the opalinids; and (3) Sarcodina (SAR-ko-DI-na; G., *sarkos*, flesh + G., *ina*, belonging to), the amebas.

Subphylum Mastigophora

The most primitive protozoans belong to the subphylum Mastigophora. The flagellum is the primary locomotory organelle. Flagellates are mostly free living and often are divided into plantlike and animal-like forms.

Class Phytomastigophorea

These plantlike flagellates have a cell wall, cellulose, chloroplasts, and stigma. The usual type of nutrition is holophytic (autotrophic). Representatives of the following three orders will be emphasized: order Dinoflagellida (*Ceratium*, *Glenodinium*, and *Gymnodinium*), order Euglenida (*Euglena* and *Peranema*), and order Volvocida (*Chlamydomonas* and *Volvox*).

Order Dinoflagellida. Dinoflagellates are widely distributed in marine, brackish, and freshwater habitats. Most are free living. Dinoflagellates have an unequal pair of flagella. One is ribbonlike and typically lies in a transverse surface furrow called a **cingulum** or girdle (Fig. 1.2). The second flagellum is directed posteriorly and usually in a longitudinal furrow, the **sulcus**. The region above the cingulum in armored forms is the **epitheca** (epicone in unarmored ones); below the cingulum is the **hypotheca** (hypocone in unarmored forms). Unarmored or naked species (e.g., *Gymnodinium*) have a cell covering of membranes, whereas thecal plates of cellulose or other polysaccharides occur on armored species (e.g., *Ceratium* and *Glenodinium*).

Many dinoflagellates (e.g., *Noctiluca*, *Gonyaulax*, and *Pyrocystis*) are bioluminescent and their bluish-green glow can be seen for some distance in the oceans. Certain genera, such as *Gonyaulax* and *Gymnodinium*, may occur in huge numbers; these blooms are often called "red tides." Blooms of *Gymnodinium breve* produce toxins that directly kill fish and some invertebrates. The toxins of other species may be concentrated in the bodies of clams, oysters, and scallops, rendering these organisms unsafe for human consumption. Reproduction is by binary fission and syngamy.

Observational Procedure: Order Dinoflagellida

Make a small ring, using a thick suspension of methyl cellulose, on a glass slide. Add a drop of the culture into the center of the ring, cover with a coverslip, and observe under low power of a microscope. After the organism is located, change to high magnification. Describe the locomotion. Attempt to locate the cingulum and sulcus in a stationary individual. Can you see flagella beating? What variations in the body shape, size, and structure do you see, especially in *Ceratium*, which has hornlike extensions (Fig. 1.2)? What might cause these variations?

Examine prepared slides of several species of dinoflagellates (e.g., *Ceratium*, *Glenodinium*, and *Gymnodinium*). Note the nature of the body covering. *Ceratium* has one apical horn and one to three antapical horns. The nucleus should be evident. Locate the cingulum and sulcus.

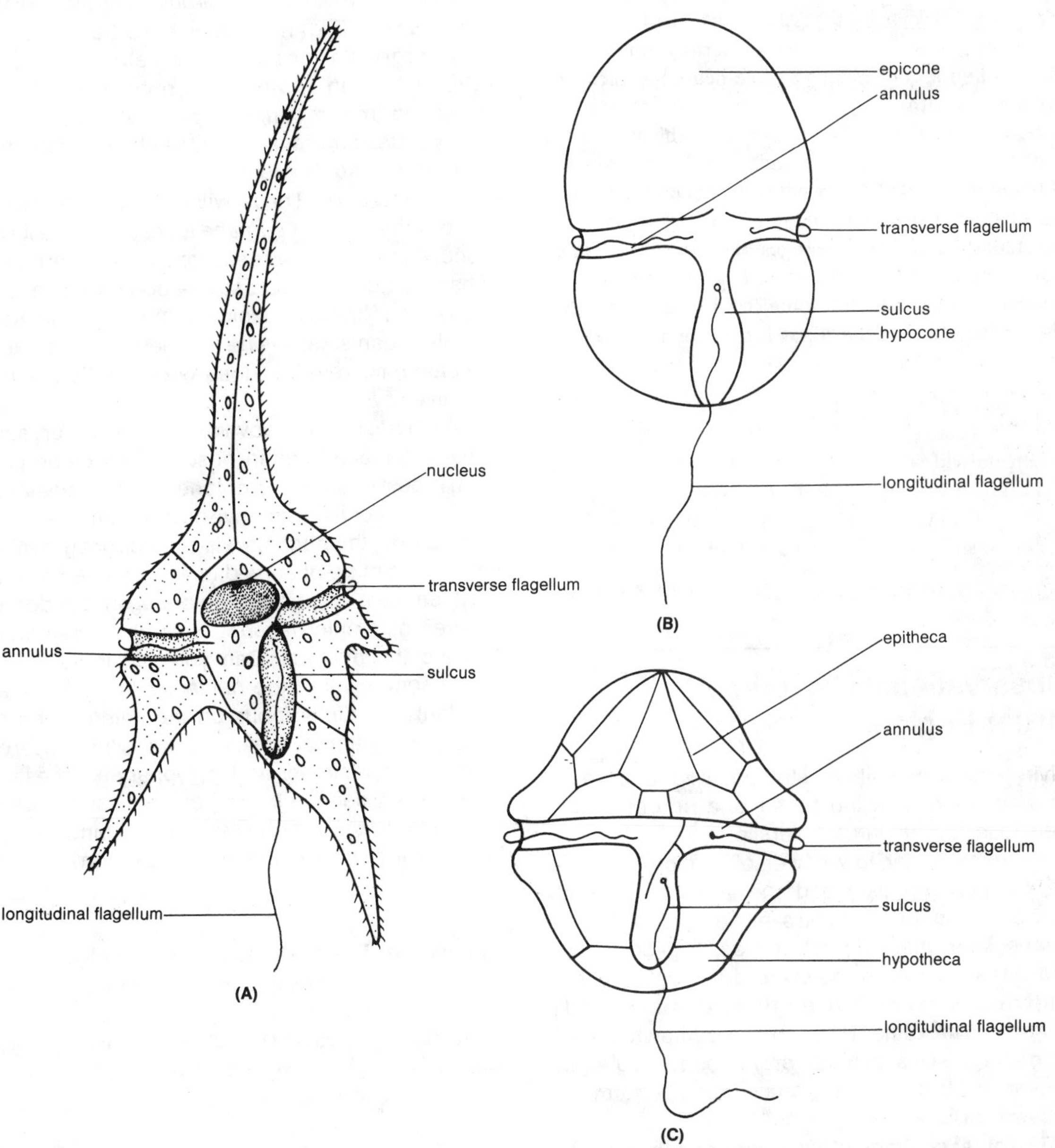

Figure 1.2. Dinoflagellates. **(A)** *Ceratium.* **(B)** *Gymnodinium.* **(C)** *Glenodinium.*

Order Euglenida. Euglenoid flagellates are primarily freshwater organisms. Over 800 species in at least 37 genera have been described. Most species are bilaterally symmetrical with bodies elongated, ovoid, spherical, or leaflike. Both chloroplast-containing and colorless forms are common. Chloroplasts (organelles with chlorophylls) vary in number and shape according to the species.

The anterior end of the cell is invaginated, forming a flask-shaped **reservoir** (Fig. 1.3). The narrow part of the reservoir is sometimes called the **cytopharynx** and its opening the **cytostome** (cell mouth). Most euglenoid flagellates have two flagella (generally unequal in length) and a contractile vacuole near the reservoir. All species with an eyespot (stigma) have the photoreceptor or paraflagellar swelling near the base of the longer, emergent flagellum and opposite the eyespot. Surrounding the cell is a **pellicle**, consisting of interlocking proteinaceous strips and microtubules beneath the cell membrane. Rapid changes of body shape, called euglenoid movement, are seen in many euglenoids. A single nucleus with **endosome** (nucleolus) is present. Reproduction is primarily by longitudinal fission.

Euglena. *Euglena* is a common, solitary phytoflagellate in well-vegetated freshwater streams, ponds, and pools. When individuals occur in great numbers the water has a greenish hue. *E. viridis* (40 to 80 μm long), *E. spirogyra* (45 to 250 μm), and *E. oxyuris* (500 μm) are common species often studied in biology courses.

Observational Procedure: Order Euglenida

Living material will be studied. Add a drop of *Euglena* culture in the center of a ring of methyl cellulose. Cover with a coverslip.

Examine under low power of a microscope to locate the organism and then change to high power. Describe *Euglena's* movement and body shape. Does the cell rotate, creep, or glide? Do all individuals move in the same direction? Are any individuals showing euglenoid movement? Euglenoid movement in *E. viridis* is somewhat jerky. *Euglena gracilis* exhibits pronounced euglenoid movement, but not all *Euglena* species perform this movement.

The outer covering of the organism is the pellicle. The forward end is generally less pointed than the trailing end. Carefully focus the forward end with the fine adjustment to see movement of the whiplike flagellum. You might want to make another wet mount and add a drop of Lugol's iodine solution before applying the coverslip. The solution will stain the flagellum, but will kill the organism. Near the reservoir is a contractile vacuole and a reddish eyespot (stigma). The eyespot is more easily seen in strains lacking pigment.

Is your specimen greenish in color? The shape and position of the chloroplasts vary with the species. For example, *E. spirogyra* has many discoid chloroplasts (Fig. 1.3). In *E. gracilis* the chloroplasts are large, flattened, and shield-shaped. The chloroplasts in *E. viridis* are ribbonlike structures radiating from a common **paramylon center** (carbohydrate storage area). In the chloroplasts, there may be **pyrenoids**, proteinaceous bodies containing starch reserves. These will not be observed. The chlorophyll allows *Euglena* to manufacture its own food. *Euglena* also has saprozoic nutrition; that is, it absorbs nutrients across the body surface in the absence of light. Holozoic nutrition, the ingesting of whole organisms, has never been demonstrated in *Euglena;* however, it does occur in *Peranema*, a relative of *Euglena*.

Attempt to locate the ovoid nucleus of the organism. This organelle is often obscured by chloroplasts and is best observed in colorless phytoflagellates. If a drop of acidified methyl-green stain is added to the culture, the nucleus will stain bright green. The nucleus can be observed in the prepared slide.

Examine a prepared slide of *Euglena* under high power of a microscope. Note the body shape. Locate the nucleus. Can you see the nucleolus (endosome) within the nucleus? Are chloroplasts evident? Carefully focus on the anterior end of a specimen. Attempt to see the flagellum, reservoir, eyespot, cytostome, and cytopharynx. These are more easily seen in the larger species of *Euglena*. *Euglena* reproduces by longitudinal binary fission. Are any individuals undergoing division?

Peranema. This freshwater euglenoid phytoflagellate is similar to *Euglena*, but is colorless, lacks the eyespot, has a long trailing flagellum, is holozoic in nutrition, and bears two rodlike structures **(rod organ)** in the reservoir for food catching (Fig. 1.3).

Observational Procedure: *Peranema*

Mount a drop of the culture on a slide as before. Observe the shape and movement of *Peranema*. Can you see striations in the pellicle? *Peranema* undergoes pronounced euglenoid movement. Observe a prepared slide of *Peranema* and compare the organism with *Euglena*.

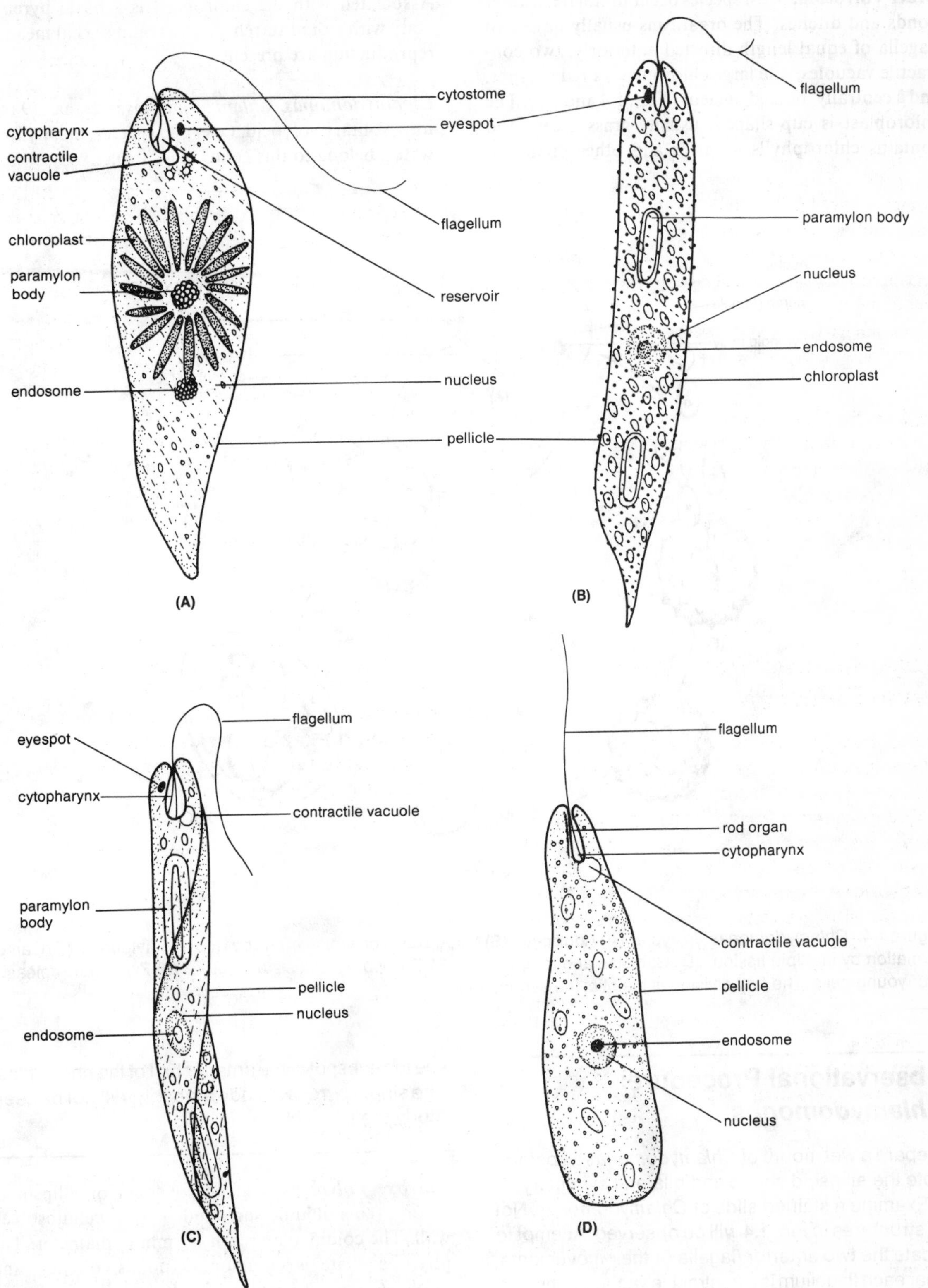

Figure 1.3. **(A)** *Euglena viridis.* **(B)** *E. spirogyra.* **(C)** *E. oxyuris.* **(D)** *Peranema.*

Order Volvocida. Most species occur in still freshwater ponds and ditches. The organisms usually have two flagella of equal length directed anteriorly, two contractile vacuoles, one large chloroplast, a red eyespot, and a centrally located nucleus (Figs. 1.4 and 1.5). The chloroplast is cup-shaped, usually grass green, and contains chlorophylls *a* and *b* and other pigments. Associated with the chloroplast is a basal pyrenoid body with stored starch. Asexual and sexual means of reproduction are present.

Chlamydomonas. *Chlamydomonas* is a small (9 to 16 μm), solitary volvocid. Over 500 species, mostly freshwater, belong to this genus (Fig. 1.4).

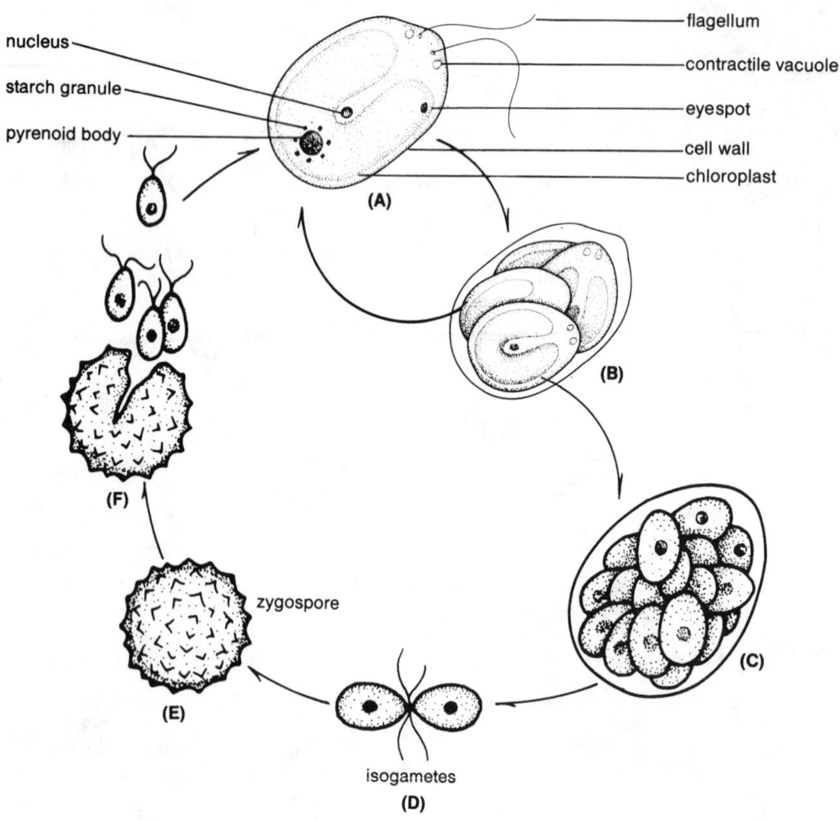

Figure 1.4. *Chlamydomonas* life cycle. **(A)** Adult cell. **(B)** Mitotic division which produces more individuals. **(C)** Gamete formation by multiple fission. **(D)** Gametes and fertilization. **(E)** Zygospore, a resistant stage. **(F)** Zygospore releasing four young cells. The first division is meiotic; the second mitotic.

Observational Procedure: *Chlamydomonas*

Prepare a wet mount of *Chlamydomonas* as before. Note the ellipsoid shape and bilateral symmetry.

Examine a stained slide of *Chlamydomonas.* Not all structures in Fig. 1.4 will be observed. Attempt to locate the two anterior flagella or their movements. Near each flagellum is a contractile vacuole. Observe the single large chloroplast and nucleus. Attempt to see the eyespot in the anterior half of the chloroplast. The single pyrenoid body probably will not be seen. Study the life cycle in Fig. 1.4.

Volvox. *Volvox* is a green spherical or ellipsoidal colony (**coenobium**) surrounded by a cellulose cell wall. The colony may reach 1.5 mm in diameter. The phytoflagellate is common in ponds, ditches, and pools. Colony formation in *Volvox* is similar to embryonic development in some metazoans.

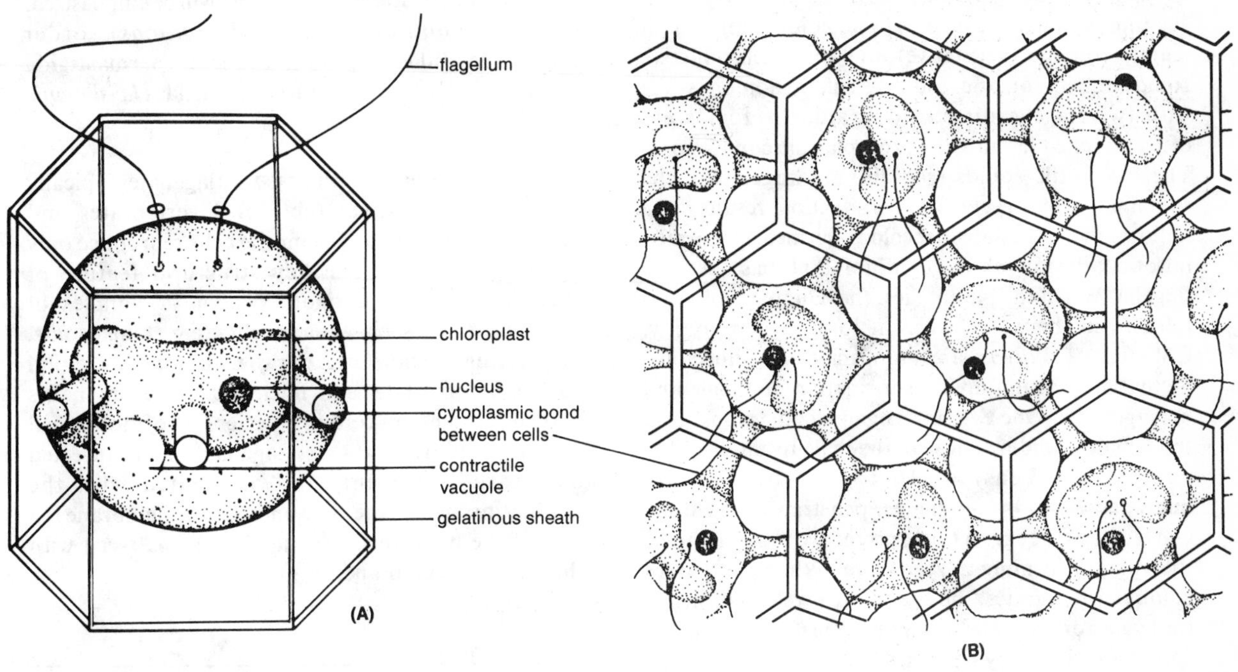

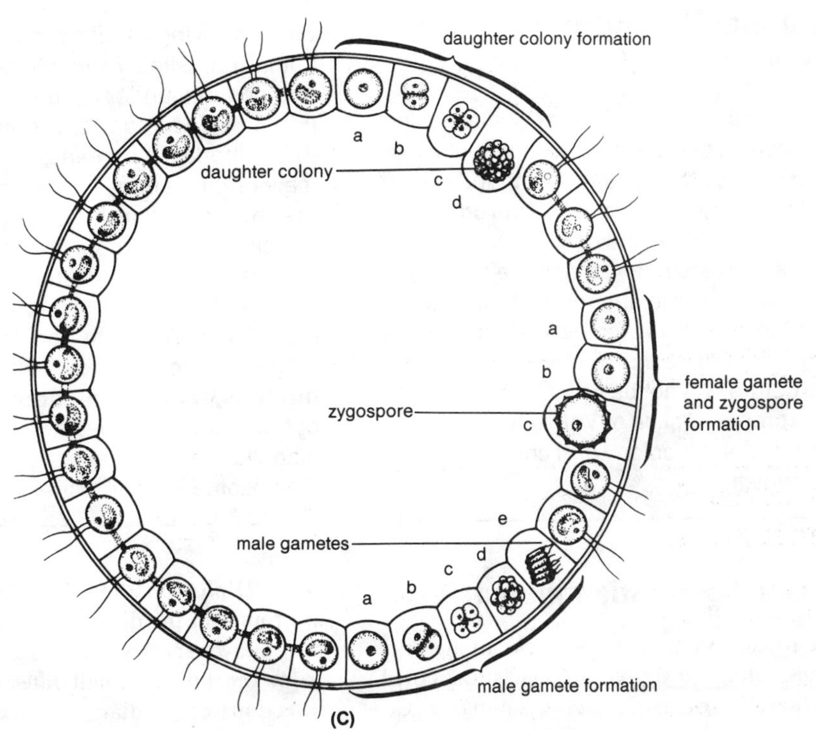

Figure 1.5. *Volvox*. **(A)** One zooid (cell). **(B)** Surface view of coenobium, showing position of zooids in relation to cytoplasmic bonds. **(C)** Daughter colony formation, gametogenesis, and zygospore formation (diagrammatic).

The organism consists of many cells or **zooids** (500 to 60,000) embedded in a single surface layer in a gelatinous matrix (Fig. 1.5). The colony consists of mainly somatic (vegetative) and some reproductive zooids. Each somatic zooid is chlamydomonas-like with two flagella directed outward; the zooids are often attached laterally by protoplasmic connections. The few fertile zooids **(gonidia)** are larger than the somatic ones. Vegetative reproduction results in a hollow sphere, the **daughter colony**, formed by repeated mitotic division of a gonidium that has lost the flagella. When the parent dies, the daughter colony is released in the water, where it forms a new colony. Colonies of *Volvox* are monoecious or dioecious. Sexual maturation involves formation of female and male gonidia. One egg (**macrogamete**) develops from the female gonidia. Those of the male result in sperm packets (**plakeas**) of 16 to 512 biflagellate **microgametes.** One sperm cell penetrates the colony and fertilizes the egg. A thick, protective wall surrounds the zygote, forming a **zygospore.** When the parent colony disintegrates, the zygospore is released. Inside the zygospore repeated division occurs and during the spring a new colony is formed.

Observational Procedure: *Volvox*

Prepare a wet mount of the *Volvox* culture using methyl cellulose. Raise the coverslip with clay or other supports. Study the organism under low power and then change to high power of a microscope. Describe the movement of the colony. Observe the daughter cells if present. How many are there? Do they move about inside the colony or remain stationary? Can you tell if the colony possesses sexual stages? The zooids are very small and details of structure will not be observed.

Examine a prepared slide of *Volvox.* Observe the general anatomy of the colony and compare it with the living organism.

Class Zoomastigophorea

In contrast to the phytoflagellates, the animal-like zooflagellates belong to the class Zoomastigophorea (ZOO-mas-ti-GOF-or-e-a; G., *zoon*, animal + G., *mastix,* whip + G., *phora*, to bear). These organisms lack chromoplasts, eyespots, cellulose coverings, and other plantlike features. Four orders will be emphasized: order Trichomonadida (*Trichomonas*), order Diplomonadida (*Giardia*), order Hypermastigida (*Trichonympha*), and order Kinetoplastida (*Leishmania* and *Trypanosoma*).

Order Trichomonadida. These zooflagellates typically inhabit the digestive tracts of invertebrates and vertebrates (Fig. 1.6). The motile cells reproduce only by binary fission. The key structural feature of trichomonad is a mastigont system—a complex of flagellum and associated organelles (e.g., **kinetosome, undulating membrane, parabasal body** or **Golgi apparatus, axostyle,** and **nucleus**). Some organelles may be absent in certain species. The motile cell typically has four to six flagella; one is directed backward (recurrent) and associated with the undulating membrane. The undulating membrane is a fold of the body surface's membrane coalesced with the flagellar membrane.

Observational Procedure: Order Trichomonadida

Examine living specimens or prepared slides of the human parasites, *Trichomonas vaginalis* or *T. tenax,* and observe under oil immersion of a microscope (Fig. 1.6). *Trichomonas vaginalis,* the most common *Trichomonas* in humans, occurs in the vagina and male urethra and prostate gland; *T. tenax* occurs in the human mouth. Cysts are unknown and transmission occurs in the trophozoite stage.

Locate the four anterior flagella and the single recurrent flagellum attached to the undulating membrane. An axial rod (axostyle) extends the length of the parasite. Note the location of the nucleus, parabasal body (Golgi body), and cytostome. In well-stained slides a **rhizoplast** (flagellar rootlet) may be seen extending from the kinetosome.

Order Diplomonadida. These zooflagellates generally are found in the digestive system of their hosts. The best known species is *Giardia intestinalis* (= *lamblia*) from the human small intestine (Fig. 1.6). Heavy infections cause diarrhea. Transmission is by cysts passed in the feces.

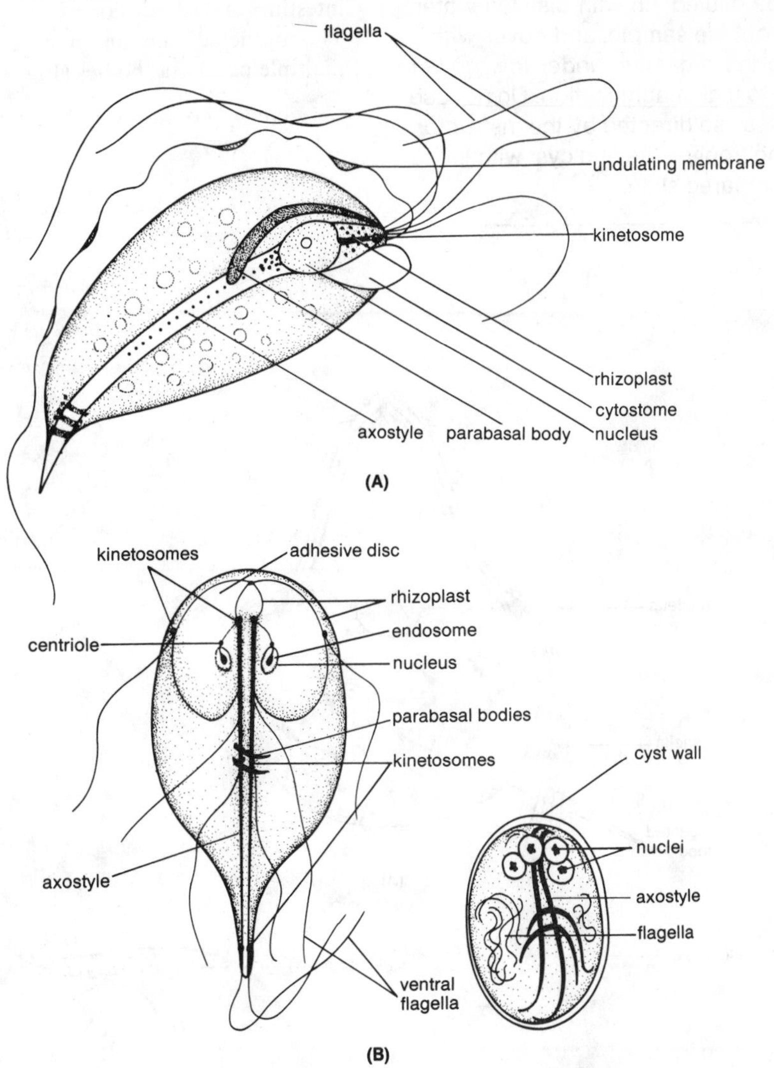

Figure 1.6. **(A)** *Trichomonas.* **(B)** *Giardia intestinalis*, trophozoite and cyst.

Observational Procedure: Order Diplomonadida

Examine a prepared slide of *Giardia intestinalis* under oil immersion of a microscope (Fig. 1.6). The trophozoite is pear-shaped, with two parallel rods (axostyles) running longitudinally in the middle of the organism. Note the two large, ovoid nuclei with endosomes and eight flagella. *Giardia* has a large ventral **adhesive disc** for attachment to the host's intestinal mucosa. Are cysts present? They have distinct cell walls, usually four nuclei (two in young cysts), axostyles, and flagella.

To observe the living organism, collect a very small amount of fresh feces on an applicator stick and transfer onto a glass slide. Be careful not to contaminate anything that might contact food; wash your hands thoroughly after handling the fecal material. Add a drop of normal saline solution and

Lugol's iodine solution to the sample. The iodine solution should be diluted 1:5 with distilled water. Mix well, spread out the sample, and cover with a coverslip. Locate the organism under low power and then change to high magnification. Do not use oil immersion unless so directed by the instructor. Compare the living trophozoite and cyst with those observed in the prepared slide.

Order Hypermastigida. These zooflagellates live in the intestines of termites, cockroaches, and wood-roaches. The motile cell has one nucleus, many flagella, and multiple parabasal bodies (Fig. 1.7).

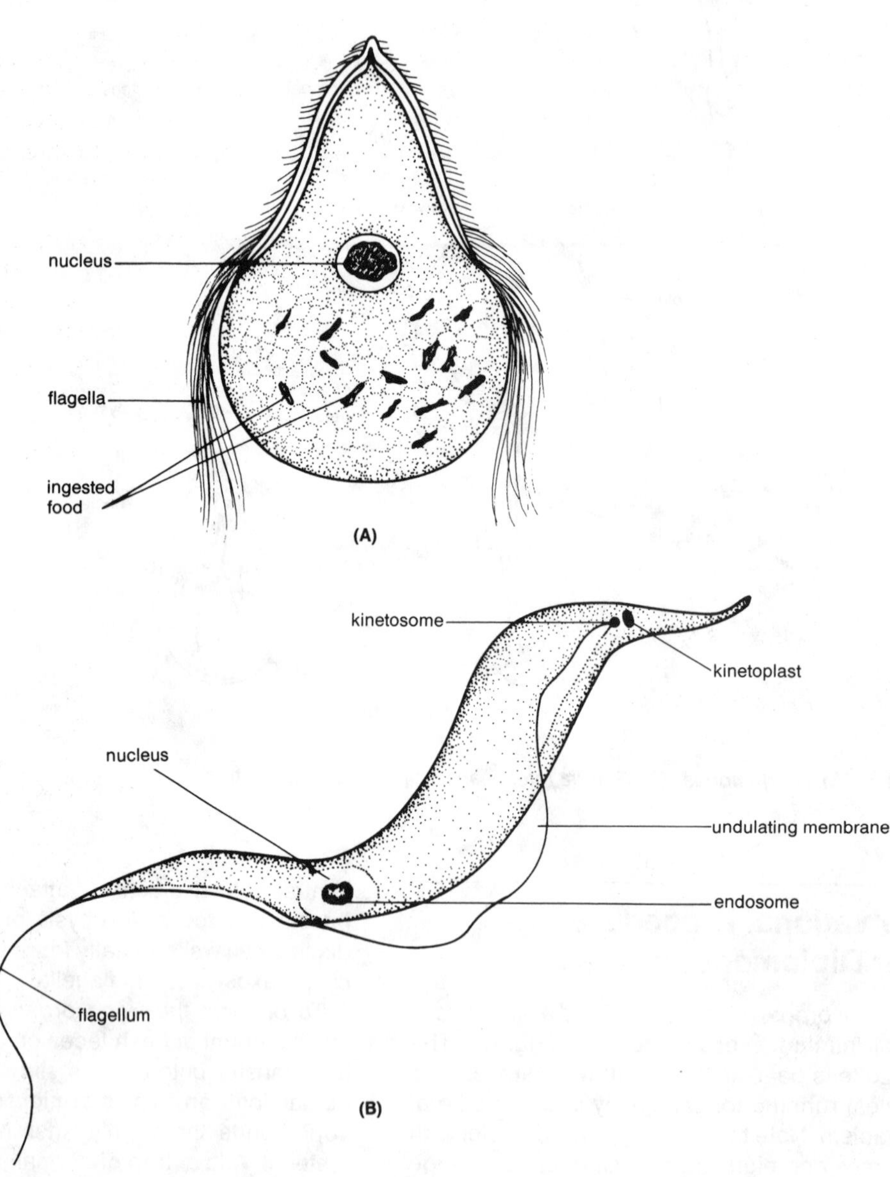

Figure 1.7. **(A)** *Trichonympha.* (From R. M. Cable.) **(B)** Trypanosome.

Observational Procedure: Order Hypermastigida

Obtain a slide of *Trichonympha.* Note the body shape, position of nucleus, and arrangement of flagella (Fig. 1.7). Living specimens may be seen by examining the gut contents of a termite. Place the termite on a glass slide. Hold the anterior end of the termite's body with forceps. Pull the other end of the body posteriorly with another pair of forceps until the gut contents emerge onto the slide. Spread the contents thinly, add a small drop of water, and examine under low and high powers with a compound microscope. There may be several species and genera represented. Note the shape of the body and arrangement of the flagella. *Calonympha* (order Trichomonadida) may be in the gut contents; it has an axostyle and several anterior nuclei which are lacking in *Trichonympha.*

Order Kinetoplastida. Parasites of blood and tissue of vertebrates as well as free-living forms are in this order. Several species of *Trypanosoma* and *Leishmania* are dreaded human pathogens in tropical regions, causing diseases such as **sleeping sickness, Chagas' disease,** and **kala azar.** The motile cells have one or two flagella. A key feature of these organisms, for which the order is named, is the **kinetoplast** (Fig. 1.7). This is a large stainable mass of mitochondral DNA located near the base of the locomotory apparatus. The kinetoplast is usually larger than the kinetosome.

Observational Procedure: Order Kinetoplastida

Observe prepared slides of *Trypanosoma* and *Leishmania* under oil immersion. Note the body shape and single flagellum that is free or attached to the undulating membrane, depending on the species and stage in the life cycle (Fig. 1.7). Carefully study the locational relationship of the nucleus, basal body, and kinetoplast.

Living *Trypanosoma lewisi* may be obtained from an infected rat by examining its blood. Put a drop of blood on a glass slide and quickly cover with a coverslip. Observe under low power of a microscope and search for movement. A moving trypanosome causes the red blood cells to move. Examine the parasite under high power. Do you see the undulating membrane moving?

B. Subphylum Opalinata

Opalinids are found primarily in the intestines of amphibians; a few species occur in fish. Their flattened leaf-shaped bodies are covered with numerous cilia (Fig. 1.8). The fibrillar associates of the kinetosomes are unlike those of ciliates. There are at least two monomorphic nuclei. A cytostome and cytopharynx are lacking; ingestion is by pinocytosis and egestion is by exocytosis.

Observational Procedure: *Opalina*

Examine a prepared slide of *Opalina.* Note the body shape, body ciliation, and nuclei (Fig. 1.8).

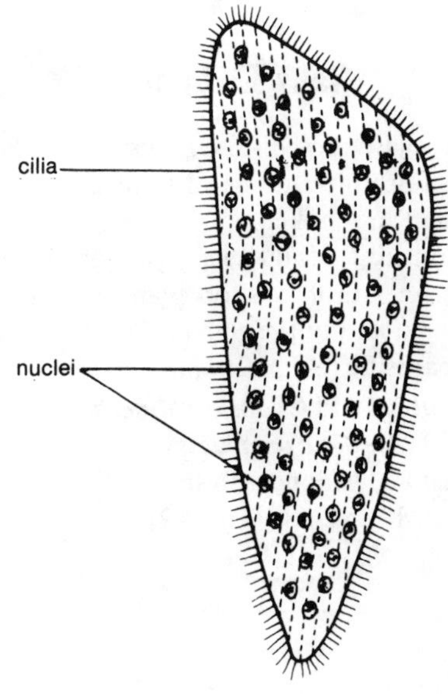

Figure 1.8. *Opalina.* (From R. M. Cable.)

C. Subphylum Sarcodina

These protozoans use pseudopodia, protoplasmic extensions of the motile cell, as the primary means for locomotion and feeding. These organisms are commonly called amebas. The body is naked (e.g., *Amoeba proteus*) or provided with a **test** or outer covering (e.g., *Difflugia*). Most species are free living; reproduction by binary fission is common.

The subphylum is usually divided into two superclasses based primarily on the type of pseudopodia. Members of the superclass Rhizopodea (ri-ZOP-o-de-a; G., *rhiza*, root + G., *podos*, foot) have **pseudopodia** called **lobopodia**, **filopodia**, and **reticulopodia** (granuloreticulopodia or rhizopodea), or by protoplasmic flow without production of pseudopodia (e.g., slime molds). Actinopodea (AK-tin-OP-o-de-a; G., *aktinos*, ray + G., *podos*, foot) includes amebas with pseudopodia called **axopods**. Representatives from both superclasses will be studied.

Superclass Rhizopodea

This large group of amebas includes five classes differentiated primarily on the type of pseudopodia present. Included in the superclass Rhizopodea are slime molds (class Mycetozoea), which have both animal and plant features. Slime molds will not be studied.

Class Lobosea

The Lobosea (LO-bo-SE-a; G., *lobos*, lobe) includes amebas that produce lobose to filiform pseudopods. Their bodies are naked or housed in a secreted proteinaceous test. Habitats are varied: in fresh and salt waters, in moist soils, on plants, in manure, in sewage waters, and in the guts of plants and animals. Some are parasitic (e.g., *Entamoeba histolytica*). ***Amoeba proteus*** and ***Chaos carolinense*** (= *Pelomyxa carolinensis*). *Amoeba proteus* is a carnivorous lobose ameba occasionally found in fresh water at temperatures ranging from 10 to 24°C (Fig. 1.9).

Observational Procedure: *Amoeba proteus*

To prevent crushing the ameba, small pieces of a broken coverslip or bits of clay can be placed on the slide around the sample of culture to be examined. Place a drop of the ameba culture, taken from the bottom of the jar, on a glass slide and cover with a coverslip. Locate the organism under low power of a microscope and then change to high magnification.

Body size ranges from 100 to 500 μm, depending on whether the organism has assumed a round or extended shape. Describe the symmetry of your specimen. Observe movement of the organism. *Amoeba proteus* has large pseudopodia called **lobopodia** used in food capture and locomotion. The body area opposite the advancing pseudopodia is the **uroid end**. Note the streaming of the cytoplasm within the body and as a pseudopodium is formed. The cytoplasm consists of an outer, clear, firmer (gel) **ectoplasm** and an inner, granular, more liquid **endoplasm** (sol) that contains the various organelles. These areas represent different colloidal states and have been used to explain in part amoeboid movement. Observe an extending pseudopodium. Note the large clear area of ectoplasm, the **hyaline cap**, at the end of the pseudopodium. This is the region of sol-to-gel conversion. Do other pseudopodia have the cap? Although ameboid movement has been studied for years, many details explaining exactly how the movement occurs are unknown. Refer to your textbook for current ideas on amoeboid movement.

Do you see a large, clear spherical area in the endoplasm? This is the **contractile vacuole**, a water-regulatory organelle. The vacuole forms near the advancing end, but discharges its contents at the uroid end. At the time of expulsion, the vacuole is about 30 μm in diameter. Make an effort to see the vacuole discharge.

Scattered throughout the endoplasm are **food vacuoles** (5 to 20 μm in diameter), where intracellular digestion occurs. A food vacuole consists of food material and fluid surrounded by a membrane.

Examine a prepared slide showing whole mounts of the organism. Compare the shape and size of the stained amebas to that of the living organism. The nucleus should be evident. Are food vacuoles present?

Amoeba proteus reproduces by binary fission. Clones of the organism have been maintained for many years by this mitotic type of reproduction. Observe prepared slides showing fission.

Chaos carolinense is a large ameba (over 300 μm) that can be seen with the unaided eye. The ameba is found in swampy fresh waters, where it feeds on protozoans and small invertebrates. Examine the living ameba first with low power and then under high power of a microscope. Compare its size and shape to that of *A. proteus*. Describe the shape of the pseudopodia. Are hyaline caps present? *Chaos* has 1000 or more nuclei, and several contractile vacuoles that range in size from 30 to 40 μm.

Reproduction is by plasmotomy, a type of multiple fission found in multinucleated one-celled organisms. The nuclei exhibit mitosis following division.

Examine a prepared slide and observe the pseudopodia, ectoplasm, and endoplasm. Look for food vacuoles, contractile vacuoles, and nuclei.

***Entamoeba histolytica*.** This widely distributed intestinal parasite of humans, apes, monkeys, dogs, rats, cats, and other animals causes **amebic dysentery** or amebiasis. The parasite produces large abscesses in the intestinal lining. Sometimes the ameba causes death in its host when it penetrates the gut, enters the bloodstream, and invades vital organs such as the brain, liver, and spleen. Other *Entamoeba* species in humans are *E. gingivalis*, a commensal in the mouth, where it feeds on leucocytes, and *E. coli*, a commensal in the colon. Fecal contamination via cysts (except *E. gingivalis*, which lacks cysts and is transmitted by kissing) is the mode of transmission and infection.

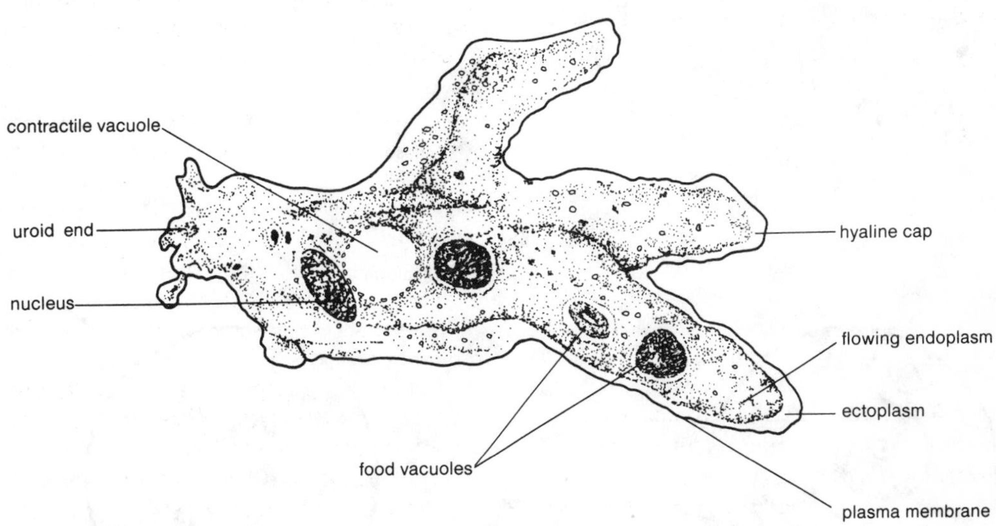

Figure 1.9. *Amoeba proteus*.

Observational Procedure: *Entamoeba histolytica*

Examine prepared slides of *E. histolytica* containing trophozoites and cysts (Fig. 1.10). Locate the organism under high power and then under oil immersion of a microscope. The trophozoite is 15 to 40 μm long, depending on whether it is in the round or extended state. There are usually four nuclei in the mature ameba. The structure of the nucleus is important in identifying *E. histolytica*. A nucleus stained with hematoxylin has fine, beadlike peripheral chromatin and a small (less than 1.0 μm), centrally placed endosome. In *E. coli* the trophozoite usually has eight nuclei, each with a large eccentric endosome and coarse peripheral chromatin. Both species lack contractile vacuoles. Food vacuoles are small (2 to 3 μm in diameter).

Look carefully for cysts. The mature cyst (10 to 20 μm in diameter) is round, smooth, and has four nuclei (3.5 to 6 μm in diameter). Eight amebas develop from a single cyst. In hematoxylin-stained specimens, dark-staining barlike rods may be seen. These are **chromatoid bodies** which represent an accumulated mass of RNA. The number varies from one to four. As the cyst ages, these bodies disappear. In contrast, *E. coli* has more chromatoid bodies that resemble a bundle of splinters. Reproduction is by binary fission.

To study the living organism, prepare a fecal sample as was done for *Giardia intestinalis*. Be careful not to contaminate anything that may contact food and wash your hands thoroughly after handling the material. First locate the trophozoite or cyst with low power and then change to high power of a microscope. Do not use oil immersion unless directed by the instructor. Does the trophozoite move quickly? Describe the appearance of the nucleus. How does it differ from that observed in the stained slide? The chromatoid bodies appear as whitish refractile bodies instead of dark-stained structures observed in the prepared slide. Does your *E. histolytica* contain erythrocytes as a result of having recently fed on the intestinal mucosa?

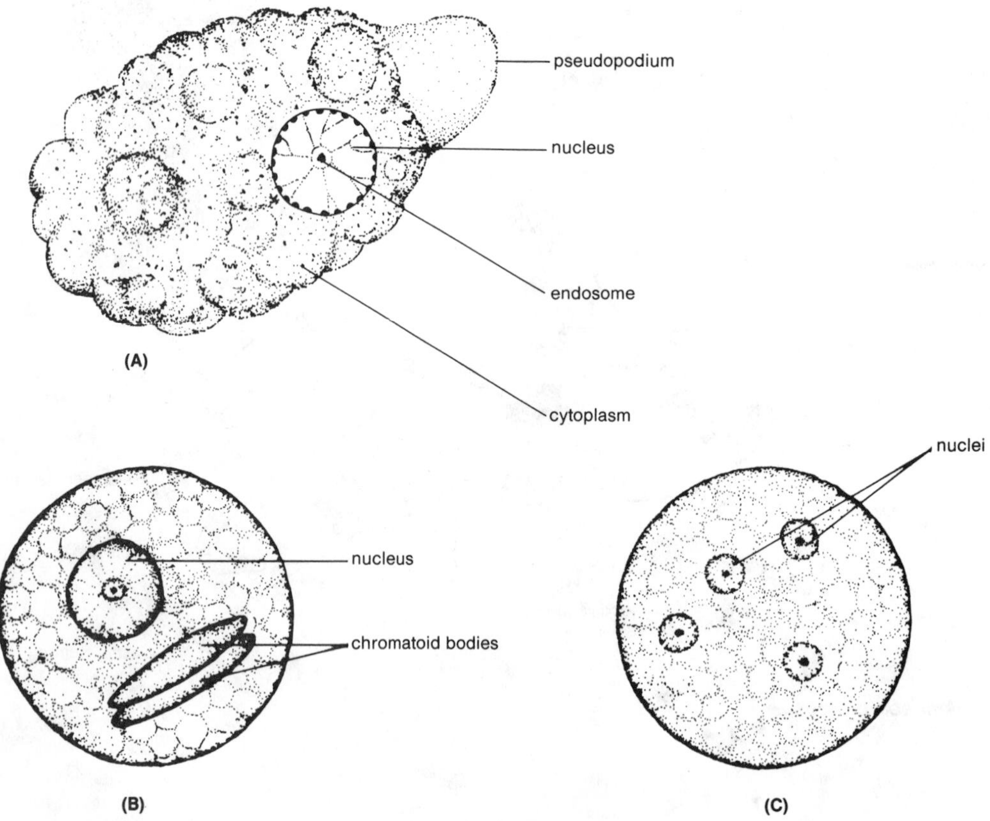

Figure 1.10. *Entamoeba histolytica.* **(A)** Trophozoite. **(B)** Immature cyst. **(C)** Mature cyst.

Arcella vulgaris. *Arcella* is a beautiful testate ameba often found on aquatic plants (Fig. 1.11). The rigid, chitinoid test is variable in shape and size, depending on the species. The opening, or **pseudostome**, into the test is ventral and round.

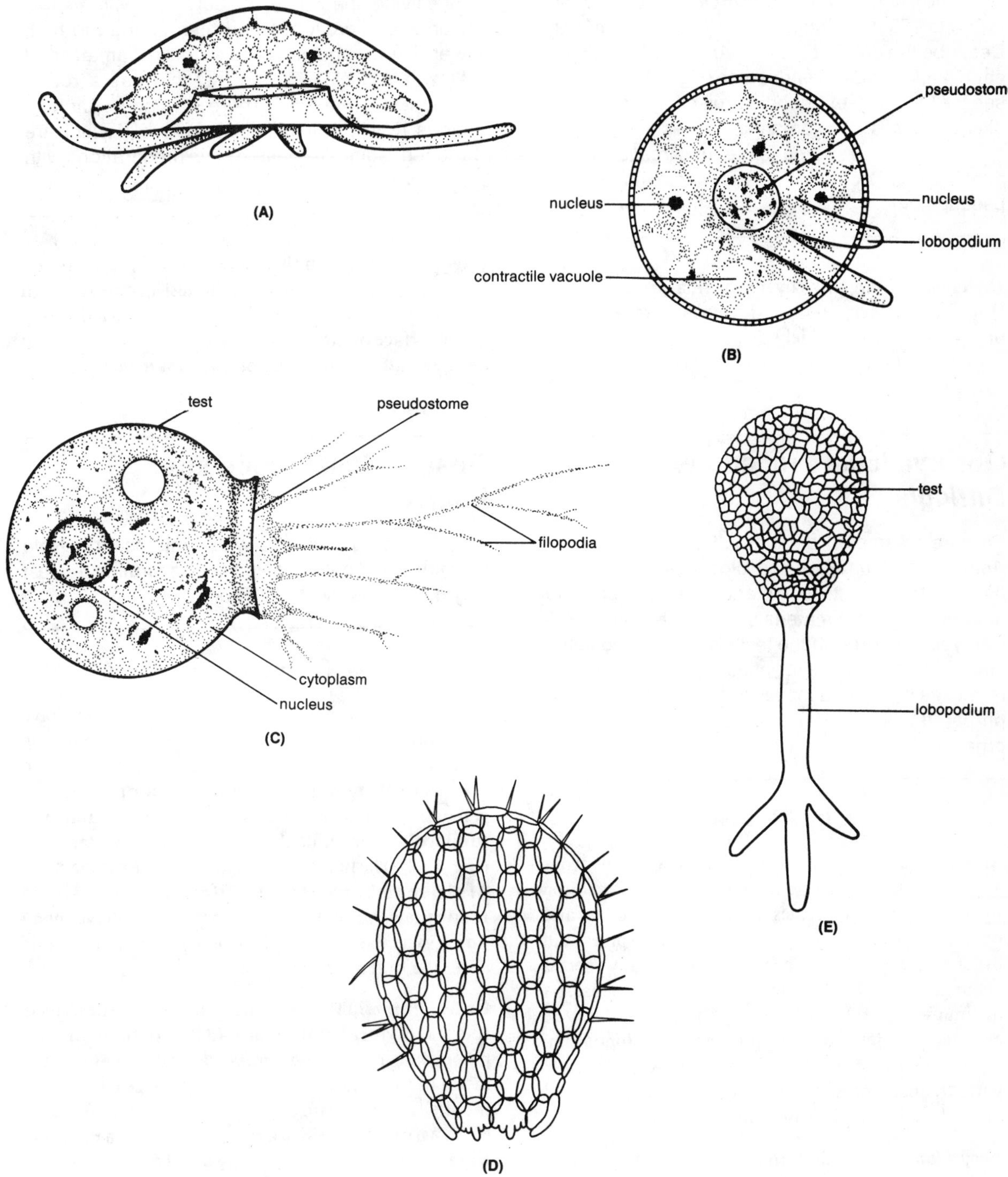

Figure 1.11. **(A)** *Arcella*, side view. **(B)** Apical view. (After Barnes.) **(C)** *Lecythium*. **(D)** *Euglypha* test. (After Barnes.) **(E)** *Difflugia*.

Observational Procedure: *Arcella vulgaris*

Examine living specimens of *Arcella*. The organism is 100 to 150 μm in diameter and 50 to 80 μm high. Describe the color of the test. *Arcella* is binucleate; each nucleus has an endosome (Fig. 1.11). Can you see the lobose pseudopodia projecting through the pseudostome as the organism moves and feeds?

Examine a prepared slide of *Arcella* with a microscope. Locate the pseudopodia, nuclei, and pseudostome.

Difflugia. *Difflugia* is covered with a protective test (Fig. 1.11). Members of the genus are beautiful organisms and are widely distributed on freshwater plants.

Observational Procedure: *Difflugia*

Prepare a sample from the culture as you did for *Amoeba proteus*. Examine closely the ovoid test (Fig. 1.11). Note its color and shape; attempt to determine of what materials the test is composed. Can you see through the test? Is the pseudostome circular? Can you see the pseudopodia extend and retract as the ameba moves and feeds? Examine a prepared slide and compare it with the living organism.

Class Filosea

Most amebas in this class are herbivorous or bactivorous and have bodies protected in a test. The tests are often formed of siliceous scales with attached spines and other debris. Scales may be lacking, but spines and other particles are attached to the test. Many species occur in fresh water, often with lobose amebas. A distinguishing feature of these amebas, from which the class name is dervied, is the presence of a **filopodium**. This long pseudopodium is filamentous, generally without anastomoses or a central axial rod. It is used in food capture and locomotion.

Lecythium. This herbivorous ameba is found in fresh water. Individuals typically occur in groups.

Observational Procedure: *Lecythium*

Locate living specimens of *Lecythium* with the low power of a microscope and then change to high power. The organism is 30 to 40 μm in diameter and lives within a clear, ovoid test that has a round pseudostome (Fig. 1.11). There is a single spherical nucleus. Note the clear filopodia, some of which are branched. Compare stained slide preparations with the living material.

Euglypha. Species in this large genus are often found living on freshwater plants. The test is covered with internally secreted scales that are transported to the outer surface of the ameba (Fig. 1.11). The scales are in definite patterns and may or may not have spines.

Observational Procedure: *Euglypha*

Examine living *Euglypha* with a microscope. Note the shape and pattern of the scales forming the test (Fig. 1.11). Observe the filopodia.

Class Granuloreticulosea

This large class typically contains amebas that have filamentous pseudopodia with branching anastomosing networks of protoplasm. This type of pseudopodium is a **reticulopodium** (granuloreticulopodium or rhizopodium). Most species have a single-chambered (**unilocular**) or multichambered (**multilocular**) test (Fig. 1.12). The first-formed chamber is the **proloculum**. Multichambered foraminiferidans are single individuals, not colonies. Perforate shells have many pores, whereas imperforate shells generally have one (e.g., the oral aperture) pore.

Order Foraminiferidia. There are more described species of foraminiferidans (34,000) than of any protozoan group. *Globigerina* is a predominate genus; much of the ocean's ooze consists of large deposits of the shells from *Globigerina* species. Foraminiferidans are ancient, primarily marine amebas. Foraminiferidans are large in size (0.5 mm to 1 cm), but larger ones (12 cm) are known from the fossil record. The composition (mostly calcium carbonate), arrangement of chambers, and openings of the test vary. The test is covered by a portion of the cytoplasm (Fig. 1.12). In multilocular forms the cytoplasm and rhizopodia extend through surface pores as well as through the oral aperture. In unilocular species, the pseudopodia extend from the oral opening of the shell. Multilocular species add chambers in definite patterns as they grow. The test is absent in a few species. The complex life cycle involves multiple division and alternation of haploid and diploid generations.

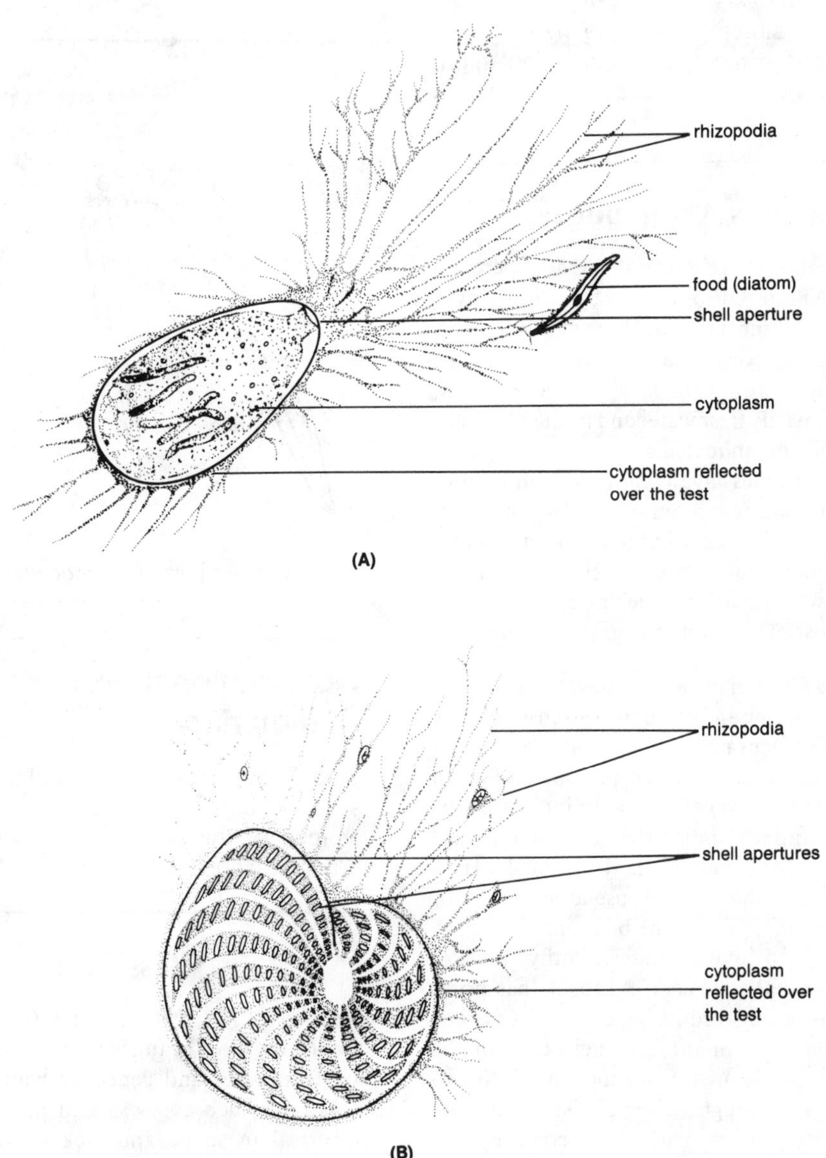

Figure 1.12. Two foraminiferidans. **(A)** *Allogromia* with unilocular shell. (After Schulze.) **(B)** *Elphidium* with multilocular shell. (After Jahn.)

Observational Procedure: Foraminiferidans

Obtain prepared slides and unmounted specimens of several species of foraminiferidans. Examine the organisms with the compound and dissecting microscopes. Note the shapes of the specimens and how chamber formation has taken place. Are the surfaces of the tests adorned with ribs, keels, spines, or other structures? Is the perforate or imperforate test the more abundant? You might want to examine bulk specimens against a black background.

Superclass Actinopodea

Four classes are recognized currently in the superclass Actinopodea (AK-tin-OP-o-de-a; G., *aktinos*, ray + G., *podos*, foot). Members of the three marine classes (e.g., Acantharea, Polycystinea, and Phaeodorea) are commonly called radiolarians. The fourth class, Heliozoea, is primarily freshwater and its members are commonly called **sun animalcules**.

Members of all classes have a type of pseudopodium called an **axopodium** (actinopodium). This straight, slender pseudopodium possesses an axial rod (**axoneme**) composed of microtubules. Axopodia are radially arranged and used primarily in feeding rather than in locomotion. Most of these amebas have tests of silica.

Radiolarians. Radiolarians are among the oldest of animals. All are marine and currently assigned to three classes. Most radiolarians are planktonic and pelagic. Their beautiful skeletons are composed primarily of silica and strontium sulfate or calcium aluminum silicate, with radially arranged spines that extend outward from the center of the body (Figs. 1.13 and 1.14). A **central capsule**, a double or single pseudochitinous or mucinoid membrane, divides the body into an inner, granular endoplasm and an outer, frothy mass of ectoplasm (**calymma**). The central capsule has pores (**fusules**) that allow outward passage of the axial rods of the axopodia. One or more nuclei occur in the endoplasm. Axopodia arise from the capsule membrane and extend through the ectoplasm. They are sticky and capture prey which is carried by the streaming cytoplasm to the central capsule, where digestion occurs within a food vacuole. The ectoplasm on one side of the axopodium moves toward its tip, while the other side moves toward the test.

Reproduction is incompletely known. Binary fission, budding, and sporogenesis have been observed.

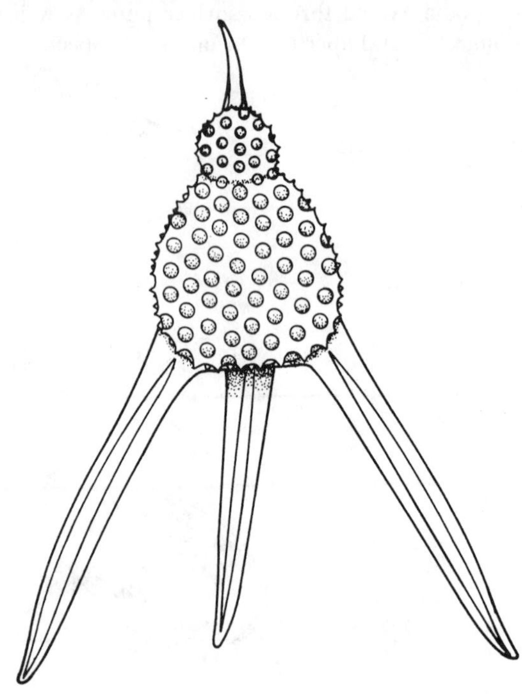

Figure 1.13. Radiolarian, *Podocyrtis*.

Observational Procedure: Radiolarians

Study prepared slides of radiolarian skeletons from several genera and species. Note the sizes and shapes of the skeletons as well as the varied ornamentations such as spines, thorns, and hooks.

Class Heliozoea

Heliozoans (G., *helios*, sun + G., *zoon*, animal) are among the most beautiful of protozoans. Most species are freshwater and generally benthic or attached to substrates such as rocks and plants. Heliozoans are spherical in shape and lack the central capsule of radiolarians, but have long, slender, granule-studded axopodia (Figs. 1.15 and 1.16).

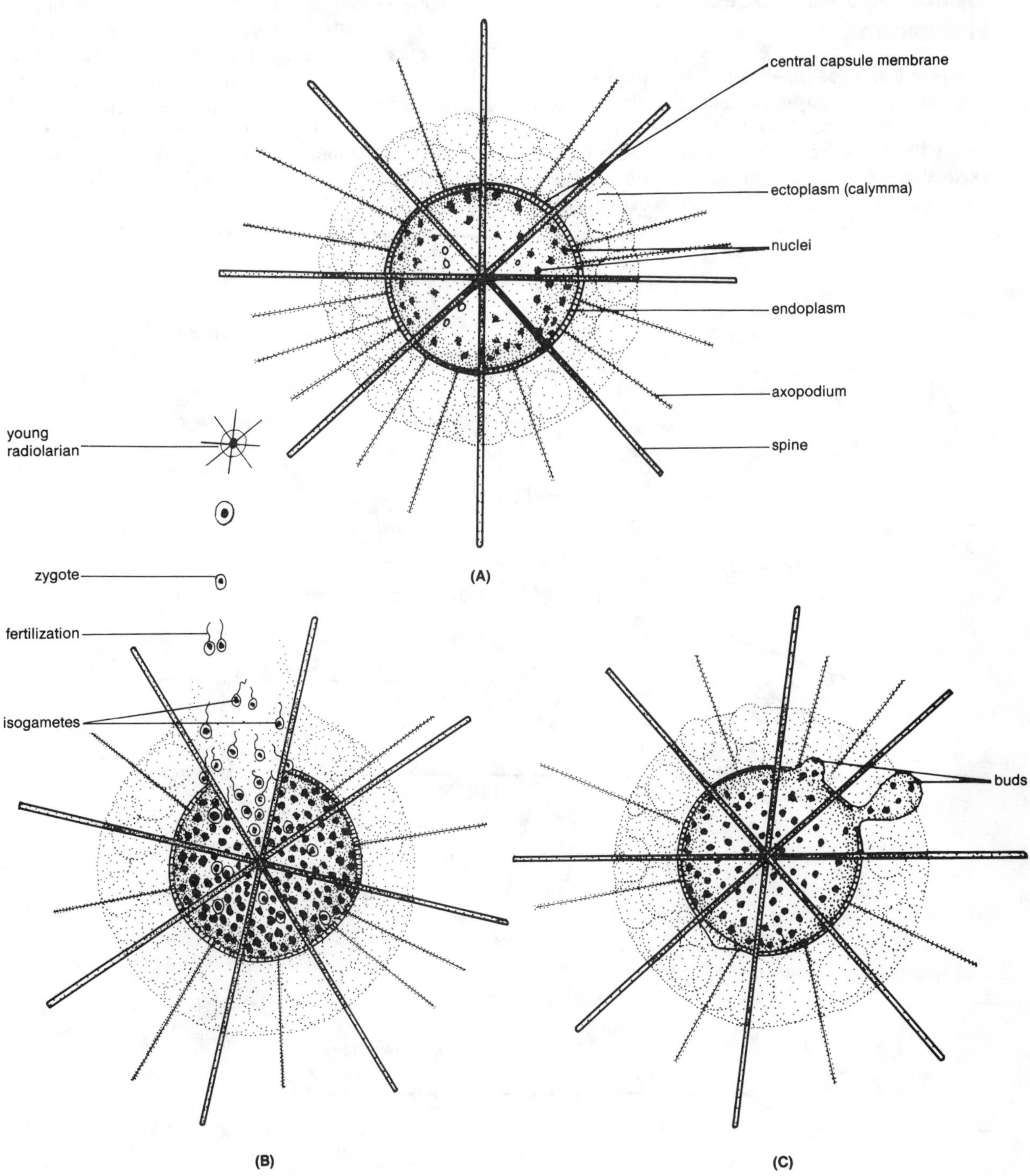

Figure 1.14. Radiolaria. **(A)** Mature individual. **(B)** Gametogenesis and syngamy. **(C)** Budding.

Observational Procedure: Heliozoans

Examine living specimens of *Actinosphaerium* and *Actinophrys* in depression slides under low and high powers of a compound microscope (Figs. 1.15 and 1.16). Note their spherical shape, the lack of skeletal spicules, and the radiating axopodia. Observe the streaming ectoplasm as it moves over the axial rods. Which of the two organisms is larger? Note that the ectoplasm of *Actinosphaerium* is highly vacuolated with a frothy appearance, whereas the endoplasm is a dense, granular central mass. The multinucleated endoplasm has many axial rods originating from its outer portion. The endoplasm of *Actinophrys* is uninucleate and the axial rods attach to the nuclear membrane. Are contractile vacuoles present?

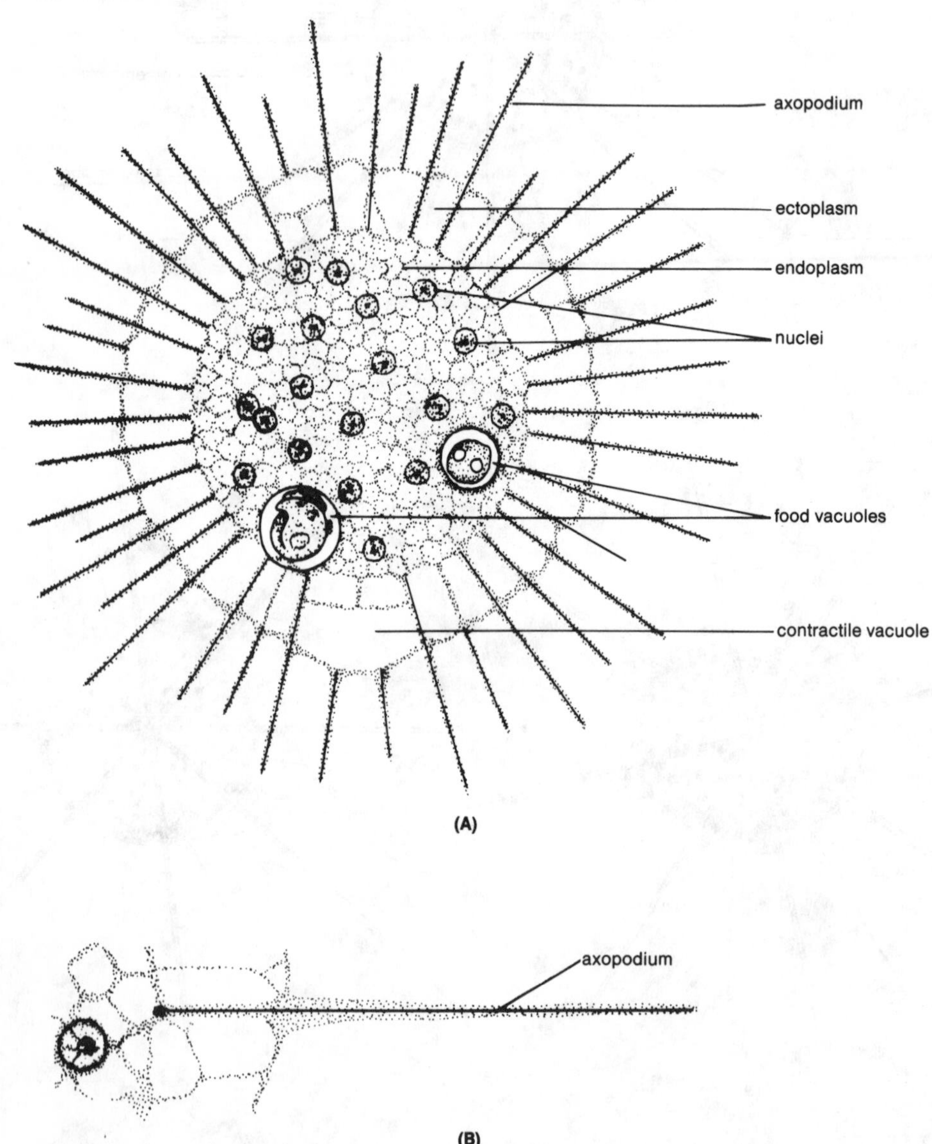

Figure 1.15. **(A)** *Actinosphaerium*. **(B)** Enlarged axopodium.

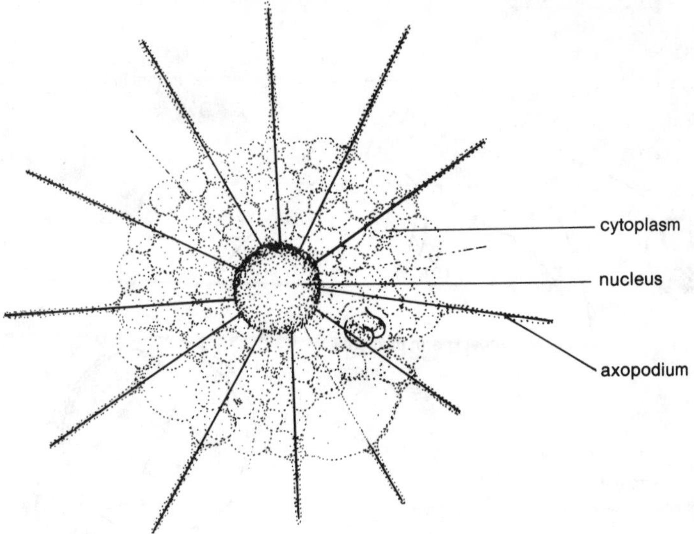

Figure 1.16. *Actinophrys.*

D. Phylum Apicomplexa

Members of the Apicomplexa (A-pi-com-PLEX-a; L., *apex*, tip + L., *complex*, twisted around) have an apical complex. All members are parasitic. The phylum contains about 4000 species and over 300 genera.

The somewhat simple-structured cell always has a vesicular nucleus. Cilia or flagella are lacking; except in flagellated microgametes in some groups.

Sporozoea is the largest and more important class of the two present in Apicomplexa. The class contains the **gregarines** (Gregarinasina), **coccidians** (Coccidiasina), and **piroplasmids** (Piroplasmasina) as three separate subclasses.

Generalized Life Cycle

The life cycle is complex, usually involving multiple fission, gamete production, and one or more hosts (Fig. 1.17). All stages in the life cycle, except the $2n$ zygote, are haploid.

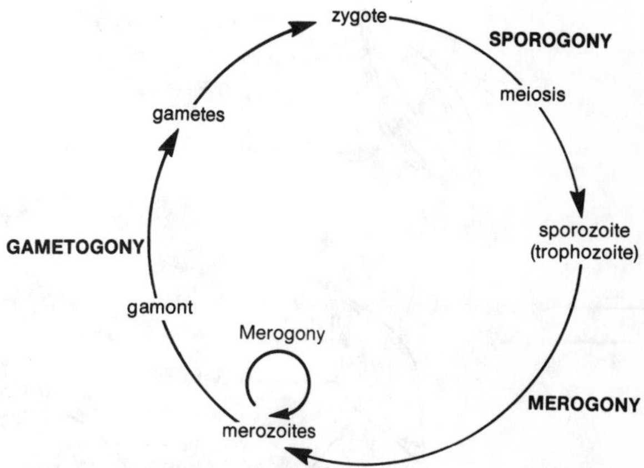

Figure 1.17. Generalized life cycle of Apicomplexa.

Gregarines. There are two types of gregarines. (1) Aseptate (acephaline) forms lack septa dividing the body into segments (Fig. 1.18). These gregarines are found especially in the coeloms of polychaetes, oligochaetes, sipunculans, nemerteans, and mollusks.

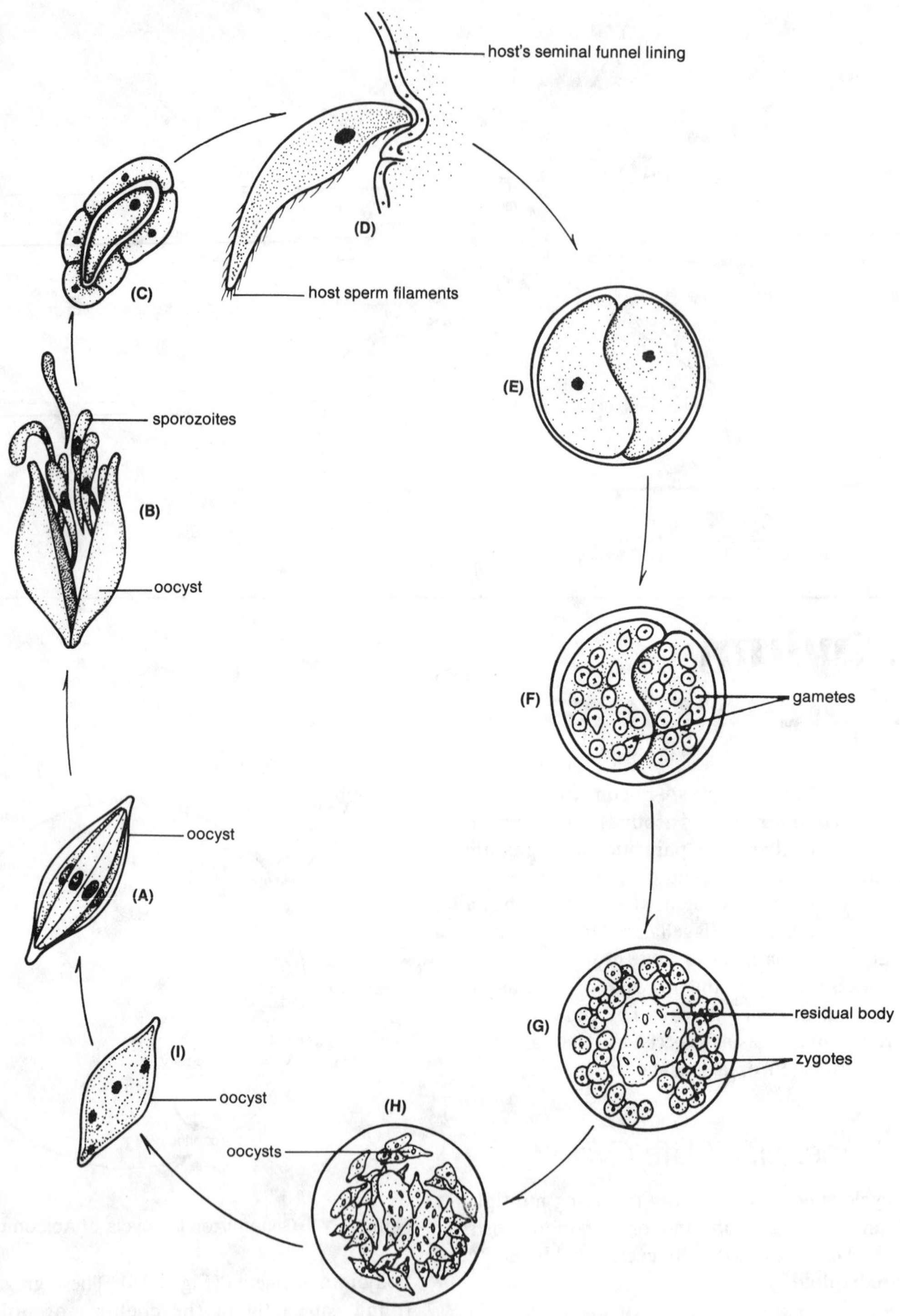

Figure 1.18. *Monocystis* life cycle. **(A)** Oocyst containing and releasing **(B)** sporozoites. **(C)** Young trophozoite. **(D)** Mature trophozoite with sperm tails clustered about the outer surface. **(E)** Pair of encysted trophozoites. **(F)** Production of isogametes. **(G)** Zygotes within a gametocyst. **(H)** Oocyst developing. **(I)** One oocyst.

(2) Septate (cephaline) gregarines have septa dividing the body (Fig. 1.19). The anteriormost segment is a modified holdfast or **epimerite** that usually breaks off when the trophozoite detaches from the host. The middle segment is the **protomerite** and the posterior **deutomerite** contains the nucleus. Septate gregarines are common in the guts of insects, millipedes, and crustaceans.

Monocystis: An Aseptate Gregarine. *Monocystis* is a large genus with about 72 species. The trophozoites (109 to 244 μm long) of most species live in the seminal vesicles of earthworms, where they feed on developing earthworm sperm cells (Fig. 1.18). Tiny filaments, seen on the surface of the trophozoite, are tails of disintegrated spermatozoa. The life cycle of *Monocystis* is similar to that of most gregarines (Fig. 1.18). The trophozoites pair off and unite together in a process called syzygy (SIZ-i-ge; G., *zygon*, yoke). A cyst wall forms around the trophozoites. Each trophozoite, now a **gamont**, undergoes multiple fission, producing isogametes that undergo syngamy with isogametes from the other gamont, so forming zygotes within a gametocyst. Each zygote is surrounded by its own wall and develops into a biconical **oocyst** (sporocyst). Each oocyst (17 to 25 μm by 8 to 10 μm) forms eight **sporozoites** by meiosis and mitosis (sporogony). Oocysts are released when the earthworm dies and decays or perhaps from feces of the live earthworm. Eight sporozoites are released in the next earthworm's gut from each oocyst ingested. The sporozoites traverse the gut wall and enter the seminal vesicles, where they mature into trophozoites.

Observational Procedure: *Monocystis*

Examine prepared slides (section and smear) of *Monocystis agilis* showing developmental stages in the seminal vesicle of earthworms. Identify as many stages as possible, using Fig. 1.18 as a guide. Is any one stage more abundant than the rest?

To obtain living *Monocystis*, put an earthworm in 7 percent alcohol for about 30 minutes. When the

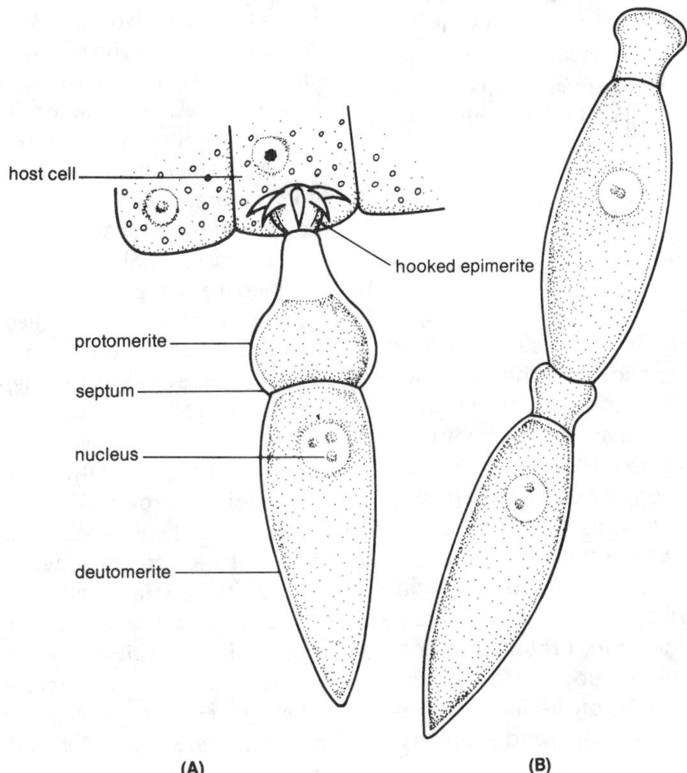

Figure 1.19. **(A)** Cephaline gregarine attached to host. **(B)** Two gregarines in syzygy. Note the lack of the epimerite.

worm is fully anesthetized, take sharp-pointed scissors or a scalpel and make a shallow, dorsal cut in the body wall from segments 10 to 15. Extend two cuts each through both sides of the body wall. The cream-colored **seminal vesicles** should be evident. Excise a small part of the vesicle and place it on a clean slide. Add a drop of water to the vesicle and tease apart with dissecting needles to make a very thin smear. Cover with a coverslip and examine under low and high magnifications of a compound microscope. Do the stages differ from those observed in the prepared slides?

***Gregarina*:** A Septate Gregarine. *Gregarina* is the largest genus of septate gregarines. The mature trophozoite (350 to 500 μm by 185 to 200 μm) parasitizes the guts of insects (Fig. 1.19). With maturity, the epimerite detaches from the body and two trophozoites unite head-to-tail in syzygy. A wall forms about the two, but each trophozoite maintains its individuality. Each trophozoite develops into a spherical gamont within its own cyst wall. Multiple fission occurs inside each gametocyst, resulting in many gametes; the gametes are released with breakdown of the gametocyst wall. Syngamy results in zygotes that develop into oocysts. Infection to the host is by ingestion of oocysts. Meiosis inside each oocyst results in eight sporozoites. The sporozoites emerge in the host after the occyst is ingested and soon develop into trophozoites.

Observational Procedure: *Gregarina*

Obtain an insect such as a grasshopper or cockroach and cut off the head. Open the body with sharp-pointed scissors or scalpel. Locate the tube-shaped intestine in the ventral area of the body and remove carefully with forceps. Place the intestine on a glass slide, tease it open, and add a drop or two of water. Cover the gut contents with a coverslip and examine with low power of a microscope. Change to high magnification once the organism has been located. Locate the immature trophozoites. Can you distinguish the **epimerite**, **protomerite**, and **deutomerite**? Based on your study of *Monocystis*, can you identify other stages of the life cycle of *Gregarina*? Study prepared slides and compare with the living organism.

Malarial Coccidians. The coccidians are perhaps the most important infectious disease-causing parasites of man. The suborder Haemospororina (subclass Coccidiasina) contains the malaria-causing parasites (*Plasmodium*) and allies (e.g., *Haemoproteus*). Over 3.5 million humans die annually from malaria; some workers believe that malaria is responsible for over one-half of all human deaths worldwide. The Haemospororina contains blood parasites of vertebrates, transmitted by a blood-sucking dipteran such as a mosquito, blackfly, or midge. The vector of *Plasmodium* in humans is the female *Anopheles* mosquito. Four species of *Plasmodium* cause malaria in humans (*P. falciparum*, *P. malariae*, *P. ovale*, and *P. vivax*) and these species differ in details related to frequency of multiple fission, morphology of schizonts (merozoites), and gametocytes (gamonts). A brief study of *P. vivax* will be made (Fig. 1.20).

***Plasmodium vivax*.** Of the four malarial diseases in humans, the type caused by *Plasmodium vivax* is the most common and widespread, but not the most severe. When an infected female *Anopheles* mosquito takes a blood meal (only females bite), the spindle-shaped sporozoites leave her salivary glands and enter the victim's bloodstream. Within a few minutes, the sporozoites enter primarily liver cells. Here they grow and multiply by several schizogonous divisions, producing **merozoites**. These merozoites are sometimes called metacryptozoites; however, they are morphologically similar to other merozoites. The merozoites enter the bloodstream, invade the erythrocytes, and transform into **trophozoites**. A young trophozoite inside an erythrocyte resembles a signet ring: the single nucleus is ruby red and is perched on one side of a blue ring of cytoplasm. This is often called the **ring stage**. Inside the erythrocytes, the ring-shaped trophozoites grow and undergo several nuclear divisions. The erythrocytes increase in size to accommodate the growing trophozoites. The cytoplasm of an infected erythrocyte contains fine spots called Schuffner's dots. In older trophozoites a brownish iron-containing residue from metabolized hemoglobin may be observed. The multinucleated trophozoite is an immature **schizont** (merozoite) distinguished by having many nuclei throughout its cytoplasm. With maturity, the cytoplasm of the schizont divides into distinct smaller bodies each with a nucleus. The mature schizont is called a **segmenter**. When the erythrocyte ruptures, the small nucleated merozoites are released into the bloodstream. Chills and fevers, so characteristic of malaria, occur when toxic wastes are released in the bloodstream with

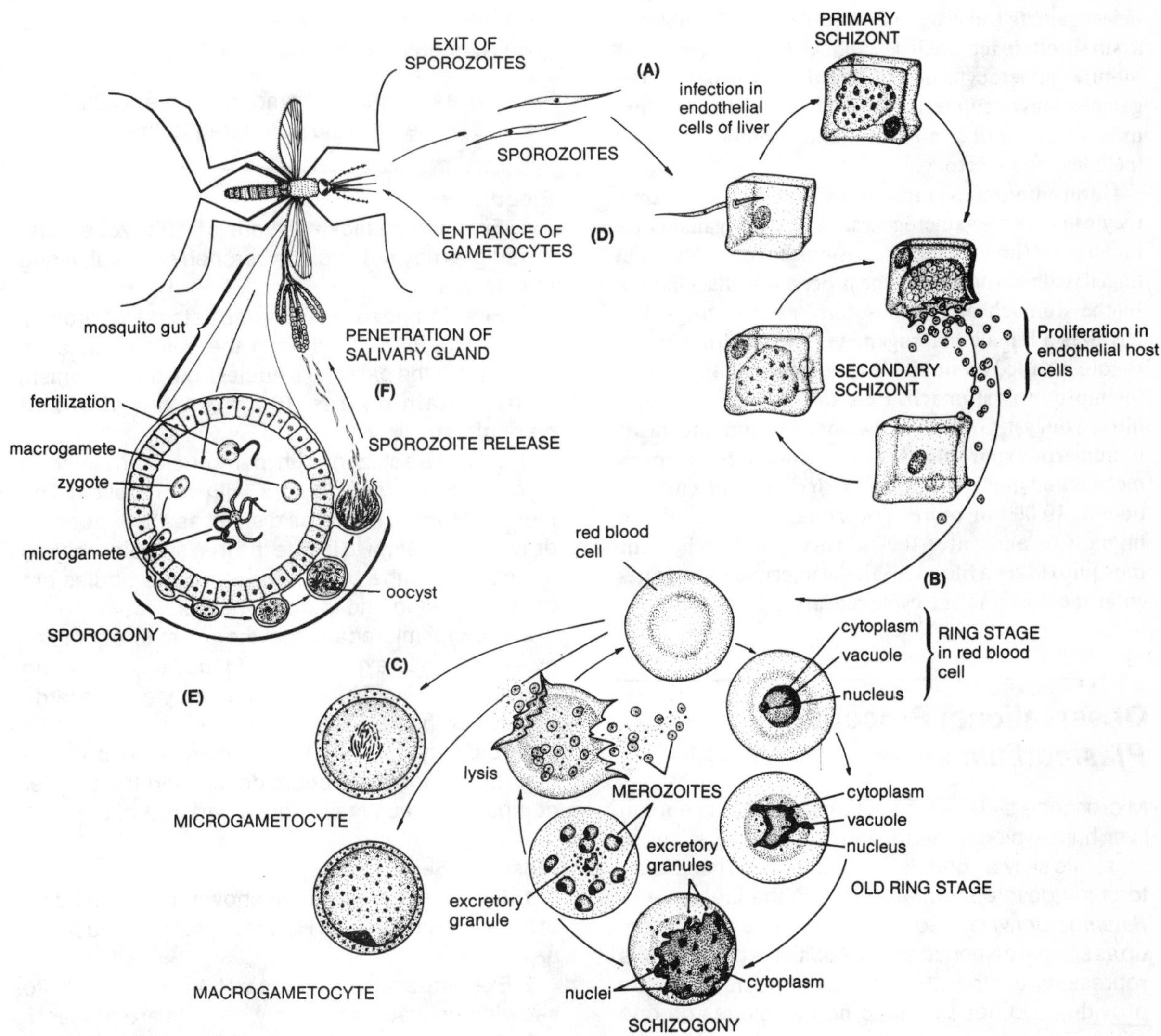

Figure 1.20. Life cycle of *Plasmodium.* **(A)** Infective stage (sporozoites) enter human, invade liver cells, undergo multiple fission (merogony), and become merozoites. **(B)** Instead of reinvading liver cells, some merozoites enter erythrocytes. These merozoites increase in numbers by multiple fission. **(C)** Some merozoites become microgametocytes (potential male gametes) and macrogametocytes (potential female gametes). **(D)** Gametocytes enter mosquito with a blood meal and become gametes. **(E)** Fertilization of macrogamete and microgamete results in a motile zygote (ookinete). The ookinete encysts on the mosquito's stomach wall and becomes an oocyst. Inside the oocyst the sporozoites develop. **(F)** With rupture of the oocyst, the released sporozoites enter the salivary glands of the mosquito and remain there until a blood meal is taken.

rupture of erythrocytes. The released merozoites invade other erythrocytes. Some merozoites develop into male and female **gametocytes** (gamonts). The male **microgamete** (microgamont) has a pale cytoplasm and a large nucleus with irregular granules. The female **macrogamete** (macrogamont) has a dense cytoplasm, a small compact nucleus, and a dark nucleolus. A mature gametocyte nearly fills the erythrocyte. The gametocytes circulate in the blood and further development ceases until a mosquito takes a blood meal that includes the gametocytes.

Upon entering the mosquito's stomach, each gametocyte undergoes gametogenesis. One microgametocyte undergoes three nuclear divisions to produce eight flagellated microgametes, the process of **exflagellation**. In the stomach of the mosquito, each microgamete fertilizes a female macrogamete. The resulting motile zygote is called an **ookinete** that lodges on the wall of the mosquito's stomach. Here the ookinete develops into an **oocyst**. Growth of the oocyst occurs and inside it, numerous **sporoblasts** (future sporozoites) form by meiosis and multiple fission. With rupture of a mature oocyst, 10,000 or more sporozoites are released and migrate to and enter the salivary gland. When the mosquito takes a blood meal, the infective sporozoites enter the host and the cycle repeats.

Observational Procedure: *Plasmodium vivax*

Microscope slides with sections of infected human liver, human blood smears, and preparations showing mosquito salivary glands and stomach are necessary to study developmental stages in the life cycle of *Plasmodium vivax*. Some of the stages may be set up as special demonstrations. Additional preparations representing other *Plasmodium* species may be provided. Do not expect to find all stages on one slide. Examine the slides under high power of a compound microscope and then under oil immersion. Extreme care must be exercised when using oil immersion to prevent breaking the slide and causing damage to the lens. Refer to Fig. 1.20 as you make your observations.

Liver Section

Find the large schizonts among the liver cells. Note the size of the parasite compared to that of a liver cell.

Blood Smear

1. Find a ring stage of a young trophozoite. Note the single ruby red nucleus perched on a bluish ring of cytoplasm.

2. Find trophozoites in various stages of growth. Observe Schuffner's dots in the cell's cytoplasm. Distinguish the parasite's nuclei from the brownish, iron-containing residue from metabolized hemoglobin.

3. Locate a schizont with many nuclei in a uniform cytoplasm and segmenters with merozoites. The merozoites may not be as distinct as they appear in drawings. Attempt to locate the free spindle-shaped merozoites in the blood. These minute bodies are often difficult to find in most smears.

4. Look for microgametocytes and macrogametocytes. You may experience difficulty differentiating between the two. The microgametocyte has a large nucleus with irregularly distributed granules and a pale blue cytoplasm. The blue cytoplasm of the macrogametocyte appears dense, and the smaller compact nucleus may show a dark red nucleolus.

Mosquito Sections

1. Observe a preparation showing exflagellation of the microgametocyte. How many tail-like structures are present?

2. Examine a stomach preparation of the mosquito showing oocysts. How many oocysts are present?

3. Observe a preparation showing sporozoites. Describe their shape.

E. Phylum Myxozoa

These are obligate, extracellular parasites in body cavities, gills, and various tissues of poikilothermic vertebrates (especially fish) and annelids. Entire fish populations may be destroyed by these parasites. Unlike the Apicomplexa and microsporans (phylum Microspora), a spore with one or two ameboid

sporoplasms and one to six (usually two) **polar capsules** and **coiled filaments** are present (Fig. 1.21). The polar filaments are holdfast structures. Each uninucleate ameboid sporoplasm enters the host when a spore is ingested and grows into a multinucleate trophozoite. The trophozoite feeds on the host's tissues and undergoes multiple fission (schizogony), thereby increasing the infection. Eventually, sporogony occurs and one or more spores are produced.

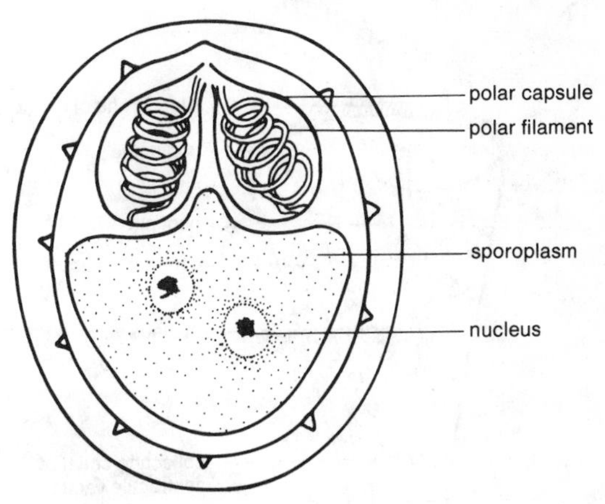

Figure 1.21. *Myxosoma*.

Observational Procedure: *Myxosoma*

Examine prepared slides of *Myxosoma* or other myxozoans with a microscope. Specimens can often be found in the gills of certain fish. Note the shape and size of the parasite. If spores are available, locate the polar filaments, polar capsules, sporoplasms, and nuclei.

F. Phylum Ciliophora

The approximate 8000 species of Ciliophora (SIL-e-OF-or-a; G., *cilium*, eyelash + G., *phora*, to bear) occur in all major ecological habitats. Most are free living and phagotrophic. Ciliates are the most structually complex of protozoans. Unique features include cilia, **dikaryotic nuclei**, complex cortex and associated organelles, and modes of reproduction. Most ciliates have a cytostome and cytopharynx for ingestion of food, at least one fixed contractile vacuole to regulate cytoplasmic water content and ion concentration, and a **cytoproct** (cytopyge) for egestion of digested food from food vacuoles. A variety of organelles collectively called **extrusomes** release materials from the cell. These include **mucocysts** that coat the body and help form cysts, **toxicysts** that eject toxins to capture prey, **haptocysts** for prey capture, and **trichocysts,** which are nontoxic explosive organelles whose exact function is unknown.

The dikaryotic nuclei distinguish ciliates from other protozoans. The small, compact **micronucleus** is involved primarily in genetic and sexual recombination, undergoing mitosis and meiosis. The larger, polyploid **macronucleus** regulates the metabolic processes of the cell. The macronucleus develops from the micronucleus, but can usually replicate once formed.

Certain ciliates have **membranelles** and **cirri**, specialized ciliary organelles used in locomotion and food getting (Fig. 1.22). Membranelles, composed of two to three rows of short cilia, appear as a thin membrane, whereas cirri are ciliary tufts that function as a unit. Cilia usually cover most of the cell. The outer portion of the ciliate body is called the **cortex**. This includes the outer plasma membrane and the membrane-lined alveoli (together called the **pellicle**) and the **infraciliature** (Fig. 1.23). The latter is a complex network consisting of **kinetosomes** and associated microtubules and fine, striated kinetodesmal fibrils. The kinetosomes plus associated fibrils form kinetids, which may be arranged in a row called a kinety. Kinetids may have one (**monokinetid**), two (**dikinetid**), or more (**polykinetid**) kinetosomes. Kinetid distribution and specializations over the ciliate body are used in taxonomy.

Reproduction is primarily by transverse binary fission; the plane of division is across the long axis of the body (homothetogenic fission). This is unlike

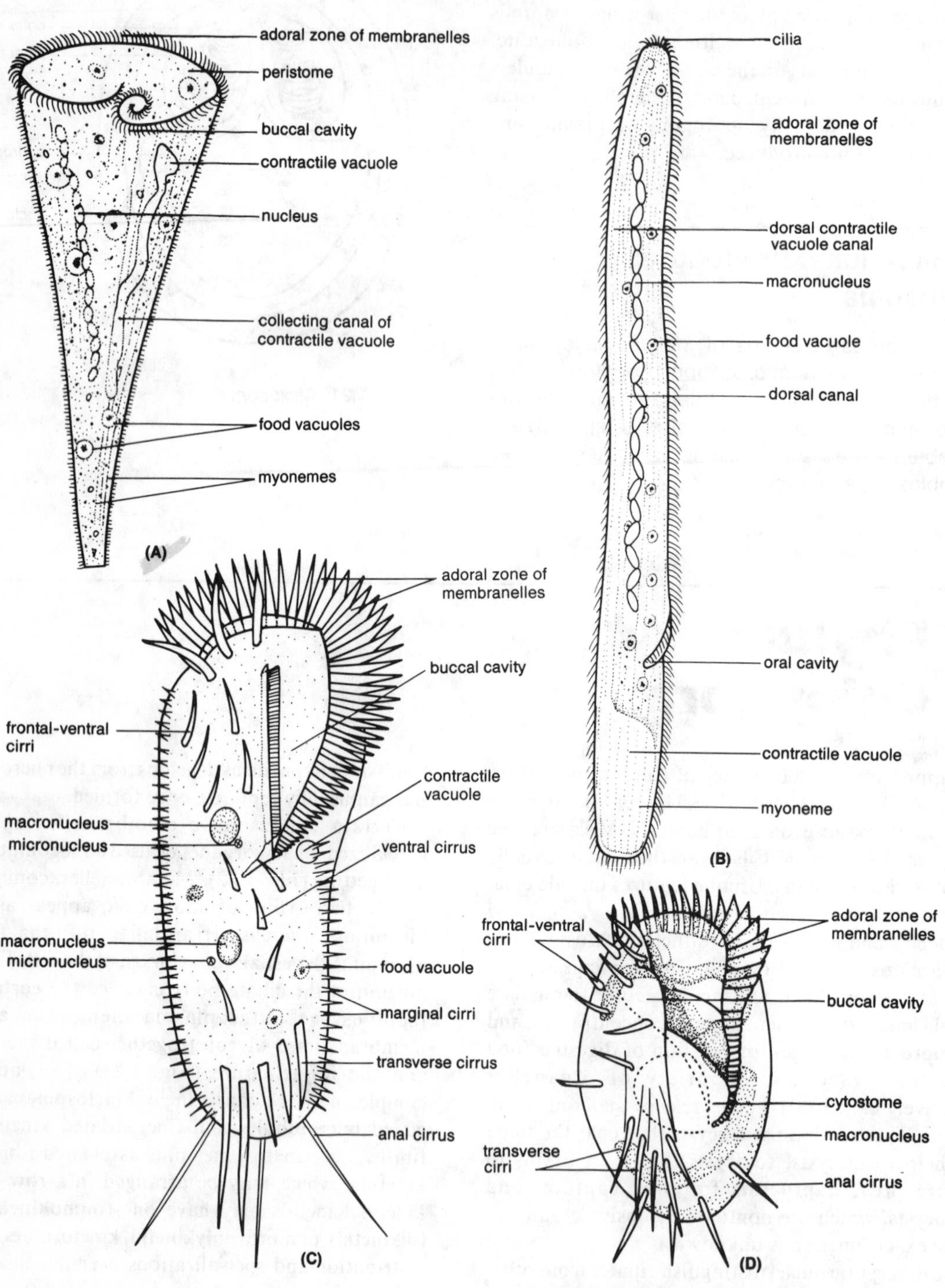

Figure 1.22. **(A)** *Stentor.* **(B)** *Spirostomum ambiguum.* **(C)** *Stylonychia.* **(D)** *Euplotes.*

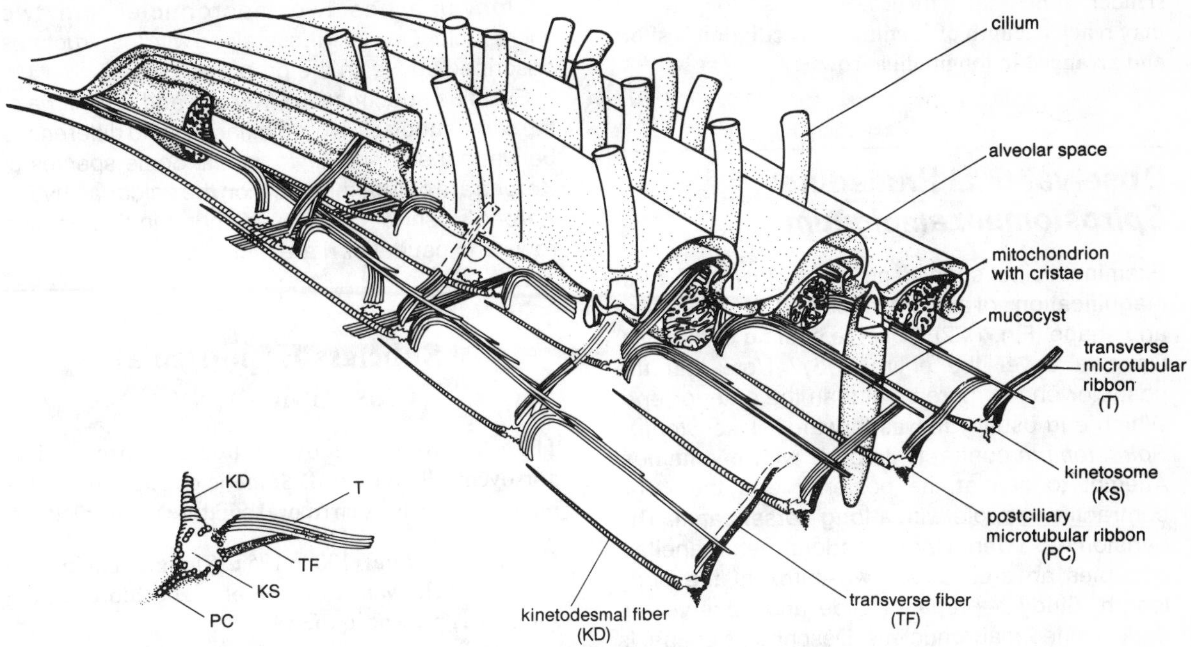

Figure 1.23. Structure of cortex and associated infraciliature of a ciliate. (From Peck 1977.)

symmetrogenic fission of flagellates, where the longitudinal furrow forms between rows of kinetosomes, if present. Budding occurs in some ciliates.

Free gametes are not formed during the sexual process; rather, micronuclei are exchanged between two sexually compatible individuals in a process called conjugation.

The recent classification scheme of Small and Lynn (in Lee et al. 1985) published in *An Illustrated Guide to the Protozoa* by the Society of Protozoologists (1985) divides the phylum Ciliophora into eight classes: Karyorelictea, Spirotrichea, Prostomatea, Litostomatea, Phyllopharyngea, Nassophorea, Oligohymenophorea, and Colpodea. Representatives of six classes will be studied in this exercise.

Subclass Heterotrichia (Class Spirotrichea)

These usually large, contractile species have conspicuous oral ciliature, well-defined myonemes, and body cilia uniformly distributed. The body shape is variable.

Stentor coeruleus. *Stentor coeruleus* is a large (1 to 2 mm long), freshwater ciliate with a conspicuous adoral zone of membranelles around the ciliated peristome. The body is like a vase or trumpet in shape.

Observational Procedure: *Stentor coeruleus*

Place a drop of the *Stentor* culture in a ring of methyl cellulose and cover with a coverslip. Observe with both low and high magnifications of a compound microscope. Note the size, shape, and bluish color derived from a cytoplasmic pigment called **stentorin** (Fig. 1.22). Which end moves forward in a swimming specimen? Locate an individual that has become attached. Observe the active-moving adoral membranelles around the peristome. In what direction are these fused rows of cilia beating? *Stentor* is a suspension feeder. Can you see the area where the membranelles lead to a spiral-shaped buccal cavity (oral cavity)? Your specimen should extend and contract its body. Longitudinal contractile filaments called **myonemes** shorten the body; microtubular arrays lengthen the body by sliding. Myonemes lie in the pellicle and may be observed in stained preparations. Locate the contractile vacuole with its long collecting canal. Is its position constant or variable? Attempt to see the moniliform-shaped (beadlike) macronucleus. The macronucleus and other structures can be seen in the prepared slide.

Spirostomum ambiguum. This freshwater ciliate is slender, somewhat flattened, and very elongated. It may reach a length of 3.5 mm. Body ciliation is short and arranged in longitudinal rows.

Observational Procedure: *Spirostomum ambiguum*

Examine living specimens under low and high magnifications of a microscope and note their size and shape (Fig. 1.22). Observe beating of the cilia. Do cilia cover the entire body? Note that the posterior end is more blunt than the anterior end. Which end usually moves forward? Like *Stentor*, *Spirostomum* contracts its body with **myonemes**. Attempt to see at the posterior end the large contractile vacuole with a long dorsal canal. The peristome, bordered by short adoral membranelles, occupies an area about two-thirds of the body length. Study a prepared slide and observe the dark-stained macronucleus. Describe its shape. Is the micronucleus evident?

Subclass Stichotrichia (Class Spirotrichea)

The body shape of these ciliates is usually ovate and dorsoventrally flattened. Oral membranelles for food capture and cirri for locomotion are characteristic of the group.

Stylonychia. *Stylonychia* (100 to 300 μm long) is a common freshwater genus, but does occur in marine waters.

Observational Procedure: *Stylonychia*

Examine living specimens under low and high magnifications of a microscope. Describe their size and shape. Observe how the specimen darts about. Locate the three anal (caudal) **cirri**, eight frontal-ventral cirri, five ventral cirri, five transverse cirri, and several marginal cirri on the ventral surface of the body (Fig. 1.22). These fused tufts of cilia serve as leglike structures. Can you see membranelles around the wide buccal (oral) cavity? Attempt to see the contractile vacuole near the membranelles. *Stylonychia* has two macronuclei and two micronuclei. Observe these, along with the structures described above, in the prepared slide.

Examine specimens of *Kerona* and *Oxytricha* of the same subclass. Note similarities and differences between these and *Stylonychia*. Some species of *Kerona* are ectocommensal on the cnidarian hydra. *Kerona's* ventral cirri are arranged in five oblique rows and caudal cirri are lacking.

Subclass Hypotrichia (Class Spirotrichea)

These ciliates have ovoid bodies more or less dorsoventrally flattened. Somatic or body ciliation is reduced, usually as cirri; oral ciliature is conspicuous.

Euplotes. *Euplotes* (100 to 195 μm long) occur in fresh and brackish waters. Symbiotic zoochlorellae are sometimes present in their bodies.

Observational Procedure: *Euplotes*

Observe living *Euplotes*; note its size, body shape, and way of movement (Fig. 1.22). Does the organism creep, crawl, or dart about? Observe the nine to ten frontal-ventral cirri, the five transverse cirri, the four anal (caudal) cirri, and the adoral zone of membranelles. Are these involved in locomotion? Examine a stained slide and locate the large C-shaped macronucleus as well as the membranelles and cirri. The buccal cavity is broadly triangular in shape. Is a posterior contractile vacuole present?

Subclass Haptoria (Class Litostomatea)

These ciliates usually have uniformly ciliated bodies. The cytopharynx inverts on ingestion and the **proboscis** or oral dome, when present, is supported by microtubules

Didinium nasutum. This species (125 μm long) is a freshwater raptorial ciliate feeding almost exclusively on *Paramecium*. Several didiniums may attack and kill one paramecium with discharged toxicysts.

Observational Procedure: *Didinium nasutum*

Place together on a depression slide a small drop each from cultures of *Didinium* and *Paramecium*. Cover with coverslip and observe under low and high powers of a compound microscope. Note the barrel-shaped body with the posterior end round and the cone-shaped anterior end (Fig. 1.24). The cytostome, located at the tip of the proboscis, can be greatly expanded to swallow a paramecium. Describe the feeding process if *Didinium* is seen taking a paramecium. Is a contractile vacuole evident? Note the girdle of cilia around the middle of the body and another circle around the anterior end. These cilia can be seen in a prepared slide along with the macronucleus and contractile vacuole. Describe the shape of the macronucleus.

Subclass Trichostomatia (Class Litostomatea)

Most members of this subclass are endosymbionts in vertebrates. The oral region forms a densely ciliated depression called a vestibulum.

Balantidium coli. This medium-sized ciliate (50 to 100 μm long) is a common commensal in the cecum and large intestine of pigs. In man, *Balantidium coli* is a pathogenic parasite that invades the intestinal mucosa.

Observational Procedure: *Balantidium coli*

Obtain a prepared slide of *Balantidium coli* and locate the ovoid trophozoite under low and then high power of a microscope (Fig. 1.24). The anterior end is more pointed than the posterior end. Body cilia occur in oblique longitudinal rows. Attempt to locate the **cytoproct** (cell anus) at the posterior end and the **cytostome** and long **vestibulum** in an apicoventral position near the anterior end. Note the large sausage-shaped macronucleus and two contractile vacuoles, one at the posterior and one at the anterior end. The vesicular micronucleus may not be evident.

Balantidium coli is transmitted via contaminated food or water containing the ovoid cysts (40 to 60 μm in diameter) passed in the host's feces. Cysts often stain poorly and the large macronucleus and one contractile vacuole are the most conspicuous organelles observed. Look for cysts in a prepared slide. Note the size, shape, and structure of the cyst wall. Locate the large macronucleus and contractile vacuole. The micronucleus will probably not be observed. Are food vacuoles evident?

Subclass Suctoria (Class Phyllopharyngea)

Suctorians are carnivorous ciliates found in both marine and freshwater habitats. Suctorians get their name from the way they obtain their food. Mature suctorians are usually sessile, stalked, lack cilia, and have one or more tentacles. Immatures are ciliated and lack the stalk and tentacles.

Ephelota. *Ephelota* (250 μm by 220 μm; stalk up to 1.5 mm) is a marine suctorian and is often found attached to the marine hydroid *Obelia*.

Observational Procedure: *Ephelota*

Examine living specimens or prepared slides with a microscope. Note the short, thick, flat-tipped tentacles and their arrangement (Fig. 1.24). The stalk enlarges distally. Do the tentacles and stalk extend or contract in the living organism? Do you see myonemes or striations in the stalk? Suctorians have **haptocysts** at the tip of the tentacles used in feeding. Can you see a contractile vacuole? Examine a prepared slide and locate the curved, elongated macronucleus.

Subclass Nassophoria (Class Nassophorea)

These ciliates have an oral cavity that is either shallow or deep. The body ciliation is typically uniform and trichocysts are present.

Paramecium caudatum. This is the most widely distributed species of paramecium. It is 150 to 300 μm long, has a single large macronucleus, and has one compact micronucleus (Fig. 1.24). The species occurs in freshwater streams and ponds containing plants and decaying organic matter. *P. caudatum* is holozoic, living on bacteria, algae, and small animals.

Paramecium aurelia with two micronuclei and *P. multimicronucleatum* with four or more micronuclei and three to seven contractile vacuoles are similar to *P. caudatum*. *Paramecium busaria*, unlike the above-mentioned species, has unicellular green **zoochlorellae** as symbionts in its body.

Observational Procedure: *Paramecium caudatum*

1. Place a drop of the paramecium culture on a glass slide containing a ring of methyl cellulose. Cover with a coverslip and locate an individual under low power of a compound microscope.

2. Observe locomotion in the paramecium. Note the organism's forward spiral movement as it rotates on its long axis. Does the organism change its direction of movement by reversing the beat of its cilia? Can the organism bend its body?

3. Locate an individual that has ceased moving rapidly and observe under high power. The anterior end is narrower and more blunt than the broad, pointed posterior end (Fig. 1.24). Can you see beating of the cilia?

4. Locate the oblique depression on the right side of the body. This is the **oral groove**; it gives the organism an asymmetrical shape.

5. The oral groove leads to a buccal or oral cavity.

6. Note the longitudinally arranged cilia covering the entire body and lining the oral groove. The buccal cavity has on its inner surface numerous cilia arranged in two major bands. These cannot be observed with light microscope. Food trapped in the oral groove travels through the buccal cavity, then through the cytostome, where a food vacuole is formed. Formation and circulation of food vacuoles can be observed by dusting a few grains of powdered carmine on the under surface of a coverslip before it is placed over the culture sample.

7. The outermost membrane layers of the body constitute the pellicle. If a fresh drop of culture is allowed to dry on the slide uncovered, a hexagonal pattern on the pellicular surface can be seen with the compound microscope.

8. In the ectoplasm are numerous rodlike **trichocysts** that alternate with the bases of the cilia. Each trichocyst is discharged through a pellicular pore and is not reused once discharged. Trichocysts are quickly regenerated. An electron micrograph of a discharged trichocyst resembles a golfer's tee attached to a long, striated shaft. If a small drop of acetic acid, methylene blue, or Dahlia stain is added to a paramecium sample, the organisms should discharge numerous trichocysts.

9. Behind the posterior end of the oral groove is a small permanent opening through which solid wastes are voided. This is the **cytoproct** (cytopyge); it is difficult to observe in the living organism.

10. A contractile vacuole with radiating canals occurs at each end of the paramecium. These are in fixed positions and open to the outside via a pellicular pore. If another preparation is made and some of the water removed by placing a blotter or piece of filter paper near the edge of the coverslip so that the organism is flattened, the vacuoles can be observed.

11. The macronucleus and micronucleus are located near the center of the paramecium. Observe these in a prepared stained slide since they are difficult to see in the living organism. The larger macronucleus is kidney-shaped and the smaller micronucleus lies in a depression of the macronucleus.

12. Paramecium reproduces asexually by transverse binary fission (across the kineties), while sex occurs by conjugation. Examine prepared slides showing fission and conjugation in *Paramecium.*

Subclass Peritrichia (Class Oligohymenophorea)

The oral apparatus, when present, is distinct. The oral structures are located in a ventral oral cavity or deeper infundibulum. The oral region can be contracted. Sessile forms are usually stalked.

Vorticella. *Vorticella* is a solitary, but often gregarious ciliate with a long stalk (up to 4150 μm). The stalk can retract and extend because of a large fiber called a **myoneme** (spasmoneme). Members of the genus occur in fresh and marine waters. *Vorticella* is a suspension feeder on bacteria.

Observational Procedure: *Vorticella*

Place a drop of culture containing *Vorticella* in a ring of methyl cellulose. Cover with a coverslip and observe under low power of a microscope. Note the inverted bell-shaped body and stalk (Fig. 1.25). The stalk can appear like a coiled spring as it retracts

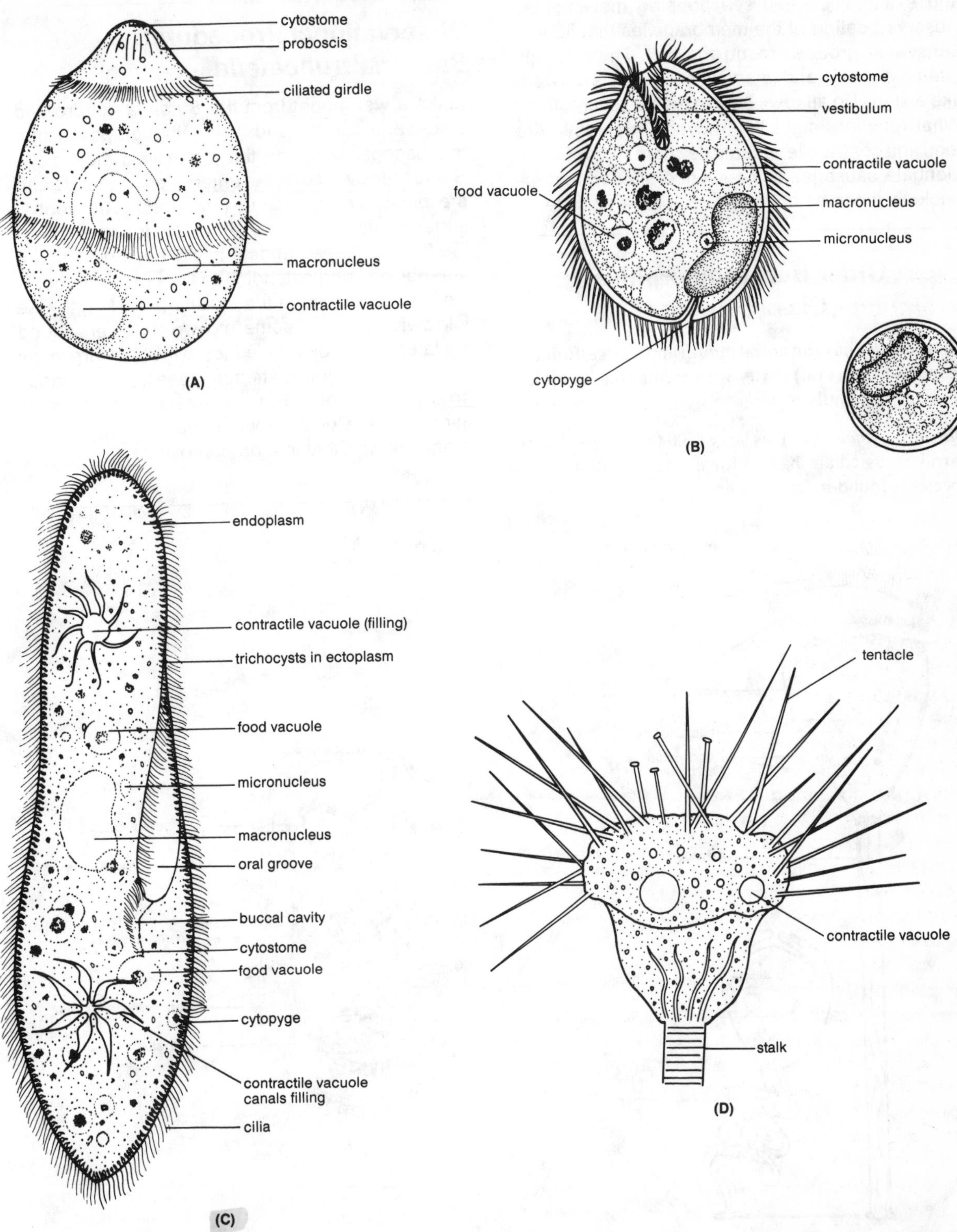

Figure 1.24. **(A)** *Didinium nasutum.* **(B)** *Balantidium coli*, trophozoite and cyst. (After Kudo.) **(C)** *Paramecium caudatum.* **(D)** *Ephelota.*

and extends. Do you see popping movements? Observe beating of the membranelles that lie in a peristomal groove around the bell. There is one inner set of cilia and one row that projects outward like a shelf. Do all rows beat in the same direction? What function might this movement serve? Are food and contractile vacuoles evident? Look for the elongate sausage- or horseshoe-shaped macronucleus. It can be observed in a prepared slide.

Order Bursariomorphida (Class Colpodea)

These ciliates have an apical mouth and a large, funnel-shaped buccal (oral) cavity with membranelles on its left and right walls.

Bursaria truncatella. This large (500 to 1000 μm long) carnivorous ciliate has uniform body ciliation. The species is found in fresh water.

Observational Procedure: *Bursaria truncatella*

Make a wet mount from the *Bursaria* culture and observe under low and then with high power of a microscope. Describe the organism's movement. *Bursaria* is ovoid with the anterior end truncate and the posterior end round (Fig. 1.25). The ventral surface is flat, but the dorsal surface is convex. Note the funnel-shaped buccal cavity beginning at the anterior end and extending to the end of the body, where it gives rise to the cytostome. A lengthwise fold divides the peristome into two chambers. Along the lateral and posterior edges of the body are many contractile vacuoles. Attempt to see the long, band-shaped macronucleus. *Bursaria* has several micronuclei. Compare what you see in the living organism with that in a prepared slide.

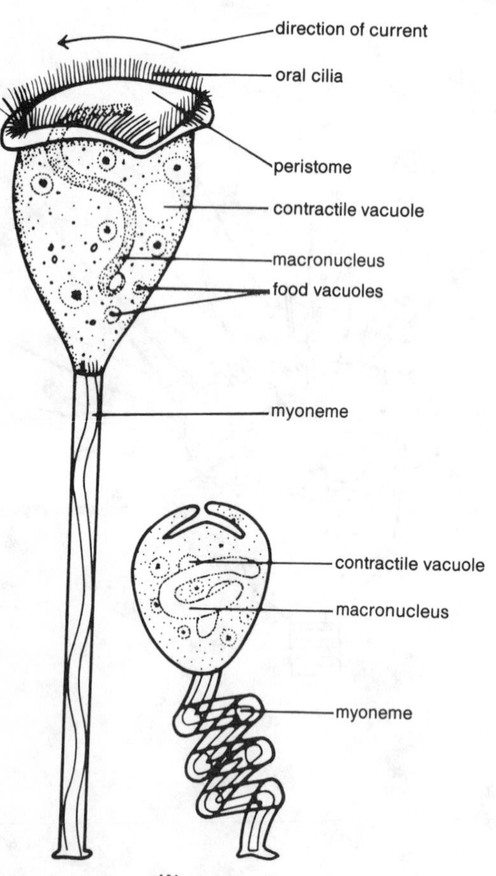

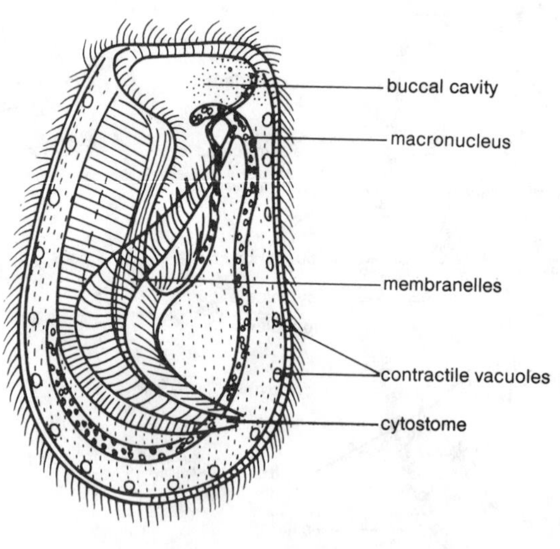

Figure 1.25. **(A)** *Vorticella.* **(B)** *Bursaria.*

Supplemental Reading

Anderson, O.R. 1983. Radiolaria. Springer, New York.

Bannister, D.H. 1972. The structure of trichocyst in *Paramecium caudatum*. J. Cell Sci. 11:899-929.

Bonner, J. T. 1983. Chemical signals of social amoebae. Sci. Am. 248:114-120.

Borror, A. C. 1973. Protozoa: Ciliophora. Marine flora and fauna of the northeastern U.S. NOAA Technical Report. NMFS circular 378. U.S. Gov't Printing Office, Washington, DC.

Bovee, E. C. 1953. Morphological identification of free-living Amoebida. Proc. Iowa Acad. Sci. 60:599-615.

Bovee, E. C., and T. L. Jahn. 1973. Locomotion and behavior of amoebae. *In*: K. W. Jeon (ed.). Biology of *Amoeba*. Academic Press, New York, pp. 249-290.

Bovee, E. C., and T. K. Sawyer. 1979. Protozoa: Sarcodina: Amoebae. Marine flora and fauna of the northeastern U.S. NOAA Technical Report. NMFS Circular 419. U.S. Gov't Printing Office, Washington, DC.

Buetow, D. E. (ed.). 1967. The biology of *Euglena*. Vols. 1 and 2. Academic Press, New York.

Buetow, D. E. (ed.). 1982. The biology of *Euglena*. Vol. 3. Academic Press, New York.

Chen, Y. T. 1950. Investigations of the biology of *Peranema trichophorum* (Euglenineae). Q. J. Microsc. Sci. 91:279-308

Coleman, A. W., and P. Heywood. 1981. Structure of the chloroplast and its DNA in chloromonadophycean algae. J. Cell Sci. 49:401-409.

Corliss, J. O. 1979. The ciliated Protozoa: characterization, classification, and guide to the literature. 2nd ed. Pergamon Press, Oxford.

Corliss, J. O. 1983. Consequences of creating new kingdoms of organisms. BioScience 33:314-318.

Corliss, J. O. 1984. The kingdom Protista and its 45 phyla. BioSystems 17:87-126.

Edds, K. T. 1981. Cytoplasmic streaming in a heliozoan. BioSystems 14:371-376.

Farmer, J. N. 1980. The protozoa: introduction to protozoology. C.V. Mosby, St. Louis, MO.

Giese, A. C. 1973. *Blepharisma*: the biology of a light-sensitive protozoan. Stanford University Press, Stanford, CA.

Grell, K. G. 1973. Protozoology. Springer-Verlag, New York.

Hall, R. P. 1953. Protozoology. Prentice-Hall, Englewood Cliffs, NJ.

Hammond. D. M., and P. L. Long (eds.). 1973. The Coccidia. University Park Press, Baltimore.

Haynes, J. R. 1981. Foraminifera. Wiley, New York.

Jahn, T. L., E. C. Bovee, and F. F. Jahn. 1979. How to know the protozoa. 2nd ed. Wm. C. Brown, Dubuque, IA.

Jeon, K. W. (ed.). 1973. The biology of *Amoeba*. Academic Press, New York.

Katz, M., D. D. Despommier, and R. Gwadz. 1983. Parasitic diseases. Springer-Verlag, New York.

Kudo, R. R. 1966. Protozoology. 5th ed. Charles C Thomas, Springfield, IL.

Leadbeater, B. S. C. 1983. Observations on the life history and ultrastructure of the marine choanoflagellate *Proterospongia choanojuncta*. J. Mar. Biol. Assoc. U.K. 63:135-160.

Lee, J. J., S. H. Hutner, and E. C. Bovee (eds.). 1985. An illustrated guide to the protozoa. Society of Protozoologists, Lawrence, KS.

Leedale, G. F. 1967. Euglenoid flagellates. Prentice-Hall, Englewood Cliffs, NJ.

Leedale, G.F. 1974. How many are the kingdoms of organisms? Taxon 23: 261-270.

Leedale, G.F. 1978. Phylogenetic criteria in euglenoid flagellates. Biosystems 10: 183-187.

Levine, N.D. 1973. Protozoan parasites of domestic animals and man. 2nd ed. Burgess, Minneapolis, MN.

Levine, N. D., J. O. Corliss, F. E. G. Cox, G. Deroux, J. Grain, B. M. Honigberg, G. F. Leedale, A. R. Loeblich III, J. Lom, D. Lynn, E. G. Merinfeld, F. C. Page, G. Poljansky, V. Sprague, J. Vavra, and F. G. Wallace. 1980. A newly revised classification of the protozoa. J. Protozool. 27: 37-58.

Long, P. L. (ed.). 1982. The biology of the Coccidia. University Park Press, Baltimore.

Margulis, L. 1974. Five-kingdom classification and the evolution and origin of cells. Evol. Biol. 7: 45-78.

Molyneux, D., and R. W. Ashford. 1983. The biology of *Trypanosoma* and *Leishmania*, parasites of man and domestic animals. Taylor & Francis, London.

Nanney, D. L. 1980. Experimental ciliatology. Wiley, New York.

Noble, E. R., and G. A. Noble. 1983. Parasitology. The biology of animal parasites. 5th ed. Lea & Febiger, Philadelphia.

Ogden, C. G., and R. H. Hedley. 1980. An atlas of freshwater testate amoebae. Oxford University Press, Oxford.

Ormerod, W. E. 1982. The life cycle of the sleeping sickness trypanosome compared with the malaria life cycle. *In*: E. U. Canning (ed.). Parasitological topics. A presentation volume to P. C. C. Garnham F.R.S. on the occasion of his 80th birthday 1981. Allen Press, Lawrence, KS, pp. 191-199.

Peck, R. K. 1977. Cortical ultrastructure of the scuticociliates *Dexiotricha media* and *Dexiotricha colpidiopsis* (Hymenostomata). J. Protozool. 24:122-134.

Pitelka, D. R. 1963. Electron-microscopic structure of protozoa. Macmillan, New York.

Pitelka, D. R. 1970. Ciliate ultrastructure; some problems in cell biology. J. Protozool. 17:1-10.

Poag, C. W. 1981. Ecologic atlas of benthic foraminifera of the Gulf of Mexico. Hutchinson Ross. Stroudsburg, PA.

Rudzinska, M. A. 1973. Do suctoria really feed by suction? BioScience 23:87-94.

Sadun, E. H., and A. P. Moon (eds.). 1972. Basic research in malaria. Proc. Helminthol. Soc. Wash. 39 (special issue).

Sarjeant, W. A. S. 1974. Fossil and living dinoflagellates. Academic Press, London.

Schmidt, G. D., and L. S. Roberts. 1985. Foundations of parasitology. 3rd ed. C.V. Mosby, St. Louis, MO.

Silva, P. C. 1980. Names of classes and families of living algae. Regnum Veg. 103:1-156.

Sleigh, M. A. 1962. The biology of cilia and flagella. Macmillan, New York.

Sleigh, M. A. 1973. The biology of Protozoa. American Elsevier, New York.

Steidinger, K. A., and E. R. Cox. 1980. Free-living dinoflagellates. *In*: E. R. Cox (ed.). Phytoflagellates. Developments in marine biology 2. Elsevier North-Holland, New York, pp 407-432.

Sweeney, B. M. 1979. The bioluminescence of dinoflagellates. *In* M. Levandowsky and S. H. Hutner (eds.). Biochemistry and physiology of Protozoa. Academic Press, New York, pp. 288-306.

Tartar, V. 1961. The biology of *Stentor*. Pergamon Press, Oxford.

Van Wagtendonk, W. J. (ed.). 1974. Paramecium; a current survey. Elsevier, New York.

Vickerman, K., and F. E. G. Cox. 1967 . The protozoa. Houghton Mifflin, Boston.

Whittaker, R. H. 1969. New concepts of kingdoms of organisms. Science 163:150-160.

Wichterman, R. 1953. The biology of *Paramecium*. McGraw-Hill, New York.

THE METAZOA

Metazoans, unlike protozoans, are composed of numerous specialized cells arranged in layers. In primitive metazoans the outer layer serves for protection and for sensory response, while the inner layer functions in digestion. Cells between these layers function in support, conduction of nutrients, reproduction, and muscular contraction. There is an increasing complexity in cellular organization, from primitive to advanced metazoans. In the least complex metazoans, cells having similar functions form aggregations known as **tissues** (tissue grade). In more advanced metazoans, two or more tissues combine to form **organs** (organ grade). In the most advanced forms, combinations of specific organs form specialized **organ systems** (system grade). Although these grades do not have taxon status, they may be thought of as a logical way of organizing the metazoan phyla.

Branch Parazoa (G., *para*, beside + G., *zoon*, animal) includes two phyla, Porifera and Placozoa. Both phyla have a cellular (or poorly defined tissue) grade of construction. In the Porifera the body has numerous pores, canals, and chambers lined with specialized flagellated feeding cells. The phylum Placozoa contains one species, *Trichoplax adhaerens*. This enigmatic organism resembles a microscopic pancake and is composed of a loose mesenchyme surrounded by layers of epithelial cells on the upper and lower sides.

Branch Mesozoa (G., *meso*, middle) contains the phylum Mesozoa, an enigmatic phylum of minute parasitic and commensal marine animals composed of a single outer cell layer and an inner reproductive layer of one or more cells. Mesozoa has been thought of as a link between protozoans and metazoans, but some workers believe that their primitive characteristics are secondary, resulting from a long evolution as obligate parasites.

Branch Eumetazoa (G., *eu*, good) includes the remainder of the metazoan phyla. These approximately 30 phyla are of the tissue, organ, or system grade of construction. See Field et al. (1988) for a phylogeny of the Metazoa based on evolutionary distances as determined by sequencing 18S ribosomal RNA from about 20 classes of animals.

Field, K. G., G. J. Olsen, D. J. Lane, S. J. Giovannoni, M.T. Ghislin, E. C. Raff, N. R. Pace, and R. A. Raff. 1988. Molecular phylogeny of the animal kingdom. Science 239:748-753.

EXERCISE 2

Phylum Porifera

Phylum Porifera (po-RIF-er-a; L., *pori*, pore + L., *ferre*, bearing), or sponges, comprises a group of simple sessile metazoans considered to have a cellular grade of construction. Sponges are either radially symmetrical or asymmetrical and have a body consistency varying from hard and stony to friable, rubbery, or gelatinous, depending on the nature and arrangement of skeletal elements. They range in size from a few millimeters to about 1 m tall and are economically important as members of the marine-fouling community and because of continued commercial importance of bath sponges.

In their simplest form, sponges resemble a tube closed at one end surrounding an inner cavity called the **spongocoel** or **atrium** (Fig. 2.1). The tube wall consists of two layers of cells arranged as concentric cylinders separated by a thin, noncellular gelatinous layer, the **mesohyl**. The outer cell layer (**pinacoderm**) is made up of flattened cells called **pinacocytes** that do not secrete a basement membrane. The inner layer consists of flagellated feeding cells called **choanocytes** or **collar cells** (Fig. 2.2). Sponges are supported internally by a calcareous or siliceous skeleton of **spicules**, or by protein **spongin** fibers. Some sponges have a combination of siliceous spicules and spongin. A large variety of different sizes (**mega-** and **microscleres**) and shapes of spicules has been described (Fig. 2.3). Spicules are absent in some sponges and a massive or reticulate calcareous skeleton is present.

The name "pore bearer" refers to the fact that sponges are perforated by numerous minute incurrent pores (**ostium**, plural **ostia**) and by one or more large excurrent pores (**osculum**, plural **oscula**). Water currents, generated by flagellar action of choanocytes, enter the sponge through ostia, pass into choanocyte (flagellated) chambers, and exit the sponge via the osculum. Viewed under high magnification of a light microscope, choanocytes appear as globular cells with long flagella surrounded at the base by a collarlike extension of the cell membrane (Fig. 2.2). Sponges feed by screening incurrent water through a progressively finer series of filters (canals to choanocyte chambers) until the finest particles are removed by the choanocytes.

The structure of sponges may be divided into three body types, based on anatomy of choanocyte chambers and incurrent and excurrent canals. These types, in increasing order of complexity, are as follows: **asconoid**, **syconoid**, and **leuconoid** (Figs. 2.1, 2.4, and 2.5).

Both asexual and sexual reproduction occur in sponges. Asexual reproduction is accomplished by fragmentation and budding, and in freshwater and a few marine sponges by production of resistant structures called **gemmules** (Fig. 2.6). Some sponges have remarkable powers of regeneration and reaggregation. Apparently, all sponges are capable of sexual reproduction. Both hermaphroditic (monoecious) and dioecious species occur in the phylum. Fertilization and development are internal, except for some species that release fertilized eggs. Two different larvae occur in sponges. Most sponges have a parenchymella larva and a few have an amphiblastula (Fig. 2.7). After a mobile phase of a few hours to several days, larvae attach to a suitable surface and metamorphose into young sponges.

Classification

Sponge taxonomy is based primarily on the chemical nature and morphology of inorganic and organic skeletal elements. However, reproductive features, comparative biochemistry, and ultrastructural histology are also important. Four classes of sponges are recognized. Reiswig and Mackie (1983) separate sponges

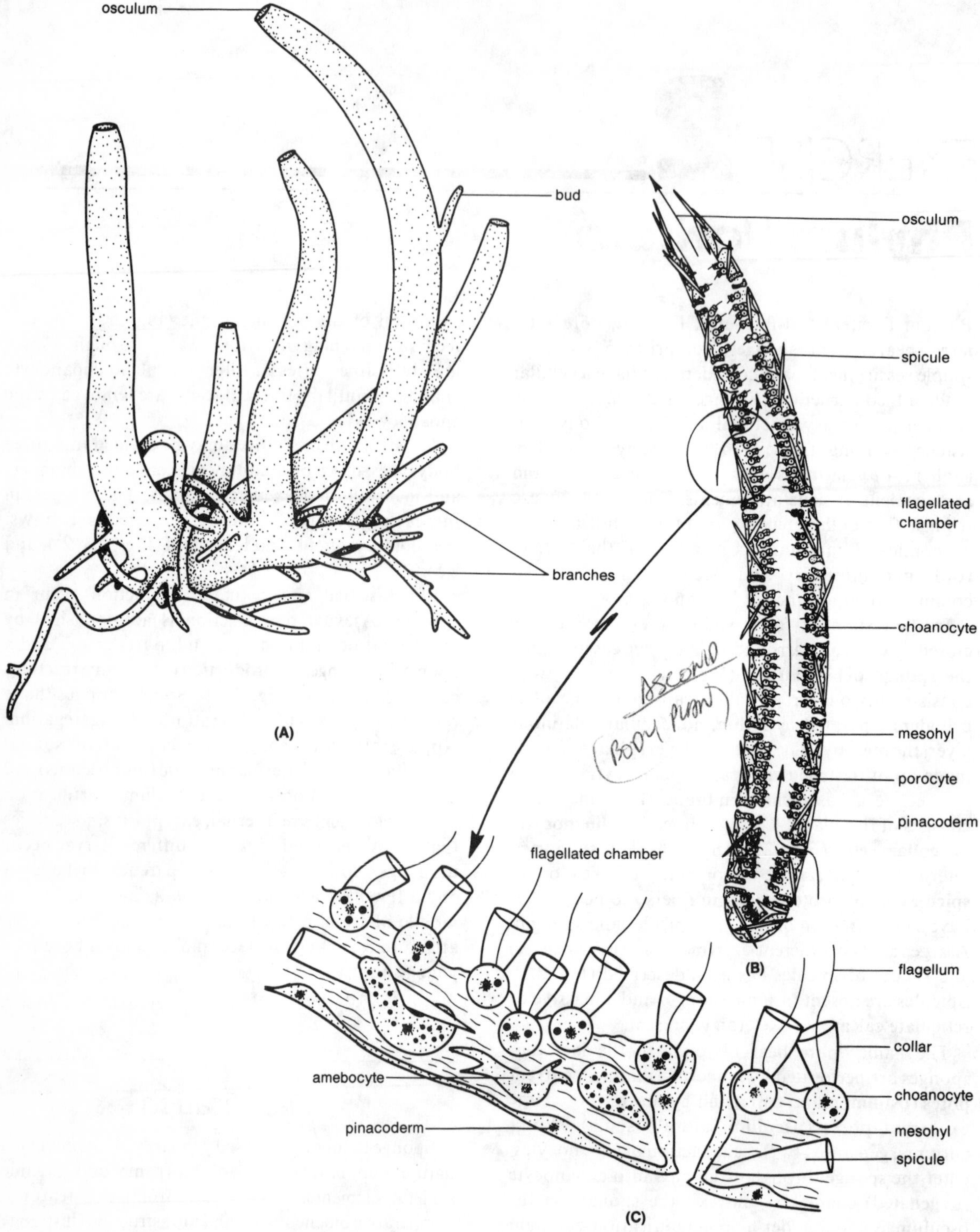

Figure 2.1. **(A)** Asconoid sponge, *Leucosolenia*. **(B)** A small portion of *Leucosolenia* colony. **(C)** A diagrammatic longitudinal section of *Leucosolenia* with an enlargement of the body wall is shown.

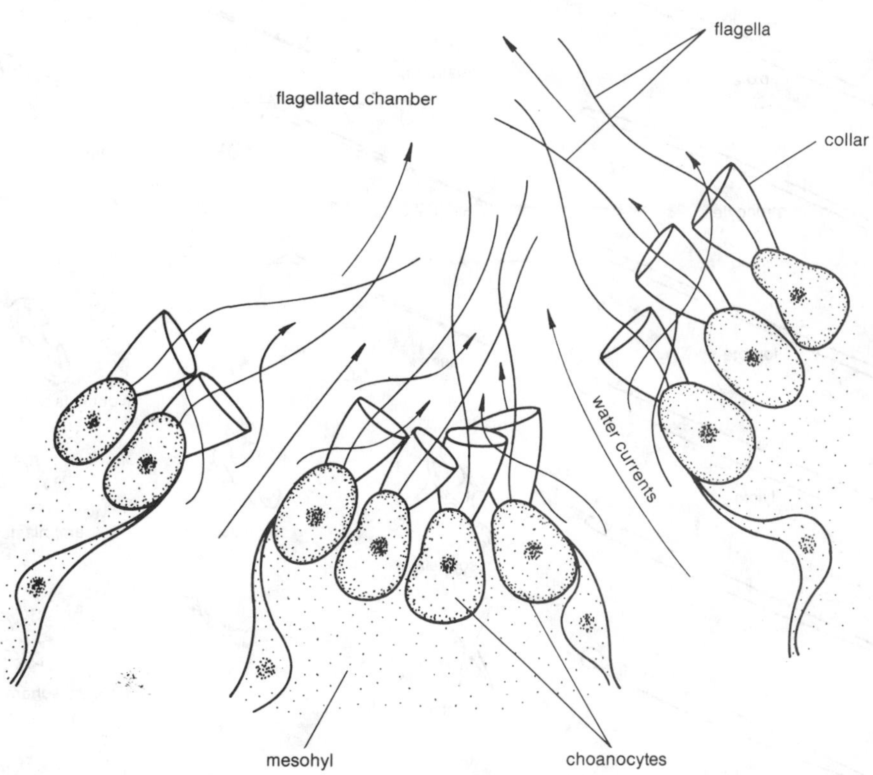

Figure 2.2. Choanocytes within flagellated chambers.

into two subphyla: **Symplasma**, comprising one class (Hexactinellida), and **Cellularia**, comprising the remaining three classes (cf. Vacelet 1985 and Bergquist 1985). Most of the 5000 species of sponges are marine; about 150 species inhabit fresh waters.

1. Class Calcarea. Sponges with spicules of calcium carbonate; massive skeletal structures also are present. Microsclere spicules lacking: entirely marine, usually located in shallow water; all three body types occur. Examples include *Leucosolenia* and *Sycon*.

2. Class Demonspongiae. Approximately 90 percent of all extant poriferans; siliceous spicules (megascleres and microscleres), spongin fibers, or both; one family, Oscarellidae, lacks any skeleton; freshwater sponges (family Spongillidae) are members of this class; only the leuconoid body type occurs. Examples include the freshwater genera *Ephydatia* and *Spongilla* and the marine genera *Cliona* (boring sponge), *Spheciospongia* (loggerhead sponge), and *Spongia* (bath sponge).

3. Class Sclerospongiae. Recently constructed class of tropical to subtropical sponges usually found in cryptic habitats and possessing massive or reticulate calcareous skeletons. Only the leuconoid body type occurs in this class. Examples are *Calcifibrospongia* and *Merlia*. Vacelet (1985) suggests that these few species (ca.15) are members of the Calcarea and Demospongiae.

4. Class Hexactinellida. Deep-water marine sponges with siliceous six-pointed spicules and long siliceous fibers forming a fused lattice structure; some hexacts have a syconoid-like body type. A prominent example is *Euplectella*, or Venus's flower basket. Because of their syncytial organization, it has been argued that hexacts should be a distinct phylum, Symplasma (Bergquist 1985).

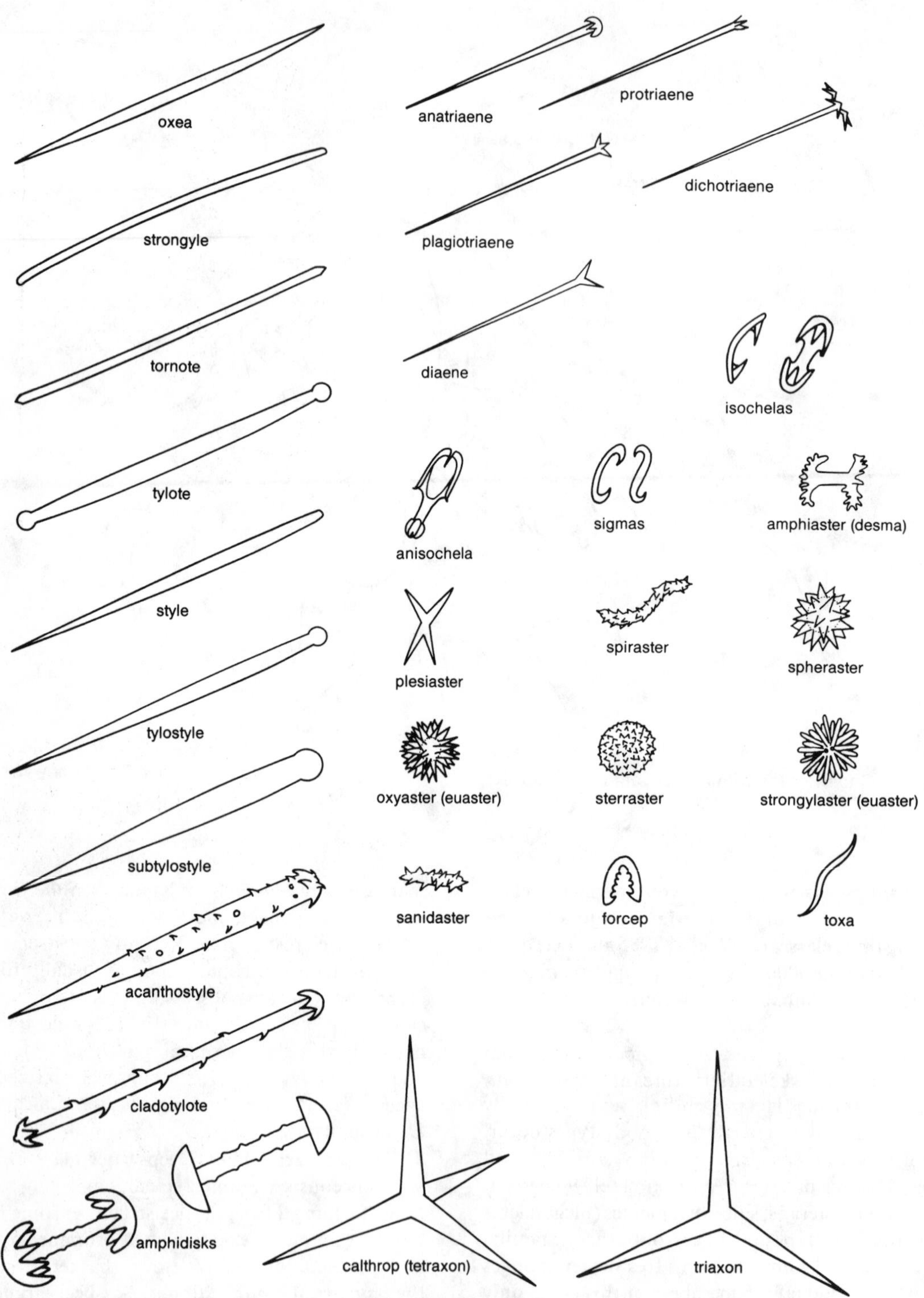

Figure 2.3. Common spicule types.

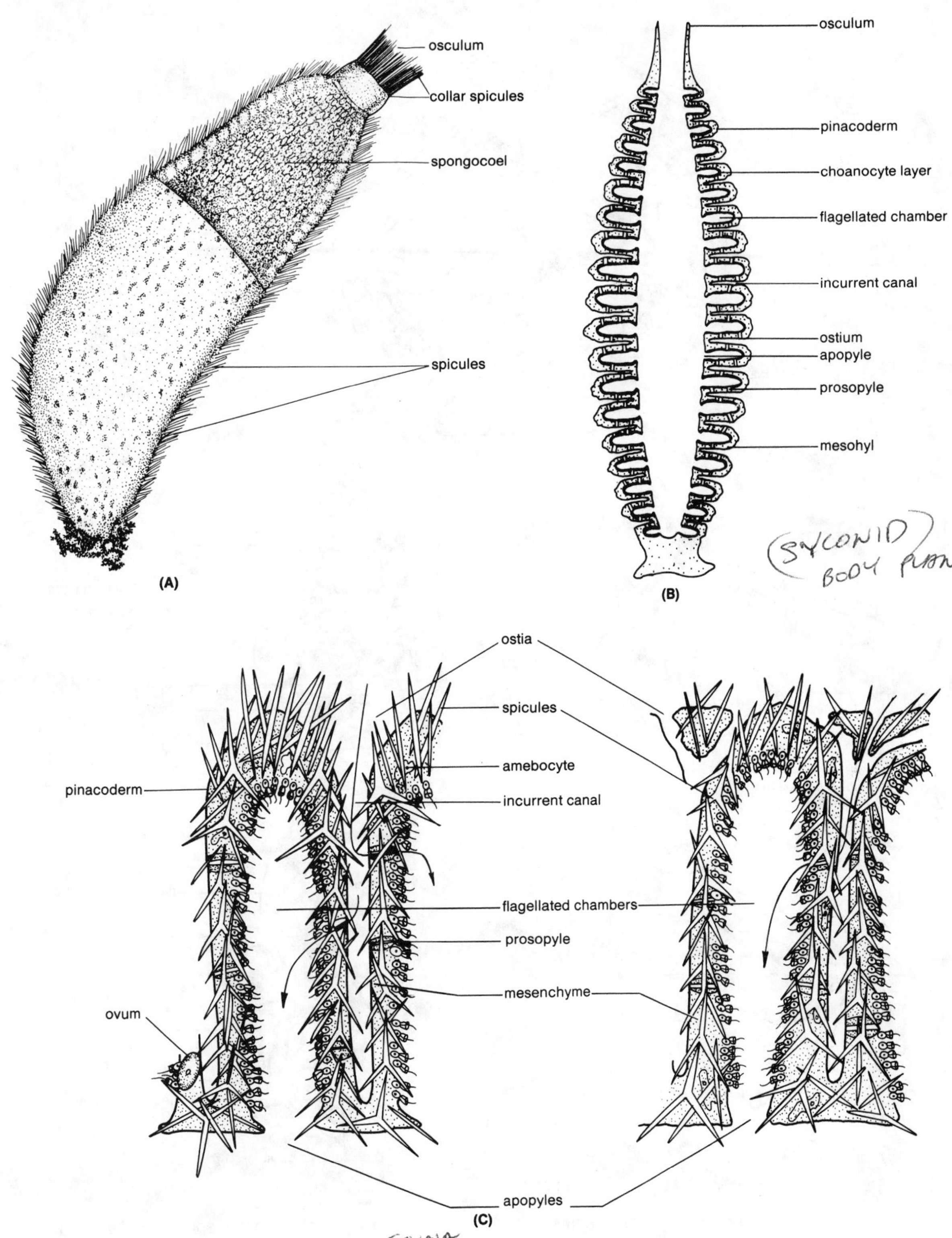

Figure 2.4. The syconoid sponge body plan of *Sycon*. **(A)** Whole specimen with a portion of the body wall removed. **(B)** Longitudinal section. **(C)** Two cross sections. The latter depicts the two stages of the syconoid canal system.

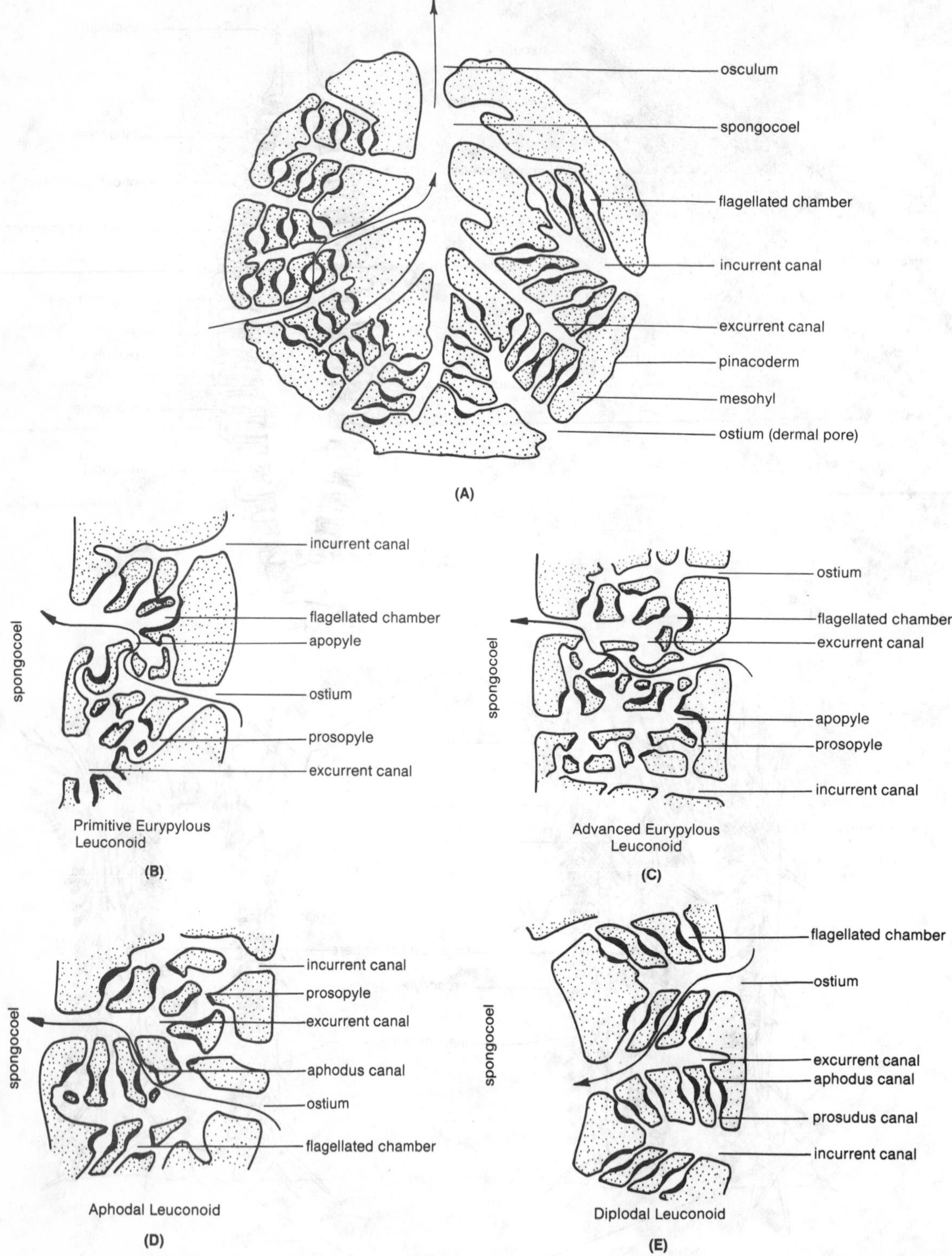

Figure 2.5. Diagrammatic views of leuconoid sponges. **(A)** Cross section of a typical leuconoid sponge and sections of the three levels of leuconoid complexity are shown: **(B)** primitive and **(C)** advanced eurypylous, **(D)** aphodal, and **(E)** diplodal. Arrows indicate water currents.

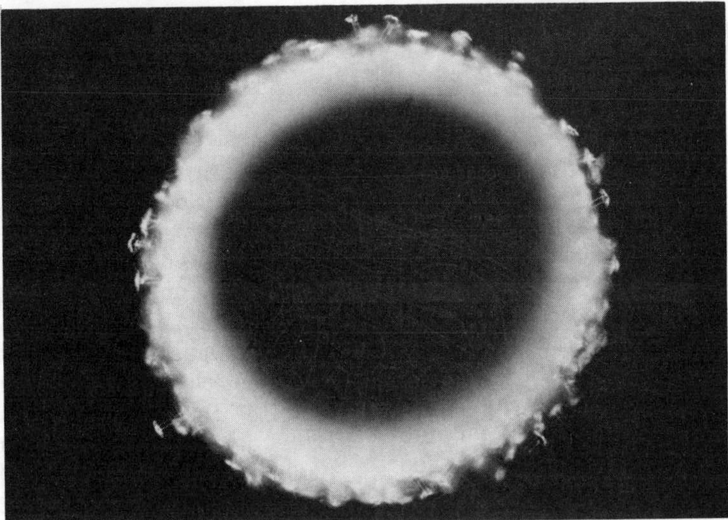

Figure 2.6. Gemmules of *Spongilla*, a freshwater sponge.

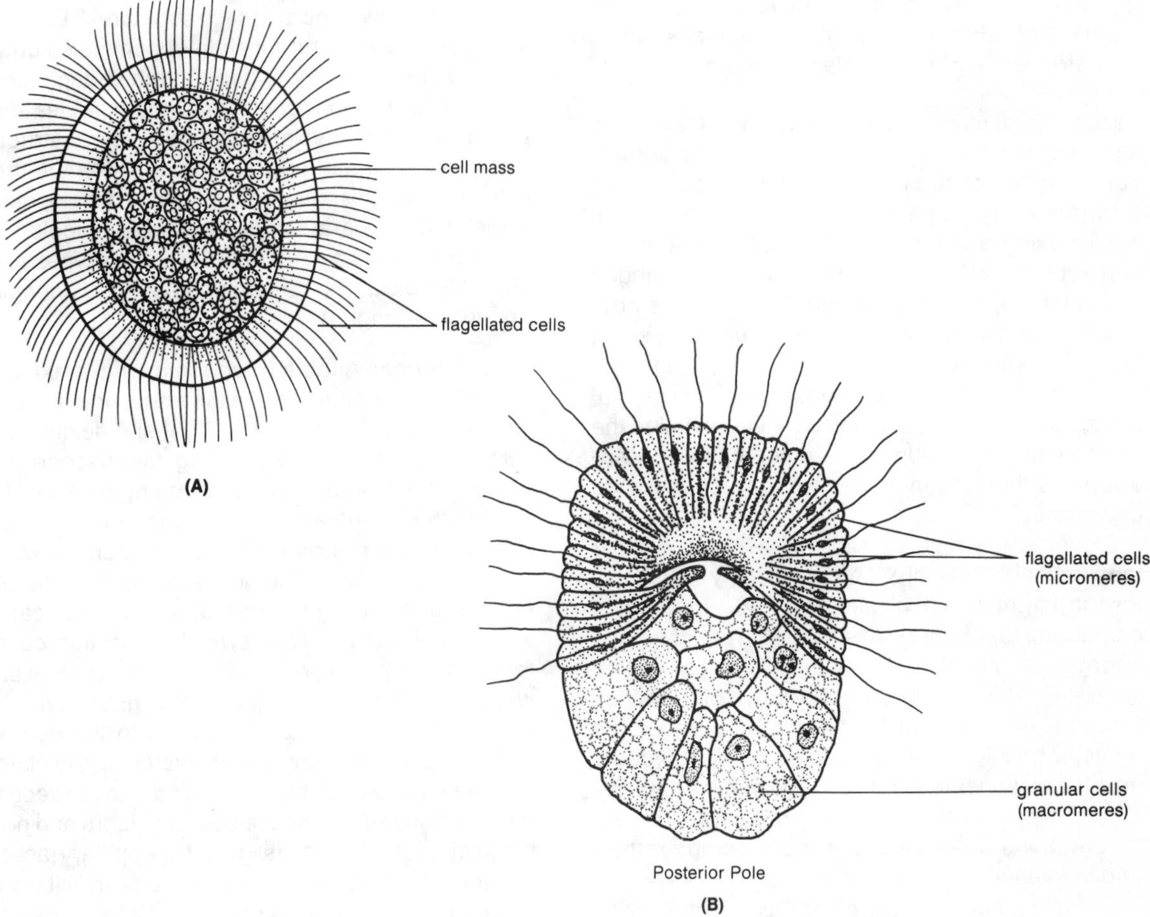

Figure 2.7. **(A)** Parenchymella and **(B)** amphiblastula larvae. (After Hyman.)

Observational Procedure

Class Calcarea

Leucosolenia: An Asconoid Sponge. Using a dissecting microscope, observe a preserved specimen of *Leucosolenia*. Keep the specimen immersed in fluid while making your observations. This small, marine sponge lives attached to rocks in the lower intertidal zone as a group of clustered, tubular structures in varying stages of growth (Fig. 2.1). The clusters result from budding. The body wall is quite thin and the surface has a rough texture, due to spicules that project through the pinacoderm. Locate the osculum, flagellated chamber, and buds. Two types of spicules may be seen in *Leucosolenia*. Place a portion of a specimen on a glass slide, tease apart with dissecting needles, and apply a small drop of bleach to the sponge. Cover the preparation with a coverslip and examine. The bleach will dissolve organic material and only spicules will remain, allowing identification of spicule types (Fig. 2.3). Observe the demonstration of birefringence in *Leucosolenia* spicules.

***Sycon* (= *Scypha*):** A Syconoid Sponge. *Sycon* is also a marine sponge. Small buds occasionally appear at the base of mature specimens and may separate from the parent. Examine a specimen of *Sycon* using a dissecting microscope and locate the osculum and the large monaxial spicules arranged around the opening as a **collar** (Fig. 2.4). Also note that the surface is interrupted by numerous small projections (Fig. 2.4). These contain the choanocyte chambers. Using a sharp scalpel, bisect the animal along its longitudinal axis and carefully examine the spongocoel. The numerous inner openings are **apopyles** that communicate with the choanocyte chambers.

Examine a prepared slide of a cross and/or longitudinal section of *Sycon* and identify the following structures: (1) pinacoderm, (2) choanocyte chambers, (3) choanocytes, (4) incurrent canal, (5) apopyle, (6) mesohyl, (7) spicules, (8) ostia, and (9) spongocoel (Fig. 2.4). The prosopyles will be indistinct. Choanocytes will not be as neatly arranged as depicted in Fig. 2.4. In a prepared cross section, note the arrangement of spicules within the sponge. Ova or embryos may be present in some preparations. How is *Sycon* more complex than *Leucosolenia*?

Make a spicule preparation of *Sycon* as described above and identify the spicule types seen (Fig. 2.3). Observe the demonstration of birefringence in *Sycon* spicules.

Class Demospongiae

Spongilla: A Leuconoid Sponge. Examine a preserved specimen of *Spongilla* under a dissecting microscope. Remember to keep the specimen covered with fluid. You should see numerous spicules projecting through the pinacoderm. Obtain a prepared thin section of *Spongilla* and observe the distribution of spicules in the sponge mass. Note the concentrations of spicules, called **fascicles**, that project through the pinacoderm (Fig. 2.8). The pinacoderm should also be visible. Choanocyte chambers are not located at the outer periphery of the sponge mass but, rather, beneath the subdermal cavity (Fig. 2.8). The choanocytes are very small and it will require great care to identify them. Compare the body form of *Spongilla* to the other species studied.

Observe a preserved specimen of *Spongilla* with gemmules. How large are these structures? Examine a prepared whole-mount slide of gemmules under high magnification, noting the tough spongin coat, in which are embedded numerous spicules (Fig. 2.6). If live gemmules are available, your instructor will provide directions for observation of their hatching and subsequent early sponge development.

Make a preparation of *Spongilla* spicules as described above and identify the spicule types seen (Fig. 2.3). Note that the siliceous spicules of *Spongilla* are not birefringent.

Bath Sponge: Although you may have used a natural bath sponge before, examine one this time with a critical eye. Note how light and flexible it is. Use a hand lens or dissecting microscope and observe the minute, interwoven spongin fibers. This is all that remains after the tissues have decayed. Bath sponges are noted for the amount of water they can hold. Since the fibers themselves cannot imbibe significant amounts of water, how can a sponge hold water? To answer this question, do the following simple exercise. Obtain a dry sponge and weigh it to the nearest tenth of a gram. Wet the sponge, wring it out, and weigh it to the nearest tenth of a gram. How much water is the damp sponge capable of holding? Under a dissecting microscope reexamine the spongin fibers and note the film of water that is held by capillary action between adjacent fibers. How many milliliters of water are held per gram dry weight of sponge?

***Cliona*:** The Boring Sponge. Examine a mollusk shell that has been colonized by *Cliona* and locate the irregularly spaced holes throughout the shell. Observe a broken or cut edge of the shell and note the extensive network of cavities within the shell. These holes were produced by sponge boring (Cobb 1969, Rützler and Rieger 1973) and represent ostia, oscula, and internal chambers. How is *Cliona* important ecologically?

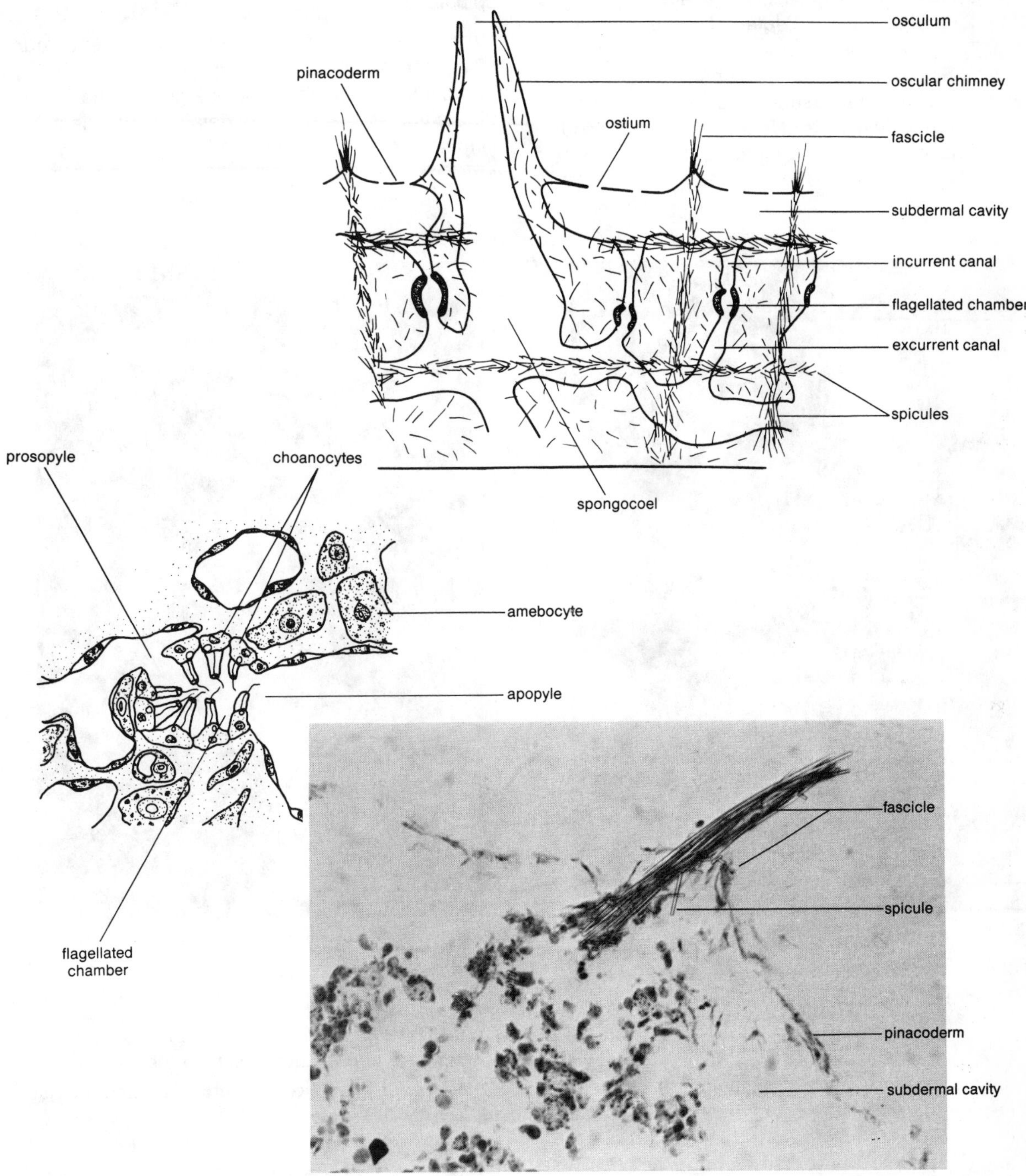

Figure 2.8. A diagrammatic view of the cross section of a freshwater sponge (*Spongilla*) with a photomicrograph of the region of the subdermal cavity. (Drawings after Pennak.)

Class Hexactinellida. *Euplectella* (Venus's flower basket) is a marine glass sponge found in waters greater than 200 m deep. Most people who have seen this sponge have viewed only the delicate skeleton. Extreme care should be taken when handling an unprotected specimen.

Observe the following aspects of general morphology (Fig. 2.9). The body of *Euplectella* is tubular and slightly curved. Internally, note the large spongocoel; remains of the commensal shrimp (*Spongicola*) often are present in the spongocoel. The osculum is covered by a **sieve plate** of fused spicules that adds additional strength to the skeleton. Carefully examine the skeleton using a dissecting microscope. Note that the skeletal framework of *Euplectella* is composed of spicules united by crossbars of silica. Note the tuft of spicules at the base. Examine these spicules using a dissecting microscope; is their surface smooth? What function might these elongate spicules serve?

Other Specimens. Observe other specimens your instructor may provide. Compare the size, shape, and skeletal elements of these specimens to the others you studied in more detail. Determine if the spicules are birefringent.

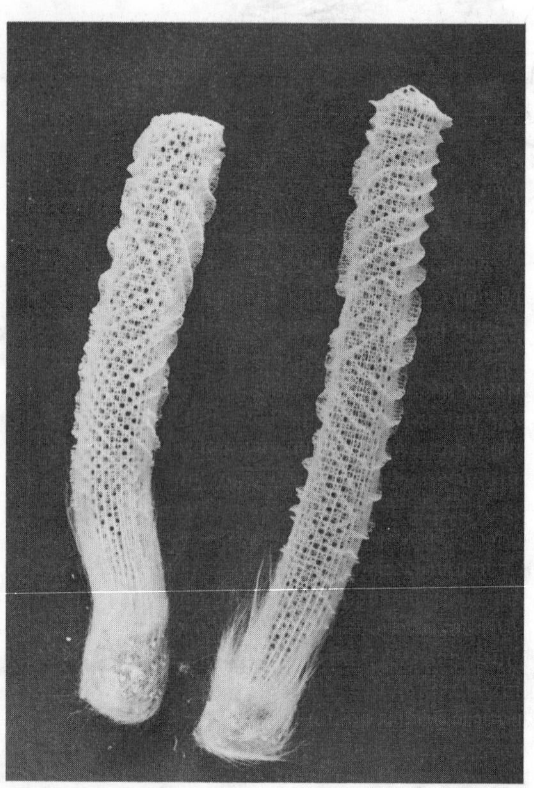

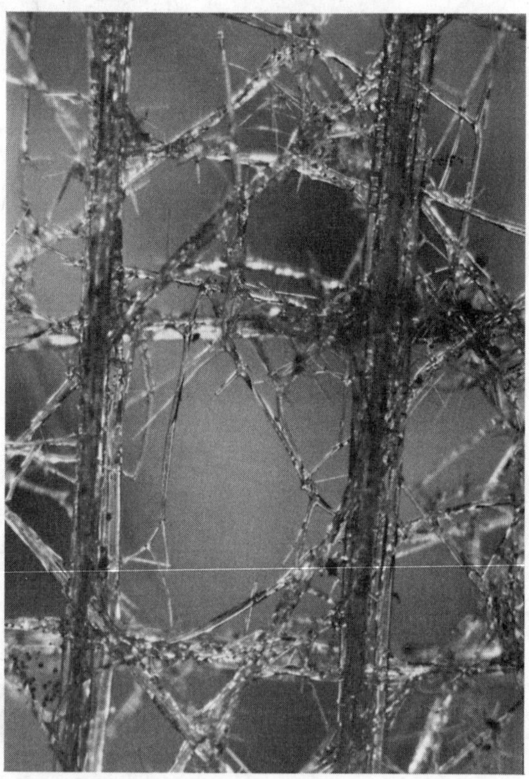

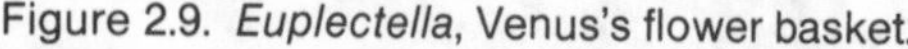

Figure 2.9. *Euplectella*, Venus's flower basket.

Fossil Forms

Many zoologists believe that sponges arose in the **Precambrian Era** from a group of protozoans similar to the modern choanoflagellates (Fig. 2.10). By the **Cambrian period**, sponges had developed an extensive fauna. However, despite their auspicious start, they are unique and difficult to relate to other phyla.

Because of the delicate nature of poriferan tissue, sponges do not preserve well and most fossil sponges are not easily recognizable. However, *Cliona* is easily recognizable as a fossil and fossil mollusk shells are often found riddled with cavities similar to those produced by this genus in modern shells.

Another interesting fossil organism is the sponge-like phylum **Archaeocyatha** (Fig. 2.11). This group of pore-bearing organisms was unlike other sponges. **Stromatoporoids**, classified by some as colonies of fossil cyanobacteria (bluegreen algae) or cnidarians, may be another extinct spongelike group.

Compare fossil and modern shells that have been colonized by *Cliona*. Examine specimens of archaeocyathids and/or other fossil sponges (Fig. 2.12).

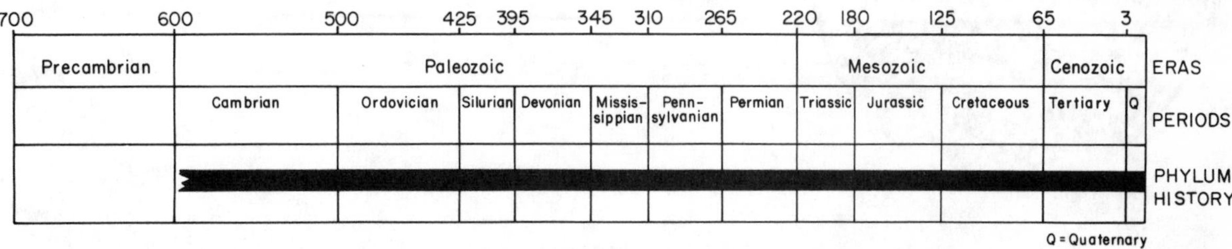

Figure 2.10. Geologic history of the phylum Porifera.

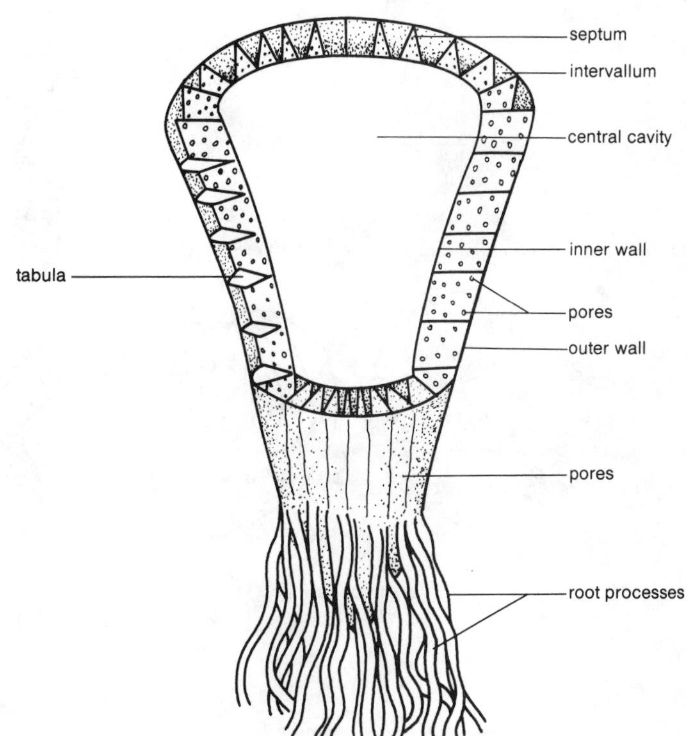

Figure 2.11. Sectioned diagrammatic view of a fossil archaeocyathid. (After Moret, from Bergquist.)

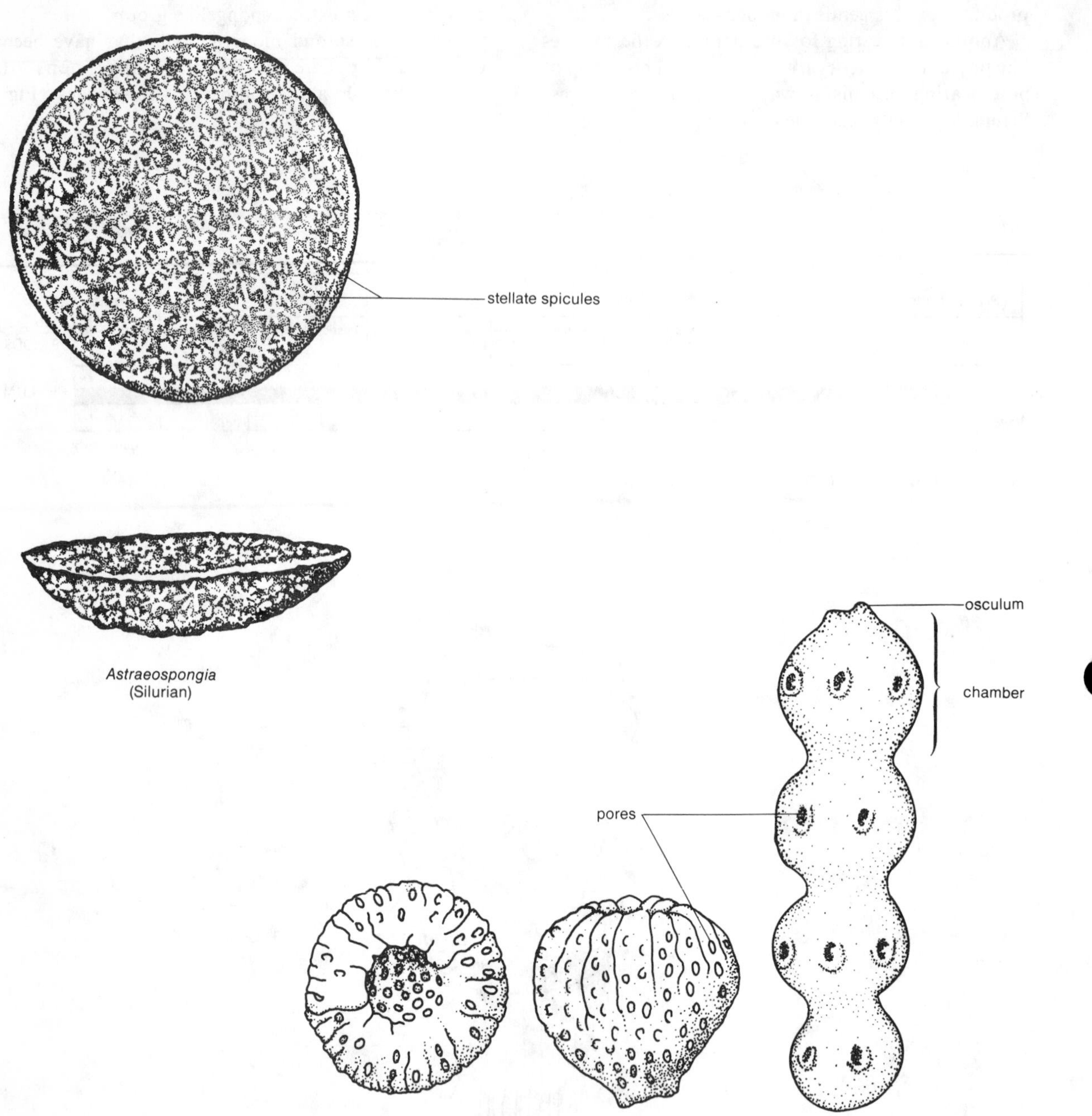

Figure 2.12. Examples of fossil sponges.

Supplemental Reading

Bergquist, P. R. 1978. Sponges. University of California Press, Berkeley, CA.

Bergquist, P. R. 1985. Poriferan relationships. *In*: S. Conway Morris et al. (eds.). The origins and relationships of lower invertebrates. Systematics Association special volume 28. Clarendon Press, Oxford, pp. 14-27.

Cobb, W. R. 1969. Penetration of calcium carbonate substrates by the boring sponge, *Cliona*. Am. Zool. 9:783-790.

Finks, R. M. 1970. The evolution and ecologic history of sponges during Palaeozoic times. Symp. Zool. Soc. Lond. 25:3-22.

Frost, T. M. 1978. In situ measurements of clearance rates of the freshwater sponge *Spongilla lacustris*. Limnol. Oceanogr. 23:1034-1039.

Gilbert, J. J. 1975. Field experiments on gemmulation in the freshwater sponge *Spongilla lacustris*. Trans. Am. Microsc. Soc. 94:347-356.

Gilbert, J. J., and H. L. Allen. 1973. Chlorophyll and primary productivity of some green, freshwater sponges. Int. Rev. Gesamten Hydrobiol. 58:633-658.

Hartman, W. D., J. W. Wendt, and F. Wiedenmayer. 1980. Living and fossil sponges: notes for a short course. Sedimenta VIII. Comparative Sedimentology Laboratory, Division of Marine Geology and Geophysics, Rosenstiel School of Marine and Atmospheric Science, University of Miami, Miami, FL.

Humphreys, T. 1970. Species specific aggregation of dissociated sponge cells. Nature (Lond.) 228:685-686.

Lawn, I. D., G. O. Mackie, and G. Silver. 1981. Conduction system in a sponge. Science 211:1169-1171.

Palumbi, S. R. 1984. Tactics of acclimation: morphological changes of sponges in an unpredictable environment. Science 225:1478-1480.

Rasmont, R. 1970. Some new aspects of the physiology of freshwater sponges. Symp. Zool. Soc. Lond. 25:415-422.

Reiswig, H. M., and G. O. Mackie. 1983. Studies on hexactinellid sponges. III. The taxonomic status of Hexactinellida within the Porifera. Philos. Trans. R. Soc. Lond. B 301:419-428.

Rützler, K., and G. Rieger. 1973. Sponge burrowing: fine structure of *Cliona lampa* penetrating calcareous substrata. Mar. Biol. 21:144-162.

Simpson, T. L. 1984. The cell biology of sponges. Springer-Verlag, New York.

Vacelet, J. 1985. Coralline sponges and the evolution of Porifera. *In*: S. Conway Morris et al. (eds.). The origins and relationships of lower invertebrates. Systematics Association special volume 28. Clarendon Press, Oxford, pp. 1-13.

THE RADIATA

The Eumetazoa is divided into two groups, one with radial symmetry called the **Radiata** and one with bilateral symmetry called the **Bilateria** (cf. Field et al. 1988). Radiata includes the phyla Cnidaria (Coelenterata) and Ctenophora. These phyla have well-developed tissues that may form organs, but they do not develop well-defined organs systems. The radiate body consists of a body wall composed of an outer epidermis, an inner gastrodermis, and a thin-to-thick gelatinous layer between called the **mesoglea**. The latter may be a noncellular gelatinous matrix or possess amoebocytes, connective tissues, and muscle cells. The body cavity (**gastrovascular cavity** or **coelenteron**) ranges from a simple sac in the lower Cnidaria to a complex canal system in the Ctenophora, with pharynx, stomach, and other canals. In Cnidaria, the body cavity opens to the exterior through only a mouth, whereas in the Ctenophora both a mouth and anal pores are present.

Field, K. G., G. J. Olsen, D. J. Lane, S. J. Giovannoni, M. T. Ghislin, E. C. Raff, N. R. Pace, and R. A. Raff. 1988. Molecular phylogeny of the animal kingdom. Science 239:748-753.

EXERCISE 3

Phylum Cnidaria (Coelenterata)

Hydras, jellyfishes, anemones, and corals are members of the phylum Cnidaria (ni-DAR-e-a; G., *cnide*, nettle) or Coelenterata (se-LEN-ter-a-ta; G., *koilos*, hollow + G., *enteron*, gut). With the exception of freshwater hydras and a few jellyfishes, the approximately 9000 species of cnidarians are marine. These carnivorous, tentaculate, multicellular animals represent the tissue level of organization; that is, organs or organ systems are lacking. The phylum contains some of the most beautiful, interesting, and perhaps most dangerous of all marine invertebrates. The sea wasp (*Chironex fleckeri*), a cubomedusa of the Indo-Pacific waters, may be the most venomous invertebrate. Human fatalities have been reported from the noxious stings of this animal. The Portuguese man-of-war (*Physalia*) is less dangerous, but often becomes a nuisance when currents carry large numbers into shallow waters.

Cnidarians possess stinging organelles called **nematocysts**, the most distinguishing feature of the phylum. Each nematocyst arises from a specialized cell, the **cnidocyte** (Fig. 3.1). Unlike sponges, cnidarians have two basic metazoan features, a **mouth** and a **gastrovascular cavity** (coelenteron or enteron) (Fig. 3.2). There is no anus, and all materials enter and exit via the mouth. The gastrovascular cavity is a large internal space for digestion and distribution of nutrients and other materials.

Two basic types of body forms are found in cnidarians: (1) **polyp (hydroid)** and (2) **medusa (jellyfish)**. Both forms are saclike and contain a gastrovascular cavity (Fig. 3.2). The sessile polyp is tubular with a mouth in the **hypostome** at the distal end that is usually surrounded by tentacles. The medusa is typically free swimming and umbrella-shaped, with tentacles extending from the outer margin of the bell. The mouth is located at the end of the **manubrium**, which hangs from the center of the subumbrellar surface somewhat like a clapper hanging from a bell. The manubrium is comparable to the hypostome of the polyp. Both the polyp and medusa are composed of the same body layers and are radially symmetrical. The medusa possesses a much thicker mesoglea than does the polyp.

Two types of medusae are found in cnidarians based on presence (**craspedote** medusa) or absence (**acraspedote** medusa) of the **velum**, an inward-projecting membrane around the margin of the bell. Rhythmic contractions of the velar musculature propel the jellyfish through the water.

The cnidarian body (polyp and medusa) is composed of three layers: (1) an outer protective dermal epithelium (epidermis), (2) a middle mesoglea, and (3) an inner gastrodermis (endoderm) that serves in digestion and absorption. The mesoglea varies from a thin noncellular layer to a thick fibrous jellylike mass with or without wandering amoebocytes. Some authors prefer the term **mesolamella** for the middle layer when it is thin and noncellular and the term **collenchyma** when the mesoglea is cellular. Regardless, the mesoglea is derived from both the epidermis and gastrodermis.

Polymorphism (G., *poly*, many + G., *morph*, shape) is common in cnidarians and often associated with their colonial existence. Various zooids function in food procurement, reproduction, defense, or protection of the colony. A large vocabulary has been developed to describe the structure and function of zooids.

Cnidarians are monoecious and dioecious. Reproduction is both asexual and sexual. Asexual budding or fission is generally associated with the polyp form, whereas sexual reproduction occurs in the medusa and in some polyps (e.g., *Hydra*). In the life cycle of many cnidarians, asexual reproduction in the polyp alternates with sexual reproduction in the medusa, a phenomenon termed **alternation of generations** or **metagenesis** (G., *meta*, back again + G., *genesis*, origin).

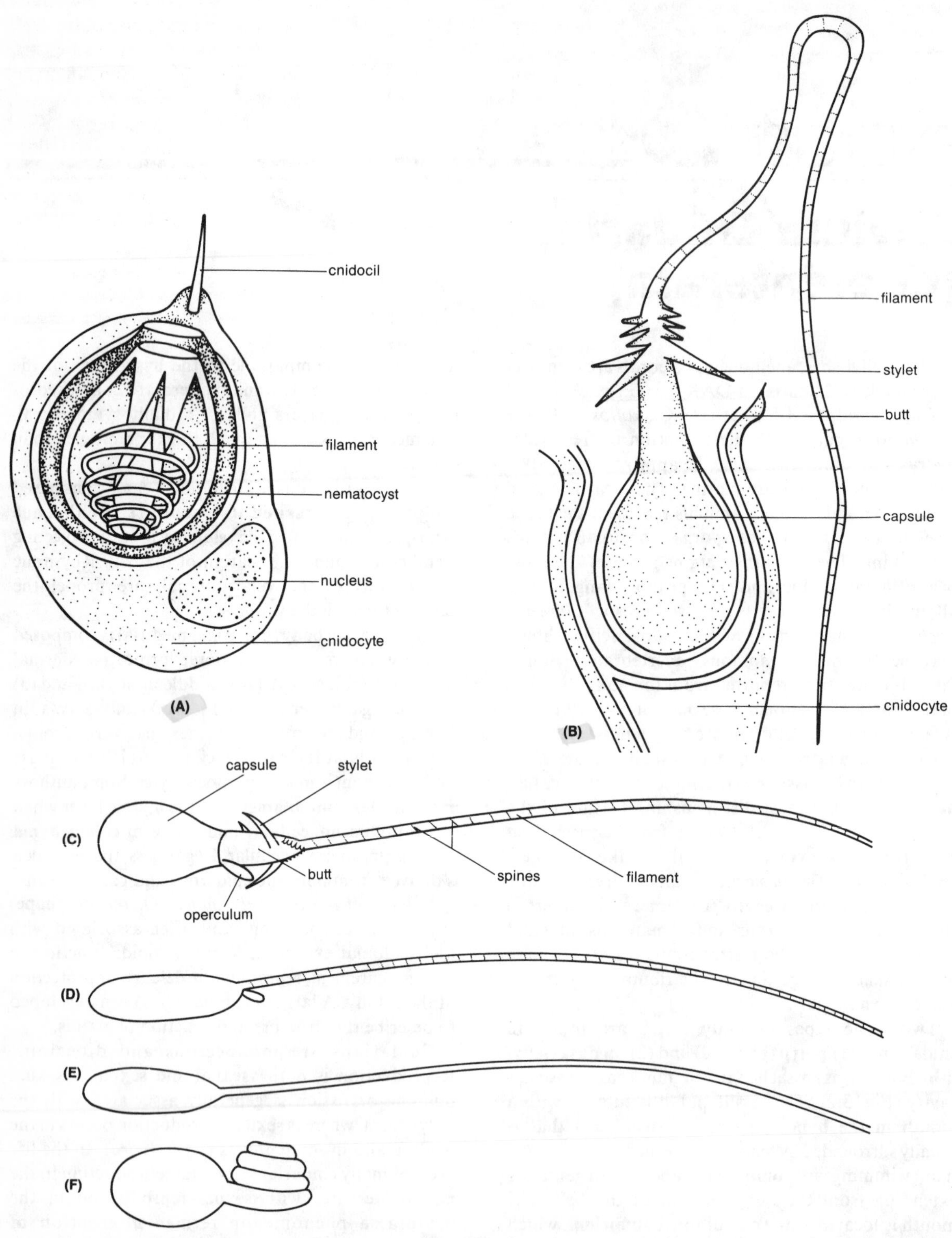

Figure 3.1. Cnidarian nematocysts. **(A)** Undischarged. **(B)** Discharged. **(C)** Penetrant (stenotele). **(D)** Streptoline glutinant (holotrichous isorhiza). **(E)** Stereoline glutinant (atrichous isorhiza). **(F)** Volvent (desmoneme).

Cnidarians in sexual reproduction typically produce a free-swimming, ciliated larva called the **planula** (Fig. 3.7). This nonfeeding, radially symmetrical larva is composed of a ciliated epidermis, mesoglea, and gastrodermis. It swims about for a time and then settles on a substrate and metamorphoses into a new individual or colony.

Classification

The phylum Cnidaria is divided into four classes.

1. Class Hydrozoa. Both polyp and medusa present, noncellular mesoglea, epidermal gonads, and medusa with velum; marine and freshwater; four orders are emphasized: *Hydroida* [polyp stage dominate, medusa present or absent, solitary and colonial; examples are the freshwater jellyfish *Craspedacusta*, *Gonionemus*, *Tubularia*, *Clava*, *Halocordyle* (= *Pennaria*), *Hydractinia*, *Polyorchis*, freshwater hydras, *Obelia*, *Campanularia*, and *Plumularia*]; *Chondrophora* [polymorphic, pelagic, and polypoid colonies; multichambered pneumatophore; examples are *Velella* and *Porpita*]; *Siphonophora* [polymorphic, pelagic, and polypoid and medusoid individuals; pneumatophore not multichambered; colonies with float or swimming bells; examples are *Physalia* (Portuguese man-of-war) and *Stephalia*]; and *Hydrocorallina* [colonial polypoids with calcium carbonate skeletons; examples are *Millepora* (stinging coral) and *Stylaster*].

2. Class Scyphozoa. Medusa dominant, no velum, cellular mesoglea, and gastrodermal gonads; all marine; three orders are emphasized: Stauromedusae (Lucernariida) (sessile, trumpet-shaped body attached by a stalk; examples are *Haliclystus* and *Lucernaria*); Semaeostomae (medusae bowl- or saucer-shaped with scalloped margins; rhopalia present and oral tentacles extend from manubrium; examples are *Aurelia*, *Cyanea*, and *Pelagia*); and Rhizostomae (tentacles lacking around the bell; oral arms of manubrium branched and fused; filter feeders; examples are *Cassiopeia*, *Stomalophus*, and *Rhizostoma*).

3. Class Cubomedusae. (Cubozoa). Medusa cuboidal in shape with unscalloped margins; four rhopalia and velum present; examples are *Chironex* and *Carybdea*.

4. Class Anthozoa. Medusa absent, cellular mesoglea, gastrovascular cavity partitioned by septa (mesenteries), endodermal gonads; solitary, or colonial; all marine; two subclasses are recognized: Zoantharia (Hexacorallia) (polyps with more than eight, paired mesenteries; unbranched tentacles; solitary or colonial; examples are sea anemones (solitary polyps without a skeleton) *Metridium*, *Calliactis*, *Cerianthus*, and *Ceriantheopsis* and hard (stony) corals [colonial polyps with calcareous skeleton; no siphonoglyphs; examples are *Astrangia* (eyed coral), *Porites*, and *Acropora* (staghorn coral)]); and Alcyonaria (Octocorallia) [polyps with eight pinnate tentacles and eight, complete, unpaired mesenteries; mostly colonial; examples are *Tubipora* (organ pipe coral), *Clavularia*, *Gorgonia* (sea fan), *Leptogorgia* (sea whip), *Stylatula* (sea pen), and *Renilla* (sea pansy)].

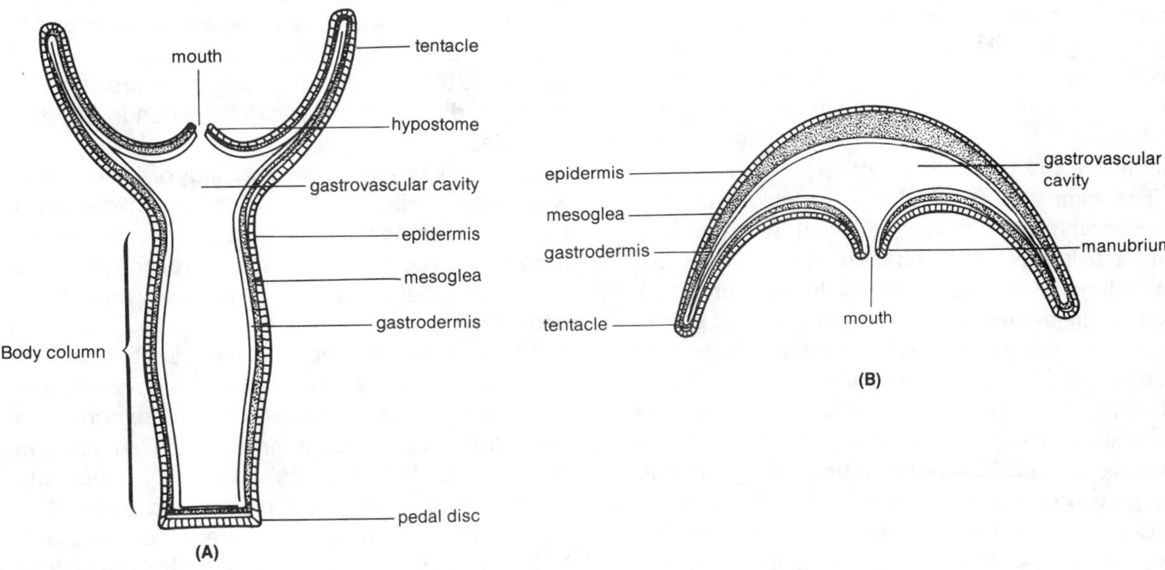

Figure 3.2. Two body forms of cnidarians. **(A)** Polyp (hydroid). **(B)** Medusa (jellyfish).

A. Class Hydrozoa

The simpler cnidarians belong to the class Hydrozoa (hy-dro-ZO-a; G., *hydra*, water serpent + G., *zoon*, animal). Seven representatives of the class are described in the following exercises. One or more of these may be omitted from your study to conserve time.

Hydra

Hydras are cosmopolitan inhabitants of fresh waters, often attached to the underside of plants and twigs in the water. Three common North American species are *Hydra littoralis, H. oligactis* (= *H. fusca* and *Pelmatohydra oligactis*), and *H. viridis* (= *Chlorohydra viridissima*). The green color of *H. viridis* is due to numbers of the symbiotic alga, **zoochlorella** (*Chlorella*), living within its gastrodermis. The description below will apply to any of these three hydras.

Hydras are small animals (0.3 to 5 cm long) with four major body regions: (1) **hypostome,** (2) **tentacles,** (3) **body column,** and (4) **pedal disc** (Fig. 3.3). The hypostome, a conical elevation of cells at the distal end, surrounds the mouth opening. Numerous mucus-secreting glands cells in the gastrodermis of the hypostome aid the hydra in swallowing food such as water fleas (*Daphnia*), cyclops (*Cyclops*), annelids, and insect larvae. When the mouth is closed and at rest, it is stellate-shaped, but it can be greatly distended when food is swallowed.

The hypostome is surrounded by 4 to 10 tentacles. Tentacles are hollow and can be extended more than twice the length of the body column. Numerous clumps or batteries of cnidocytes with nematocysts inside impart a bumpy appearance to the outer surface of each tentacle. Four kinds of nematocysts occur in hydra (Fig. 3.1): penetrants, streptoline and stereoline glutinants, and volvents.

The main part of a hydra is the body column, a cylindrical tube that can be greatly expanded or contracted. In *H. oligactis* the column can be divided into a large gastric region (**stomach**) and a proximal, narrow, light-colored stalk (**peduncle**). The gastrodermis of the gastric region contains flagellated epithelionutritive cells and enzyme-secreting gland cells. Most digestion occurs in the stomach. The gastrodermis of the stalk is composed of vacuolated cells that do not produce enzymes. In *Hydra*, but not all hydrozoans, the gastrovascular cavity extends into the tentacles. Because the cavity is filled with water, it serves as a **hydrostatic skeleton** as well as in digestion and circulation.

The base of the pedal disc forms the proximal (or aboral) end of the body. Tall epidermal cells of the disc secrete a sticky mucus for attachment and form a gas bubble that allows the animal to float.

Observational Procedure: *Hydra*

Living Organism. A small dish containing hydras will be provided. Examine the specimen with a dissecting microscope. A hydra in the extended state may reach 5 cm in length. Locate the tentacles, hypostome, body column, and basal disc (Fig. 3.3). Count the number of tentacles on your specimen. Can you see the mouth? Observe hydra capture prey in the small dish provided. If your hydra has a lateral growth from the body, it is most likely an asexual **bud**.

Observe your specimen extending and contracting. Do the tentacles extend and contract at the same time? If disturbed, the organism will contract and remain in a nonextended state for several minutes. When contracted, a hydra resembles a jellylike ball.

Microscopic Study

1. With a pipette, remove a hydra from the dish and place it in a depression slide, adding a drop of water if needed. Cover with a coverslip and observe under low power of a compound microscope.
2. Locate the central, light-colored gastrovascular cavity and trace the cavity as it extends into the tentacles.
3. Compare the external morphology of the tentacles with that of the body column. Locate the batteries of cnidocytes along the tentacles. Can you determine if the cnidocytes are more numerous on the tentacles then on the body column? Are cnidocytes on the pedal disc?
4. Place a small drop of weak acetic acid or methylene blue near the edge of the coverslip and observe what occurs in the tentacular region when the solution contacts the specimen. Can you see discharged nematocysts? Carefully focus the nematocysts under high power and attempt to identify the different types by reference to Fig. 3.1.
5. Examine prepared longitudinal and cross sections of a hydra under high power of a compound

microscope. Determine the three body layers: outer **epidermis**, middle **mesoglea** (mesolamella), and inner **gastrodermis** (Fig. 3.4). The mesoglea appears as a line wedged between the epidermis and gastrodermis. The noncellular mesoglea is secreted by both epidermis and gastrodermis; it is thinnest in the tentacles and thickest in the body column.

Study the cell types and their distribution in the epidermis and gastrodermis by examining the slide preparations under oil immersion. With the description below and using Fig. 3.4 as a guide, locate the following cells and understand their functions.

1. Epitheliomuscular cells. Most of the epidermis is composed of these cuboidal to columnar-shaped cells. The basal region is expanded longitudinally and contains contractile fibers. The fibers when contracted shorten the hydra, therefore functioning as longitudinal muscles. The cells also support and protect the body.

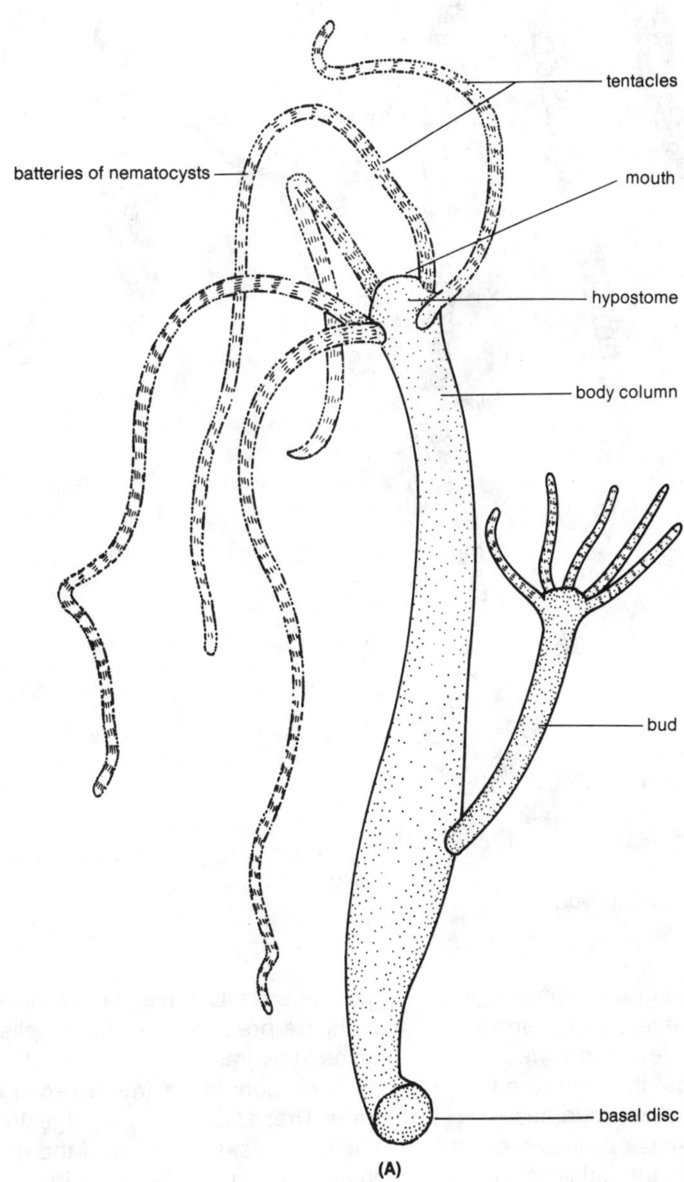

Figure 3.3. **(A)** *Hydra* with a bud showing the gross external features.

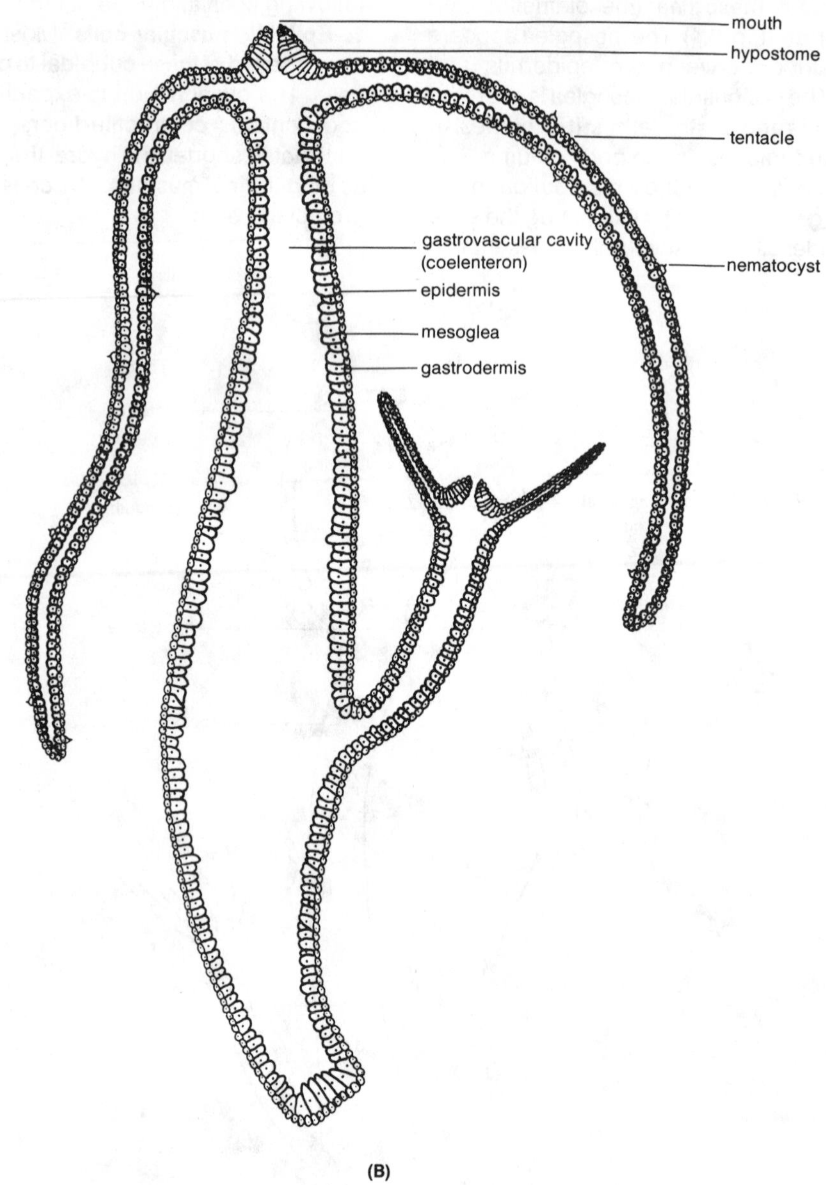

Figure 3.3. **(B)** Longitudinal section of *Hydra*.

2. Cnidocytes. These cells contain nematocysts. They occur in groups or batteries and are most abundant on the tentacles. A few cnidocytes occur in the stalk of *H. oligactis*, but they are rare in the pedal disc. None occur in the gastrodermis.

3. Interstitial cells. These **I-cells**, as they are often called, are found primarily in the epidermis of the gastric region, wedged between epitheliomuscular cells. I-cells occur singly or in clusters (nests). Interstitial cells have dark-stained nuclei. These cells are precursors or stem cells of all other cell types in hydras.

4. Epithelionutritive (**nutritive or nutritive-muscular**) **cells**. These tall, columnar digestive cells with food vacuoles make up most of the gastrodermis. Like their counterparts in the epidermis (i.e., epithelio-muscular cells), the base of each nutritive cell is elongated and contains contractile fibers. The fibers

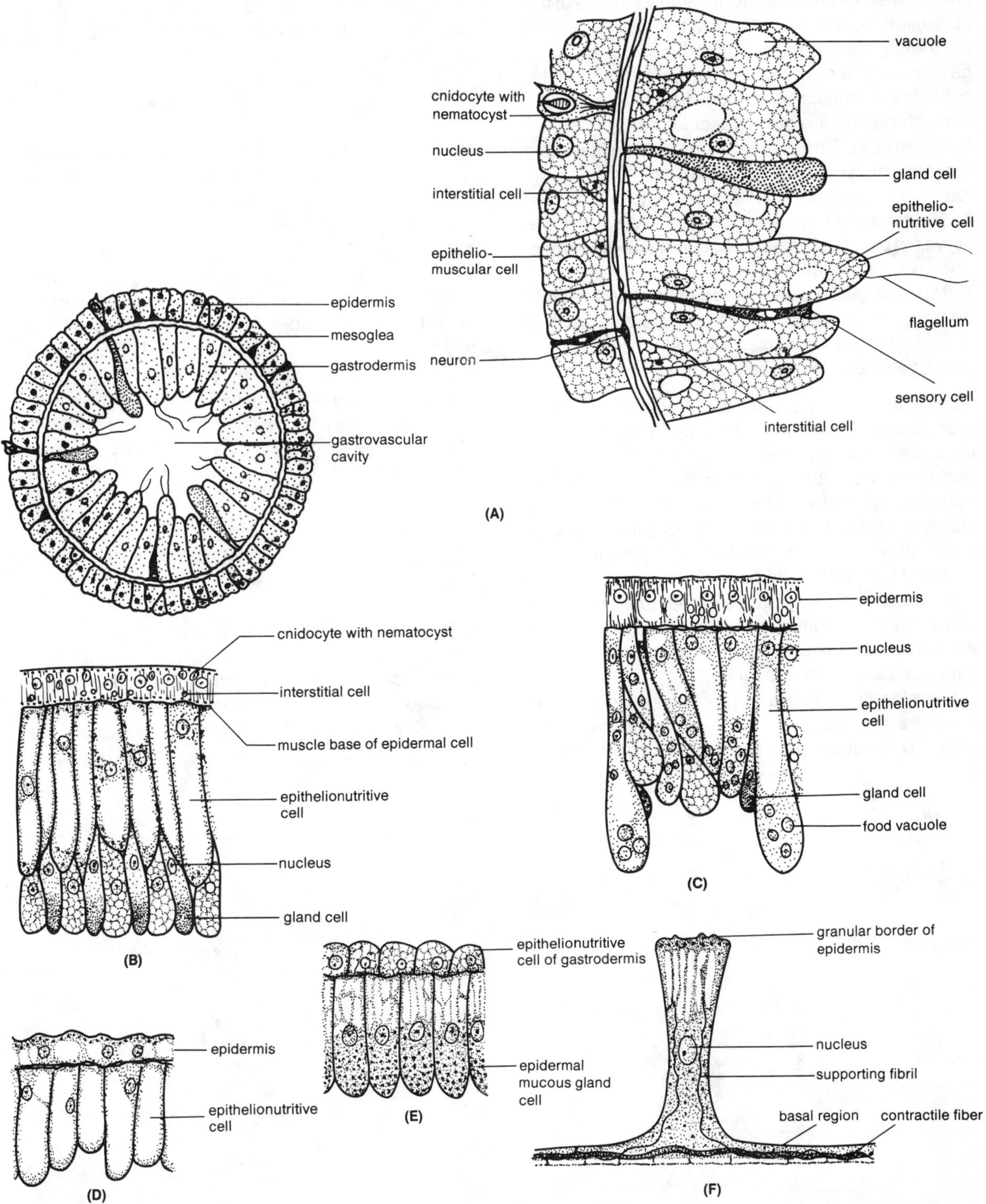

Figure 3.4. **(A)** *Hydra* cross section. **(B)** Section of body wall in the region of the hypostome. **(C)** Section of the stomach region. **(D)** Section of the stalk region. **(E)** Section of the pedal disc. **(F)** Epitheliomuscular cell. (Figures B-F after Hyman).

are oriented circularly at right angles to those of the epitheliomuscular cells. Contraction of the nutritive cells elongates the hydra and reduces the size of its gastrovascular cavity. Unlike the epitheliomuscular cells, the epithelionutritive cells bear two flagella that extend into the gastrovascular cavity.

5. Gland cells. These large, vacuolated cells secrete digestive enzymes and mucus. They are commonly found in the gastrodermis and hypostome.

6. Neurons and sensory cells. These form the nerve net plexus of both epidermis and gastrodermis. They will be difficult to see.

7. Mucous gland cells. These tall epidermal cells do not produce enzymes. The cells are especially abundant in the pedal disc and hypostome. Those in the pedal disc produce mucus for attachment.

Reproduction. Asexual (**budding**) and sexual reproduction occur in hydras. Dividing epitheliomuscular and nutritive cells in the stalk initiate budding. The young polyp or bud grows until it detaches from the parent hydra in 2 to 4 days. Observe budding in living hydras or from prepared slides (Fig. 3.3). What is the general shape of the bud; is it identical to the parent?

Most species of *Hydra* are dioecious. The ovary and testis arise from aggregated interstitial cells in the epidermis. Unlike most cnidarians, hydras lack a medusa and a planula larva.

Examine living hydra or prepared slides (whole mounts and cross sections) showing the testis and ovary. Distinguish between the two reproductive structures. Are the testis and ovary true organs such as found in higher animals? Support your conclusion.

Pennaria

Pennaria is a colonial, shallow-water, marine hydroid that forms a minute plantlike growth on pilings, rocks, and seaweed. The pinnate colony, which may reach 15 cm in height, is attached to the surface of objects by hollow, rootlike **hydrorhiza** (stolons or rhizomes). Stemlike structures called **hydrocauli** grow upright from the hydrorhiza. The hydrocaulus buds off alternating side branches called **hydrocladia** from which are budded the terminal feeding polyps called **hydranths** (gastrozooids). The type of growth in *Pennaria* is **monopodial** because the original or primary hydrocaulus elongates throughout the colony's life. Therefore, the oldest polyp occupies the tip of the original hydrocaulus (Fig. 3.5).

Observational Procedure: *Pennaria*

1. Obtain specimens of a preserved or living colony and prepared slides (whole mounts and cross sections) for study with a dissecting and a compound microscope, respectively. Can you determine the monopodial growth form from the preserved or living specimen?

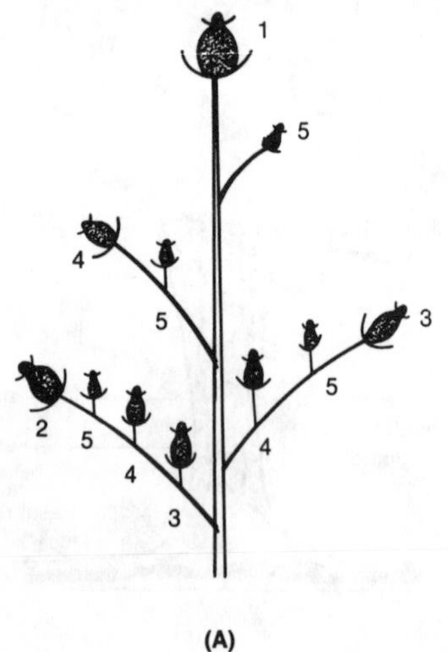

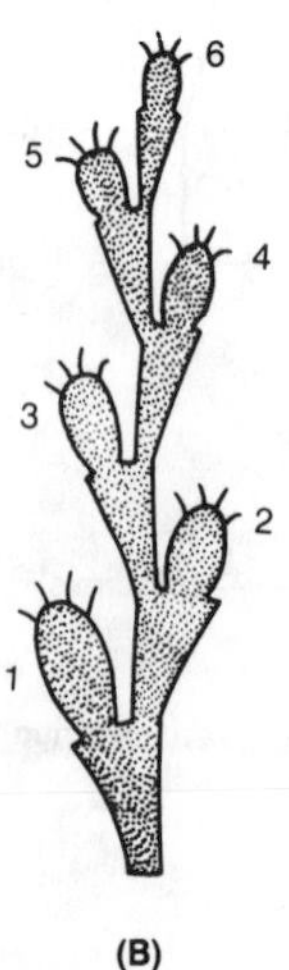

Figure 3.5. Monopodial **(A)** and sympodial **(B)** growth patterns (diagrammatic). Age of the polyps indicated by relative size and numbers, with the original polyp numbered 1.

2. Locate a hydranth and observe the large conical-shaped hypostome containing the mouth (Fig. 3.6). Note the **capitate tentacles** on the hypostome and the ring of **filiform tentacles** that circles the base of the gastral region.

3. Examine the tentacles for cnidocytes.

A living hollow tube, the **coenosarc**, extends throughout all branches of the colony (Fig. 3.6). The coenosarc consists of an outer epidermis, a middle mesoglea (mesolamella), and an inner flagellated gastrodermis that lines the gastrovascular cavity. The cavity is continuous throughout the colony, and at the base of the hydranth the cavity enlarges as a stomach region. The epidermis of the coenosarc of the hydrocaulus and hydrocladium secretes an outer, protective chitinous covering, the **perisarc** (periderm). In *Pennaria*, the perisarc adheres closely to the outer surface of the colony and is annulated at the base of each stem. *Pennaria* is an **athecate** (**gymnoblastic**) hydrozoan because the hydranths and gonophores (medusa buds) lack the perisarc covering.

Pennaria is dioecious. Locate the small, thimble-shaped gonophores that arise by budding off the hydranth just distal to the filiform tentacles. The medusa lacks a mouth but has a velum and four short tentacles.

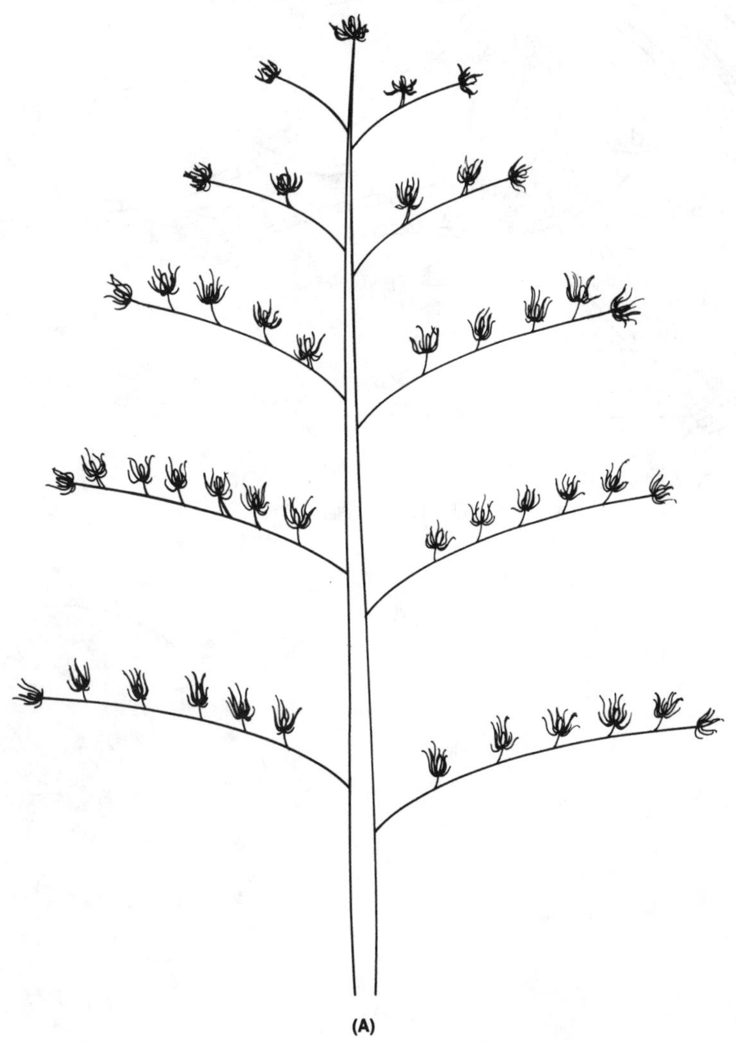

(A)

Figure 3.6. **(A)** Colony of *Pennaria*.

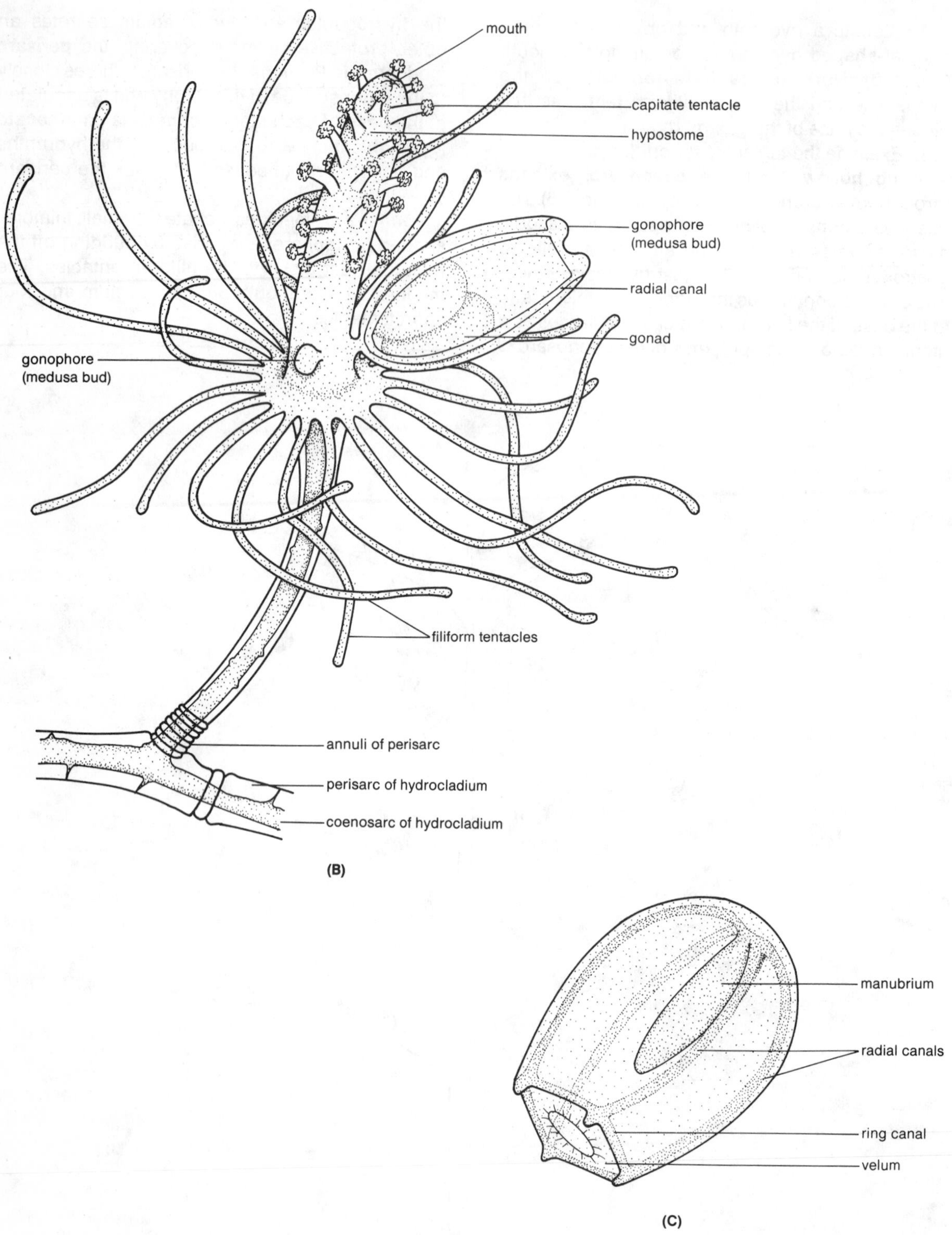

Figure 3.6. **(B)** Enlarged hydranth with medusa bud of *Pennaria*. **(C)** *Pennaria* medusa.

Obelia

Obelia is another colonial marine hydroid that forms a minute plantlike growth on rocks, pilings, and other substrates. *Obelia* is similar to *Pennaria* in a number of ways; however, there are important differences between the two hydrozoans.

Observational Procedure: *Obelia*

Examine preserved or living specimens of the *Obelia* colony with a dissecting microscope and whole-mount slide preparations with a compound microscope.

1. Locate the two kinds of polyps occurring on the *Obelia* colony: **hydranth** (**gastrozooid**) and **gonangium** (**gonozooid**) (Fig. 3.7). Both polyps are supported by upright, stemlike, branched **hydrocauli** and rootlike **hydrorhiza** attached to a substrate. The type of growth in *Obelia* is **sympodial** because the original hydrocaulus does not elongate throughout life. Therefore, the youngest polyps occur at the tips of the branches and the oldest polyps near the base of the colony (Fig. 3.5). Can you distinguish the youngest polyps from the oldest ones on the preserved or living materials supplied?

2. Observe that the entire colony is covered with a nonliving, chitinous **perisarc**. The perisarc covering each hydranth is bell-shaped and called the **hydrotheca**, whereas the perisarc of the gonangium is the **gonotheca**. Because the perisarc extends onto the polyps, *Obelia* is a **thecate** (**calyptoblastic**) hydroid, unlike *Pennaria* which is athecate (gymnoblastic). The perisarc of the hydrocauli is annulate at various places, which allows flexibility of movement to the colony. Like *Pennaria*, the living **coenosarc** consists of epidermis, mesoglea, and gastrodermis; the hollow gastrovascular cavity is continuous throughout the colony.

3. Examine a hydranth (feeding polyp) and locate its hypostome, which bears the mouth. The filiform tentacles of *Obelia* are solid and provided with nematocysts. The tentacles can be retracted into the hydrotheca. Note that the tentacle arrangement in *Obelia* is different from that of *Pennaria* and that capitate tentacles are absent.

4. Study prepared cross and longitudinal sections of *Obelia* made in the regions of the hypostome and hydrocaulus. How do these sections differ with comparable regions of hydras?

5. Examine a **gonangium** or reproductive polyp. Note that it lacks tentacles, a mouth, and hypostome. Observe the asexually produced discoid medusa buds (**gonophores**) on the club-shaped **blastostyle** of the gonangium. The blastostyle is a modified polyp and represents a continuation of the coenosarc into the gonangium. The medusa, termed a **leptomedusa**, escapes into the water from the blastostyle via the gonopore at the terminal end of the gonangium. A pelagic existence occurs before the medusa reaches sexual maturity.

6. Observe a prepared slide of *Obelia's* medusa (Figs. 3.7 and 3.8). Note its bell or umbrella shape. The upper,convex surface is the **exumbrella** and the lower, concave surface is the **subumbrella**. The **manubrium**, comparable to the hypostome of the hydranth, hangs down from the center of the subumbrella surface. Can you see the mouth in the manubrium?

7. Observe the solid tentacles around the bell's margin. Count the tentacles. Does each medusa have the same number of tentacles? Are cnidocytes on the tentacles?

8. Locate the four **radial canals** and four **gonads** attached to them. Radial canals meet the **ring canal** that encircles the margin of the bell. Eight **statocysts** are located on the margin of the bell in the interradial areas. The gastrovascular cavity of the medusa is lined with gastrodermal cells. The cavity includes the spaces of the manubrium, radial canals, and the ring canal. *Obelia's* medusa is atypical, compared to most hydromedusa such as *Gonionemus* because a velum is lacking.

Obelia is dioecious. Gametes are fertilized in the water and the zygote develops into a planula larva which settles onto a substrate and metamorphoses into a colony. *Obelia* is a good example for understanding metagenesis, the alternation of asexual and sexual reproduction (Fig. 3.7).

Plumularia

Plumularia is a marine, calyptoblastic colonial hydroid with creeping hydrorhiza. Arising from the hydrorhiza are hydrocauli with attached plumelike branches (hydrocladia), each bearing on one side a series of bell-shaped hydrothecae containing the gastrozooids. The sessile hydrotheca is so small that the gastrozooid with its filiform tentacles cannot be retracted within the encasement.

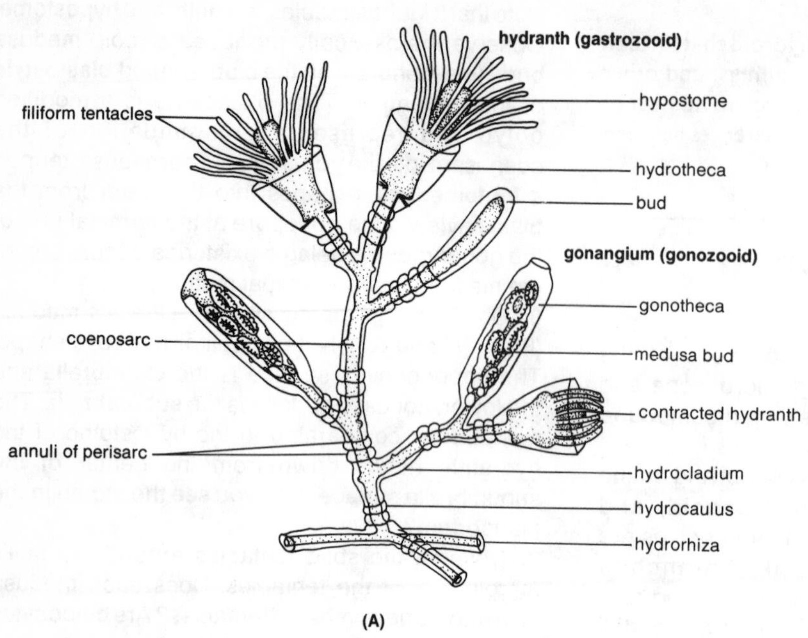
hydranth (gastrozooid)
filiform tentacles
hypostome
hydrotheca
bud
gonangium (gonozooid)
gonotheca
coenosarc
medusa bud
contracted hydranth
annuli of perisarc
hydrocladium
hydrocaulus
hydrorhiza

(A)

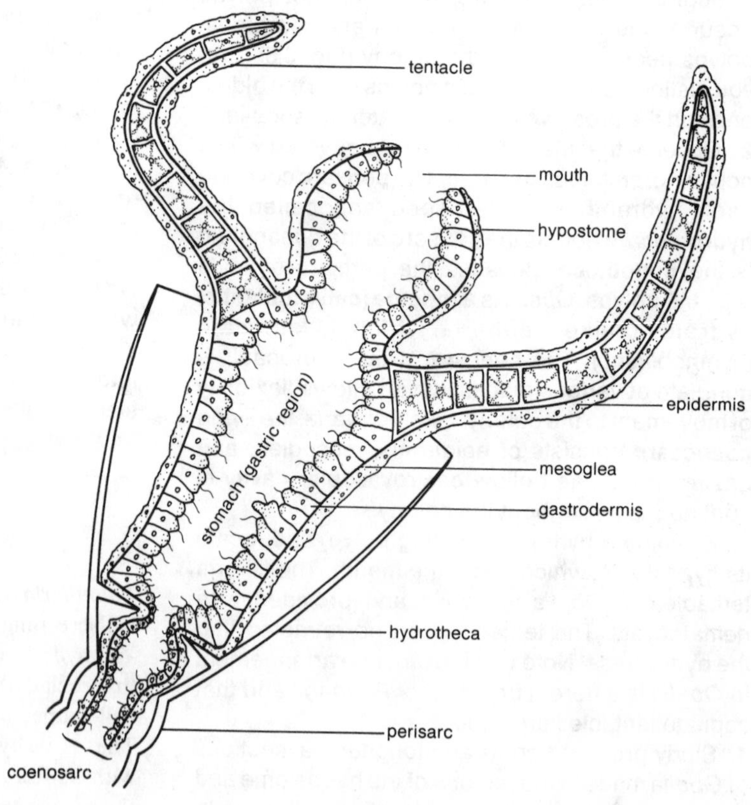
tentacle
mouth
hypostome
stomach (gastric region)
epidermis
mesoglea
gastrodermis
hydrotheca
perisarc
coenosarc

(B)

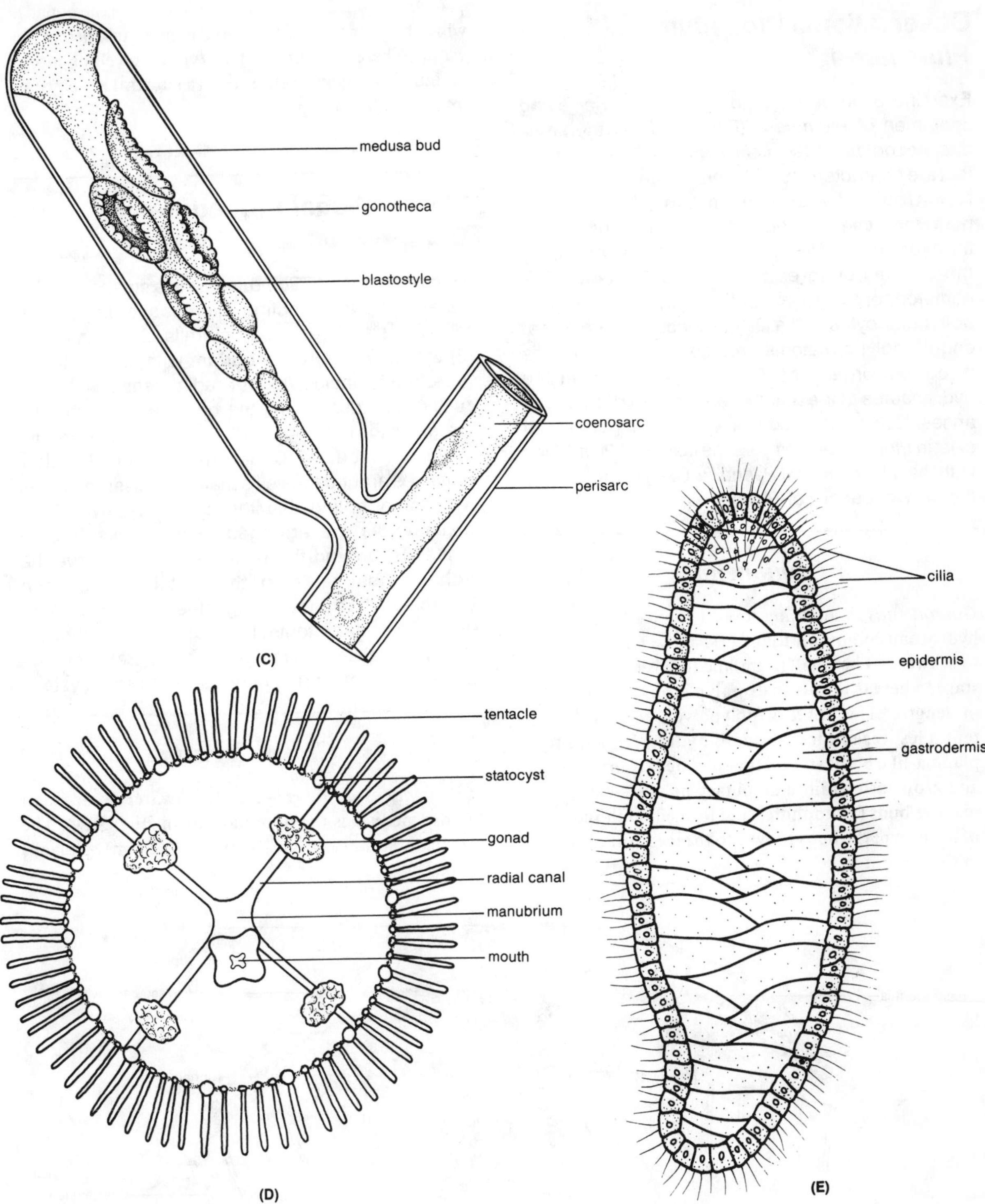

Figure 3.7. *Obelia*. **(A)** Entire colony. **(B)** Longitudinal section of hydranth. **(C)** Single gonangium. **(D)** Oral view of medusa. **(E)** Planula larva.

Observational Procedure: *Plumularia*

Examine a prepared slide or preserved or living specimen of *Plumularia* (Fig. 3.9). Note the small dactylozooids called **nematophores** (sarcostyles) that are characteristic of *Plumularia* and its relatives. Nematophores arise from tiny thecae, the **nematothecae**, located on the hydrocaulus and on the hydrotheca of the gastrozooid. There are usually three nematophores to each gastrozooid. The nematophores are protective polyps. They have both cnidocytes and a long amoeboid process that engulfs polyps, diatoms, protozoa, and larval stages of epizoic organisms. Gonangia extend from the hydrocaulus at the point where each hydrocladium arises. No special modifications for the gonotheca exist in *Plumularia*. Compare the colony of *Plumularia* with that of *Obelia* and *Pennaria*. Can you determine the growth pattern of *Plumularia*?

Gonionemus

Gonionemus, containing several species, is a marine hydrozoan commonly found in shallow water attached to seaweed. The small (2 cm) medusa is the dominant stage, whereas the athecate polyp is only about 1 mm in length. The solitary polyp with its four solid tentacles resembles a small hydra. Nonciliated, planula-like bodies called **frustules** bud from the polyp and grow into additional solitary polyps. Seasonally, medusa buds (gonophores) are formed on the column of the polyp; the buds grow into the medusae observed in the laboratory.

Gonionemus is dioecious. Gametes from the gonads, which are attached to the radial canals, are shed in the water, where they unite. The zygote develops into a planula larva which settles onto a substrate and grows into the small polyp.

Observational Procedure: *Gonionemus*

Study a medusa of *Gonionemus* in a small dish with water using a dissecting microscope. Locate the following structures: (1) exumbrella, (2) subumbrella, (3) velum, (4) short manubrium with its four frilled lobes, (5) tentacles, (6) four radial canals with their ruffled gonads, and (7) the ring canal around the margin of the bell (Fig. 3.10). Can you find the mouth in the manubrium? Count the number of hollow tentacles around the bell's margin. Closely examine a tentacle, noting that the cnidocytes with their nematocysts are arranged in a ringlike fashion (batteries) around the tentacles. Can you see the **adhesive sucker** near the distal end of each tentacle? What function might the sucker serve? At the base of each tentacle is a prominent sensory swelling called a **tentacular bulb**. Between the bulbs are balancing structures called statocysts.

Craspedacusta

Craspedacusta sowerbyi is freshwater hydrozoan whose medusa is the dominant form (Fig. 3.11). The athecate, solitary polyp is very small (2 mm) and lacks tentacles.

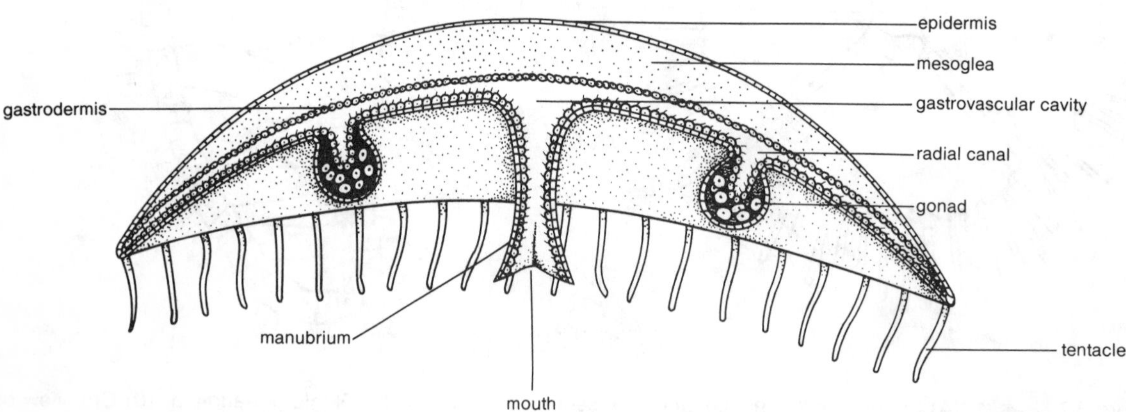

Figure 3.8. Oral-aboral section of *Obelia* medusa.

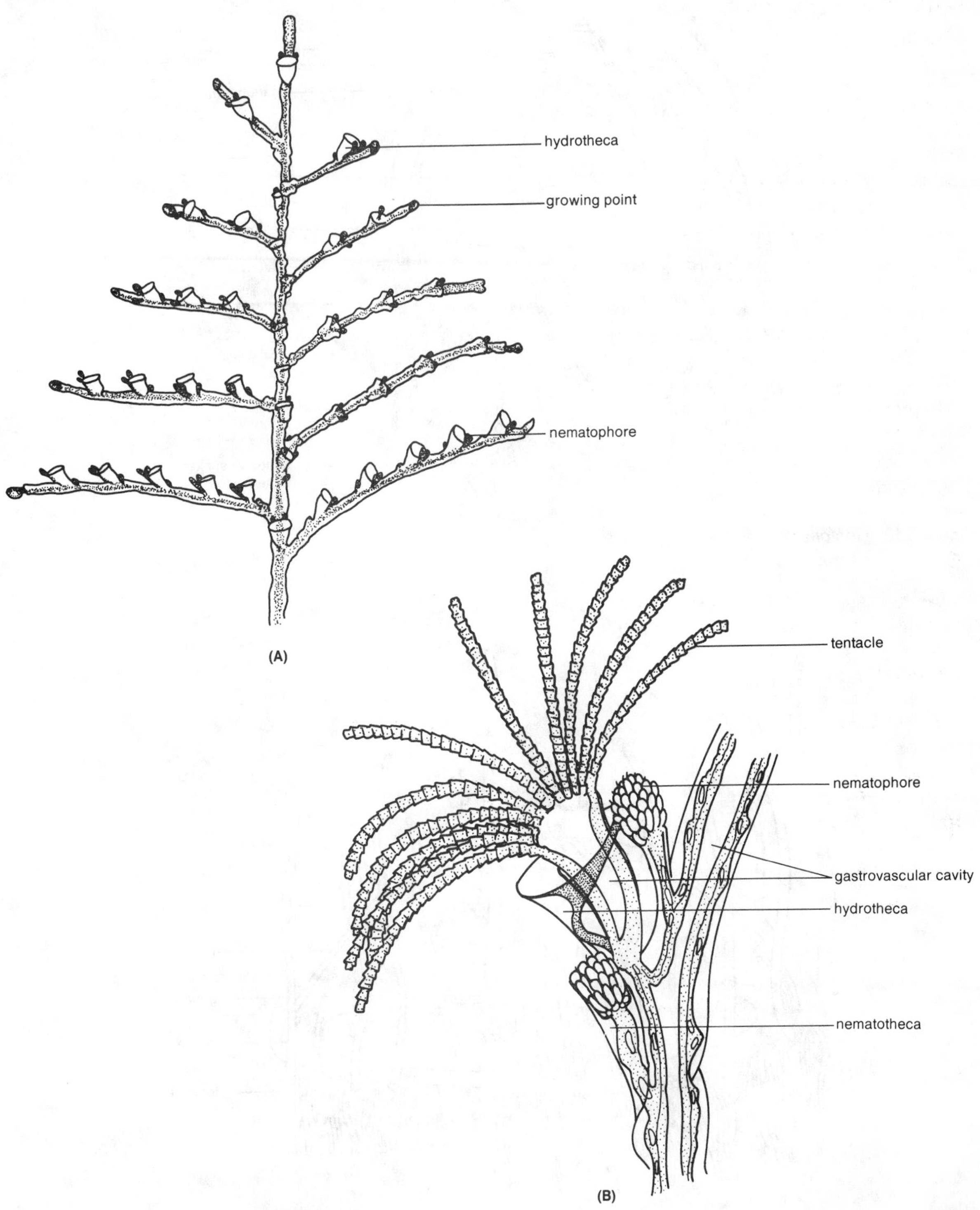

Figure 3.9. **(A)** Colony of *Plumularia*, showing monopodial growth. **(B)** Hydranth of *Plumularia*, showing two of three nematophores.

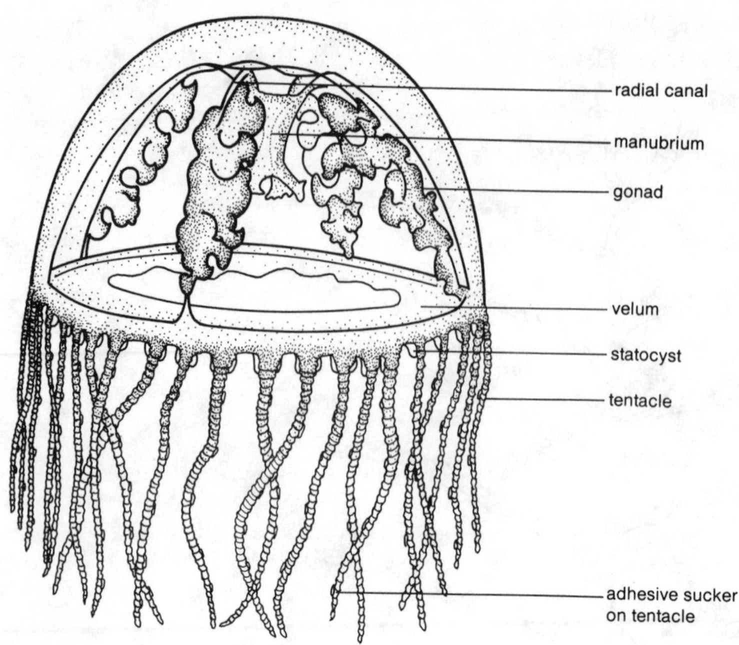

Figure 3.10. *Gonionemus* medusa.

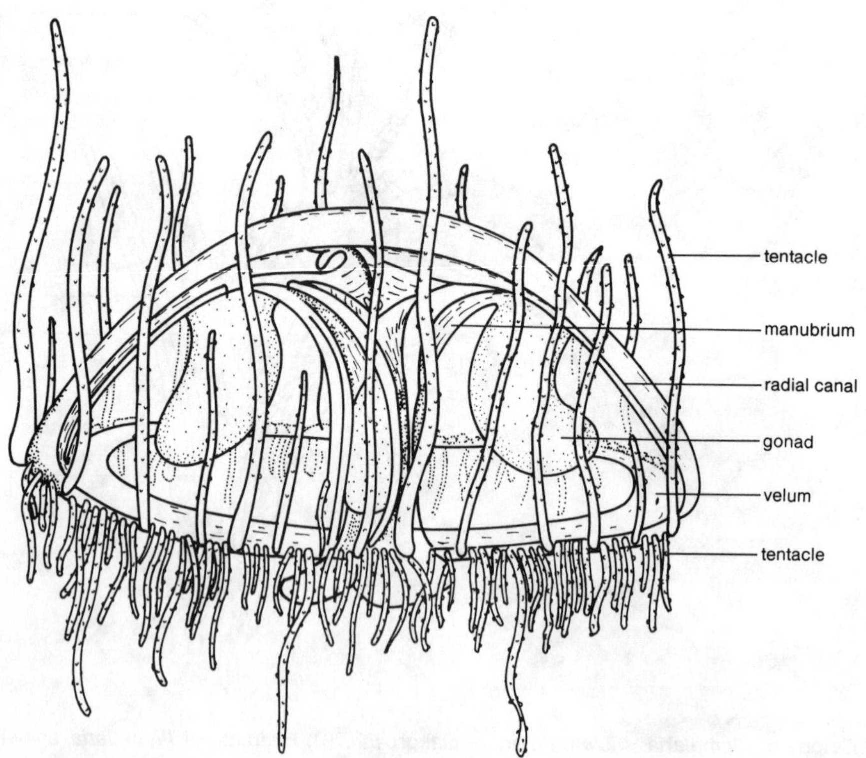

Figure 3.11. *Craspedacusta* medusa.

Observational Procedure: *Craspedacusta*

Examine a medusa of *Craspedacusta* in a small dish with water using a dissecting microscope. Compare the anatomy of the medusa to that of *Gonionemus*. How are these two organisms similar? Do you see any differences? Compare the polyp of *Craspedacusta* to hydra.

Siphonophora

Members of the order Siphonophora, such as the Portuguese man-of-war (*Physalia*), are perhaps the most complex and specialized of hydrozoans. These pelagic, colonial marine animals attain the maximum development of polymorphism found in cnidarians.

The siphonophore colony originates from a planula larva. The colony consists of both modified medusoid and polypoid individuals (zooids) specialized to perform different functions. The polypoid individuals are modified for feeding, capturing of prey, and producing the reproductive zooids. The medusoid individuals are concerned with locomotion, flotation, and reproduction. The various types of zooids that may occur in siphonophores are listed below.

1. Polypoid zooids
   a. **Gastrozooid (trophozooid)**. A feeding polyp with a mouth and one long, contractile, hollow tentacle with cnidocytes. The tentacle originates near the base of the polyp.
   b. **Dactylozooid (palpon)**. This tactile polypoid resembles a gastrozooid in having a long tentacle, but a mouth is lacking. The tentacle is never branched.
   c. **Gonozooid**. This reproductive polyp lacks a tentacle. A mouth may be present as in *Velella*. A gonozooid usually bears several gonophores (see below).
2. Medusoid zooids
   a. **Nectophore (swimming bell)**. This modified, very muscular medusa is used for flotation. The nectophore has a bell, velum, four radial canals, and a ring canal. It lacks a mouth, manubrium, tentacles, and sense organs.
   b. **Phyllozooid (bract)**. This leaflike zooid bears cnidocytes for protection.
   c. **Gonophore**. This medusoid occurs on gonozooids or branched stalks called gonodendra. Gonophores lack a mouth, tentacles, and sense organs. They have a bell, velum, manubrium, and radial canals. Gonophores occur singly or in grapelike clusters. Some are saclike, such as the sporosacs of some hydroids. Gonophores are dioecious even though the colony is monoecious. Female gonophores are medusoid and may swim free, whereas male gonophores are reduced.
   d. **Pneumatophore**. This is a float that may or may not have a pore to regulate the gas it contains. Some workers believe that the pneumatophore is a modified polyp instead of a medusa.

Physalia. The bluish pneumatophore of *Physalia* is derived from an aboral evagination from the planula larva. The float is highly muscular, double-walled, lacks mesoglea, and is lined with a chitinous epidermal secretion. The float has its greatest development in *Physalia*, where it is an ovate bladder 30 cm in length. On the floor inside the pneumatophore is a modified glandular epithelium called the **gas gland** that produces gas (mainly nitrogen and carbon dioxide) contained inside the float. *Physalia*, unlike some species, lacks a pore to regulate the amount of gas within the float. Attached to the pneumatophore is an erectile, sail-like **crest** that can be lowered and raised. In the living organism the crest is blue bordered with pink.

The various zooids of *Physalia* are budded from a ventral disc area beneath the pneumatophore. The gastrovascular cavity and the coenosarc of the zooids and pneumatophore are continuous. Each group of zooids extending from the disc beneath the float is called a **cormidium**. *Physalia* possesses several cormidia, each consisting of a gastrozooid, male and female gonozooids, and one dactylozooid with an extremely long tentacle (10 m). The dactylozooids from several cormidia form a suspended driftlike net. Contact with the tentacles results in discharge of the nematocysts. The tentacles contract and pull the prey to the gastrozooids.

Observational Procedure: *Physalia*

In a dish containing water, carefully examine the specimen provided. Locate the pneumatophore, crest, and gas gland (Fig. 3.12). The latter may be observed within the floor of the pneumatophore by holding the pneumatophore up to a bright light. Study one cormidium and identify the gastrozooid, with its mouth and long tentacle; the gonozooid, with small, grapelike, sessile gonophores; and the dactylozooid, with its long tentacle. The stemlike, branched structure bearing the gonozooid is called a **gonodendron**. A gonodendron may bear **gonopalpons** (modified dactylozooids lacking a mouth), **gelatinous zooids**, and **nectophores** (modified medusae). Each nectophore consists of a bell, velum, four radial canals, and a ring canal, but it lacks a mouth, manubrium, tentacles, and sense organs.

Velella. *Velella*, commonly called "by-the-wind sailor," is a pelagic, polymorphic chondrophoran. The ovoid body is flattened with the pneumatophore on the aboral surface as a chitinous disc (Fig. 3.13). Arising from the aboral surface of the pneumatophore is a thin, chitinous, triangular **sail**. The living organism is bluish in color and the sail is edged with purple. *Velella* lacks control over the sail and is at the mercy of the ocean currents for movement.

Beneath the pneumatophore are three types of zooids forming the bulk of the colony (Fig. 3.13). Unlike *Physalia*, zooids of *Velella* are not greatly elongated. In the center of the colony is a single, large gastrozooid that contains the mouth at its free end. The mouth leads to a large gastrovascular cavity. On both sides of the gastrozooid are several gonozooids with small, grapelike gonophores that escape into the water as free-swimming medusae. Around the margin of the disc is a fringe composed of several dactylozooids.

The cavities of gastrozooids and gonozooids communicate with the cavity of the pneumatophore via ectodermal tubes. These tubes are extensions of the multichambered pneumatophore. *Velella* lacks swimming bells.

Observational Procedure: *Velella*

Examine a *Velella* with a dissecting microscope. Locate the pneumatophore with its sail, gastrozooid, gonozooids with the gonophores, and the dactylozooids.

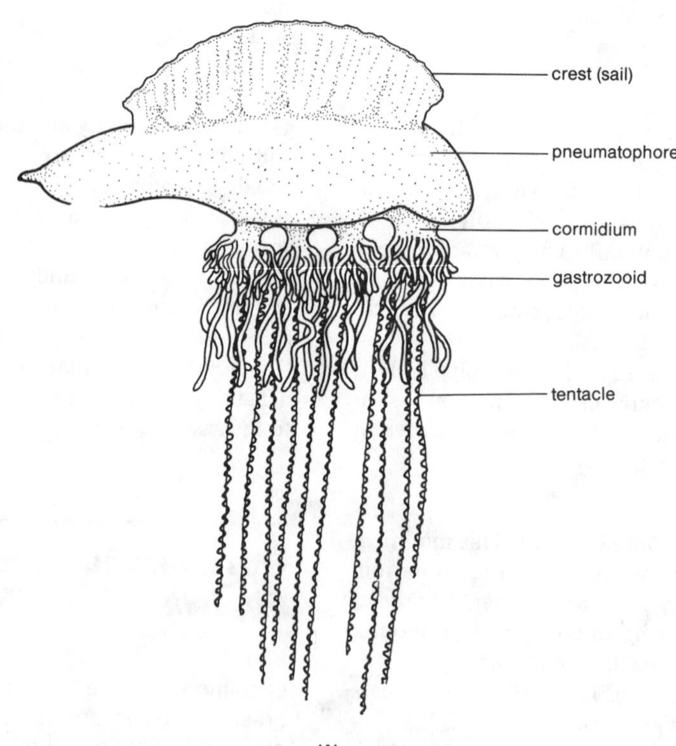

(A)

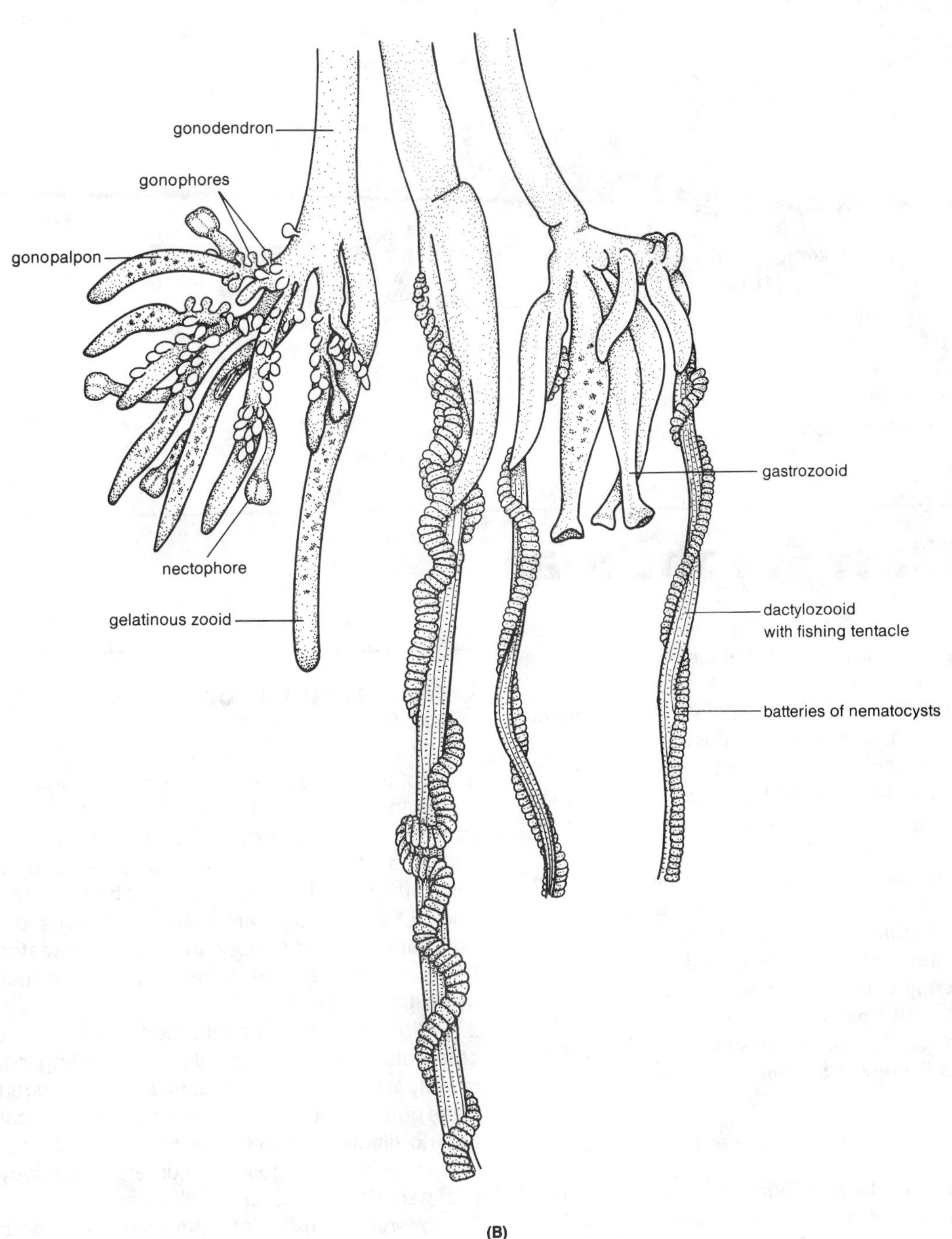

Figure 3.12. *Physalia.* **(A)** Entire colony. **(B)** Portion of a colony showing the types of zooids.

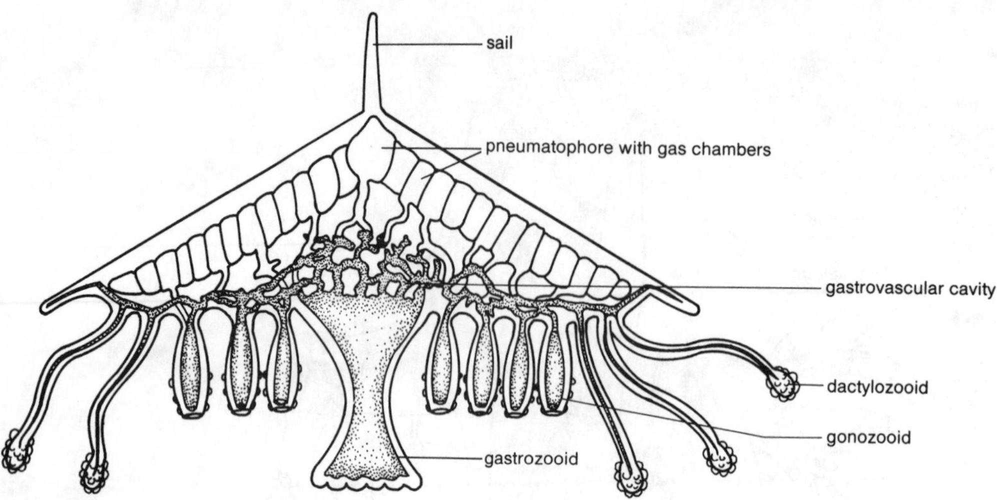

Figure 3.13. *Velella* colony.

B. Class Scyphozoa

The approximately 200 species in the class Scyphozoa (sy-fo-ZO-a; G., *skyphos*, cup + G., *zoon*, animal) are often called true jellyfish. The dominant body form is an acraspedote medusa (velum lacking) having a cellular mesoglea or collenchyma. The medusa may attain a diameter of 1 m, such as in *Cyanea capillata*. Scyphozoans are carnivorous, feeding on a variety of marine animals: fish, annelids, ctenophores, crustaceans, and planktonic forms. Most scyphozoan medusae are active swimmers propelled through the water by rhythmic pulsations of the bell; however, some members are sluggish bottom dwellers.

Two examples of the class will be studied: *Aurelia aurita*, an active pelagic scyphozoan in the medusa stage, and *Cassiopeia*, a sluggish jellyfish that spends most of its life on the bottom of shallow seas.

Aurelia aurita

The moon jelly *Aurelia aurita*, is the most common jellyfish on both American and European coasts. The tetramerous medusa is relatively flat and ranges from 7 to 10 cm in diameter (Fig. 3.14). Four **subgenital pits** occur on the subumbrella surface. Cilia cover both the exumbrella and subumbrella surfaces.

Observational Procedure: *Aurelia*

1. Obtain an adult *Aurelia* in a bowl containing water and examine the specimen with a dissecting microscope. Find the centrally located mouth at the free end of the manubrium, surrounded by four long, frilly **oral arms** bearing short tentacles (Fig. 3.14). The oral arms are extensions of the manubrium. Food trapped in mucus is transferred to the mouth by **lappets** located around the margin of the bell.

2. Follow the mouth into the central stomach that connects to four short canals, each leading to four radially arranged **gastric pouches**. The gastrodermis of the pouches bears tentacle-like projections called **gastric filaments**, chief sites of enzyme secretion and intracellular digestion. Numerous cnidocytes with nematocysts cover the filaments.

3. Observe the many radiating canals in the bell; these connect with the ring canal located around the bell's margin. The fluid-filled canals have special names depending on their location and function: **adradial, interradial,** and **perradial** (Fig. 3.14). Count

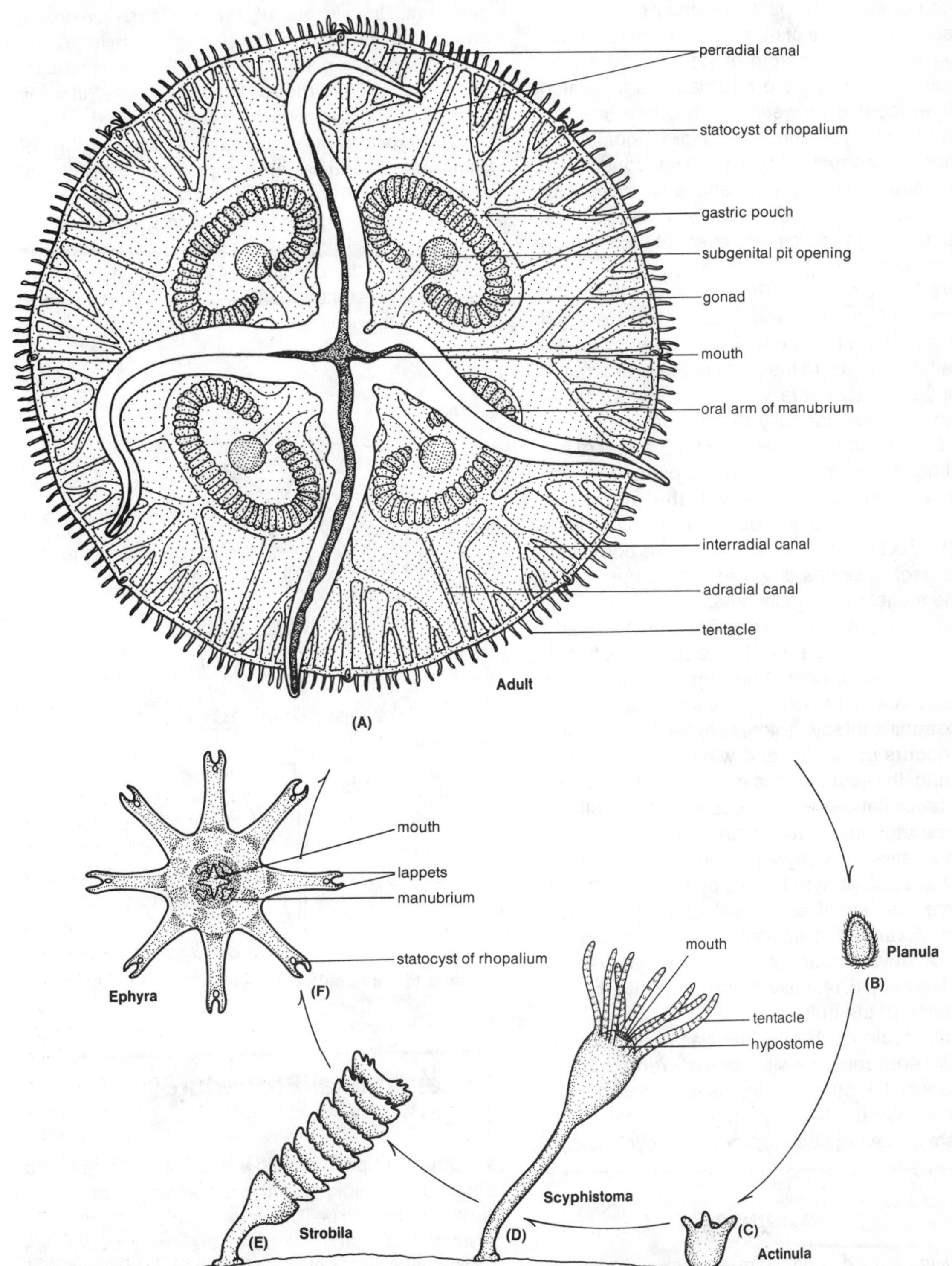

Figure 3.14. *Aurelia* life cycle.

the number of adradial canals. All canals plus the stomach constitute the gastrovascular cavity.

4. Observe the fringe of numerous, short tentacles hanging around the margin of the bell. Where the perradial canal meets the ring canal is a ciliated **rhopalium** located between two rhopalial lappets (Figs. 3.14 and 3.15). *Aurelia* has eight rhopalia and 16 rhopalial lappets. The rhopalium consists of sensory cells, sensory pits, and a statocyst, an organ of equilibrium. Unlike many of its relatives, *Aurelia* has a light-perceiving eyespot (**ocellus**) on the rhopalium.

5. *Aurelia* is dioecious. Find the four interradial gastrodermal gonads and four well-developed genital pits. Fertilization and development to the planula larva occur in the folds of the oral arms. If small opaque patches are seen on the inner surfaces of the oral arms, carefully dissect out and open some of them and examine under high power of the dissecting microscope. Do you find planulae? The mature ciliated planula leaves the adult and metamorphoses into an **actinula** or **hydratuba** stage (Fig. 3.14). This stubby, hydralike body with solid tentacles settles down and develops into a polypoid stage, the scyphistoma.

6. Examine a cross section of a **scyphistoma** and observe the four septa that partition off the gastrovascular cavity. Like the medusa, the scyphistoma is tetramerous. The scyphistoma can produce additional scyphistomae by asexual budding which occurs in autumn and winter. In late winter and spring, the scyphistoma undergoes transverse fission (**strobilation**), which results in a **strobila**, a polyp resembling a stack of saucers. The saucer-shaped bodies are **ephyrae** or immature medusae (Fig. 3.14). Each ephyra breaks off the strobila one at a time and eventually develops into an adult medusa. Examine prepared slides showing all stages of development in *Aurelia*. Compare the stages with Fig. 3.14, paying particular attention to comparison of the ephyra to the adult *Aurelia*.

The life cycle of *Aurelia* involves alternation of asexual reproduction with sexual reproduction (metagenesis). Distinguish the asexual phase from the sexual. What advantage might there be to the organism in having such reproductive cycles?

Cassiopeia

Cassiopeia, named from a mythological queen of Ethiopia, is a common rhizostome scyphozoan living in calm, shallow-water lagoons of Florida and the West Indies. The adult medusa may reach 26 cm in diameter and is unusual in its habit of living upside-down on the bottom of the seafloor. *Cassiopeia* attaches to the seafloor by having a raised circular zone around the margin of the bell, thus creating a suckerlike action on the bell's aboral surface. Pulsation of the bell creates water currents that bring oxygen and food to the animal as well as carry wastes away. Small planktonic organisms are swept over the frilly oral arms, are paralyzed by discharged nematocysts, trapped in mucus strands, and then carried into the secondary, porelike mouths. Symbiotic **zooxanthellae** also live within the mesoglea of *Cassiopeia*. If disturbed, *Cassiopeia* rises off the seafloor, swims a short distance, and settles down again.

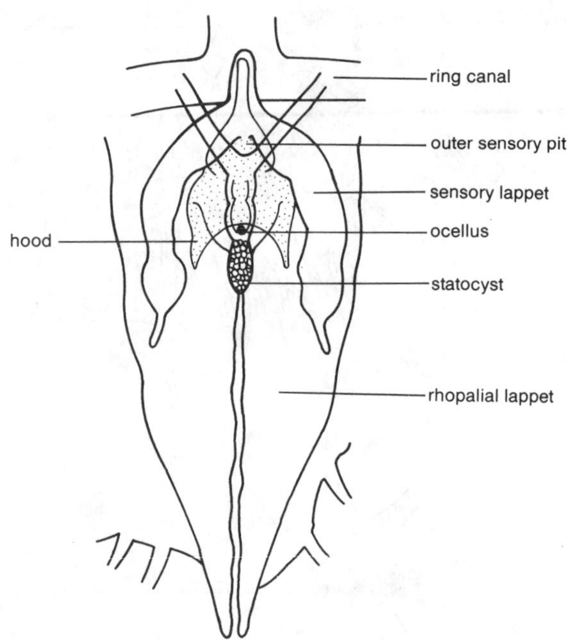

Figure 3.15. Rhopolium of *Aurelia*.

Observational Procedure: *Cassiopeia*

1. Obtain a *Cassiopeia* in a small dish of water and observe the specimen with a dissecting microscope. Note the flattened bell and lack of tentacles (Fig. 3.16). Instead of tentacles around the bell's margin as seen in *Aurelia*, *Cassiopeia* has a fringe of numerous, small **scallops**.

2. Regularly spaced between certain scallops are 16 rhopalia. The rhopalia play an important role in controlling pulsations of the bell and in regenerating missing areas of the bell. Excise a rhopalium with a sharp razor blade or needle. Place the rhopalium on a glass slide with a drop of water, cover with a coverslip, and examine with a compound microscope. A small amount of methylene blue added to the specimen may enhance your observation. Compare the rhopalium with that of *Aurelia*. Is a hood present? Unlike *Aurelia*, *Cassiopeia* lacks an eyespot.

3. Note that *Cassiopeia* lacks the distinct, central manubrium with a mouth and four separate oral arms observed in *Aurelia*. The four oral lobes are expanded and bifurcated in the adult to form eight, thick, gelatinous frilly arms. The original arm groove closes over, forming a branched **brachial canal** that opens to the oral surface by minute pores called secondary or brachial mouths. If adequate light passes through your specimen, the brachial canals can be seen. Numerous elongated appendages bearing many cnidocytes with nematocysts and mucus cells extend from the arms (Fig. 3.16). These are used in capturing food.

4. Cut off a small piece of the frilly arm at its distal end, make a wet mount, and observe under high power of the compound microscope. A small drop of methylene blue added to the specimen before the coverslip is secured may aid your observation. You should see numerous undischarged nematocysts.

5. Follow the **brachial canals** to the central stomach, where extracellular digestion occurs. Numerous radial canals extend from the stomach; the canals form an anastomosing network of channels throughout the bell. A distinct ring canal, as observed in *Aurelia*, is lacking. Examine the flattened aboral surface of the bell to see the numerous radial canals.

6. Lift up the bell and locate the four openings, called **subgenital pits**, locate in the stalklike area between the arms and flattened bell. The pits lead to a cross-shaped space, the **subgenital porticas,** which lies below the stomach.

7. With a sharp razor blade or scalpel, cut the flattened bell from the stalk to better observe the four **gonads**. Note their curved shapes. The gonads lie on the edge of the central stomach above the

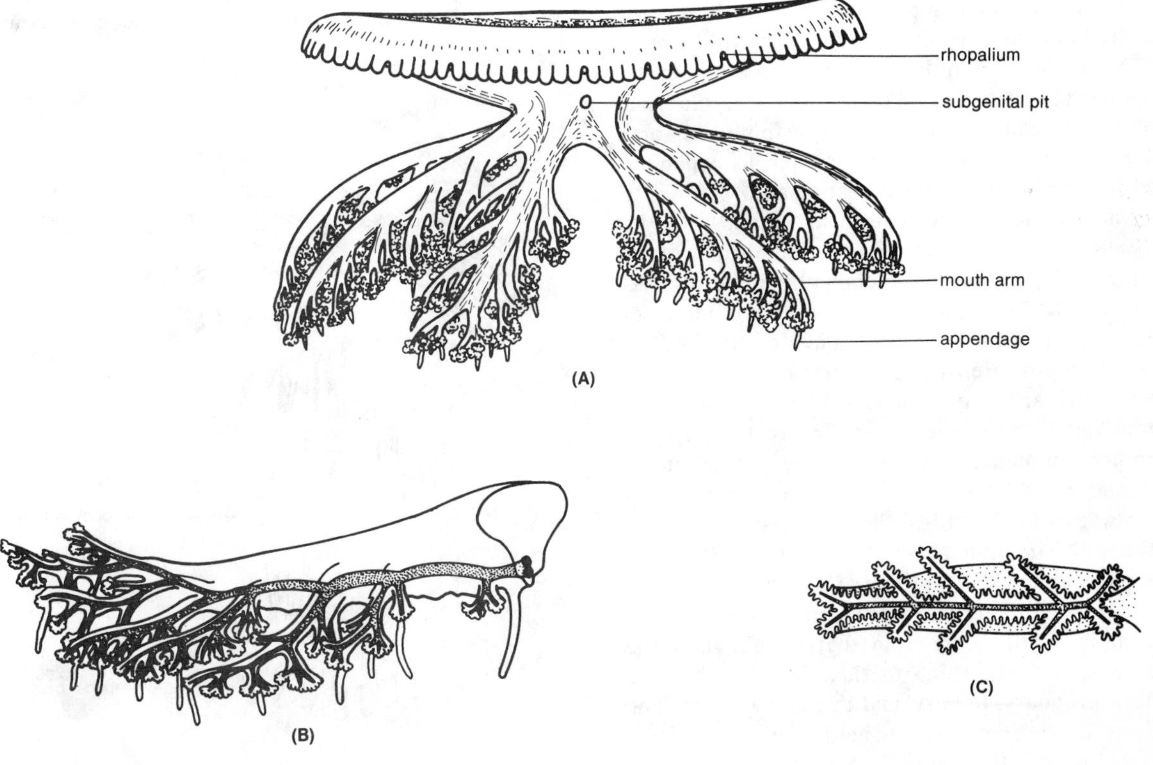

Figure 3.16. **(A)** Lateral view of adult *Cassiopeia* with oral surface down. **(B)** Lateral view of *Cassiopeia's* mouth arm, showing the gastrovascular canals. **(C)** A mouth of *Cassiopeia* closed.

subgenital pits. The pits do not connect to the gonads. The gametes are discharged into the porticus.

Cassiopeia is dioecious.The planula larva settles down and develops into a scyphistoma. One ephyra at a time buds from the scyphistoma. *Cassiopeia* lacks a strobila. The ephyra settles on the seafloor when it reaches about 2 cm in diameter, and there it developes into the adult. *Cassiopeia* also buds off small planulae from the scyphistoma.

C. Class Anthozoa

Over 6000 species, including sea anemones and corals, belong to the largest cnidarian class, Anthozoa (an-tho-ZO-a; G., *anthos*, flower + G., *zoon*, animal). The name "flower animal" is appropriately assigned to these animals since many are highly colored and resemble flowering plants. These marine invertebrates occur in both deep and shallow waters of tropical and temperate seas but reach their greatest abundance in tropical areas.

All adult anthozoans are sessile polyps; there is no medusa. Anthozoan polyps may be elongated and thin-walled or short, rather squatty, and thick-walled. The polyps occur as solitary individuals (e.g., sea anemones) or as complex colonies with several hundred interconnecting polyps (e.g., corals).

The anthozoan polyp differs in a number of important features from hydrozoans and scyphozoan polyps. Instead of having a hypostome, the oral surface of an anthozoan polyp is typically expanded and flattened as an **oral disc** bearing an ovoid mouth and hollow tentacles studded with cnidocytes (Fig. 3.17). Radial muscles in the disc permit opening of the mouth during feeding. The number of tentacles varies from 6 to >100, and the cavities of the tentacles communicate with the gastrovascular cavity.

Anthozoans are the only cnidarians with the gastrovascular cavity divided by longitudinal (oral-aboral) **septa** or **mesenteries** and in having a **pharynx** (stomodaeum, actinopharynx, esophagus, gullet) (Figs. 3.18 and 3.19). The septa are composed largely of mesoglea bounded by a layer of gastrodermis on either side. The septa that extend from the body wall to the pharynx are called **primary** (perfect or complete) septa (Fig. 3.18). In zoantharian anthozoans, there are **incomplete** (imperfect) septa that do not meet the pharynx, and according to their lengths, they are called secondary, tertiary, and quaternary septa. The free ends of the septa near and below the pharynx form coiled **septal filaments** (mesenteric filaments) containing enzyme-producing cells and cnidocytes (Fig. 3.18).

The pharynx is an ectodermally lined, ciliated tube chiefly for ingestion of food. The pharynx connects the mouth and gastrovascular cavity, and typically bears one or more **siphonoglyphs** (sulcuses)—ciliated grooves or gutters for moving water through the mouth and into the gastrovascular cavity (Fig. 3.19). Coral polyps lack the siphonoglyphs.

Two subclasses, Zoantharia (or Hexacorallia) and Alcyonaria (or Octocorallia), comprise the class Anthozoa. Zoantharians (sea anemones, stony corals) have their structures arranged in sixes (hexamerous) or multiples thereof. The several unbranched hollow

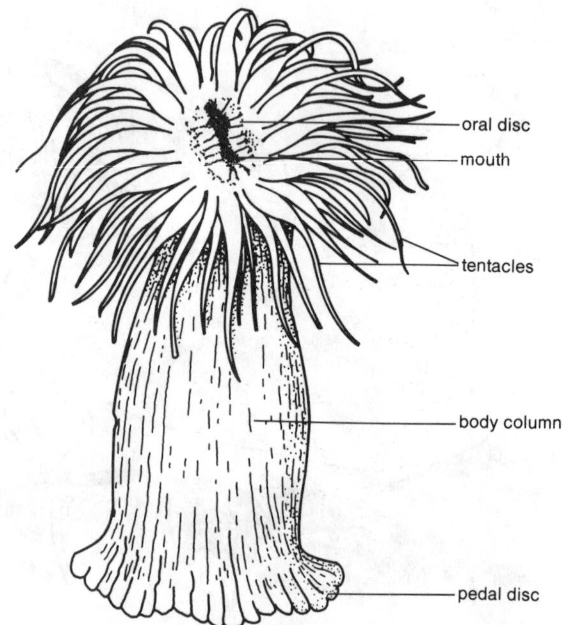

Figure 3.17. External view of sea anemone.

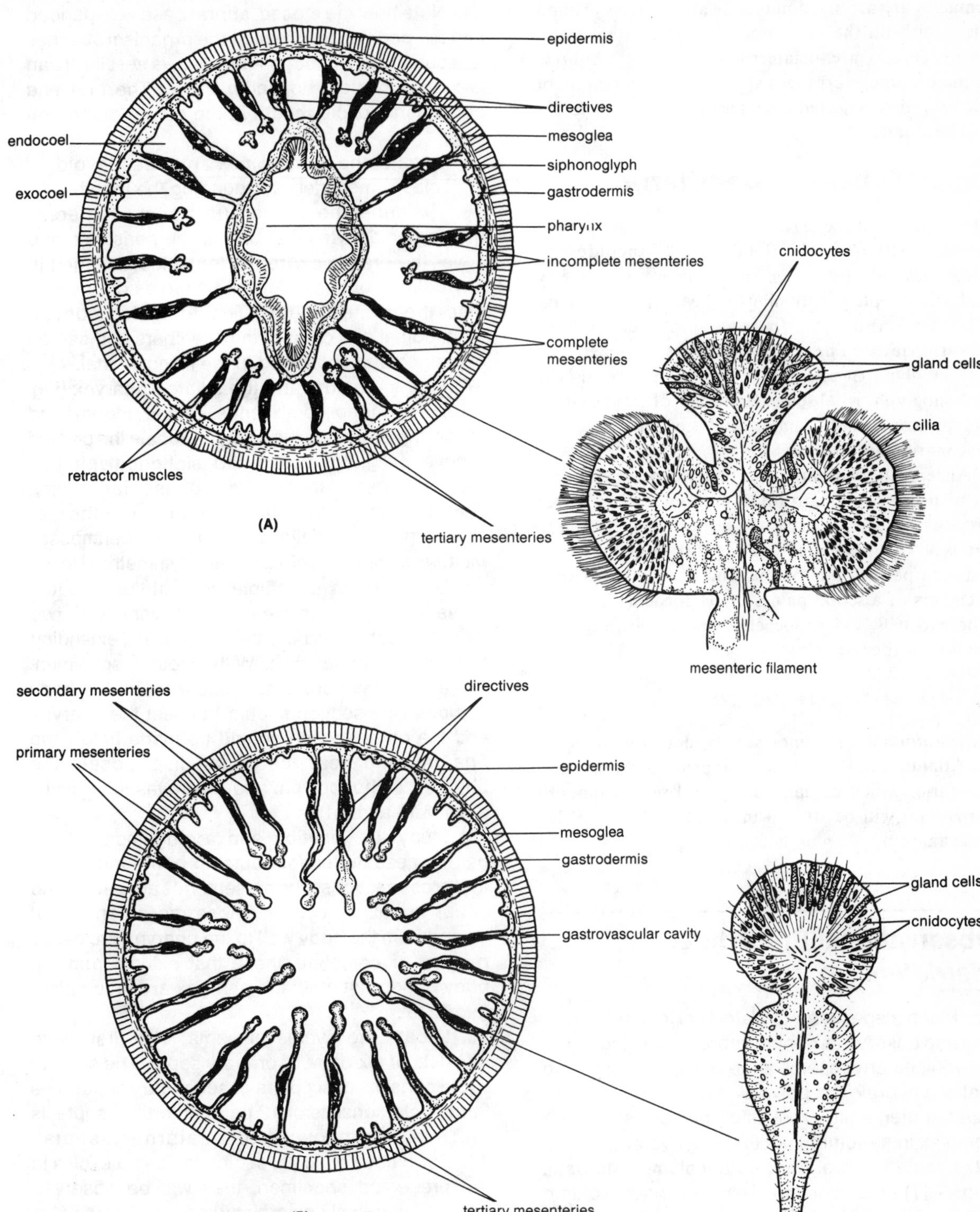

Figure 3.18. **(A)** Sea anemone cross section through pharynx, showing two siphonoglyphs. Mesenteric filament enlarged. **(B)** Sea anemone cross section below the pharynx. Mesenteric filament enlarged.

tentacles are arranged in one or more circles around the mouth on the oral disc. Alcyonarians (soft and horny corals) are octamerous with eight, hollow, pinnate tentacles arranged around the outer margin of the oral disc. Selected examples from the subclasses will be studied.

Subclass Zoantharia

Most species of the class Anthozoa are zoantharians (hexacorallians). The 3500 species include sea anemones and hard corals. The hexamerous symmetry and paired septa partitioning the gastrovascular cavity are two important distinguishing features of zoantharians. The pharynx bears at least one siphonoglyph in anemones, but in hard corals the siphonoglyph is absent. Tentacles of zoantharian polyps are usually simple and not like those of alyconarian anthozoans.

All septa in zoantharians are coupled, being arranged in a bilateral or biradial manner. Also, most zoantharian's septa occur close together in twos and are said to be paired. The gastrovascular cavity between pairs is the **exocoel** and the cavity between members of a septal pair is the **endocoel** (Fig. 3.18). The exocoels and endocoels communicate with the cavities of the tentacles.

Metridium

Metridium is a common sea anemone found along the Atlantic and Pacific coasts. It belongs to the order Actinaria, which contains all typical sea anemones. *Metridium* will be studied in detail to illustrate the basic anatomy of an anthozoan polyp.

Observational Procedure: *Metridium*

1. Place a specimen of *Metridium* in a dissection pan and observe its general morphology (Fig. 3.17). Do you see any resemblance of your specimen to that of a hydra? The polyp of *Metridium* is larger in size, stouter built, more thick-bodied, and more complex in structure than that of a hydra.

2. Locate the **oral disc**, **body column**, and **base** (Fig. 3.17). The cylindrical-shaped body column can be divided into an upper, thin-walled **capitulum** and a lower, thick-walled **scapus** (Fig. 3.19).

3. Between the capitulum and scapus is a groove, the **fosse**, bordered by a prominent fold of the body wall called the **collar** (parapet).

4. Note how the closed, aboral base is expanded into the **pedal disc** by which the organism attaches to a substrate. Although *Metridium* is sessile, it can detach and move to another site. The oral end contains an ovoid **mouth** in the center of the oral disc.

5. Locate the **peristome**, an area devoid of tentacles immediately surrounding the mouth.

6. Examine the numerous hollow tentacles bearing cnidocytes that border the peristome and cover the oral disc. Are the tentacles arranged in any definite pattern? Are all tentacles on the oral disc the same length? The tentacles contain longitudinal and circular muscle fibers and can be expanded and shortened in the living animal.

7. Find the mouth and the ciliated **pharynx** (Fig. 3.19). At one or both ends of the mouth and extending down into the pharynx, locate the ciliated groove or gutter called the **siphonoglyph** that moves water into the gastrovascular cavity. Consider how the position and number of siphonoglyphs in relation to the mouth superimposes bilateral symmetry onto a radially symmetrical form.

8. Using a sharp scalpel, longitudinally cut a preserved specimen cleanly and evenly into two equal halves beginning at the oral end and extending through the pedal disc. With another specimen, make two clean cross sections through the body column. One section should transect the pharynx and the other section should be made below the pharynx and near the pedal disc. Locate the pharynx, siphonoglyph, and gastrovascular cavity (Figs. 3.18 and 3.19).

9. Study the location and arrangement of the paired mesenteries (or septa) in both longitudinal and cross sections. The mesenteries are paired and occur in sets of cycles. Those mesenteries that extend from the body wall to the pharynx are called **primary** or **complete**; those that extend from the body but do not meet the pharynx are **incomplete** (Fig. 3.18).

10. Locate the cycles of septa. The first cycle consists of six pairs of primary septa. The second cycle consists of six pairs of secondary septa. The third cycle consists of 12 pairs of tertiary septa. Is there a fourth cycle of quaternary septa?

11. If you have difficulty seeing the septa cycles in the preserved specimen, they will be observed when prepared slides are studied.

12. In the longitudinally halved specimen, locate a longitudinal **retractor muscle** on one face of a septum (Fig. 3.19). This muscle occurs on the side of each paired septum facing the endocoel, except for the septa attached at the ends of the siphonoglyphs in the pharynx. On these paired, complete septa, called **directives**, the retractor muscles face the exocoels (Fig. 3.18). The retractor muscles retract the oral disc. The relationship between the septa, the muscles, and their spaces can best be seen in cross section on the prepared slide. Depending on how the longitudinal section was made, one or two **ostia** may be seen in the septa near the pharynx and body wall at the distal end of the animal. What function might these septal perforations serve?

13. At the free end of each incomplete septum, locate the longitudinal cord or ridge called the **septal** or **mesenteric filament** (Fig. 3.18). The filament is trilobed in the region of the pharynx; those near the base are single-lobed. The detail structure of a filament will be seen in the prepared cross section.

14. Locate the many threadlike structures called **acontia** (Fig. 3.19). These might be more easily observed in the longitudinal section. Acontia are continuations of the middle lobe of the lower septal mesenteries. They aid in defense and capture of prey, bear cnidocytes and enzyme-producing gland cells, and can be protruded through the mouth and through minute openings, **cinclides**, located in the lower body column. When the column contracts, water is expelled through the cinclides.

Metridium is dioecious. Careful dissection and observation may reveal the **gonads**, which are thickened bands located behind a mesenteric filament (Fig. 3.19).

15. Examine a prepared slide showing a cross section of *Metridium* through the pharynx under low power of a compound microscope (Fig. 3.18). Locate the outer epidermis of columnar cells, the inner gastrodermis lining the gastrovascular cavity, and the middle mesoglea containing ameboid cells and muscle fibers in the body wall. The inner wall of the pharnyx is folded and lined with epidermis, containing cnidocytes and gland cells. The outer covering of the pharynx is gastrodermis and the middle area is mesoglea.

16. Locate the siphonoglyph. Do you see more than one siphonoglyph? Observe the cilia of the pharynx and siphonoglyph under high power.

17. Study the arrangement of the mesenteries. Note that each mesentery consists of an inner layer of mesoglea surrounded by gastrodermis.

18. Study the structural relationship of a mesentery to the body wall from which it arises. How many pairs of primary and secondary mesenteries are present? Do you see any tertiary mesenteries?

19. Locate the retractor muscles on the secondary and primary septa. Note how the muscles face the endocoels. Do the muscles on the directives face the same way as those on the remaining septa?

20. Locate the clover-shaped mesenteric filament. The central lobe is nonciliated and contains gland cells that produce enzymes used in digestion and

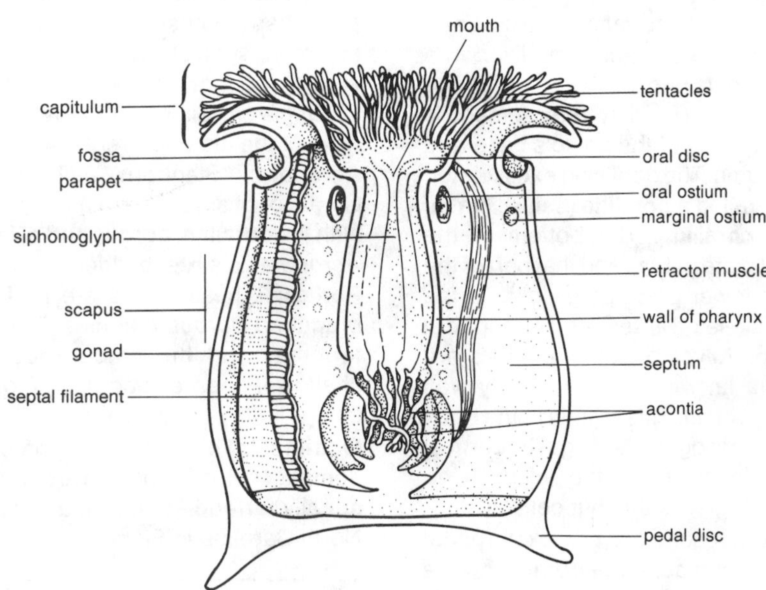

Figure 3.19. Oral-aboral section of *Metridium*, showing internal structures.

numerous cnidocytes. The two lateral lobes are ciliated and lack cnidocytes. What function might the cilia serve?

21. Study a longitudinal section on a prepared slide and identify as many of the above-mentioned structures as possible.

Stony Corals

True or stony corals are zoantharian anthozoans of the order Scleractinia (Madreporaria). There are two groups of hard corals: **hermatypic** and **ahermatypic**. Hermatypic corals are colonial reef builders, found in shallow tropical seas. Ahermatypic corals are noncolonial, nonreef builders, and may occur in rather deep and cold waters of higher latitudes. Most coral species are hermatypic with polyps 1 to 3 mm in diameter.

Growth patterns and arrangement of the polyps in a colony are varied, resulting in different skeletal configurations. Coral deposits may be upright, branched, round, or flattened masses. Selected examples of common species are described briefly below. Your instructor may provide additional examples to observe.

Observational Procedure: Stony Corals

Examine the specimens available for study, noting differences in weight (density), growth form (branching or globular), and surface patterns. Each coral polyp is a miniature sea anemone fixed in a cup of calcium carbonate (Fig. 3.20) into which the living polyp can retract. The limy encasement is secreted by the epidermis of the polyp's base and lower area of the column. The nonliving exoskeleton of a single polyp is called a **corallite (calyx)**; that of the colony is the **corallum**. The bottom of the corallite is the basal plate (**tabla**), and that part of the cup containing the lower part of the polyp is the **theca**. The limy exoskeleton is secreted throughout the life of the polyp. Polyps of colonial corals are interconnected by a lateral fold of the body wall extending from the column. This results in the body tissues (epidermis, mesoglea, and gastrodermis) and the gastrovascular cavities of the polyps being continuous. The mesoglea is thin but cellular.

Unlike sea anemones, coral polyps lack the pedal disc, siphonoglyphs, and septal perforations. The hollow tentacles occur in sets of six and the septa are usually hexamerous too. Six complete mesenteric septa typically alternate with six incomplete septa. Tertiary septa may occur. In your specimens locate the skeletal septa or ridges, called **sclerosepta**, present in each theca. These represent radial folds found between each complete septa that are secreted by the underside of the polyp; this results in the lower part of the gastrovascular cavity divided by folds of tissue.

The septal filaments are single-lobed, containing gland cells and cnidocytes. The highly convoluted acontia can be extended through the ovoid mouth in the oral disc.

Meandrina: Brain Coral. Note how the polyps are arranged close together in rows but that the rows are well separated. The resulting coral has ridges and valleys that appear somewhat like a human brain. The polyps of *Meandrina* are produced by intratentacular buddings; however, the oral discs and columns never separate after new mouths are formed. Therefore, the polyps of a new row have many mouths but share a common oral disc. *Meandrina sinuosa* occurs in Florida and the West Indies.

Astrangia: Eyed Coral. This North Atlantic coastal coral forms small, encrusting colonies. The skeleton shows a pitted appearance because the polyps are well separated. The polyps in *A. danae* cannot withdraw into the cups. This species occurs from North Carolina to Massachusetts.

Fungia: Mushroom Coral. This solitary, rather deep coral has a mushroom-shaped skeleton. It is a common species of the Great Barrier Reef; *F. elegans* occurs in the Gulf of California. Individuals of *Fungia* can slide themselves over the substratum without the aid of currents.

Acropora: Staghorn, Elkhorn, or Antler Coral. Corals of this genus are mostly branched colonies with the corallum porous and of loose construction. *Acropora*. is a reef builder.

Porites: These corals are reef builders, forming branched porous colonies. The polyps are close together and the calcareous cups are rather shallow. *P. porites* occurs in Florida and the West Indies.

Oculina: This genus forms branched, rather compact colonies; polyps are widely separated and spirally arranged. *Oculina diffusia* occurs from North Carolina to Florida.

Subclass Alcyonaria

Organ-pipe coral (*Tubipora*), sea fans(*Gorgonia*), sea pansies (*Renilla*), sea pens (*Stylatula*), sea whips (*Leptogorgia*), and their allies are alcyonarian cnidarians. All 3000 of the known living species of alcyonarians are colonial and most abundant in warm tropical areas of the Indo-Pacific.

Unlike polyps of zoantharians, each alcyonarian polyp has eight unpaired mesentaries that connect to the body wall and a relatively short pharynx (Fig.3.21). Furthermore, each polyp bears eight pinnate tentacles that are arranged symmetrically. The number of tentacles and septa and their arrangement give alcyonarians an octamerous radial symmetry. The feathery appearance of the tentacles is due to many

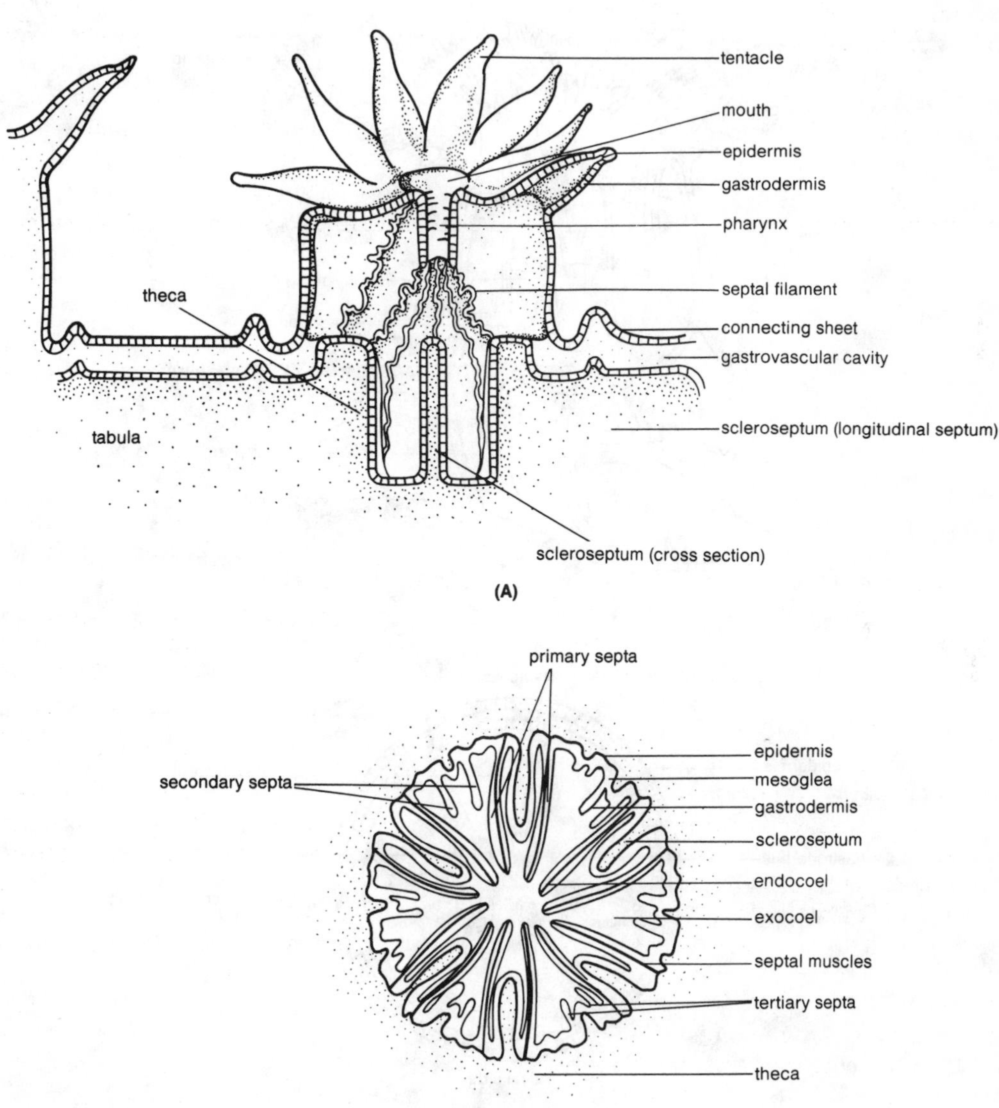

Figure 3.20. **(A)** Oral-aboral section of a hard coral polyp and its calcareous cup. **(B)** Cross section of a coral polyp below the pharynx.

fingerlike projecting pinnules that arise from the main part of the tentacle. The tentacles are hollow; their cavities are extensions of the gastrovascualr system. Cnidocytes and cilia on the tentacles capture food.

The gastrovascular cavities of the polyps are connected by gastrodermal tubes—branched ontinuations of the gastrovascualr walls of the polyps, called **solenia** (Fig.3.21). New polyps arise as buds or spouts from the solenia. Arrangement of the solenia varies. The solenia are embedded in a thick mesoglea called **coenenchyme** and the entire outer surface of the colony is covered with epidermis. The thick matrix of coenenchyme also contains ameboid cells, skeletal elements (e.g., calcareous spicules and horny protein),

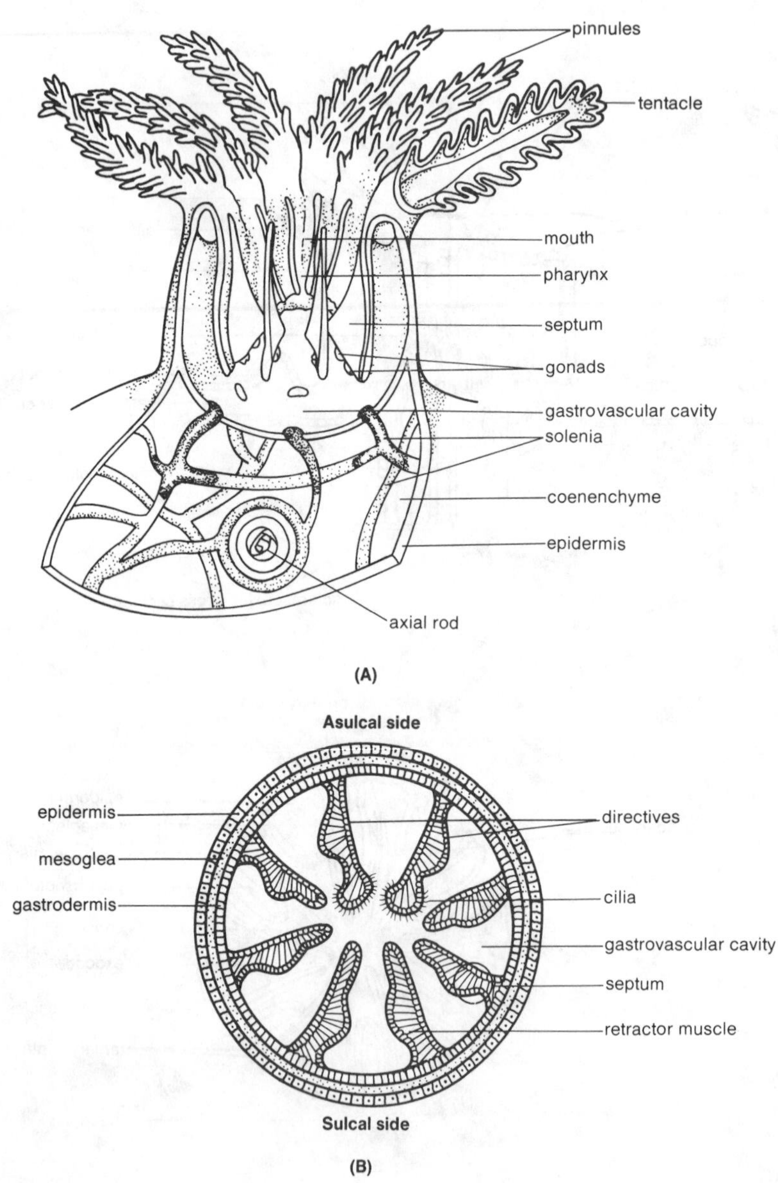

Figure 3.21. **(A)** Oral-aboral section of an alcyonarian polyp. **(B)** Cross section of the polyp below the pharynx.

and most of the body of the polyp. Only the oral ends of the polyps (**anthocodia**) prodect from the surface of the coenenchyme.

There is usually one siphonoglyph (or sulcus) located in the pharynx of the polyp. This ciliated groove, like that of other anthozoans, drives water into the gastrovascular cavity. The side of the polyp bearing the siphonoglyph is the sulcal side; the opposite side is the asulcal side.

Each septum on its sulcal side (i.e., toward the siphonoglyph) bears a strong, longitudinal retractor muscle (Fig.3.21). The free edge of each septum below the pharynx bears a septal (or mesenteric) filament.

Observational Procedure : Alcyonaria

There are several orders of Alcyonaria, which differ in the nature of the skeleton and mode of colony formation. Examples of some of these orders will be studied.

Tubipora: Obtain a specimen of the organ-pipe coral (*Tubipora*). The order to which this genus belongs (Stolonifera) is characterized by the lack of the coenenchymal mass (Fig. 3.22). The polyps arise singly from a mat of stolons. Observe that the colony consists of upright, parallel polyps supported by hard skeletal tubes of fused spicules giving an impression of the pipes of an organ. The erect tubes are united at spaced levels by transverse platforms. Iron salts in the skeleton impart a dull red color to the organism.

Gorgonia: The sea fan (*Gorgonia*) is a horny coral of the order Gorgonacea, which also includes the sea whips and sea feathers (Fig. 3.22). These highly colored, graceful colonies occur in tropical and subtropical waters, especially in the Indo-Pacific Ocean and in the subtropical Atlantic around Bermuda, the Bahamas, and West Indies. Because of their sessile nature and firm upright position, sponges, hydroids, and other invertebrates become attached to the sea fan. The most characteristic feature of these anthozoans is the horny protein material called **gorgonin**, located in their skeletons. The polyps secrete the gorgonin.

Examine a sea fan closely (Fig. 3.22). Note the short main trunk to which the colony was attached to the substratum. Extending from the trunk is a central axial rod of gorgonin. Emerging from the axial skeleton are numerous small branches united by cross connections into a lattice network. The many branches of the colony are in one plane, resulting in the fan-shaped pattern. A thin layer of coenenchyme covers the colony.

Observe the branches of the sea fan using the dissecting microscope. Locate the small polyps (**anthocodia**) scattered over the stems and branches. If you are using a dried specimen, you will see openings where the once-living polyps occurred. Can you see polyps on the main stem? Do polyps occur on both surfaces of the sea fan?

Renilla: Sea pansies (*Renilla*), of the order Pennatulacea, occur on soft bottoms of nearshore waters of the Atlantic, Gulf, and southern California coasts. The colony can uncover itself if buried by moving sediment.

Obtain a preserved or living *Renilla* and observe the specimen under the dissecting microscope. The primary axial polyp (**rachis**) is leaflike or somewhat heart- shaped, and is supported by a short, proximal stalk (**peduncle**) to which the colony was anchored to a substrate by peristaltic contractions of the peduncle (Fig. 3.22). If the organism is dislodged, it can reanchor itself. Note the bilateral symmetry of the colony. The symmetry is established by a stem (or "track") extending from the peduncle to about the middle of the rachis. Do polyps occur on the peduncle? Are there polyps on the ventral surface? Many secondary polyps are embedded in the fleshy upper surface of the rachis. The secondary polyps are dimorphic and called **autozooids** and **siphonozooids**. Autozooids are feeding polyps, containing a mouth and six pinnate tentacles. Siphonozooids are tiny, nonfeeding, warty protuberances arranged in clusters (Fig. 3.22). Tentacles of siphonozooids are rudimentary. These polyps move water through the colony. The secondary polyps are formed by budding. Sexual reproduction, as in other anthozoans, results in a free-swimming planula larva.

Other Alcyonarians. Examine specimens of sea whips (*Leptogorgia*, order Gorgonacea), sea pens (*Stylatula*, order Pennatulacea), or other examples of alcyonarians. Study these with the dissecting microscope and compare their structures with members previously studied. Note the branching patterns and locations of the polyps.

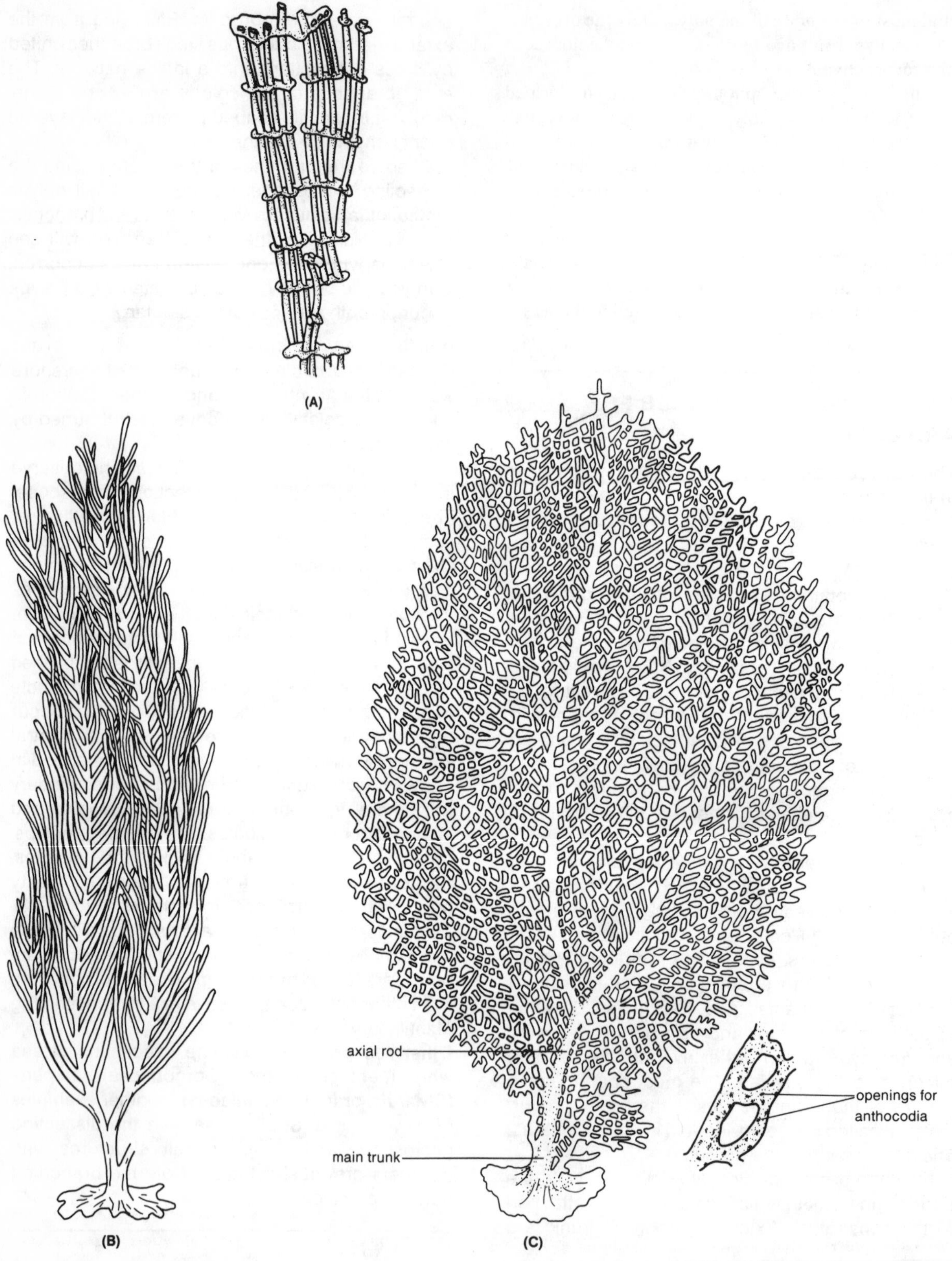
(A)
axial rod
main trunk
openings for
anthocodia
(B)
(C)

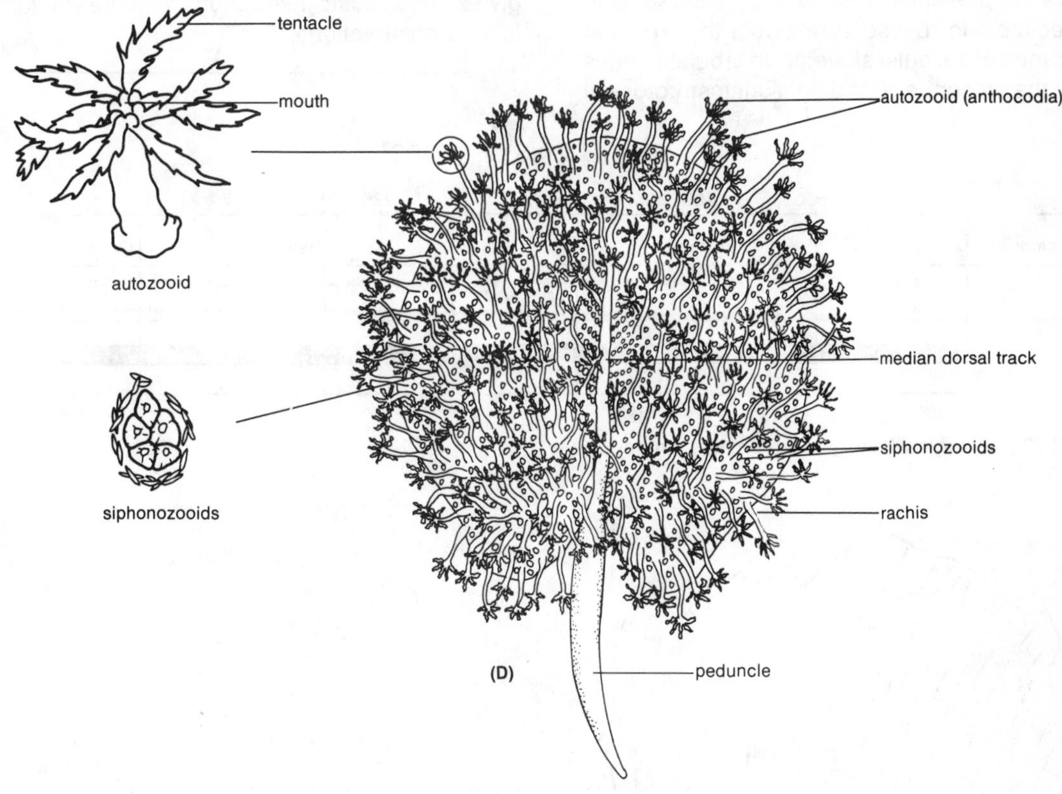

Figure 3.22. **(A)** Piece of *Tubipora* (organ-pipe coral) skeleton. **(B)** Sea whip. **(C)** Sea fan (*Gorgonia*); magnified part of fan showing holes for polyps. **(D)** Sea pansy (*Renilla*), dorsal view. (From Hyman.)

D. Fossil Cnidarians

Cnidarians represent an ancient group of animals dating from the **Cambrian Period** (Fig. 3.23). Because of the soft-bodied nature of most individuals, the fossil record of cnidarians is scant. However, individuals with chitinous perisarcs and calcareous skeletons have been preserved as fossils. Extinct individuals of Hydrozoa, Scyphozoa, and Anthozoa have been found as fossils, but the largest number of extinct cnidarians is fossil corals. The extinct **rugose** (Order Rugosa or Tetracoralla) and **tabulate** (Subclass Tabulata) corals were important reef builders throughout most of the **Paleozoic Era** (Fig. 3.24). Both groups died out in the **Permian**. The rugose corals were mostly solitary, cone-shaped corals with an outer skeleton having a wrinkled appearance as indicated by the name rugose (Fig. 3.24). The tabulate corals were colonial anthozoans with calcareous skeletal tubes containing tabulae, or horizontal platforms, on which the polyps rested. These extinct corals were replaced by scleractinian and hydrozoan corals. Many well-known living genera, such as *Acropora*, *Fungia*, *Favia*, and *Orbicella*, have a long fossil history.

Observational Procedure: Fossil Cnidarians

Examine fossil specimens provided and note the similarities and differences to the present-day forms.

Observe the presence of theca with sclerosepta in the specimen. In rugose corals note the wrinkled appearance of the outer skeleton. In tabulate corals look for the presence of tabulae. Your instructor will give some specific instructions to guide you further in your observations.

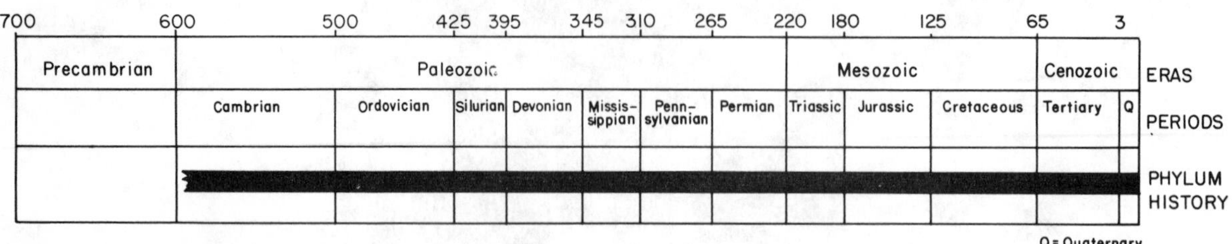

Figure 3.23. Geological history of the phylum Cnidaria.

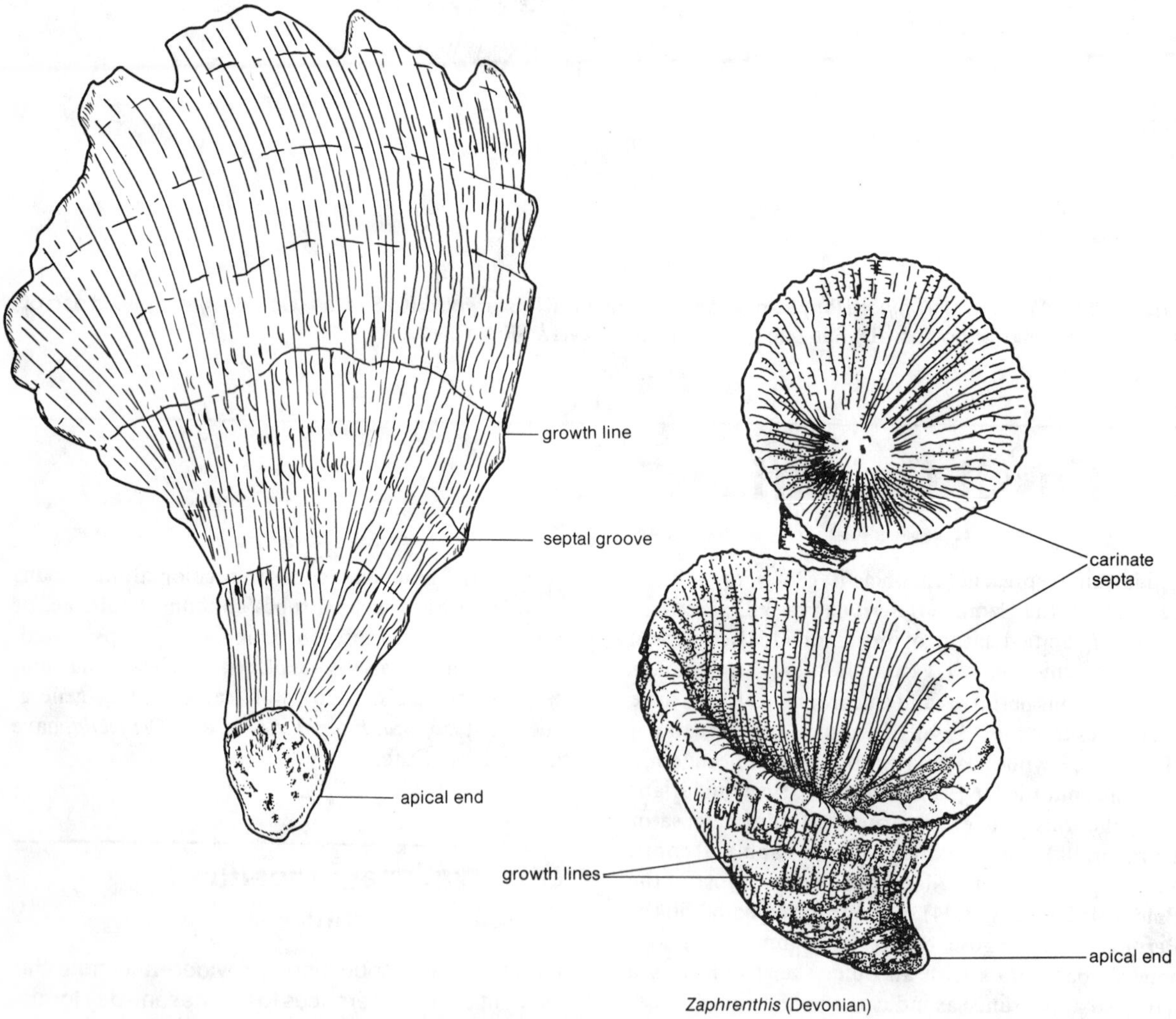

growth line

septal groove

apical end

carinate septa

growth lines

apical end

Zaphrenthis (Devonian)

Figure 3.24. Two views of *Zaphrentis*, a rugose coral.

Supplemental Reading

Barzansky, B., and H. M. Lenhoff. 1974. On the chemical composition and developmental role of the mesoglea of hydra. Am. Zool. 14:575-581.

Bayer, F. M., and H. B. Owre. 1968. The free-living lower invertebrates. Macmillan, New York.

Bouillon, J. 1968. Introduction to coelenterates. *In*: M. Florkin and B. T. Scheer (eds.). Chemical zoology. Vol. 2. Academic Press, New York.

Burnett, A. L. (ed.). 1973. Biology of hydra. Academic Press, New York.

Calder, D. R. 1971. Nematocysts of *Aurelia*, *Chrysaora*, and *Cyanea*, and their utility in identification. Trans. Am. Microsc. Soc. 90:269-274.

Calder, D. R., H. N. Cones, and E. B. Joseph. 1971. Bibliography on the Scyphozoa with selected references on Hydrozoa and Anthozoa. Va. Inst. Mar. Sci. Spec. Sci. Rep. 59.

Campbell, R. D. 1974. Cnidaria. *In*: A. C. Giese and J. S. Pearse (eds.). Reproduction of marine invertebrates. Vol. 1. Academic Press, New York.

Carre, C., and C. Carre. 1980. Triggering and control of cnidocyst discharge. Mar. Behav. Physiol. 7:109-117.

Chapman, G. 1953. Studies on the mesoglea of coelenterates. I. Histology and chemical properties. Q. J. Microsc. Sci. 94:155-176.

Faulkner, D., and R. Chesher. 1979. Living corals. Clarkson N. Potter, New York.

Florkin, M., and B. T. Scheer (eds.). 1968. Chemical zoology. Vol. 2. Porifera, Coelenterata, and Platyhelminthes. Academic Press, New York.

Gierer, A. 1974. Hydra as a model for the development of biological form. Sci. Am. 231:44-45.

Goreau, T. F., and N. I. Goreau. 1979. Corals and coral reefs. Sci. Am. 241 (2):124-136.

Hughes, T. P., and J. B. C. Jackson. 1980. Do corals lie about their age? Some demographic consequences of partial mortality, fission, and fusion. Science 209:713-715.

Hughes, T. P., and J. B. C. Jackson. 1985. Population dynamics and life histories of foliaceous corals. Ecol. Monogr. 55:141-166.

Kramp, P. L. 1961. Synopsis of the medusae of the world. J. Mar. Biol. Assoc. U.K. 40:1-469.

Lane, C. E. 1960. The Portuguese man-of-war. Sci. Am. 202:158-168.

Lang, J. 1973. Interspecific aggression by scleractinian corals: why the race is not only to the swift. Bull. Mar. Sci. 23:260-279.

Lenhoff, H. M. 1983. Hydra research methods. Plenum Press, New York.

Lenhoff, H. M., and W. F. Loomis (eds.). 1961. The biology of hydra and some other coelenterates. University of Miami Press, Coral Gables, FL.

Lenhoff, H. M., L. Muscatine, and L. V. Davis (eds.). 1971. Experimental coelenterate biology. University of Hawaii Press, Honolulu, HI.

Lenz, T. F. 1965. Hydra: induction of supernumerary heads by isolated neurosecretory granules. Science 150:633-635.

Lubbock, R. 1979. Chemical recognition and nematocyte excitation in sea anemone. J. Exp. Biol. 83:283-292.

Mackie, G. O. (ed.). 1976. Coelenterate ecology and behavior. Plenum Press, New York.

Mackie, G. O., P. A. V. Anderson, and C. L. Singla. 1984. Apparent absence of gap junction in two classes of Cnidaria. Biol. Bull. 167:120-123.

Muscatine, L., and H. M. Lenhoff (eds.) 1974. Coelenterate biology: reviews and new perspectives. Academic Press, New York.

Oliver, W. A. 1980. The relationship of the scleractinian corals to the rugose corals. Paleobiology 6:146-160.

Rosen, B. R. 1982. Darwin, coral reefs, and global geology. BioScience 32:519-525.

Schlichter, D. 1982. Nutritional strategies of cnidarians: the absorption, translocation, and utilization of dissolved nutrients by *Heteroxenia fuscescens*. Am. Zool. 22:659-669.

Sebens, K. P. 1980. The regulation of asexual reproduction and indeterminate body size in the sea anemone, *Anthopleura elegantissima*. Biol. Bull. 158:370-381.

Smith, F. G. W. 1971. A handbook of the common Atlantic reef and shallow-water corals. University of Miami Press, Coral Gables, FL.

Stoddart, D. E. 1969. Ecology and morphology of recent coral reefs. Biol. Rev. Camb. Philos. Soc. 44:433-498.

Totton, A. K., and G. O. Mackie. 1960. Studies on *Physalia physalia* (L.). Discovery reports 30. Cambridge University Press, Cambridge.

Wyttenbach, C. R. (ed.). 1974. The developmental biology of the Cnidaria. Am. Zool. 14:440-866.

Yonge, C. M. 1963. The biology of coral reefs. Adv. Mar. Biol. 1:209-261.

Yonge, C. M. 1968. Living corals. Proc. R. Soc. Lond. B. 169:329-344.

EXERCISE 4

Phylum Ctenophora

Sea walnuts, comb jellies, and sea gooseberries are common names often applied to members of the phylum Ctenophora (te-NOF-or-a; G., *ctenos*, comb + *phora*, to bear). The transverse, ciliary comb-shaped plates or **ctenes** located on eight bands called **comb rows** are the basis for the phylum name (Fig. 4.1A). The ctenes are used in locomotion.

There are approximately 100 carnivorous species in this marine group of **biradial** symmetrical invertebrates. Their biradial symmetry is established by the bilateral symmetrical digestive system, which is composed of paired canals arising from a central stomach, the radially arranged comb rows, and the bilaterally arranged tentacles when present. Ctenophores are worldwide in distribution, and occur at all depths in the open sea as well as in coastal waters.

Ctenophora and Cnidaria are the two major phyla that constitute the Radiata because their members possess primary radial symmetry. Although both phyla have several similarities (e.g. body wall construction and gastrovascular digestive system), there are important differences. Unlike cnidarians, ctenophores are not colonial and polymorphic and lack nematocysts. Instead of nematocysts, tentaculate ctenophores have peculiar cells called **collocytes** (glue or lasso cells) that secrete a sticky substance for catching and holding prey such as copepods and other small organisms (Fig. 4.1B).

Ctenophores are monoecious except for the lobate ctenophore, *Ocyropsis*. The testes and ovaries develop as two bands located within each of the eight meridional canals located beneath the comb rows. Fertilization occurs in the seawater; most forms have a free-swimming **cydippid larva** that resembles an adult ctenophore of the order Cydippida.

Classification

The phylum is divided into two classes based primarily on the presence or absence of tentacles.

I. Class Tentaculata. Tentacles present.

- **A. Order Cydippida**. Ovoid body with two branched tentacles that retract into pouches or sheaths. Examples are *Pleurobrachia* and *Mertensia*.
- **B. Order Lobata**. Compressed body in tentacular plane forming two oral lobes on either side of that plane. Tentacles are small and not located in pouches. Examples are *Mnemiopsis*, *Bolinopsis*, and *Ocyropsis*.
- **C. Order Cestida**. Belt-shaped body and compressed in tentacular plane; tentacles in pouches, but reduced and near the mouth. Examples are *Cestum* and *Velamen*.
- **D. Platyctenea**. Body flattened along an oral-aboral axis and adapted for creeping movements. Examples are *Coeloplana and Ctenoplana*.

II. Class Nuda. Tentacles absent.

- **A. Order Beroida**. Body cylindrical and compressed; mouth and pharynx enlarged. Example is *Beroë*.

Pleurobrachia

Pleurobrachia, often called the sea gooseberry, is a tentaculate ctenophore. Individuals of the genus are commonly found in coastal waters of the Atlantic Ocean.

Observational Procedure: *Pleurobrachia*

Place a preserved *Pleurobrachia* in a small dish filled with water. Using a dissecting microscope, observe the ctenophore's coloration and ovoid shape (Fig. 4.1A). The mouth is at the narrower (oral) end. It leads to the pharynx, stomach, and aboral canal that terminates in aboral anal pores. Arising off either side of the stomach are **transverse canals** that extend as **interradial canals** connecting the eight **meridional canals**. Two **pharyngeal canals** run parallel to the pharynx (Fig. 4.1A). This complex of canals constitutes the gastrovascular digestive system.

Locate the eight radially arranged comb rows located above the meridional canals. These orally-aborally arranged rows are often whitish in color. Examine closely one comb row and locate the ctenes, each composed of fused cilia arranged like teeth of a comb. Find the openings of the two **tentacular sheaths** at the aboral end (Fig. 4.1A). Preserved specimens usually have the solid tentacles retracted into the sheaths; however, in life the extended tentacles (up to 15 cm long) with many

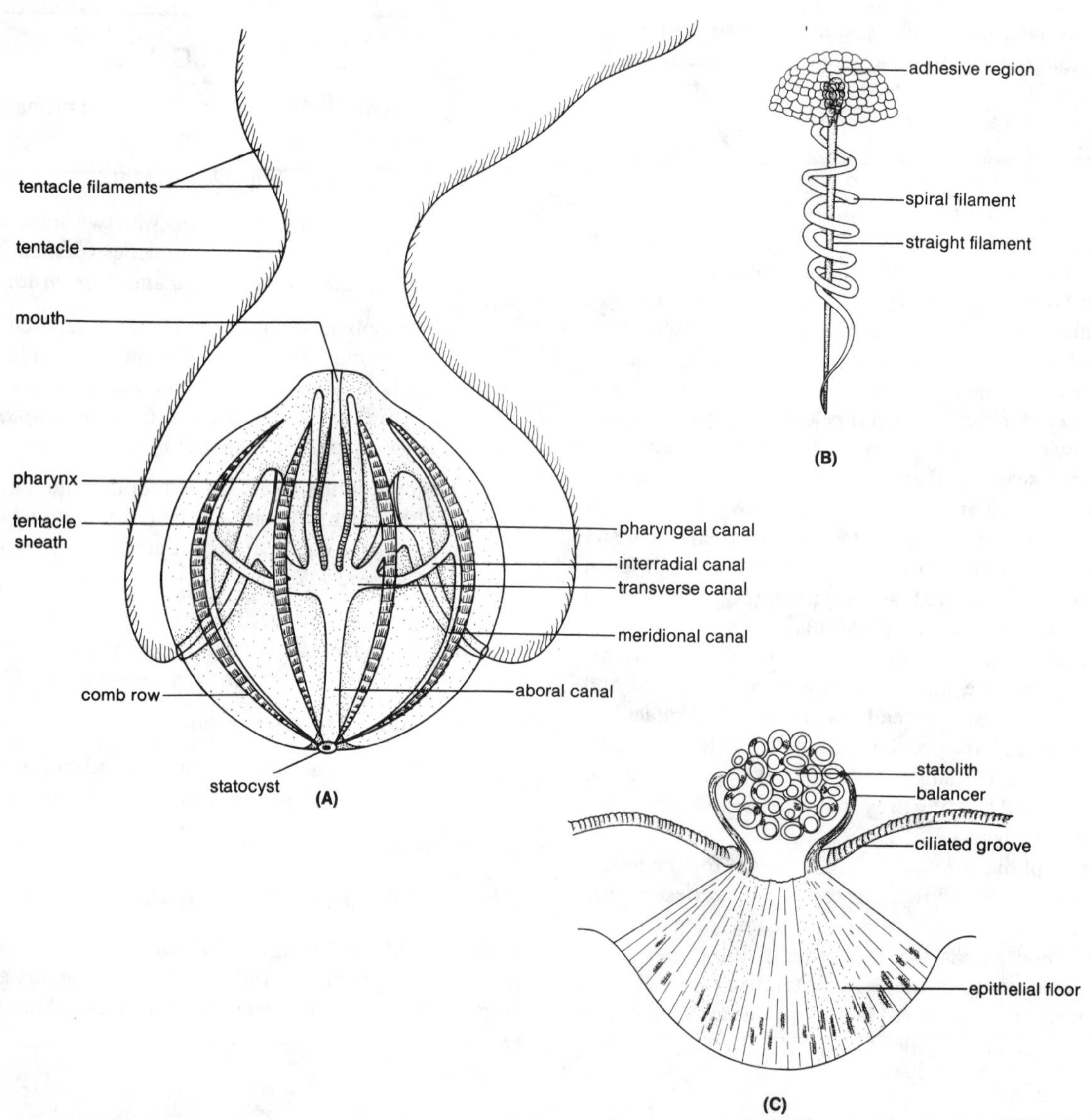

Figure 4.1. *Pleurobrachia.* **(A)** Entire organ. **(B)** Collocyte. **(C)** Aboral sense organ. (From Shelton 1982.)

lateral filaments function as a fishing net to ensnare prey.

At the aboral end is a sense organ called a **statocyst**, containing a statolith that initiates beating of the cilia on the comb rows (Fig. 4.1C). With a razor blade or sharp scalpel, cut away the aboral end and place it in a shallow dish filled with water. Examine the statocyst under high power of a dissecting microscope. Can you see the four ciliary tufts that support the calcareous statocyst in its dome-shaped covering?

Cut away the outer wall of an aboral quarter of the body. Locate the pharynx, pharyngeal canals, and meridional canals of the digestive system. Locate the tentacular sheath. If the tentacles are retracted into the sheath, cut an entire sheath away and carefully tease the tentacle from its sheath.

Beroë

Beroë, the thimble jelly, is larger (5 cm x 10 cm long) than *Pleurobrachia*. *Beroë* lacks tentacles and is pinkish in color.

Observational Procedure: *Beroë*

Obtain a specimen of *Beroë* in a small dish with water. Note its bell-shaped and laterally compressed body (Fig. 4.2). Observe the enlarged mouth. Locate the eight comb rows and highly branched gastrovascular cavity.

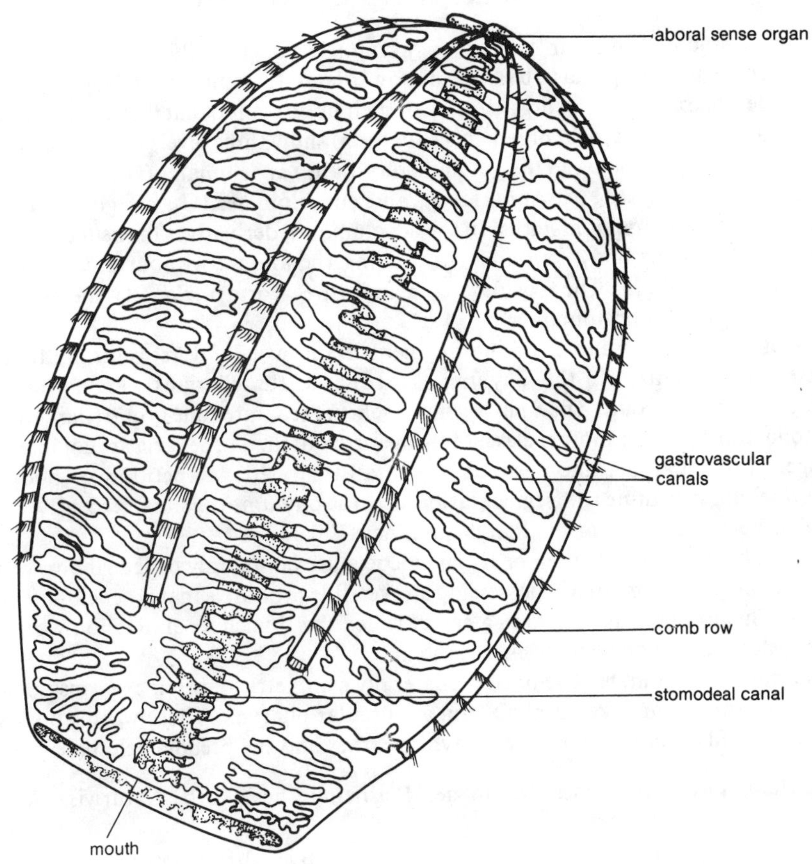

Figure 4.2. *Beroë*.

Supplemental Reading

Franc, J. M. 1978. Organization and function of ctenophore colloblasts: an ultrastructural study. Biol. Bull. 155:527-541.

Greve, W. 1978. Ctenophores. Fiches Identif. Zooplancton 146.

Harbinson, G. R., L. P. Madin, and N. R. Swanberg. 1978. On the natural history and distribution of oceanic ctenophores. Deep-Sea Res. 25:233-256.

Horridge, G. A. 1974. Recent studies on the Ctenophora. *In*: L. Muscatine and H. M. Lenhoff (eds.). Coelenterate biology. Academic Press, New York, pp. 439-458.

Pianka, H. D. 1974. Ctenophora. *In*: A.C. Giese and J. S. Pearse (eds.). Reproduction of marine invertebrates, Vol. 1. Academic Press, New York, pp. 201-265.

Reeve, M. R., M. A. Walter, and T. Ikeda. 1978. Laboratory studies of ingestion and food utilization in lobate and tentaculate ctenophores. Limnol. Oceanogr. 23:740-751.

Tamm, S. L. 1980. Cilia and ctenophores. Oceanus 23:50-59.

Tamm, S. L. 1982. Ctenophora. *In*: G.A.B. Shelton (ed.). Electrical conduction and behaviour in 'simple' invertebrates. Oxford University Press, Oxford, pp. 266-358.

Tamm, S. L. 1983. Motility and mechanosensitivity of macrocilia in the ctenophore *Beroë*. Nature (Lond.) 305:430-433.

ACOELOMATE BILATERIA

The metazoans studied thus far are either asymmetrical, radially symmetrical, or biradially symmetrical. The remaining phyla to be studied show **bilateral** symmetry, although members of several phyla have undergone additional reorganization of their body plan so that they exhibit secondary radial symmetry. Bilaterally symmetrical organisms (**Bilateria**) show three levels in the organization of their body cavity (i.e., the space between the gut and body wall): **acoelomates** (G., *a*, without + G., *coelo*, hollow), **pseudocoelomates** (G., *pseudo*, false), and **eucoelomates** (G., *eu*, good). These groupings generally are not given taxonomic status. In the acoelomates, derivatives of the mesoderm completely fill the space between the body wall and gut, leaving the animal without a coelom (but compare Turbeville and Ruppert 1985). The acoelomates include three worm phyla: Platyhelminthes, Nemertea (Rhynchocoela), and Gnathostomulida.

Members of the Platyhelminthes are nonsegmented and usually dorsoventrally flattened worms. Three classes are recognized in this phylum: Turbellaria (turbellarians), Trematoda (flukes), and Cestoda (tapeworms), although a new phylogenetic scheme separates one group (the Monogenea) from the Trematoda, thus creating another class. Most members of the Turbellaria are completely free living; the Monogenea, Trematoda, and Cestoda are totally parasitic. Repetition of body units in the tapeworms does not constitute true segmentation as in the phylum Annelida. Recently, an entirely new taxonomy has been proposed for the Platyhelminthes (Ehlers 1984).

Members of the phylum Nemertea or Rhynchocoela are commonly called ribbon worms. Ribbon worms differ from platyhelminths in possessing an eversible proboscis that is developmentally separate from the gut. The proboscis is housed in a cavity called the rhynchocoel. Ribbon worms also have a circulatory system and a complete digestive system.

Gnathostomulids are minute acoelomate, bilaterally symmetrical, marine worms possessing paired jaws and an unpaired, comblike basal plate in a muscular pharynx. These worms possess characters suggestive of an affiliation with both the Platyhelminthes and the pseudocoelomate phyla.

Ehlers, U. 1984. Phylogenetisches system der Plathelminthes. Verh. Naturwiss. Ver. Hamburg. 27:291-294.

Turbeville, J. M., and E. E. Ruppert. 1985. Comparative ultrastructure and the evolution of nemertines. Am. Zool. 25:53-71.

EXERCISE 5

Phylum Platyhelminthes

Platyhelminthes (PLAT-e-hel-MIN-thez; G., *platy*, flat + G., *helminth*, worm) are acoelomate, nonsegmented, dorsoventrally flattened worms. There are about 15,000 species of flatworms, which range in size from microscopic to an incredible 30 m. Most species in this phylum are entirely endo- or ectoparasitic.

There is no coelom in flatworms (cf. Turbeville and Ruppert 1985). The internal organs are embedded within a compact mass of **parenchymal** tissue of mesodermal origin, which functions in secretion, storage, and regeneration. Flatworms lack true segmentation. The repetition of body parts in the cestodes is not considered to be segmentation. Usually, a digestive system, which is incomplete (i.e., no anus), is present only in the turbellarians and flukes. Cestodes lack a digestive system.

Classification

Traditional classification schemes subdivide Platyhelminthes into three classes (Turbellaria, Trematoda, and Cestoda), while recent work recognizes that the Trematoda comprises some rather distantly related taxa. Therefore, that group has been subdivided by some into two classes, the Monogenea and a more restrictively defined Trematoda (Digenea + Aspidobothria) (e.g., Schmidt and Roberts 1985:203). Ehlers (1985) and Brooks et al. (1985) offer new systems of classification constructed on the basis of phylogenetic systematics and thus lack either the three- or four-class hierarchy. The traditional three-class hierarchy is retained here for teaching purposes.

1. **Class Turbellaria**. Relatively small flatworms with ciliated epidermis richly supplied with glands; distinct rod-shaped structures called rhabdoids (including rhabdites) are among the glandular secretions; possessing a blind gut with oral opening usually located midventrally; mostly free living, some commensal or parasitic; 12 orders are recognized; examples include *Dugesia*, the common pond planarian, *Bdelloura*, a commensal on the gills of horseshoe crabs, and *Stenostomum*, a common microscopic freshwater turbellarian.

2. **Class Trematoda**. Moderate-sized, leaf- to-worm-like parasitic flatworms, characterized by having one or more prominent, ventral attachment structures; simple to complex life cycles, as endo- or ectoparasites of both invertebrate and vertebrate hosts; examples include *Aspidogaster conchicola* (freshwater clam fluke), *Polystoma nearcticum* (found in the urinary bladders of toads), *Fasciola hepatica* and *Opisthorchis sinensis* (sheep and human liver flukes, respectively), and *Trichobilharzia* and *Schistosoma* spp. (duck and human blood flukes, respectively).

3. **Class Cestoda**. Elongate, obligately, endoparasitic flatworms lacking a gut; most possess an anterior knoblike holdfast called the head or scolex; elongate body or strobila consisting of many repeated units called proglottids; subclass Cestodaria includes tapeworms of fish and turtles having a leaflike body shape; true tapeworms (subclass Eucestoda) parasitize all vertebrate classes; examples include *Taenia solium* (pig tapeworm), *Taeniarhynchus saginatus* (beef tapeworm), *Dipylidium caninum* (dog tapeworm), and *Moniezia expansa* (sheep tapeworm).

A. Class Turbellaria

Turbellaria (ter-be-LAR-e-a; L., *turbella*, a little crowd) is a moderate-sized class of flatworms comprising about 3000 species. Many of them have an oval to elongate, dorsoventrally flattened shape and range in size from microscopic to over 50 cm long; most are less then 5 mm long. Turbellarians are predominantly free living, although many commensal and parasitic species are known. Members of this class are found in freshwater, marine, and terrestrial habitats. All have a simple life cycle.

Turbellarians have a single-layered epidermis which is cellular or **syncytial**. In most species the epidermis is completely ciliated all around and amply provided with glands; some have ciliation restricted to special areas of the epidermis such as a creeping sole. One prominant secretory product in the epidermis is the **rhabdoid**. There are several different types of rhabdoids, but one common type is the **rhabdite**.

A mouth opens into the gut, in most species through a simple or complex pharynx which in some forms may be protruded through the mouth to feed. The location of the mouth is highly varied within the class. The pharynx, in turn, opens into a blind gut. In some lower turbellarians a true digestive cavity is lacking. Other forms have a simple sac or complex intestine with many diverticula. Digestion begins extracellularly, and it is completed intracellularly.

Observational Procedure: Turbellaria

Preserved Specimens

Whole specimens. Obtain a prepared slide of a stained specimen of the freshwater planarian *Dugesia* (Fig. 5.1). Locate the anterior and posterior ends. The flaplike structures at the anterior end are chemosensory lobes called **auricles.** Near the middle of the worm is the muscular pharynx, which is housed in a pharyngeal cavity. The pharynx is protruded through the mouth when the animal feeds. In an animal that has been fed opaque material, trace the outline of the intestine (Fig. 5.1). How many main branches of the gut are present? Note that each main branch is highly branched. Locate the pair of **eyespots** at the anterior end. These are special photoreceptive sensory organs and are of the inverse pigment-cup type. The shape and position of the pigment cup in relation to the photoreceptive cells gives the planarian its cross-eyed appearance. Observe a cross section of a planarian at the level of the eyespots. Note the shape of the pigment cups. Which way do they face? What function does the black pigment serve?

Compare a whole-mount slide of *Bdelloura* to that of *Dugesia* (Fig. 5.2). *Bdelloura* is a marine turbellarian that is commensal on gills of the horse-shoe crab, *Limulus.* At the anterior end of *Bdelloura*, note the paired eyespots, but absence of auricles. Locate under each eye a ganglion and trace the paired ventral nerve cords which leave these ganglia and run laterally the length of the animal. In the middle of the worm is the large pharynx and just posterior to the pharynx is a copulatory organ.

Cross Sections. Examine several cross sections of *Dugesia* and make the following observations.

1. Epidermis. Observe the epidermal surfaces (Fig. 5.3). Are cilia uniformly distributed on the dorsal and ventral surfaces? Note that cilia cover the ventral surface except for a narrow band along the lateral margin of the body. In this area locate the special eosinophilic glandular cells, which produce a secretion enabling the worm to adhere to surfaces. Locate the rhabdoids in the epidermis. Do they have a uniform distribution? Find rhabdoids in the parenchyma.

2. Muscular system. Just under the epidermal basement membrane, locate the muscle layers made up of circular, longitudinal, and diagonal muscles (Fig. 5.3).

3. Parenchyma. Examine the parenchymal tissue and note that this tissue completely fills the body with a loose cellular matrix (Fig. 5.4). The only cavities present in the body of the worm are those of the gut. Passing through the parenchyma are numerous dorsoventral muscle fibers.

4. Gut. Locate the gut in a cross section. How many regions of the gut do you see in a single cross section of the planarian worm? Note the vacuolate phagocytic cells of the gastrodermis (Fig. 5.4). In the

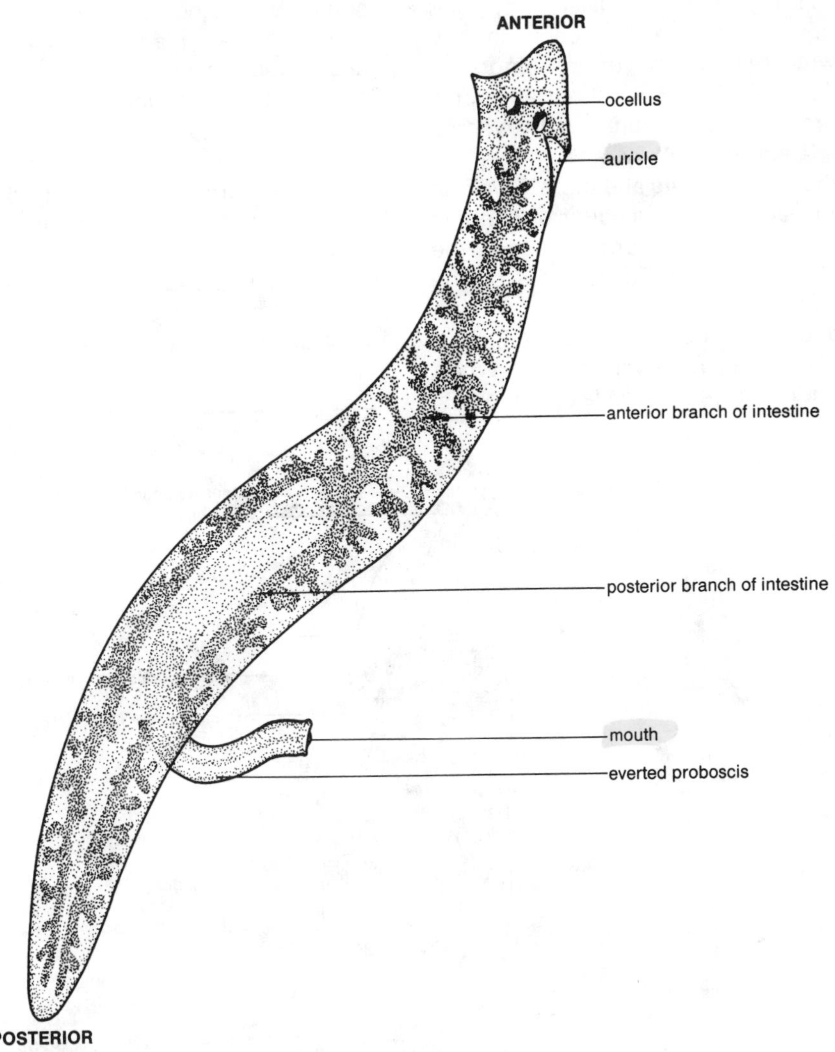

Figure 5.1. Dorsal view of the freshwater turbellarian *Dugesia*.

gastrodermis of some preparations, a gregarian parasite (*Lankesteria planariae*) may be present. This parasite may be distinguished from cells of the planarian gastrodermis by the distinct nucleation, larger size, and pear-shaped body.

The pharynx is a protrusible organ, capable of being projected through the mouth. Carefully examine a cross section of the pharynx. Observe how it is located in the pharyngeal cavity. Note the relative size of the pharynx in comparison to the rest of the body. Find the outer ciliated layer and muscle layers of the pharynx. Also examine a longitudinal section of a planarian through the pharynx and determine the anatomical relationship between the pharynx and the mouth. Does the section you are examining pass through the mouth?

Live Specimens. Carefully remove several living specimens from the holding jar and place them in a shallow dish containing some pond water. Observe the specimens under a dissecting microscope and note their general manner of locomotion and any behaviors they may exhibit when obstacles are encountered. Using a probe, turn a specimen upside down and watch how it rights itself. Place a specimen in a depression slide filled with pond water, put a cover glass over the specimen, and observe it with low power of a compound microscope. As the animal moves about, attempt to examine its dorsal and ventral surfaces. The ventral surface is ciliated and ciliary action may be observed as the animal twists and turns. To aid in this observation add a small amount of carmine powder to the depression and observe the movement of the carmine brought about by the action of the cilia.

If time and climatic conditions permit, field collections of living planarians will be made. When making collections in freshwater lakes, streams, ponds, or springs, examine the undersurface of submerged objects such as stones, old leaves, and wood. If present, planarians will be seen attached to substrates sometimes in clusters (hence the use of the Latin term *turbella*). Collect at least a dozen or so specimens and place them alive in collection bottles.

If time permits, your instructor may provide instructions for observing planarian (*Dugesia*) feeding bits or other behaviors, or provide directions for exercises in planarian regeneration. The study of other turbellarians (e.g., *Stenostomum*) also may be possible.

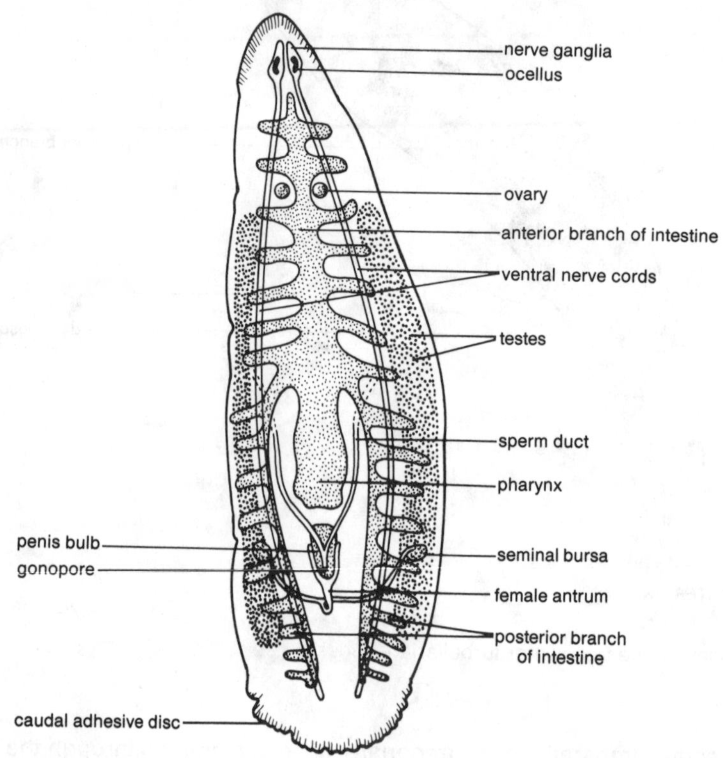

Figure 5.2. Dorsal view of the marine commensal turbellarian *Bdelloura*.

B. Class Trematoda

Trematoda (trem-a-TO-da; G., *trema*, a hole) is a large class containing about 8000 species of parasitic flatworms. Trematodes, or **flukes** as they are commonly called, are similar in size and shape to the turbellarians. Most are dorsoventrally flattened; however, some have a cylindrical body. As a group they range in size from microscopic to several meters in length. Flukes have enormous medical, veterinary, and economic importances. Together with the class Cestoda (tapeworms) and phylum Nematoda (unsegmented roundworms), flukes are the subject of the field of **helminthology** (G., *helminth*, worm + G., *ology*, study).

Trematodes possess one or more suckers, usually an oral one surrounding the mouth and one located near the ventral midline. These may be armed with many small spines or hooks. The fluke body wall differs from that of turbellarians. Although once believed to be nonliving, the **tegument** is a complex living syncytium which may be ornamented with minute spines.

Trematodes may have a simple life cycle involving a single host (subclasses Aspidogastrea and Monogenea) or a very complex one involving several larval stages

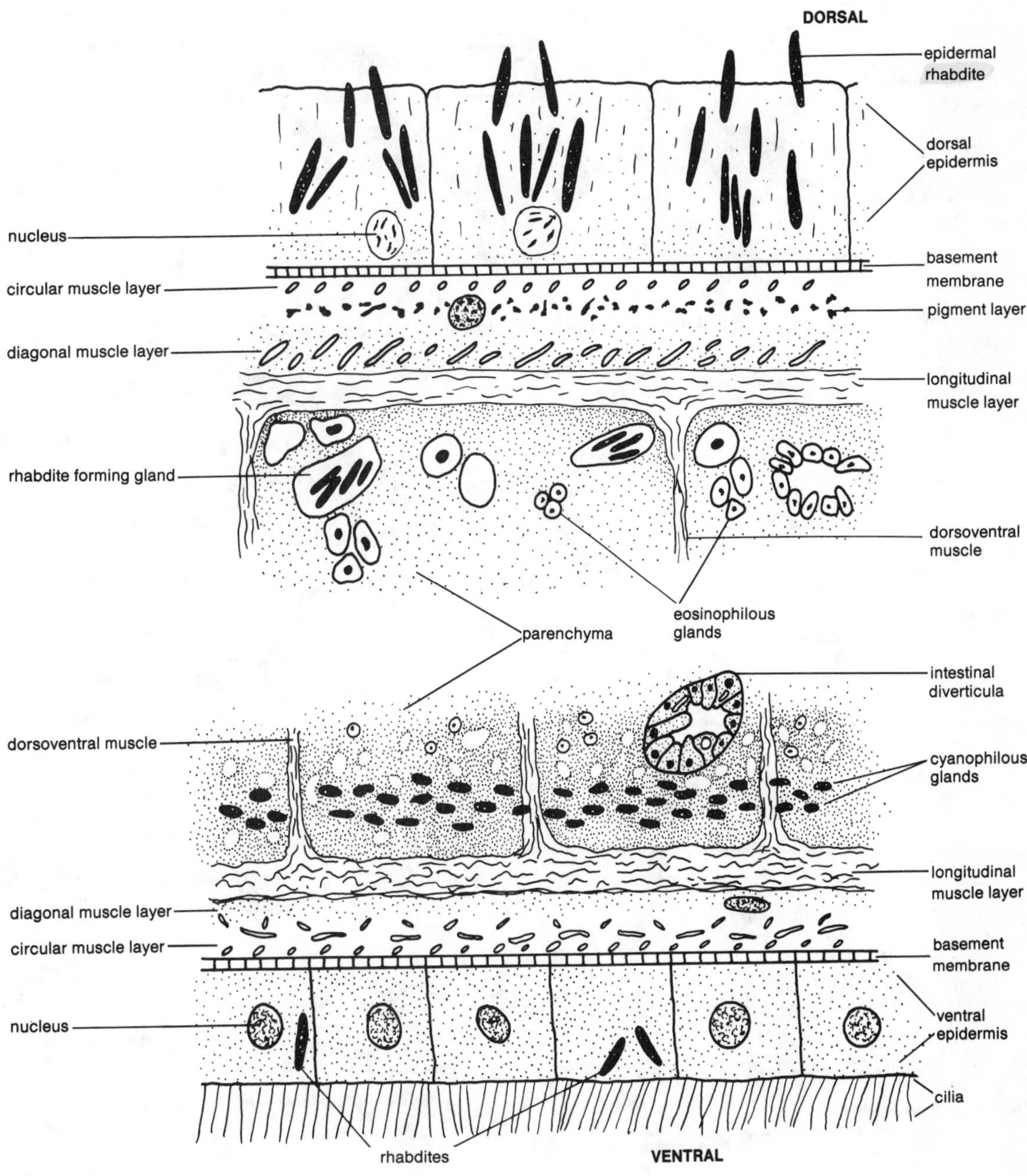

Figure 5.3. Cross sections of the dorsal and ventral body walls of *Dugesia*.

and more than one host (subclass Digenea). In the Digenea, development may proceed through as many as five different larval forms: **miracidium, sporocyst, redia, cercaria , metacercaria**. Some of these are small free-swimming stages which are involved in the location of the next host. Others are sedentary within a host or encysted on a substrate.

It is not possible to completely cover this diverse class, in an introductory laboratory; therefore, exercises for only a few selected species are provided here. At this point, it will be profitable if students review the life history of the digenetic fluke, *Fasciola hepatica* (Fig. 5.5).

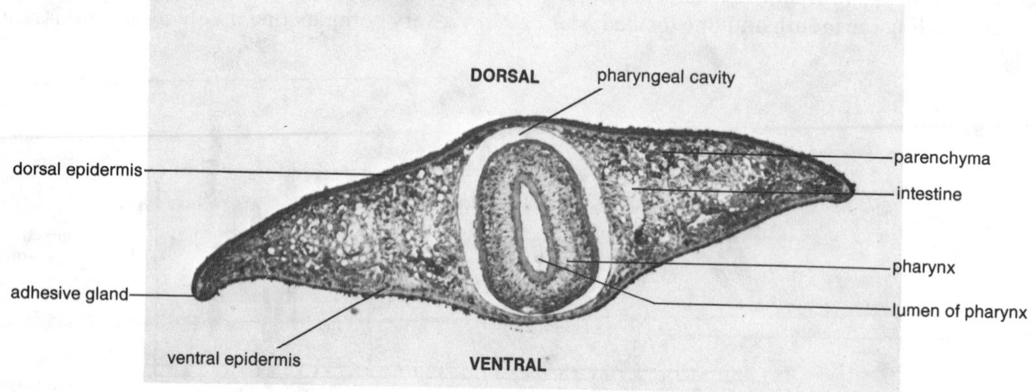

Figure 5.4. Photomicrograph of a cross section of *Dugesia* at the level of the pharynx.

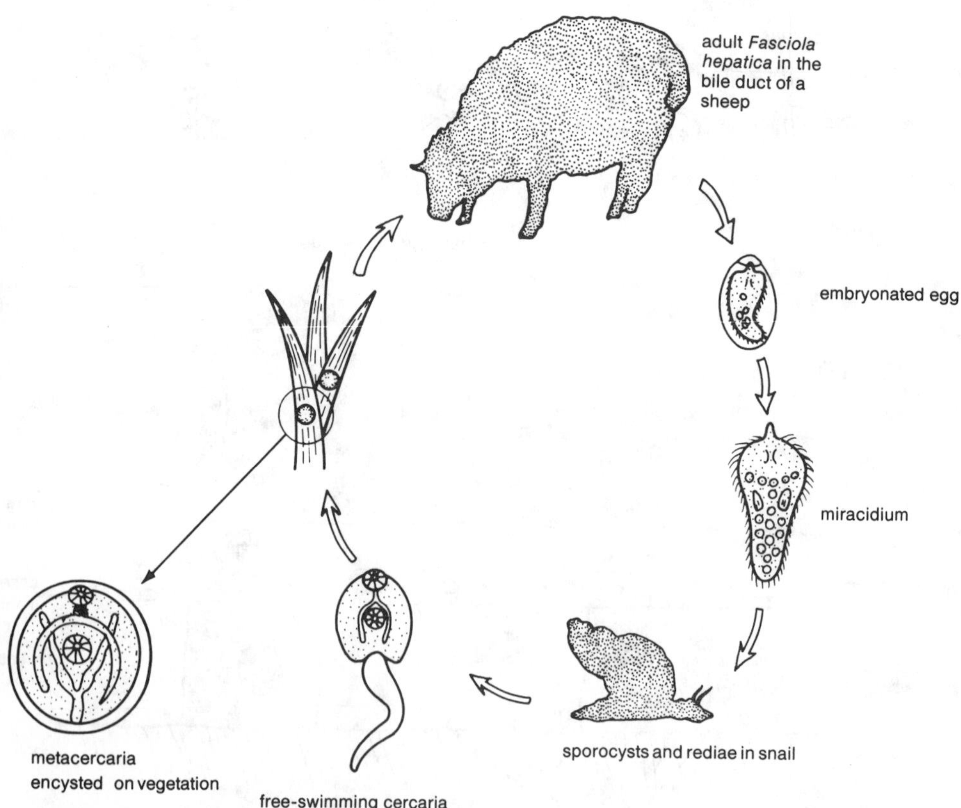

Figure 5.5. Life cycle of the digenetic trematode *Fasciola hepatica*.

Observational Procedure: Trematoda

Adults

Fasciola hepatica (Sheep Liver Fluke). *Fasciola hepatica* is a large (ca. 30 mm) leaf-shaped fluke often found in the livers of deer, goats, cattle, and sheep. Observe a slide of a stained specimen and identify the anterior and posterior ends (Fig. 5.6). The body tapers abruptly at the anterior end, giving the effect of a pointed head with shoulders. This region is called the oral cone. Locate the **oral sucker**, which surrounds the mouth. The mouth leads to the pharynx, which connects to a short esophagus and then to a biramus intestine (**intestinal ceca**) with many diverticula. The arms of the intestine run the length of the fluke. Just posterior to the pharynx locate the large ventral sucker called the **acetabulum**, and in front of it the common genital pore. The **genital pore** receives both male and female gametes. Locate the highly branched testes that lie in tandum in the central region. **Vasa efferentia** transport sperm to the seminal vesicle that together with the **cirrus sac** may be seen to one side of the genital pore. Locate the tubular **uterus** that courses posteriorly from the genital pore. It may contain many dark embryonated eggs. The uterus terminates at the shell gland, which receives yolk from the large lateral **vitellaria** (yolk glands). The vitellaria may obscure the intestinal ceca. On the right side of the fluke find the ovary that also terminates at the shell gland. It is branched, but darker than the testes and lies anterior to them. At the posterior end, locate the excretory pore that drains the excretory vesicle.

Examine a cross section slide of *Fasciola* and note the following features. The body surface is not ciliated and rhabdites, characteristic of turbellarians, are not present. However, buried in the tegument are numerous scalelike **spines.** Below the tegument find the muscle layers and note the dorsoventral

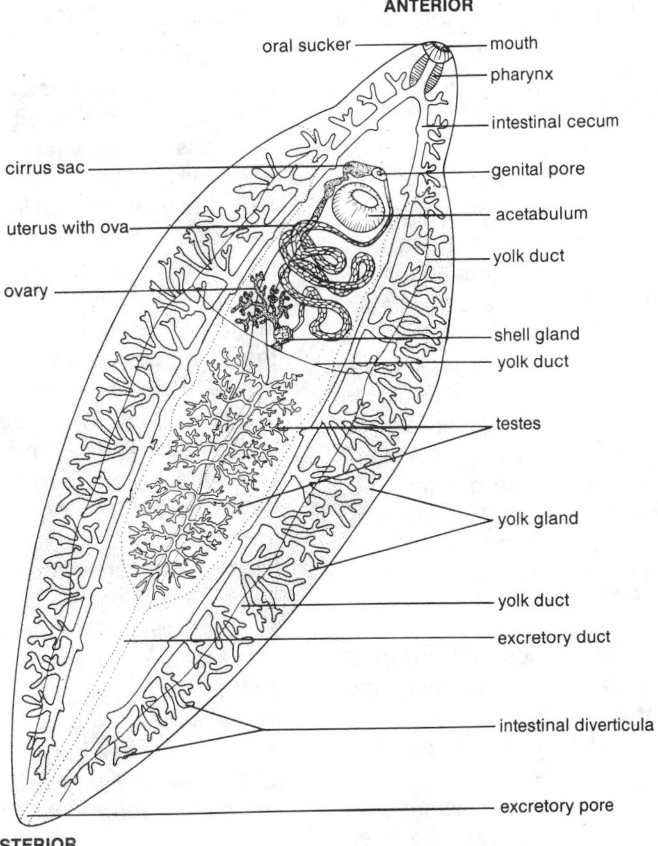

Figure 5.6. Ventral view of the sheep liver fluke *Fasciola hepatica*.

muscles which pass through the parenchyma. In thin sections from the anterior region, locate the uterus containing shelled eggs and laterally find the intestinal ceca and vitellaria. In sections from the central region, locate the testis and again laterally find the intestinal ceca and vitellaria.

Fasciolopsis buski (Human Intestinal Fluke). *Fasciolopsis buski* is an important fasciolid which parasitizes people living in Southeast Asia. It has a life history similar to that of *Fasciola* except that it inhabits the intestine. There are a few notable differences in the body plan of these species. If slides are available, compare the general anatomy of *Fasciolopsis buski* to *Fasciola hepatica*.

Note that *Fasciolopsis buski* is much larger (up to 75 mm long), lacks the oral cone, and its acetabulum is larger and closer to the oral sucker than that of *Fasciola*. The intestine of *Fasciolopsis* is biramus like that of *Fasciola*, but it lacks ceca. Considering the differences in habitat of the two species, what might be the significance of a larger (stronger) acetabulum and an intestine that lacks ceca? The reproductive systems of the two species are essentially the same. Testes lie in tandum in the central region of the body, a smaller ovary lies ahead of the testes next to the shell gland, and large vitellaria are found along the lateral margins of the fluke.

Opisthorchis sinensis (Chinese Liver Fluke). The opisthorchiid, *Opisthorchis* (= *Clonorchis*) *sinensis*, like *Fasciola*, lives in the liver of its host (Fig. 5.7). However, it is a smaller fluke (ca. 25 mm long) that lacks tegmental spines. If slides are available, compare this fluke to the fasciolids studied previously. The oral sucker is slightly larger than the acetabulum. Two large, branched testes lie in tandum near the posterior end, and just anterior to them lies a small three-lobed ovary. From the ovary the uterus courses anteriorly to the genital pore, which lies in front of the acetabulum. To either side of the uterus are the moderately sized vitellaria. As in *Fasciolopsis*, the intestine is biramus, but lacks branches.

Schistosome Flukes. Another important group of flukes are the **schistosomes** (blood flukes), including those that afflict humans: *Schistosoma mansoni* (Fig. 5.8), *S. japonicum*, and *S. haematobium*. The first two are found in veins of the intestine and the latter in veins of the urinary bladder. Schistosomiasis is endemic in Africa, in parts of the Middle East, Southeast Asia, Japan, and in eastern and northern South America.

Observe prepared slides of *Schistosoma* (male and female). Identify the anterior and posterior ends. Note the distinct cylindrical shape of this fluke and the size difference between the males and females. Find the oral and ventral suckers of the specimens. Are there differences in the sizes of the suckers? Locate the **gynecophoral canal (groove)** on the male's ventral surface. Does the female have this groove? What is the significance of this structure? If available, observe a slide that shows the sexes *in copula*.

Observe a prepared slide of *Schistosoma* eggs and note the lateral **spine**. The eggs are responsible for very serious pathologic aspects of schistosomiasis. After being liberated from the female the eggs pass through the intestinal wall, where they do considerable damage. Eggs also may be swept into the liver through the hepatic portal system, where they become lodged. If available, observe slides of *Schistosoma* eggs in the wall of the intestine and in the liver. Note any evidence of tissue damage in the liver in the region immediately surrounding the eggs (pseudotubercles).

General Study of Fluke Larval Stages. Obtain prepared slides of the five larval stages from species such as *Fasciola hepatica*, *Schistosoma* species, or any other digenetic trematode (Fig. 5.9). The basic sequence of stages is miracidium, sporocyst, redia, cercaria, and metacercaria (Fig. 5.5). However, several important modifications may be found in different species. Sporocyst and redia stages may have more than one generation (mother and daughter) or may be absent; the metacercaria stage may be absent also. The internal structures of the larval stages shown in Fig. 5.9 are often not visible in stained preserved specimens.

1. **Miracidium**. The miracidium is a small, free-swimming, ciliated larva which in a general way resembles a protozoan. This nonfeeding larva hatches from the egg and must locate the proper intermediate host (usually molluscan) within a few hours. After penetrating the host tissues it usually differentiates into a sporocyst.

2. **Sporocyst**. This larval generation absorbs nutrients directly from host tissues and reproduces another embryonic stage (redia or cercaria). Look for developing embryos in your specimens.

3. **Redia**. Rediae leave the sporocyst and relocate in the host's digestive gland or gonad. Note the elongate shape and the presence of many embryos (daughter rediae or cercariae). At the anterior end, locate the muscular pharynx, which leads to a small saccate gut.

4. Cercaria. This stage leaves the intermediate host and either locates its definitive host directly (vertebrate) or encysts, becoming a metacercaria. Note the tail used for locomotion. Find the oral and midventral suckers. A biramous intestine without lateral branches may be seen.

5. Metacercaria. In most trematodes this encysted stage precedes the adult. Encystment may take place on the surface of a plant, as in *Fasciola hepatica*, or in a second intermediate host. In *Fasciola*, metacercariae are small caplike structures. Oral and ventral suckers and the intestine may be seen.

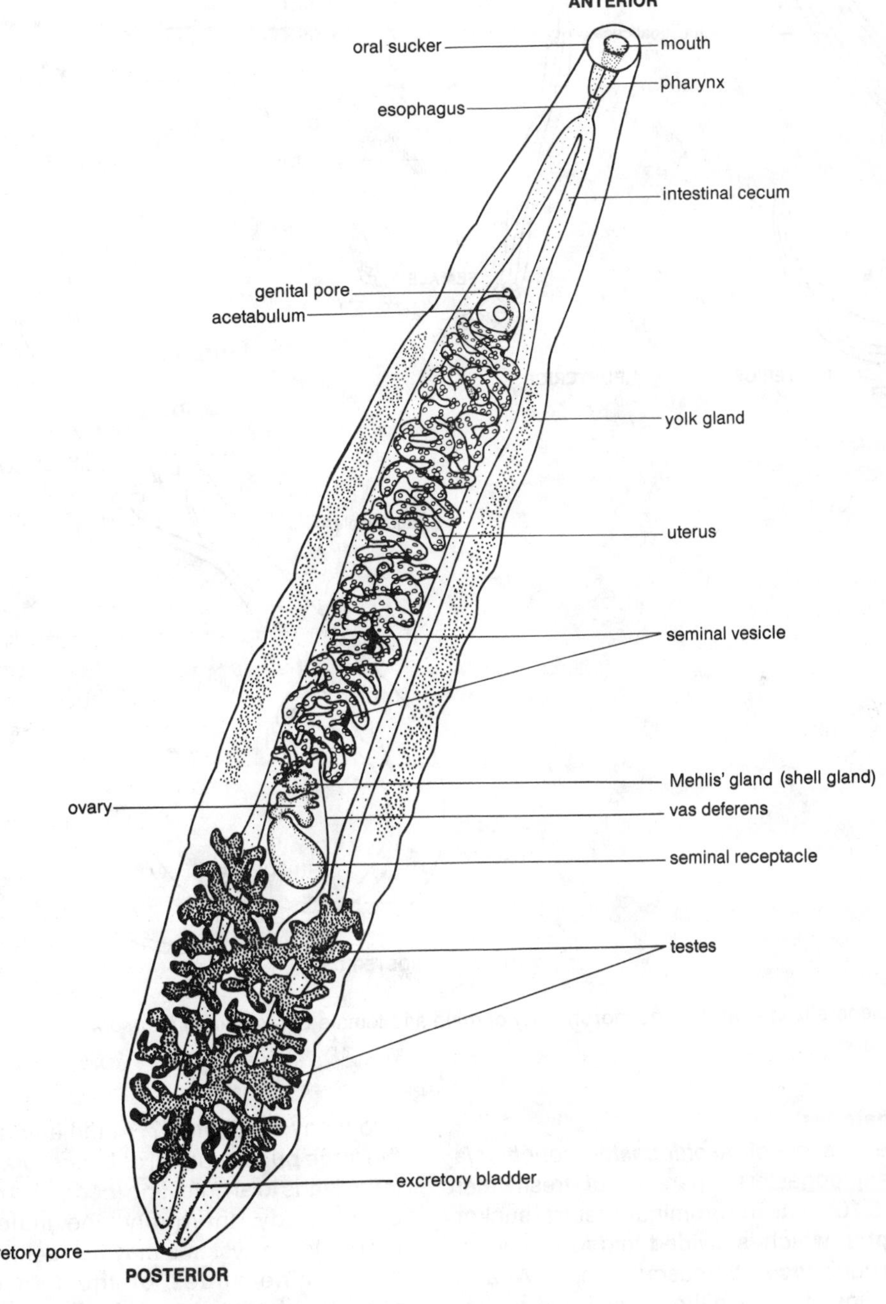

Figure 5.7. Ventral view of the Chinese liver fluke *Opisthorchis sinensis*.

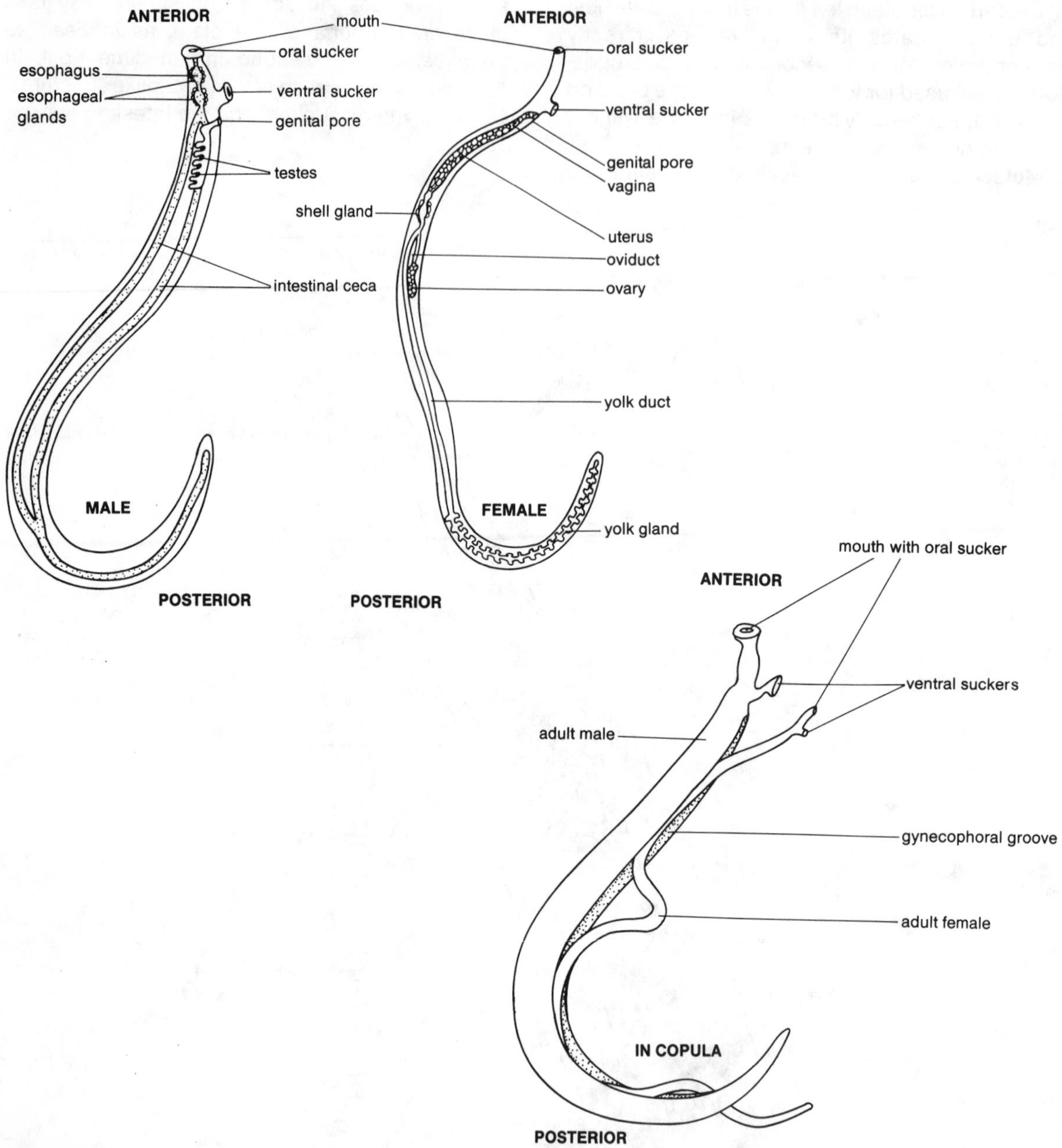

Figure 5.8. Generalized anatomy and morphology of male and female *Schistosoma mansoni*.

Other Trematodes

1. Examine a slide of *Aspidogaster conchicola* (subclass Aspidogaster), a parasite of freshwater clams (Fig. 5.10). Note the prominant ventral sucker or **opisthaptor**, which is divided into four rows of many small cups (**loculi**) by muscular septa. What is the significance of dividing the single large sucker into many small ones? At the anterior end, locate the large buccal funnel and muscular pharynx. The latter leads to an intestine (ceca) that runs the length of the body. Internally, the male and female reproductive systems may be seen.

2. Examine slides of the fish ectoparasite *Gyrodactylus* (subclass Monogenea) if they are

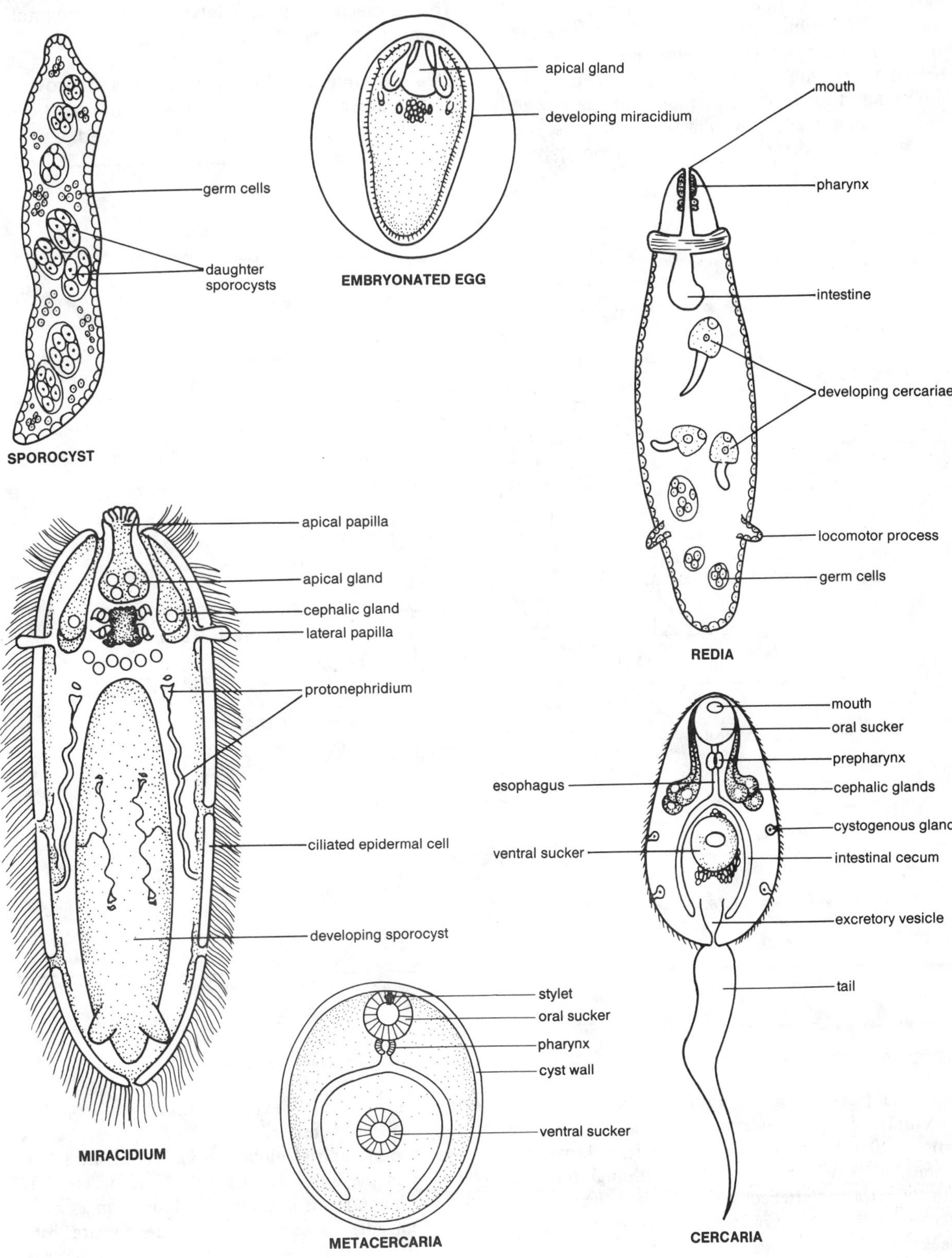

Figure 5.9. Fluke larval stages. (After several sources.)

available (Fig. 5.11). At the posterior end, note the large ventral sucker (opisthaptor), which is armed with two large, central hooks and a series of smaller hooks along the margin of the sucker. At the anterior end, locate the large muscular pharynx and biramus intestine. A juvenile worm may be seen developing internally in this viviparous monogenean. The normal monogenean larval stage (oncomiracidium) is not present in this species.

Live Specimens. Your instructor will provide instructions on retrieval of live cercariae from freshwater snails if these are to be studied.

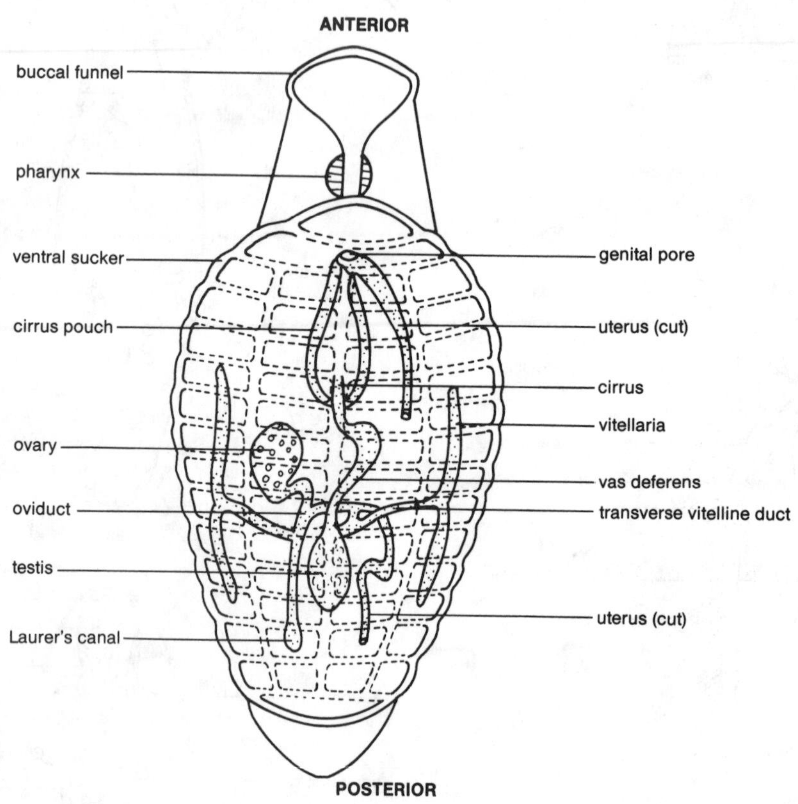

Figure 5.10. Ventral view of *Aspidogaster conchicola*.

C. Class Cestoda

Cestoda (ses-TO-da; G., *cest*, girdle + G., *oda*, resemblance) is a moderate-sized class comprising about 4000 species of endoparasitic flatworms commonly known as **tapeworms**. Although they are dorsoventrally flattened, cestodes differ from other flatworms in a number of ways. Tapeworms completely lack a mouth and digestive tract. Most have a long, ribbonlike body that may be many meters in length; however, some cestodes are less than 3 mm long. The body is usually comprised of three regions: **scolex** (head), **neck**, and **strobila** (body). Adults are found lodged in the intestine of a vertebrate, where they utilize the nutrients of their host. Larval stages often cause considerable damage to their intermediate hosts. Humans infected with tapeworms normally found in animals (a condition called **zoonosis**) may have serious complications (e.g., **hydatid cysts** of *Echinococcus granulosus*). Cestodes have enormous medical, veterinary, and economic importances.

The strobila is made up of many identical units called **proglottids**, except in the subclass Cestodaria, whose members have a short undivided body. The strobila looks as if it were segmented, but tapeworms lack true segmentation. Production of proglottids is similar to strobilization of the ephyra from the strobila stage of the scyphozoan jellyfish *Aurelia*.

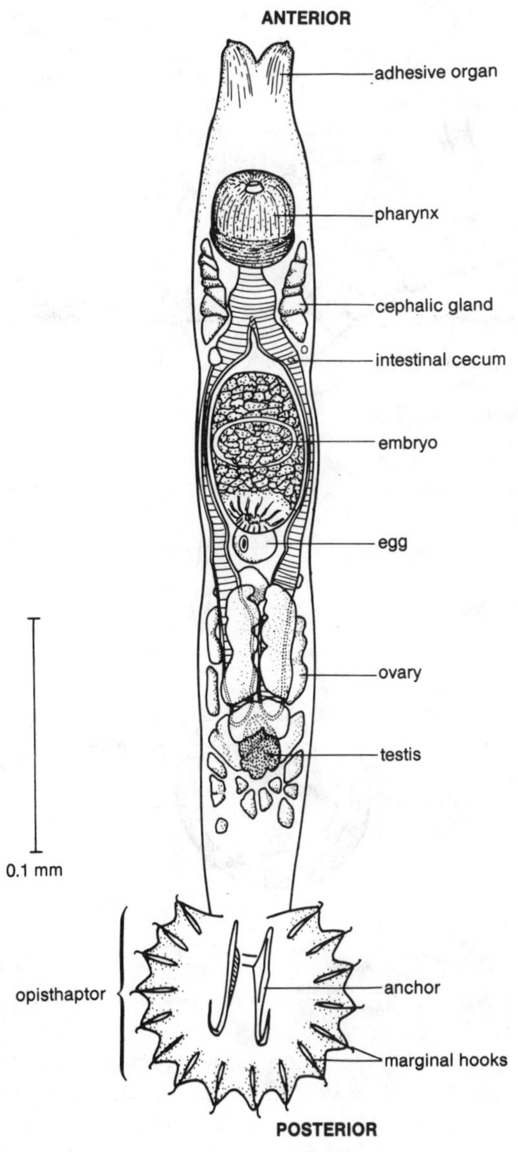

Figure 5.11. Ventral view of *Gyrodactylus*.

Proglottids are formed sequentially in the neck region and are at first relatively undifferentiated, but they mature as they grow older. Therefore, proglottids are progressively older and more mature the farther away from the anterior end they are located. Eggs of very old proglottids have been fertilized and these proglottids with expanded uteri are said to be gravid. There are no septa (walls) separating proglottids, and the tissues grade from one proglottid to the next. These characteristics are very different from the production of true segments in segmented phyla (e.g., Annelida).

The scolex acts as an anchor and is modified in different species to include various holdfast organs, with bulbous ends, spines, hooks, suckers, and several types of glands. The scolex also contains most of the animal's neural ganglia and sensory receptors. The tapeworm's body wall is a living tissue (tegument) covered by numerous microscopic projections (**microtriches**). These microvilli-like structures are responsible for absorbing nutrients from the host's gut.

Tapeworms have an immense reproductive capacity, some being capable of releasing 1 million eggs per day! Tapeworms usually have one or more larval stages, each with its own intermediate host. However, there are many variations in the life cycle and several different types of intermediate stages are recognized (Schmidt and Roberts 1985:358-360).

It is not possible to cover this diverse class completely; therefore, we will study in some detail only a few species. Other species will be introduced if they are available. At this point it will be profitable if students review the life history of a typical tapeworm (Fig. 5.12).

Observational Procedure: Cestoda

Scolices. Obtain slides of the anterior ends of several different tapeworms (e.g., *Diphyllobothrium latum*, *Echinococcus granulosa*, *Moniezia* species, *Taenia solium*, and *Taeniarhynchus saginatus*) and examine the scolices of each (Fig. 5.13). What types of structural adaptations do scolices have to act as holdfast organs? Do they all have the same type of armaments? Are there any unusual structures? Compare the scolices of *D. latum* to *T. solium*. Are they the same? Does *D. latum* have a scolex comparable to the other species studied?

Strobila and Proglottids

1. Observe a whole specimen of a tapeworm. As best you can, determine the length of the specimen. Identify the scolex, neck, and strobila of this specimen. Note the individual proglottids of the strobila. Is the worm of uniform width throughout?

2. Turn your attention to the slides of tapeworm scolices and examine the neck regions (Fig. 5.13). This is the area where young proglottids are being formed. Note that proglottids do not begin at a specific point. Their development is gradual; they get larger and more well defined toward the posterior.

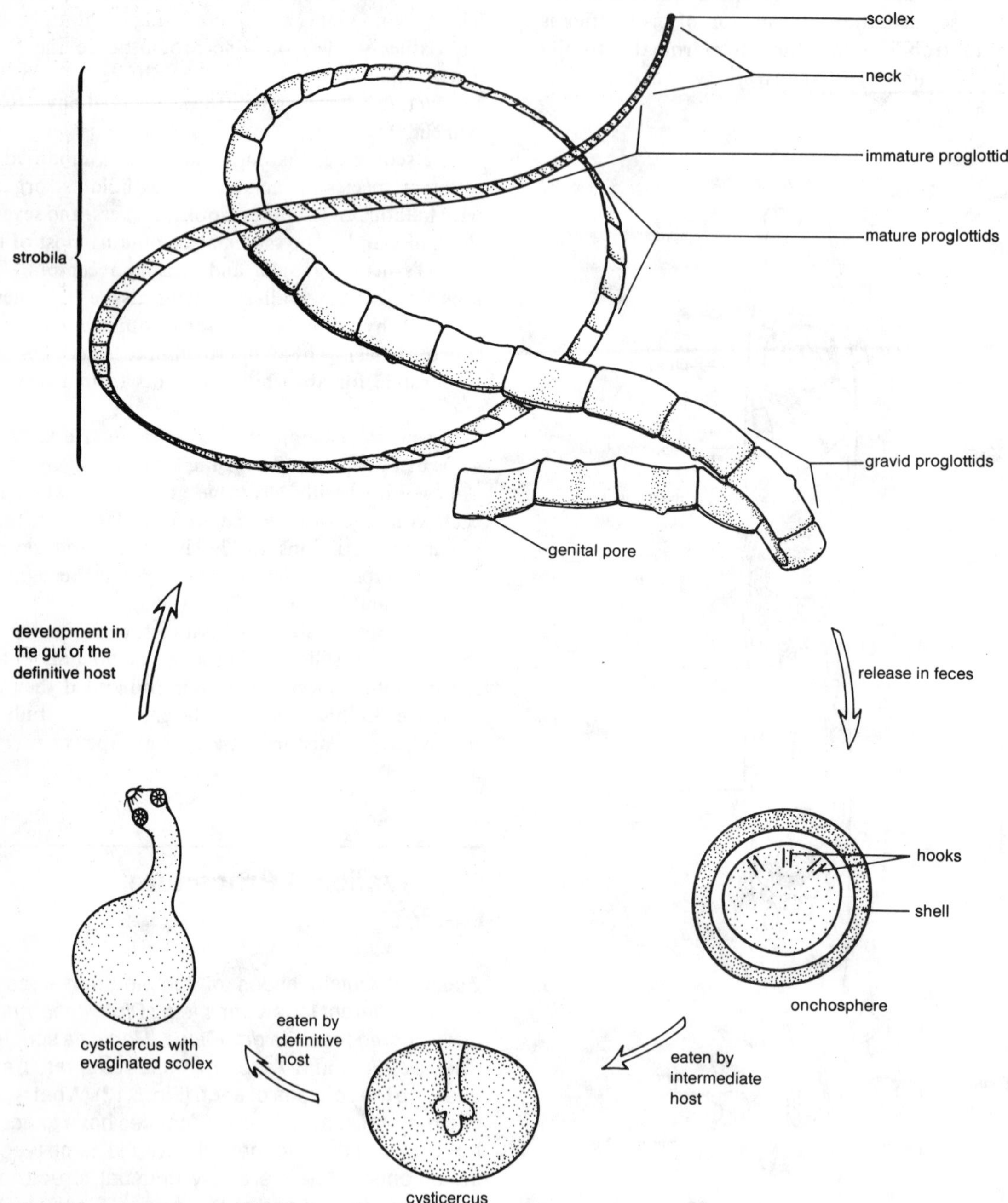

Figure 5.12. Life cycle of the pig tapeworn, *Taenia solium*.

Is there any differentiation of internal organs noticeable in the neck? Look for two strips of lighter tissue which extend posteriorly from the scolex. These are the **excretory canals**.

3. Examine a slide of a longitudinal section of *Taenia* and note there are no septa between the proglottids (Fig. 5.14). What is the significance of this fact?

4. Obtain a slide that includes several stages of development of proglottids of *Taenia* (or another tapeworm), and with the aid of Fig. 5.15, identify components of the male and female reproductive systems. In gravid proglottids, eggs fill the uterus and obscure parts of the reproductive systems. Compare mature and gravid proglottids and note how the morphology of the uterus has been changed by the storage of eggs (Fig. 5.15).

5. Compare the anatomy of *Taenia* with *E. granulosus* (Fig. 5.16). Note that in the latter species, the entire worm is only a few millimeters long and contains only three proglottids. The scolex has a **rostellum** (L., little beak) with a double row of hooks and four

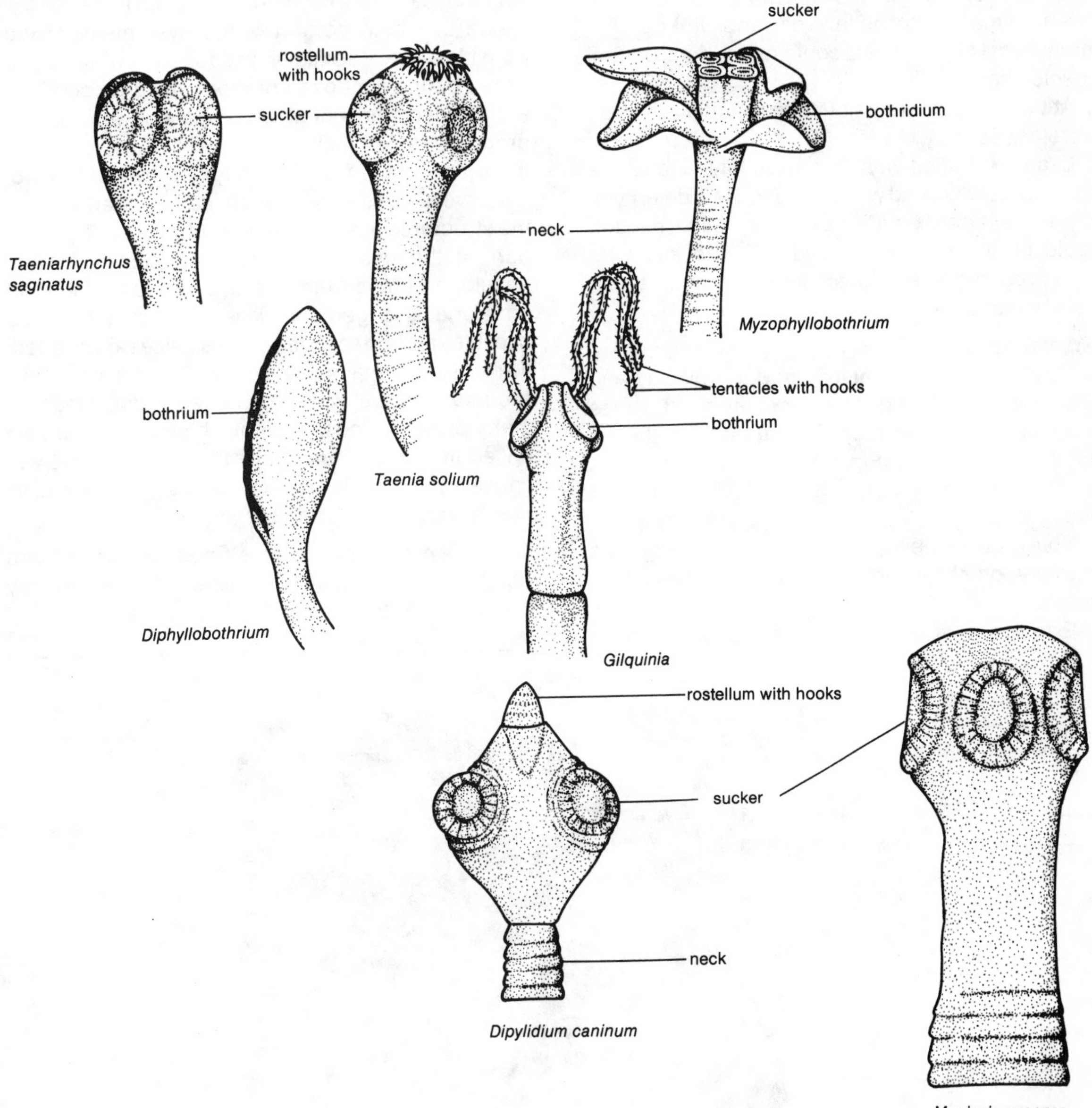

Figure 5.13. Representative tapeworn scolices.

suckers. Immediately behind the scolex is a very short neck and then three proglottids. The first is immature, the second mature, and the last gravid.

6. Observe a prepared slide with cross sections of a sexually mature proglottid of *Taenia* (Fig. 5.17). On the surface, note the tegument and just beneath it the layers of circular and longitudinal muscle tissues. The tissue from this region inward is called the **cortex** region of the parenchyma. Look for the two parallel layers of transverse muscles that surround the reproductive systems and a less dense central parenchyma called the **medulla**. Unlike the parenchyma of turbellarians and flukes, the parenchyma of tapeworms contains few dorsoventral muscle fibers.

7. At the lateral margins of the proglottid are large excretory ducts.

8. Examine a slide with serial sections of a single mature proglottid, and with aid of Fig. 5.15, determine where the approximate postion of each section would be in the proglottid. Identify the following structures in the sections: testes, ovary, uterus, vitelline gland.

Larval Stages

1. Observe a slice of **measly pork** on display and note the pits scattered throughout the meat. These are the **bladderworm** or **cysticercus** stage of the pig tapeworm (*T. solium*) (Fig. 5.12). Do you think that a single cysticercus could go undetected during meat processing?

2. Note that some cysticerci contain a bulbous structure on the inside. This is the invaginated scolex.

3. Examine a whole mount of a cysticercus from *Taenia* and locate the scolex. How does the orientation of the hooks compare to that in the adult? Why is there a difference? What process produces the reorientation of the hooks from the larval to the adult condition?

4. If available, compare cysticerci from other tapeworms (e.g., *Hymenolepis*) to that of *Taenia*.

5. The hydatid tapeworm, *E. granulosus* demonstrates asexual reproduction in its larval stage. Once the **oncosphere** has implantated in the lungs or liver of an intermediate host, it develops into a structure called the hydatid cyst (Fig. 5.16). Eventually, inside the main portion of the cyst, many **brood capsules** are asexually produced. Within each brood capsule a dozen or more tiny protoscolices are formed. Individual protoscolices and whole brood capsules may be freed within the cyst and then are known collectively as **hydatid sand**. Observe a cross section of a hydatid cyst and locate the structures indicated below (Fig. 5.16). The outer part of the cyst is a thick, noncellular, fibrous envelope formed from host tissues. Just inside the envelope is a germinal layer from which brood capsules and independent protoscolices are budded. If an animal containing the hydatid cyst is eaten by a predator, the cyst walls are broken down, releasing many protoscolices, each of which may grow into an adult worm. Fluid from the hydatid cyst will cause anaphylactic shock in its host if it is released due to trauma or during surgery.

Other Tapeworms and Live Specimens. Compare the general morphology of other species that may be available for study to those examined here.

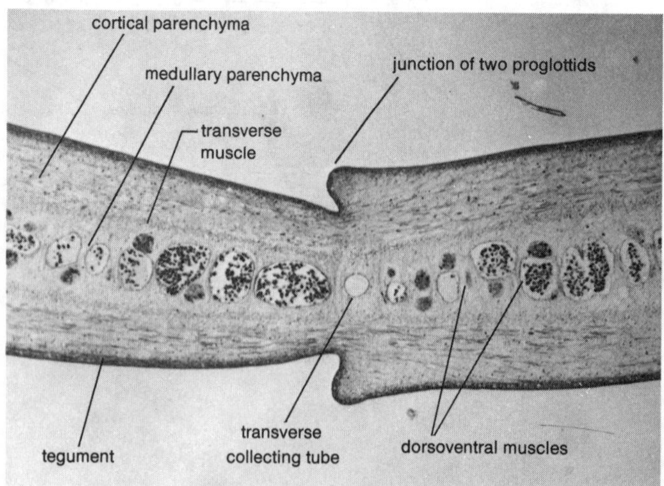

Figure 5.14. Photomicrograph of a longitudinal section of two proglottids of *Taenia*.

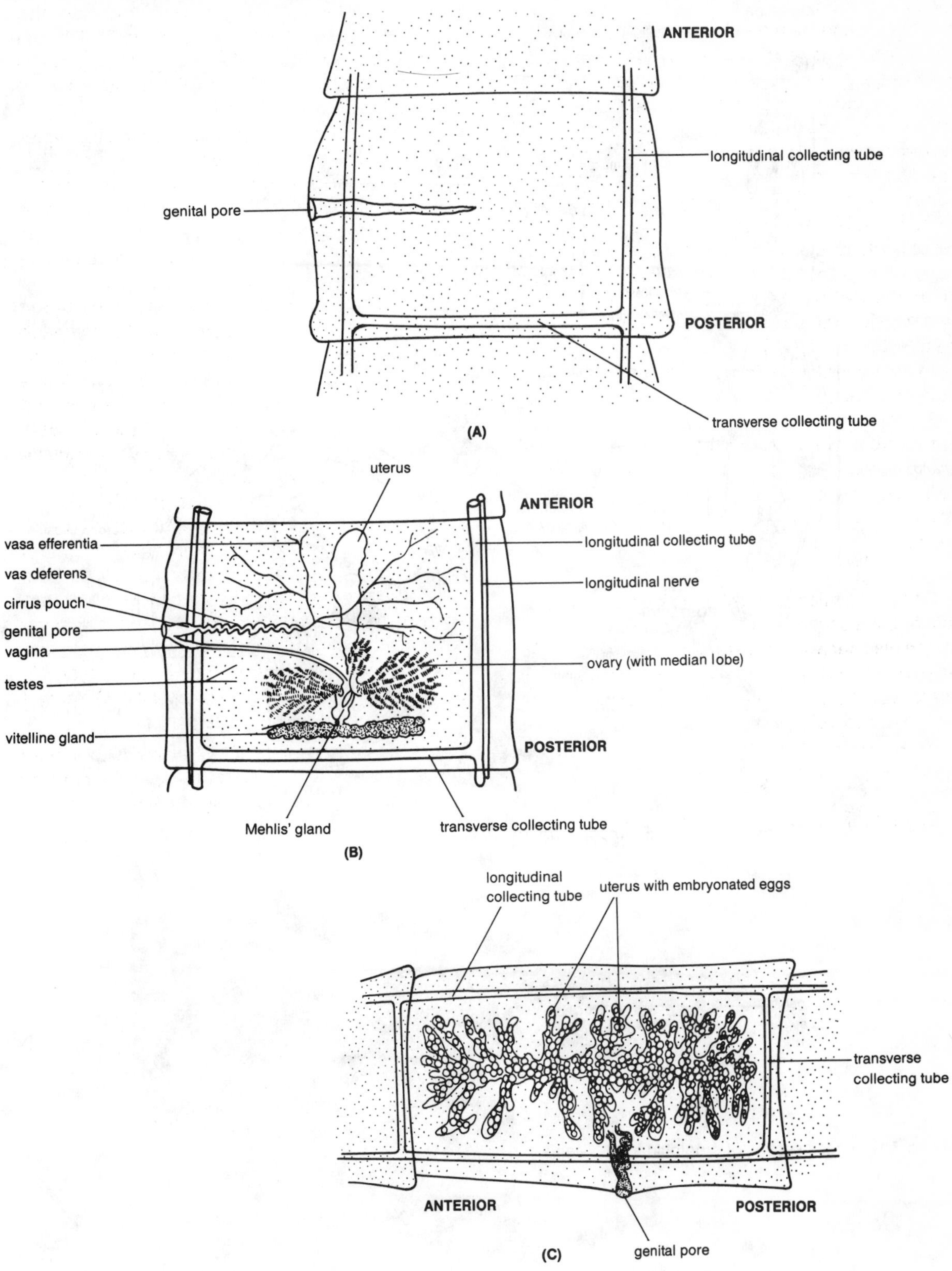

Figure 5.15. **(A)** Immature, **(B)** mature, and **(C)** gravid proglottids of *Taenia*.

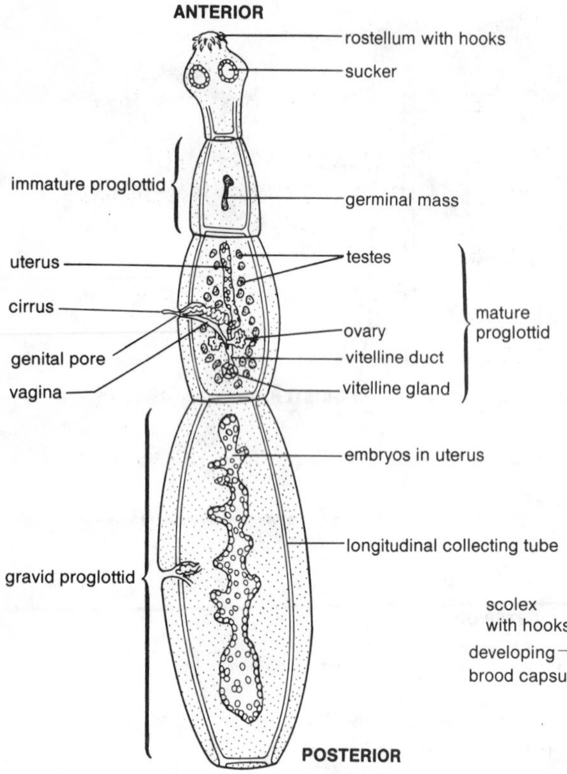

ANTERIOR
rostellum with hooks
sucker
immature proglottid
germinal mass
uterus
testes
cirrus
mature proglottid
ovary
genital pore
vitelline duct
vagina
vitelline gland
embryos in uterus
longitudinal collecting tube
gravid proglottid
POSTERIOR

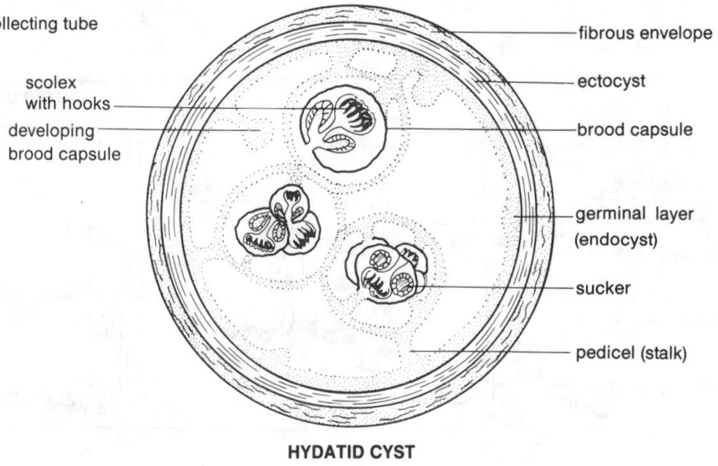

fibrous envelope
ectocyst
scolex with hooks
brood capsule
developing brood capsule
germinal layer (endocyst)
sucker
pedicel (stalk)
HYDATID CYST

Figure 5.16. Adult and hydatid cyst of *Echinococcus granulosus*.

uterus
longitudinal collecting tube
genital antrum
vagina
transverse muscle
tegument
longitudinal nerve

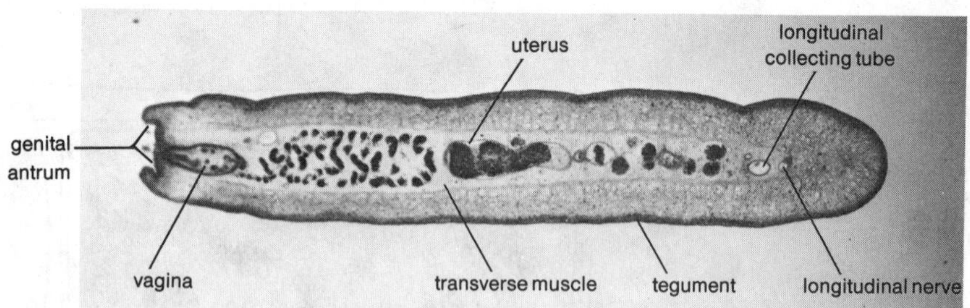

Figure 5.17. Cross section of proglottid from *Taenia*.

Supplemental Reading*

Arai, H. P. (ed.). 1980. Biology of the tapeworm *Hymenolepis diminuta*. Academic Press, New York. (C)

Bronsted, H. V. 1955. Planarian regeneration. Biol. Rev. Camb. Philos. Soc. 30:65-126. (TU)

Brooks, D. R., R. T. O'Grady, and D. R. Glen. 1985. Phylogenetic analysis of the Digenea (Platyhelminthes: Cercomeria) with comments on their adaptative radiation. Can. J. Zool. 63:411-443. (TR)

Ehlers, U. 1985. Comments on a new phylogenetic system of the Platyhelminthes. Hydrobiologia 131:1-12. (G)

Komiya, Y. 1966. *Clonorchis* and clonorchiasis. Adv. Parasitol. 4:53-106. (TR)

Loker, E. S. 1983. A comparative study of the life-histories of mammalian schistosomes. Parasitology 87:343-369. (TR)

Lumsden, R. D. 1975. Surface ultrastructure and cytochemistry of parasitic helminths. Exp. Parasitol. 37:267-339. (G)

Martin, G. 1978. A new function of rhabdites: mucus production for ciliary gliding. Zoomorphologie 91:235-248. (TU)

Montgomery, J. R., and S. J. Coward. 1974. On the minimal size of a planarian capable of regeneration. Trans. Am. Microsc. Soc. 93:386-391. (TU)

Pantelouris, E. M. 1965. The common liver fluke, *Fascicola hepatica* L. Pergamon Press, Oxford. (TR)

Reynoldson, T. B., and A. D. Sefton. 1976. The food of *Planaria torva* (Müller) (Turbellaria-Tricladida), a laboratory and field study. Freshwater Biol. 6:383-393. (TU)

Rieger, R. M. 1981. Morphology of the Turbellaria at the ultrastructural level. Hydrobiologia 84:213-229. (TU)

Schell, S. C. 1985. Trematodes of North America north of Mexico. University of Idaho Press, Moscow, ID. (TR)

Schmidt, G. D. 1970. How to know the tapeworms. Wm. C. Brown, Dubuque, IA. (C)

Schmidt, G. D. 1985. Handbook of tapeworm identification. CRC Press, Boca Raton, FL. (C)

Schmidt, G. D., and L. S. Roberts. 1985. Foundations of parasitology. 3rd ed. C.V. Mosby, St. Louis, MO. (G)

Smith, J., S. Tyler, M. B. Thomas, and R. M. Rieger. 1982. The morphology of turbellarian rhabdites: phylogenetic implications. Trans. Am. Microsc. Soc. 101:209-228. (TU)

Spiegelman, M., and P. Dudley. 1973. Morphological stages of regeneration in the planarian, *Dugesia tigrina*: a light and electron microscope study. J. Morph. 139:155-183. (TU)

Turbeville, J. M., and E. E . Ruppert. 1985. Comparative ultrastructure and the evolution of nemertines. Am. Zool. 25:53-71.

Uglem, G. L., and J. J. Just. 1983. Trypsin inhibition by tapeworms: antienzyme secretion or pH adjustment? Science 220:79-81. (C)

*The supplemental literature listed here is coded to indicate the general topic covered by the paper: G, general Platyhelminthes; TU, Turbellaria; TR, Trematoda; C, Cestoda.

Supplemental Reading*

EXERCISE 6

Phylum Nemertea (Rhynchocoela)

Nemertea (ne-MER-te-a; G., *nemerte*, unerring) is a phylum comprising about 900 species of nonsegmented, dorsoventrally flattened worms, lacking a body cavity. They are commonly known as ribbon worms because of their length and shape. Nemertines range in size from a few millimeters to an incredible 30 m in *Lineus longissimus*; the average length is less than 20 cm. Most nemertines are marine and benthic, but pelagic forms and freshwater and terrestrial species are known. Most nemertines are carnivorous; a few are parasitic.

Nemertines have several features also found in turbellarians, including a ciliated epidermis, protonephridia, and a bilateral nervous system. The space between the gut and the body wall also is filled with parenchyma, but this condition may not be acoelomate in the traditional sense. Turbeville and Ruppert (1985) suggest that both nemertines and flatworms have evolved from coelomate taxa, perhaps the annelids. Nevertheless, nemertines are unlike flatworms in three important characteristics. (1) They possess an eversible, tubular **proboscis** housed in a special, fluid-filled cavity called the **rhynchocoel**. For this reason the alternative phylum name **Rhynchocoela** (RING-ko-SE-la; G., *rhynchos*, snout + G., *coelo*, hollow) is often used. The proboscis is used in food capture, but has a different embryological origin from the mouth. Prey is seized by the proboscis, which is everted from the rhynchocoel by hydrostatic pressure created by muscular contractions. The accuracy of the proboscis is said to be unerring, hence the phylum name, Nemertea. (2) There is a complete digestive system. (3) A closed vascular system, which may be homologous with the coelom of annelids, is present (Turbeville and Ruppert 1985).

Classification

The division of nemertines into two classes is based on position of the mouth relative to the cerebral ganglia and on the nature of the proboscis. In Anopla (AN-o-pla; G., *anoplos*, unarmed) the mouth is below or posterior to the cerebral ganglia; it opens separate from the proboscis pore. In this class the proboscis is unarmed. Examples include *Cerebratulus* and *Lineus*. In Enopla (EN-o-pla; G., *enoplos*, armed) the mouth lies anterior to the cerebral ganglia and in most forms the mouth and proboscis pore are united as a common opening. Proboscises are armed. The freshwater nemertine *Prostoma* is an example.

Observational Procedure: *Cerebratulus*

1. Examine a nemertine such as *Cerebratulus* and identify the anterior and posterior ends (Fig. 6.1). Note that the body is long, somewhat dorsoventrally flattened, and nonsegmented.

2. Locate the proboscis pore at the very tip of the anterior end. Some specimens may have the proboscis partly extended. In *Cerebratulus* the large, slitlike, ventral mouth is near the anterior end.

3. On the lateral margins just anterior to the mouth are two thin longitudinal grooves called **cephalic slits**. They probably function as chemoreceptors.

4. At the very tip of the posterior end is the anus.

5. Examine several cross sections of a nemertine. Note that the body is covered by an epidermis of ciliated columnar glandular cells (Fig. 6.2). Beneath the epidermis is a layer of connective tissue called the extracellular matrix or dermis. It occasionally has extensions that go deeper into the body.

6. Just under the dermis are circular and longitudinal muscle layers.

7. Note the absence of a coelom. The body is filled with circular and longitudinal muscles and parenchymal tissue.

8. Locate the proboscis and the gut with its vacuolate phagocytic cells. The hollow proboscis is constructed of layers of circular and longitudinal muscle. How does this tubular structure function? Surrounding the proboscis is a cell-lined, fluid-filled cavity called the rhynchocoel.

9. Also present within the body of nemertines are lateral blood vessels, lateral nerves, protonephridia, and gonads.

10. Examine a slide of a helmet-shaped **pilidium larva** (G., *pilidium*, small felt cap) (Fig. 6.3). The larva is covered with cilia, and at the aboral pole is an apical sensory organ with a tuft of long cilia. Find the foregut (stomodeum) flanked on either side by the oral lobes. The foregut leads to the small blind midgut. The anus develops later by an ectodermal invagination when the larva undergoes metamorphosis. The large internal cavity is the blastocoel.

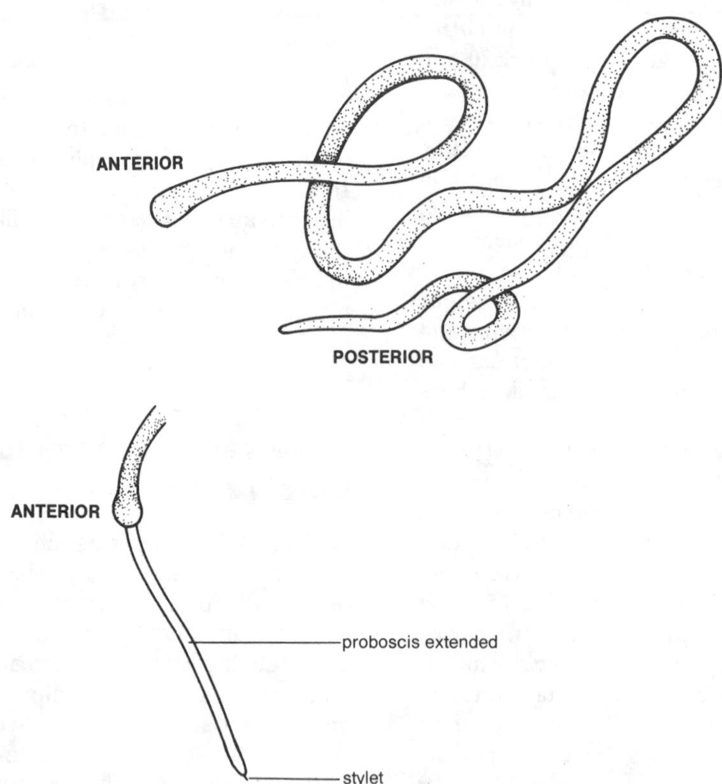

Figure 6.1. External view of a nemertine worm.

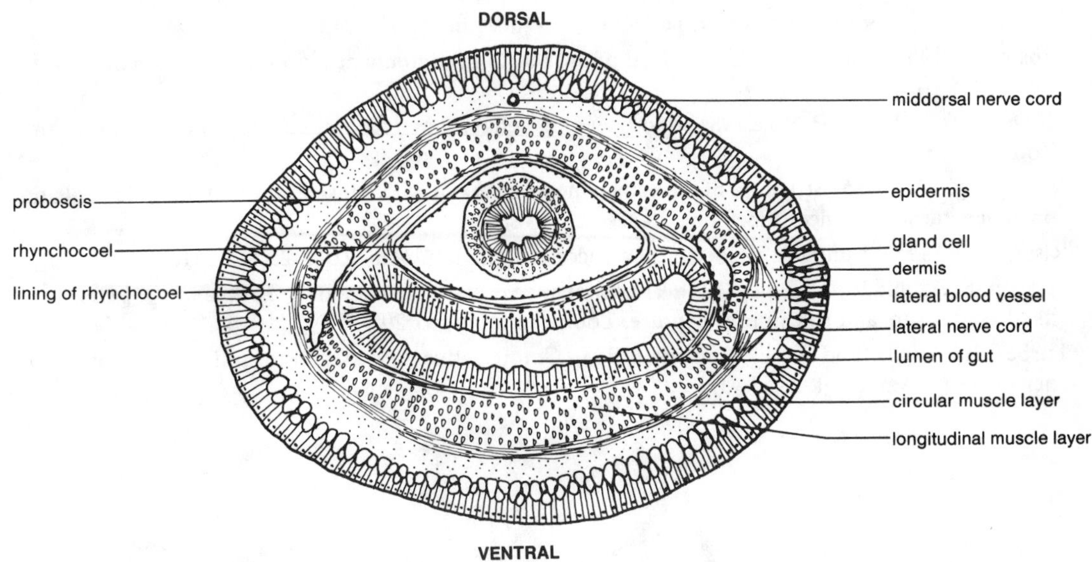

Figure 6.2. Cross section of a nemertine worm.

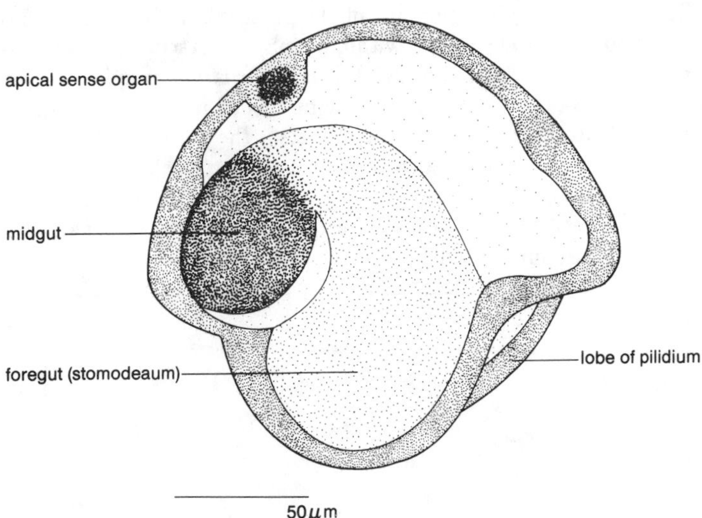

Figure 6.3. Side view of a pilidium larva.

Supplemental Reading

Gibson, R. 1972. Nemertines. Hutchinson University Library, London.

Gibson, R. 1976. Freshwater nemertines. Zool. J. Linn. Soc. 58:177-218.

Gibson, R. 1985. The need for a standard approach to taxonomic descriptions of nemertines. Am. Zool. 25:5-14.

McDermott, J. J., and P. Roe. 1985. Food, feeding behavior and feeding ecology of nemertines. Am. Zool. 25:113-125.

Norenburg, J. L. 1985. Structure of the nemertine integument with consideration of its ecological and phylogenetic significance. Am. Zool. 25:37-51.

Riser, N. W. 1985. Epilogue: Nemertinea, a successful phylum. Am. Zool. 25:145-151.

Stricker, S. A., and R. A. Cloney. 1981. The stylet apparatus of the nemertine *Paranemertes peregrina*: its ultrastructure and role in prey capture. Zoomorphology 97:205-223.

Turbeville, J. M., and E. E. Ruppert. 1985. Comparative ultrastructure and the evolution of nemertines. Am. Zool. 25:53-71.

PSEUDOCOELOMATE PHYLA

There are seven **pseudocoelomate** phyla: Acanthocephala, Gastrotricha, Kinorhyncha, Loricifera, Nematoda, Nematomorpha, and Rotifera. To this list some zoologists add Priapulida and Entoprocta, arguing that the body cavities of these phyla are not lined with a cellular peritoneum, and hence are not true coeloms. Because the pseudocoelomate phyla possess certain similar features, the phyla are considered by some zoologists to be classes within the phylum **Aschelminthes** (AS-kel-MIN-theez; G., *ascus*, bladder + G., *helmins*, worm). Although this practice is common in European publications, it is not followed by most texts of invertebrate zoology published in the United States.

The only consistent feature of the pseudocoelomates is the presence of a pseudocoelom. This body cavity lacks a mesodermal lining and is formed from a persistent blastocoel. Most pseudocoelomates are usually less than 1 cm long with an outer, noncellular cuticle. The body lacks a well-formed head and gas exchange and circulatory systems. Other important features found in many aschelminths include a complete digestive tract with muscular pharynx, a protonephridial system, and eutely.

EXERCISE 7

Phylum Gastrotricha

Gastrotricha (GAS-tro-TRIK-a; G., *gastro*, stomach + G., *trich*, hair) is a small phylum (ca. 450 species) of marine and freshwater wormlike, bilaterally symmetrical animals. Commonly, they are found gliding with their ventral surface in contact with other organisms or inanimate objects. Three characteristic features of gastrotrichs may be seen readily (Fig. 7.1): (1) the presence of a forked tail, especially in the freshwater species; (2) a modified cuticular surface with spine- or scalelike structures covering the dorsal side; and (3) the patterned distribution of cilia which are more or less restricted to the ventral side, as the phylum name suggests.

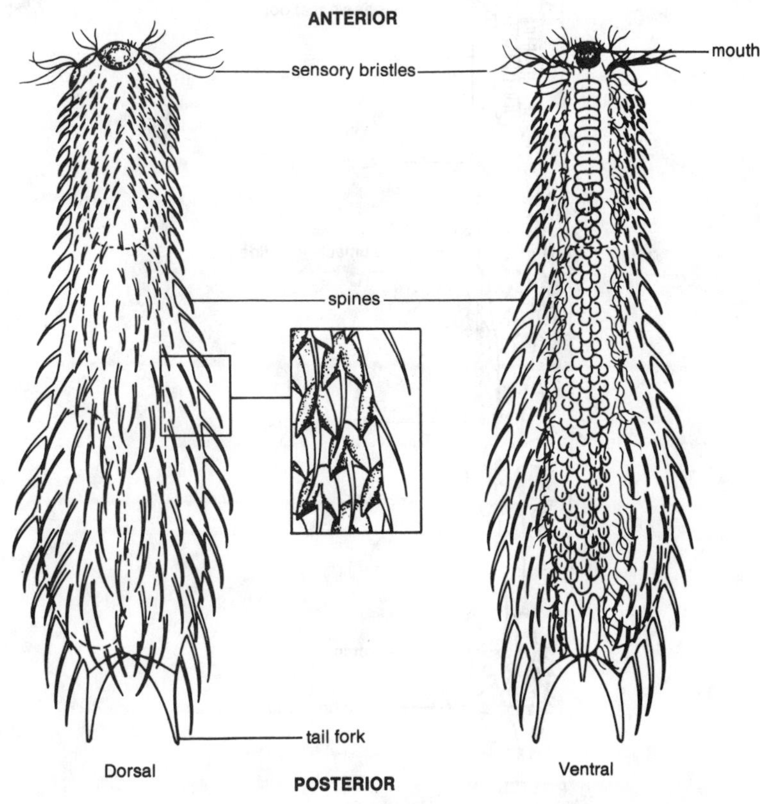

Figure 7.1. Dorsal and ventral views of *Chaetonotus*, a typical gastrotrich. (After Balsamo.)

Classification

The phylum is divided into two orders: Macrodasyida (marine, with elongate, cylindrical body; posterior end of various shapes; adhesive tubes usually present along sides of the body and at anterior and posterior ends; example includes *Macrodasys* (Fig. 7.2); Chaetonotida (marine and all freshwater gastrotrichs; body bowling pin-shaped with a neck and ending in a forked tail; adhesive tubes are found only at the posterior end; examples include *Chaetonotus* (Fig. 7.1) and *Lepidodermella* (Fig. 7.3).

Observational Procedure: Gastrotricha

1. Examine a culture of the chaetonotid gastrotrich, *Lepidodermella*, under a dissecting microscope. You should be able to see the animals gliding about on the bottom of the culture dish.

2. Using a Pasteur pipette, place 1 to 2 drops of the culture on a slide, apply a coverslip supported with small corner posts made of modeling clay, and check for the presence of gastrotrichs using low power. When you have a slide with one or more

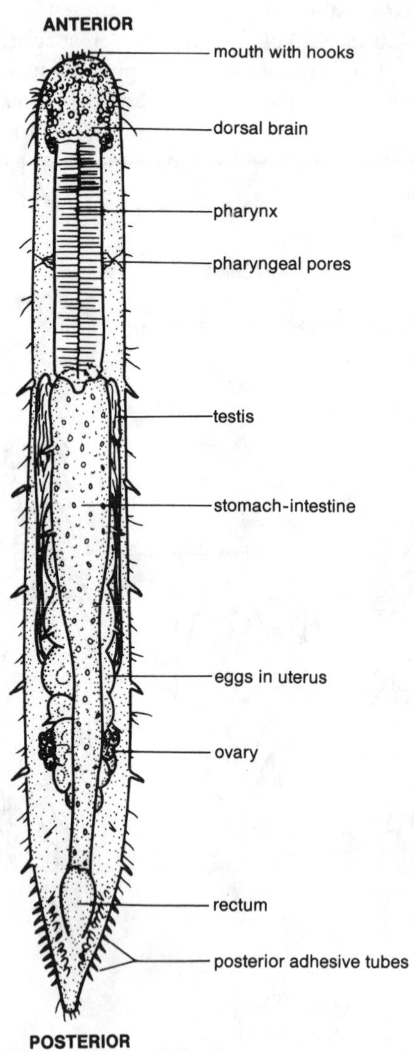

Figure 7.2. Dorsal view of a marine gastrotrich, *Macrodasys*. (From Hyman after Remane.)

specimens, withdraw water slowly using a strip of filter paper until an animal is caught in a small amount of water, then switch to higher magnification and make the following observations (Fig. 7.3). Alternatively, mount the drop of culture fluid on a slide within a ring of methyl cellulose, add a coverglass, and observe.

3. Note the distinct scaly appearance. The body consists of a rounded, anterior head region, a slightly and gradually constricted neck region, and an expanded, elongate trunk with a forked tail. With the appropriate lighting, you may be able to see the anterior sensory bristles, mouth opening, pharynx, stomach-intestine, and reproductive system.

4. The internal anatomy may be observed more easily by applying a single drop of neutral red stain to the edge of the coverslip of a sample prepared without methyl cellulose. Draw the stain across the field by touching a small piece of filter paper to the opposite edge of the coverslip. If the specimens do not stain too darkly, you should be able to differentiate many of the internal structures noted above (Fig. 7.4). Be careful not to let the slide dry out during your examination. This may be avoided by periodically adding a drop of culture fluid to the edge of the coverslip.

5. Students should examine samples taken from the field or laboratory fish tank for other specimens. One genus often present in such samples is *Chaetonotus* (Fig. 7.1). In *Chaetonotus* the dorsal side is covered with spines as well as scales. As the animal turns, you should be able to note the ventral cilia. Short sensory bristles may be seen at the anterior and lateral parts of the head region. Denser, internally located bodies at the base of each tail fork are adhesive glands.

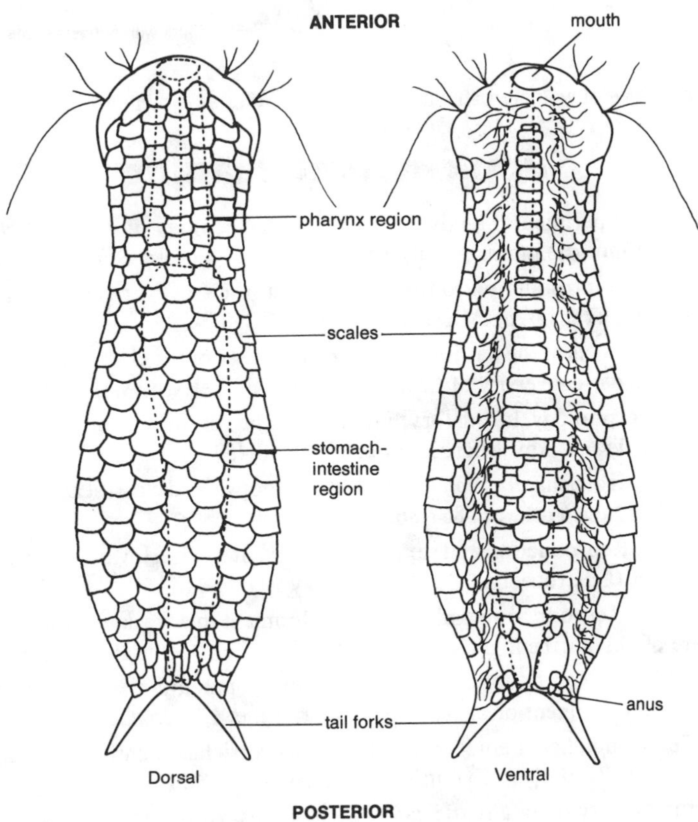

Figure 7.3. Dorsal and ventral views of *Lepidodermella*. (After Balsamo.)

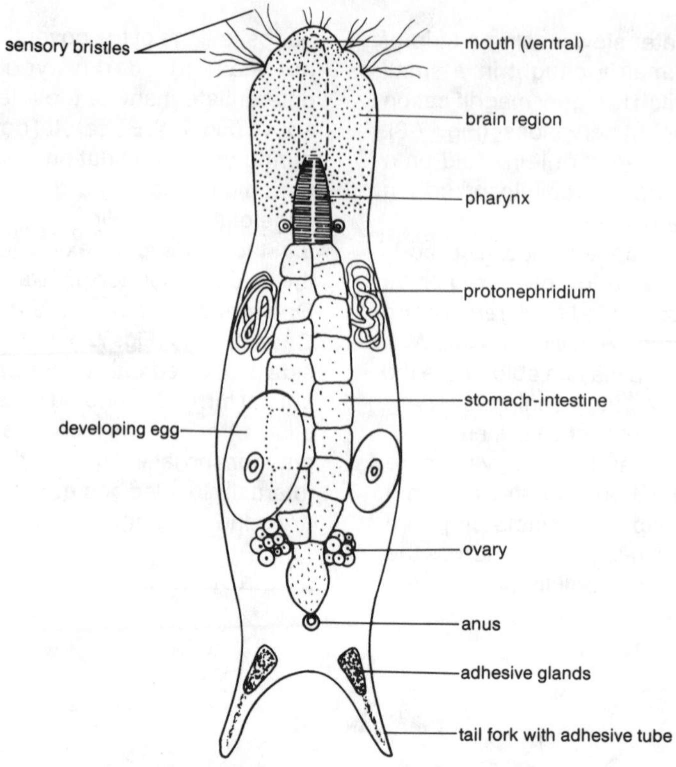

Figure 7.4. Dorsal internal view of typical freshwater gastrotrich. (After several authors.)

Supplemental Reading

Bennett, L. W. 1979. Experimental analysis of the trophic ecology of *Lepidodermella squammata* (Gastrotricha: Chaetonotida) in mixed culture. Trans. Am. Microsc. Soc. 98:254-260.

Brunson, R. B. 1950. An introduction to the taxonomy of the Gastrotricha with a study of eighteen species from Michigan. Trans. Am. Microsc. Soc. 69:325-350.

Brunson, R. B. 1963. Aspects of the natural history and ecology of the Gastrotricha. *In*: E. Dougherty (ed.). The lower metazoa. University of California Press, Berkeley, CA, pp. 473-478.

Hummon, W. 1966. Morphology, life history, and significance of the marine gastrotrich, *Chaetonotus testiculophorus* n. sp. Trans. Am. Microsc. Soc. 85(3):450-457.

Hummon, W. 1982. Gastrotricha. *In*: S. P. Parker (ed.). Synopsis and classification of living organisms. Vol. 1. McGraw-Hill, New York, pp. 857-863.

Hummon, M. R. 1984. Reproduction and sexual development in a freshwater gastrotrich. 1. Oogenesis of parthenogenic eggs (Gastrotricha). Zoomorphology 104:33-41.

Hummon, M. R. 1984. Reproduction and sexual development in a freshwater gastrotrich. 2. Kinetics and fine structure of postparthenogenic sperm formation. Cell Tissue Res. 236:619-628.

Hummon, M. R. 1984. Reproduction and sexual development in a freshwater gastrotrich. 3. Postparthenogenic development of primary oocytes and the X-body. Cell Tissue Res. 236:629-636.

Rieger, R. M. 1976. Monociliated epidermal cells in Gastrotricha: significance for concepts of early metazoan evolution. Z. Zool. Syst. Evolutionsforsch. 14 (3):198-226.

Sacks, M. 1964. Life history of an aquatic gastrotrich. Trans. Am. Microsc. Soc. 83:358-362.

Teuchert, G. 1977. The ultrastructure of the marine gastrotrich *Turbanella cornuta* Remane (Macrodasyoidea) and its functional and phylogenetical importance. Zoomorphologie 88:189-246.

Thane-Fenchel, A. 1970. Interstitial gastrotrichs in some south Florida beaches. Ophelia 7(2):113-138.

EXERCISE 8

Phylum Rotifera

Rotifera (ro-TIF-e-ra; L., *rota*, wheel + L., *ferre*, to bear) is a small phylum of some 1800 species of bilaterally symmetrical, pseudocoelomates, having two distinguishing features: (1) a ciliated, apical region called the **corona**, used in locomotion and food gathering (Figs. 8.1 and 8.2), and (2) a muscular pharynx (**mastax**) equipped with complex jaws of seven pieces called **trophi** (Fig. 8.3). The name "wheel bearer" is an allusion to the corona which resembles a rotating wheel due to the metachronal beat of its cilia. In one group of rotifers (Collothecaceae) cilia are nearly lacking and long setae are used in prey capture (Fig. 8.2).

Most rotifers are small (100 to 1500 μm) and saccate to cylindrical in shape. Typically, they are found in freshwater habitats, where they are often very abundant, although several species are exclusively marine. One class (bdelloids) inhabits films of water found in soils or covering terrestrial plants such as mosses. Most rotifers are free moving, either by swimming or crawling, but many sessile species live attached to freshwater plants (Fig. 8.2). Nearly all rotifers are solitary, but about 25 species form colonies.

Classification

Three classes are commonly recognized.

1. Class Seisonidea [separate sexes of similar size and morphology; paired gonads; single genus (*Seison*); epizoic on marine crustaceans].

2. Class Bdelloida [only females; paired ovaries; mainly swimming and crawling forms; many capable of becoming desiccated and then rehydrated; examples include *Habrotrocha* and *Philodina*].

3. Class Monogononta [separate sexes; single gonads; males rare and structurally reduced; reproduction mainly parthenogenic, sexual reproduction results in a diapausing embryo called a resting egg (Fig. 8.4); three orders (Ploima, Flosculariaceae, Collothecaceae); benthic, swimming, and sessile forms; examples include *Asplanchna*, *Brachionus*, *Filinia*, *Keratella*, *Lecane*, and *Synchaeta*].

Observational Procedure

1. Examine a culture of rotifers under a dissecting microscope and describe the locomotory activities you observe. Is the "wheel organ" evident?

2. Make a ring of methyl cellulose on a slide and using a Pasteur pipette, put a few rotifers into the center of the ring and place a coverslip over the preparation.

3. Under low power locate an animal and identify its anterior and posterior ends (Figs. 8.1 and 8.2). Note that the body may be divided into three general regions: head (with corona), body, and foot with toes (Fig. 8.1). A neck region also may be present or the foot absent. Although highly variable, the corona usually consists of two ciliated rings (**trochus** and **cingulum**) at the anterior end. Switch to higher magnification and attempt to distinguish the details of the coronal ciliation, including the trochus and cingulum (Fig. 8.1). How do the cilia appear to move? Are cirri present?

4. Turn your attention to the rest of the animal and under high power locate the stomach, intestine, ovary (ovaries in bdelloids), musculature, protonephridial flame cells, foot with pedal glands, and toes (Fig. 8.1). As you examine the organs, look for the presence of nuclei.

5. If additional species are available for study, examine them and compare their anatomy. What differences do you observe, especially between monogonont and bdelloid species (Figs. 8.1 and 8.2)? Zooplankton samples from the field may contain a variety of species, including some with **resting eggs** (Figs. 8.2 and 8.4).

6. Make another slide without methyl cellulose and place a small drop of food suspension (yeast or algae) or inert material (carmine powder or latex beads) at the edge of the coverglass. Draw the suspension across the slide using a piece of toweling and observe the rotifers under low power. Track any particles that become caught in the coronal currents and follow them as they enter the mouth. Describe the activity of the mastax during feeding. Can you see the particles accumulate in the gut?

7. If time and materials permit, make another slide preparation of one or more of the available species, but add a small drop of bleach to the fluid before putting the coverslip in place. Quickly locate a rotifer on the slide and observe the bleach dissolve the soft tissues. Examine the trophi that remain and locate the various pieces illustrated in Fig.8.3.

8. Observe a prepared slide of preserved rotifers and compare them to the live preparations just examined. Locate as many of the features mentioned above as you can.

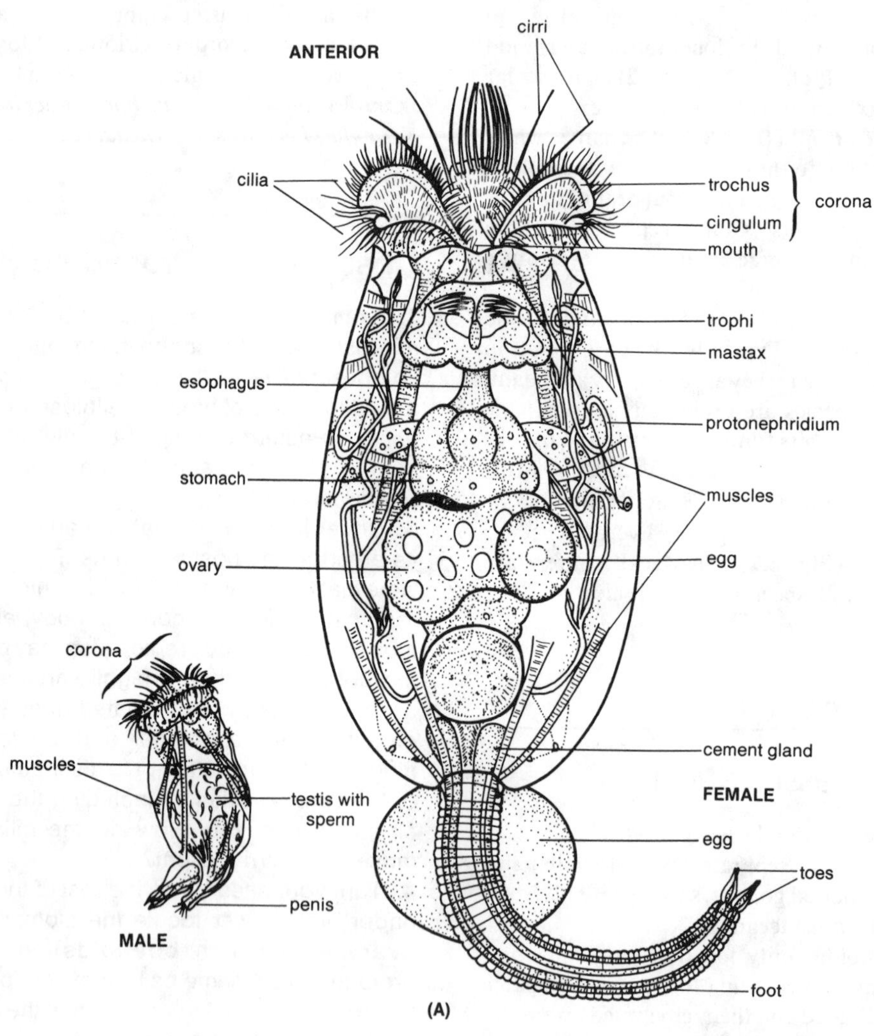

(A)

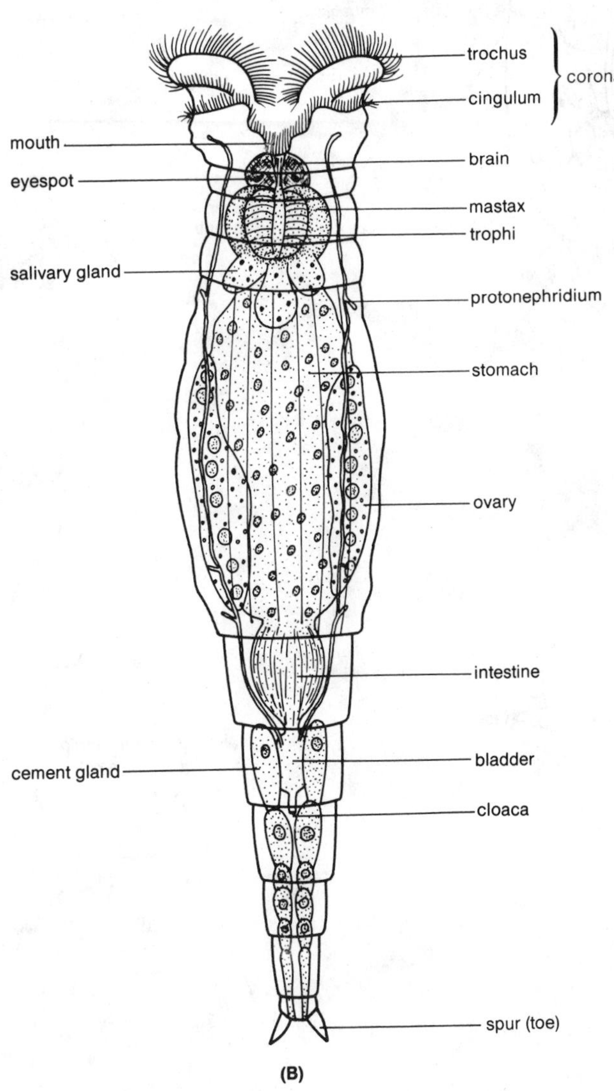

Figure 8.1. Views of two common rotifers. **(A)** *Brachionus plicatilis*, amictic female and male (Monogononta). **(B)** *Philodina roseola* (Bdelloida). Each about 300 μm in length. (After several sources.)

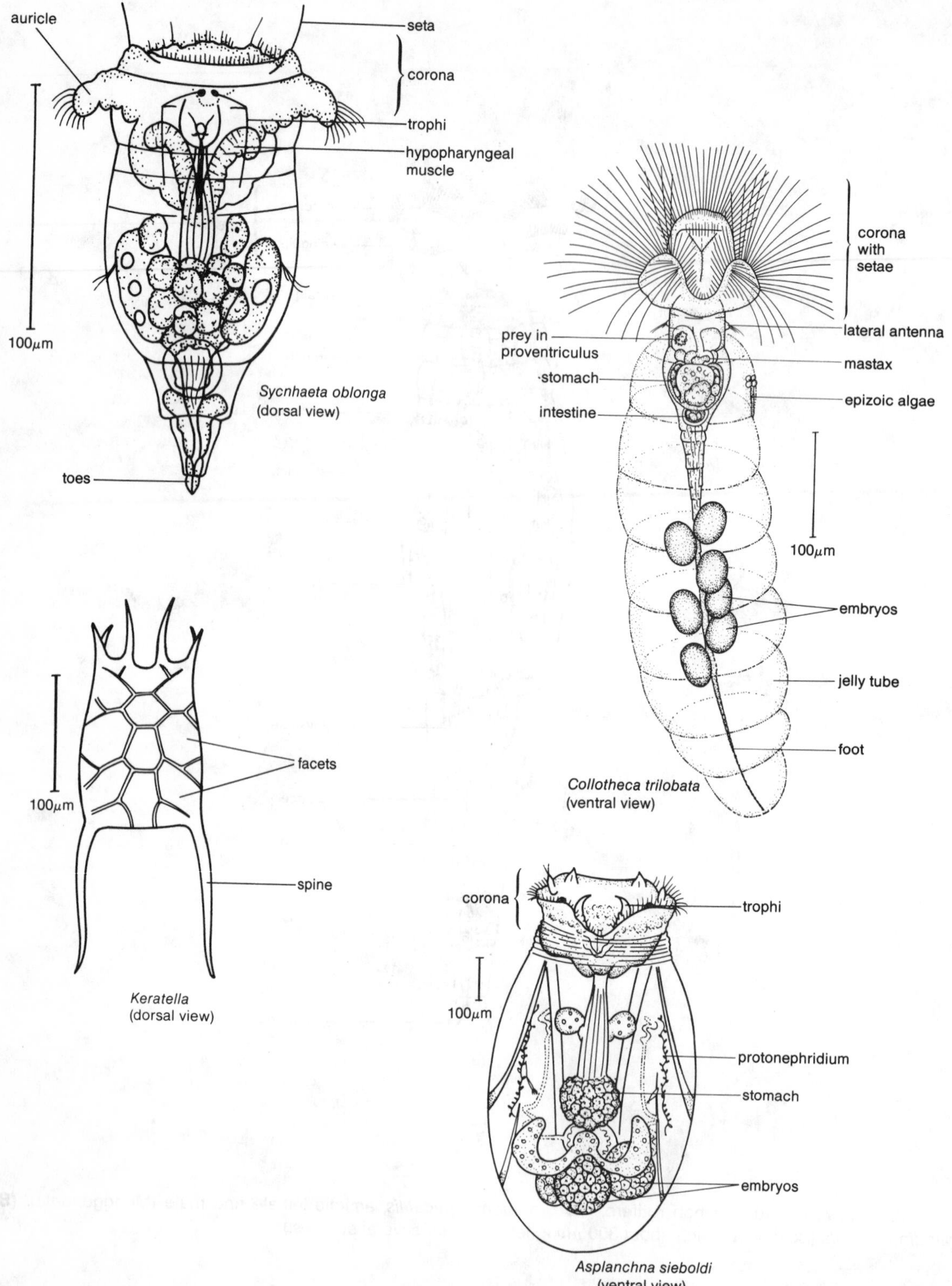

Sycnhaeta oblonga (dorsal view)

Collotheca trilobata (ventral view)

Keratella (dorsal view)

Asplanchna sieboldi (ventral view)

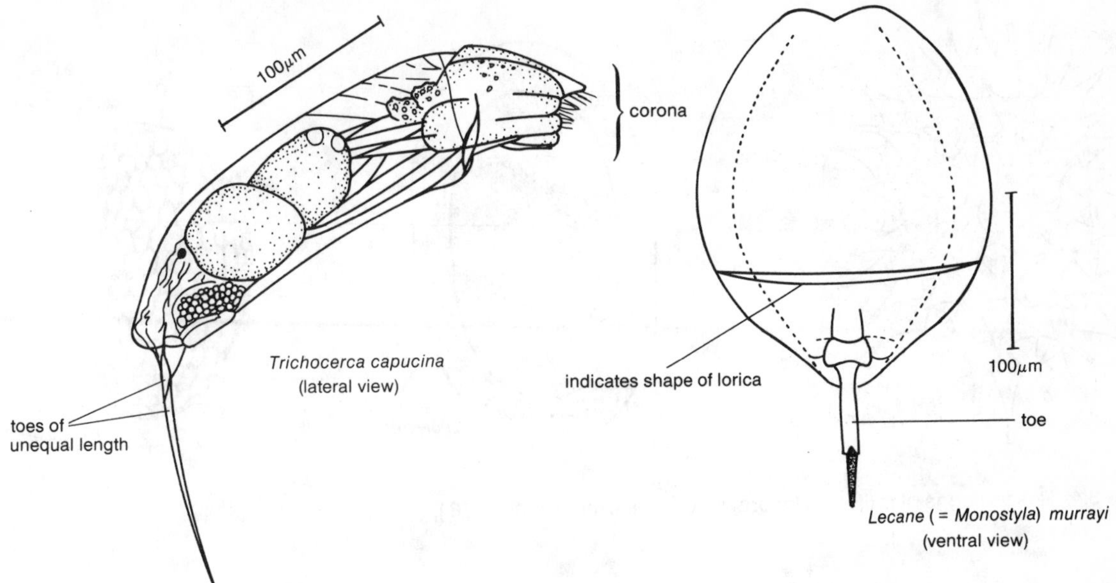

Figure 8.2. Several common freshwater rotifers. (After Koste 1978.)

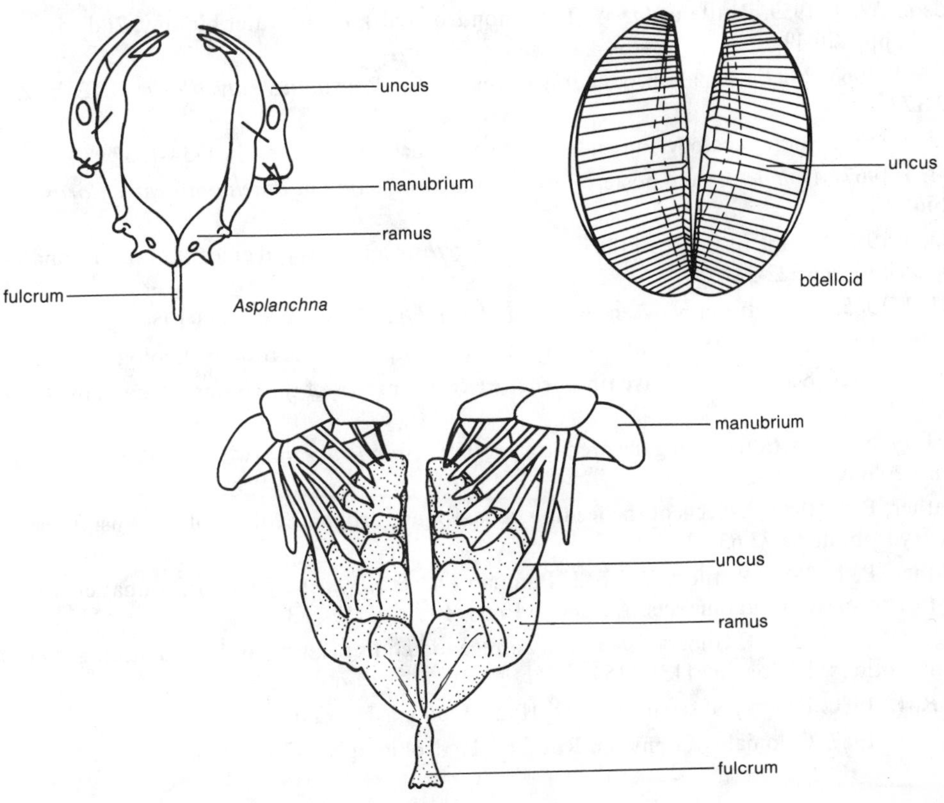

Figure 8.3. Rotifer trophi. (Not drawn to scale. After several sources.)

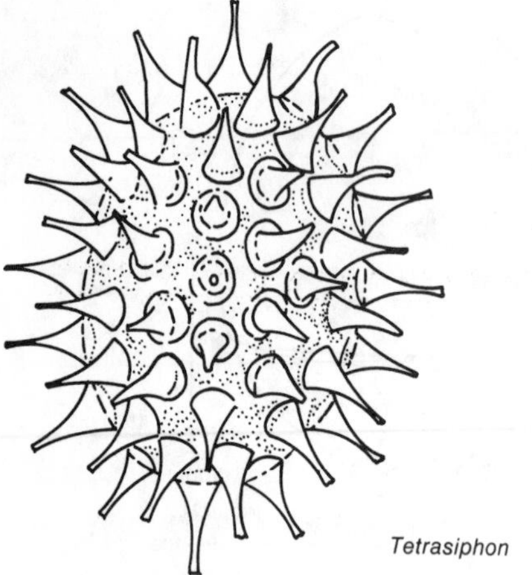

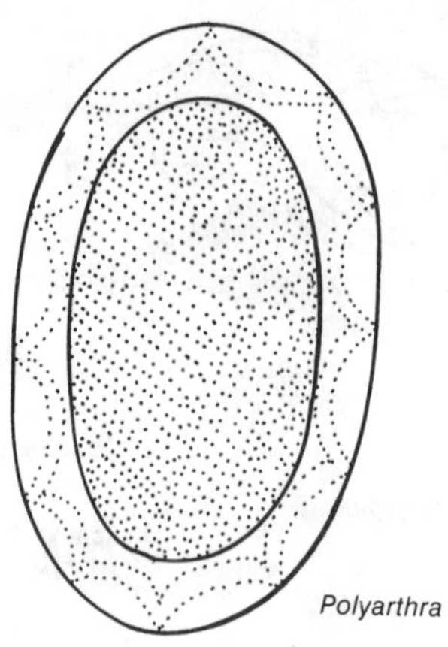

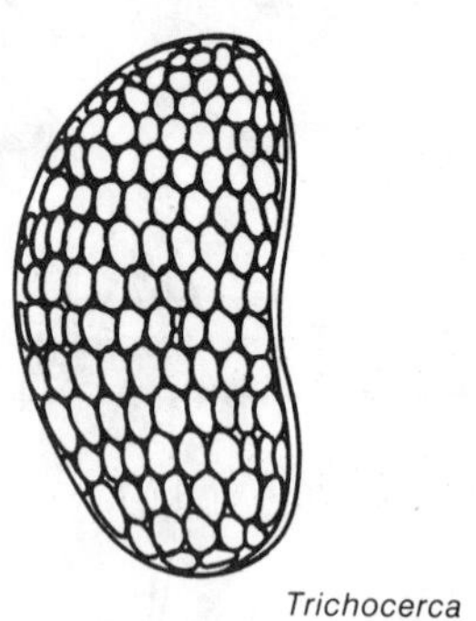

Figure 8.4. Resting eggs of rotifers. (Not drawn to scale. After Koste 1978.)

Supplemental Reading

Edmondson, W. T. 1959. Rotifera. *In*: W. T. Edmondson (ed.). Fresh-water biology, 2nd ed. Wiley, New York, pp. 420-494.

Gilbert, J. J. 1963. Mictic female production in the rotifer *Brachionus calyciflorus*. J. Exp. Zool. 153:113-123.

Gilbert, J. J. 1966. Rotifer ecology and embryological induction. Science 151:1234-1237.

Gilbert, J. J. 1967. *Asplanchna* and postero-lateral spine production in *Brachionus calyciflorus*. Arch. Hydrobiol. 64:1-62.

Gilbert, J. J. 1985. Escape response of the rotifer *Polyarthra*: a high-speed cinematographic analysis. Oecologia (Berl.) 66:322-331.

Gilbert, J. J. 1985. Competition between rotifers and *Daphnia*. Ecology 66:1943-1950.

King, C. 1967. Food, age, and the dynamics of a laboratory population of rotifers. Ecology 48:111-128.

Koehler, J. K. 1966. Some comparative fine structure relationships of the rotifer integument. J. Exp. Zool. 162:231-244.

Maly, E. J. 1975. Interactions among the predatory rotifer *Asplanchna* and two prey, *Paramecium* and *Euglena*. Ecology 56:346-358.

Starkweather, P. L. 1980. Aspects of the feeding behavior and trophic ecology of suspension feeding rotifers. Hydrobiologia 73:63-72.

Starkweather, P. L. 1987. Rotifera. *In*: T. J. Pandian and F.J. Vernberg (eds.). Animal energetics, volume 1: Protozoa through insects. Academic Press, FL. pp. 159-183

Stemberger, R. S., and J. J. Gilbert. 1985. Body size, food concentration and population growth in planktonic rotifers. Ecology 66:1151-1159.

Wallace, R. L. 1980. Ecology of sessile rotifers. Hydrobiologia 73:181-193.

Wallace, R. L. 1987. Coloniality in phylum Rotifera. Hydrobiologia 147:141-155.

EXERCISE 9

Phylum Nematoda (Nemata, Nema)

Phylum Nematoda (nem-a-TOD-a; G., *nema*, thread) is a large group of some 15,000 species of threadlike pseudocoelomates known commonly as the roundworms. Free-living species have been found in every habitat examined, including Arctic ponds, hot springs, mountain and desert soils, deep-sea marine muds, and unpasteurized vinegar. Diversity of animal and plant parasitic nematodes is tremendous. Every vertebrate species may serve as a host to one or more nematodes. On a global scale, humans are infected commonly by about 15 species of nematodes, but an additional 15 species may occasionally parasitize humans.

Despite their breadth in nutritional habits and geographic distribution, all nematodes have the same basic body structure. Typically, they are cylindrical worms which taper at both ends (Fig. 9.1), ranging in size from less than 1 mm to several meters long. Nematodes are covered by a resistant **cuticle**, unique to this group, which is composed of three layers (**cortical**, **matrix**, and **fibrous**). Nematodes have a straight digestive tract consisting of a mouth, buccal cavity, muscular esophagus, intestine, rectum, and anus. The buccal cavity, esophagus, and rectum are lined with cuticle. The nematode mouth is surrounded by up to six **lips**, but these may be fused into three lips or absent totally. Some nematodes, especially those that parasitize plants, are armed with a long stylet, which is located in the buccal cavity and used in feeding (Fig. 9.1).

The young worms that hatch from the eggs are often called larvae, but they possess most adult structures, except for an immature reproductive system. As juvenile worms grow, they undergo four molts. During these molts the entire cuticle is shed. However, unlike arthropods, which also molt, individuals grow in size during intermolt periods. Larvae of some parasitic forms are given specific names: **rhabditiform, filariform,** and **microfilaria**. Parasitic nematodes may have a life cycle which is direct (**monoxenous**), having only one host, or indirect (**heteroxenous**), having two or more hosts.

The main objective of the following study is to direct student attention to the general structure, biology, and diversity of free-living and parasitic nematodes, and to review aspects of some species parasitic on humans. Owing to time constraints, it is probable that only a few of the following studies may be done in your laboratory.

Classification

Two classes and 20 orders of nematodes are recognized.

1. Class Adenophorea (= **Aphasmida**): predominantly free-living nematodes lacking phasmids, with amphids located behind the "head" region; males lacking lateral extensions of the tail; 12 orders; examples of animal parasites include *Trichinella spiralis* (trichina worm) and *Trichuris trichiura* (whipworm); members of the orders Dorylaimida and Mermithida are important plant and insect parasites, respectively; the free-living genus *Metoncholaimus* is found in mud on North American coasts.

2. Class Secernentea (= **Phasmida**): predominantly parasitic or free-living terrestrial nematodes having both phasmids and amphids, the latter being located forward in the head; males often have lateral extensions of the tail; eight orders; examples of animal parasites include *Ascaris lumbricoides* (intestinal roundworm),

Dracunculus medinensis (Guinea worm), *Dirofilaria immitis* (dog heartworm), *Enterobius vermicularia* (pinworm), *Necator americanus* (American hookworm), *Rhabditis maupasi* (earthworm parasite), and *Wuchereria bancrofti* (filarial worm); members of the order Tylenchida are important parasites of plants and insects; the vinegar eel, *Turbatrix aceti*, is a commonly studied free-living species.

Observational Procedure

Free-Living Nematodes. Using a dissecting microscope and a Pasteur pipette, remove some of the nematodes from a culture containing the vinegar eel, *T. aceti* (Fig. 9.2), or *Chiloplacus* sp. Place your sample on a concave microslide and examine it for

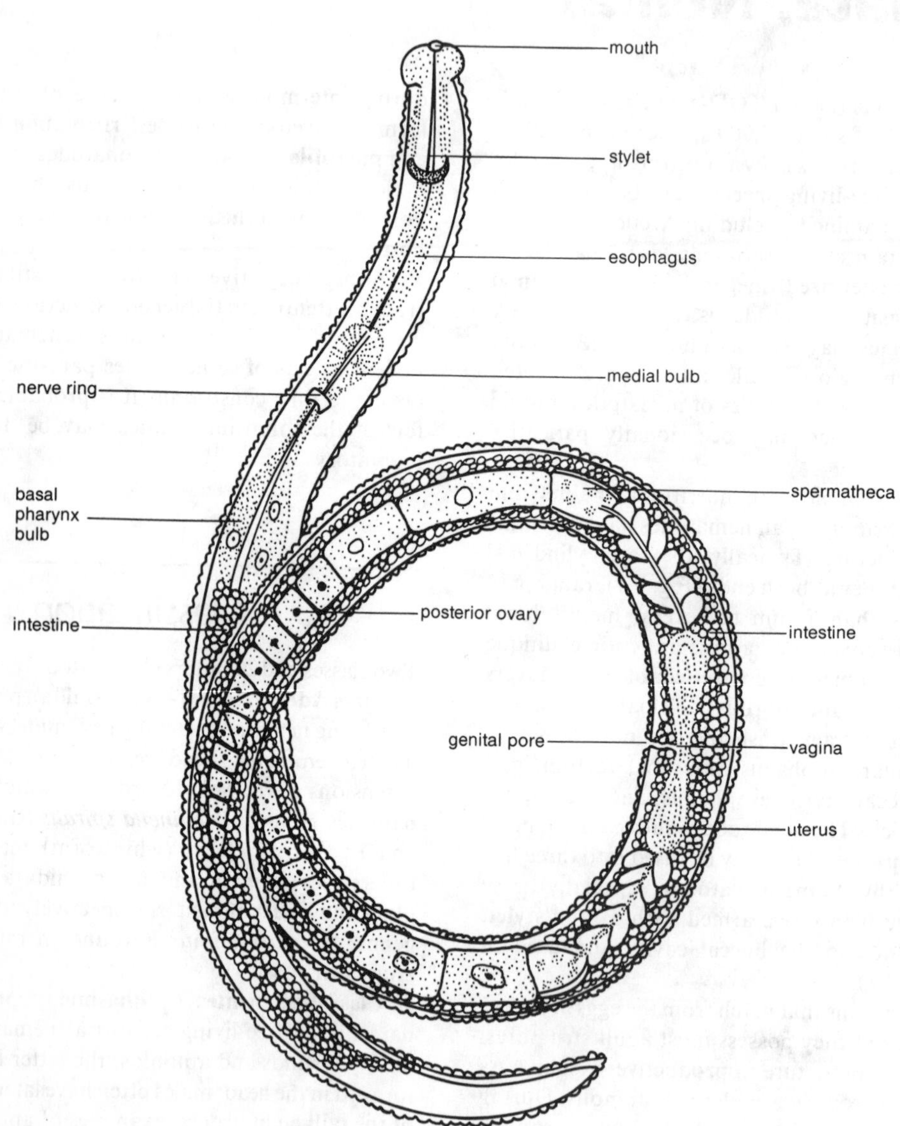

Figure 9.1. Typical female nematode, *Tylenchorhynchus cylindricus*.

specimens. Note the distinctive sinuous body movements. Remove a few specimens, place them on a microscope slide, add a drop of water and a coverslip, and examine under low and high magnifications. All developmental stages should be present in these cultures. Each successive stage is larger than the previous one. Are any of the worms molting? Identify the anterior and posterior ends. Occasionally, you may see copulating pairs of worms. What region of his body does the male use to grasp the female? What region of her body does the male grasp? How does this species feed? Can you see the muscular esophagus or any other internal organs? The thick cuticle makes identification of internal organs difficult. Examine a prepared slide of *Chiloplacus* and/or *Turbatrix* and locate the esophagus, the remainder of the gut, and reproductive system. Attempt to trace the digestive tract from mouth to anus. Measure the length and diameter of these specimens.

Parasitic Nematodes. Nematodes are some of the most important parasites affecting the lives and general welfare of humans, including domestic and game animals and crops. Only a few of the better known forms have been selected for this exercise. For each species studied, a brief description of the life cycle will be given followed by directions for the observations to be made. If live specimens are studied, be very careful not to infect yourself with infective stages.

Pinworm (Fig. 9.3). The pinworm (***Enterobius vermicularis***) or seat worm is a cosmopolitan oxyurid parasite of humans having a direct life cycle. Fertilized females migrate from the colon (usually at night) and lay thousands of eggs in the perianal region. The embryonated eggs contain rhabditiform larvae which may enter the digestive tract either by inhalation of dust contaminated by the egg or through direct ingestion. Once embryos reach the intestine, they hatch and after several successive molts the adults mate and become attached to the colon wall. It is estimated that 500 million people worldwide have enterobiasis.

Examine a prepared slide of pinworms and with the aid of Fig. 9.3, identify the anatomical structures of the specimen. Determine its size. Examine a slide of pinworm eggs. Note their oval shape and flattened side. Your instructor may discuss the clinical method used for obtaining eggs in order to diagnose enterobiasis.

Hookworms (Fig. 9.4). One of the most devastating cosmopolitan diseases of human is **ancylostomiasis**, more commonly known as "hookworm disease." Several species of hookworms may be studied with equal educational value; *Ancylostoma duodenale* (Old World hookworm), *Ancylostoma caninum* (domestic dog and cat hookworm), and *Necator americanus* (New World or American hookworm) are examples. The New World hookworm is believed to have been brought to the Americas by the slave trade. Most hookworms have a direct life cycle. Adults live attached to the mucosal lining of their host's intestine and feed on blood and tissue fluids. Embryonated eggs are passed in the feces and the first two developmental stages are passed in the soil. Third-stage larvae burrow into skin and are transported via the bloodstream to the lungs, where the larvae burrow into the alveoli. They are then carried up the trachea and are swallowed. Final molt and sexual maturity are achieved in the small intestine.

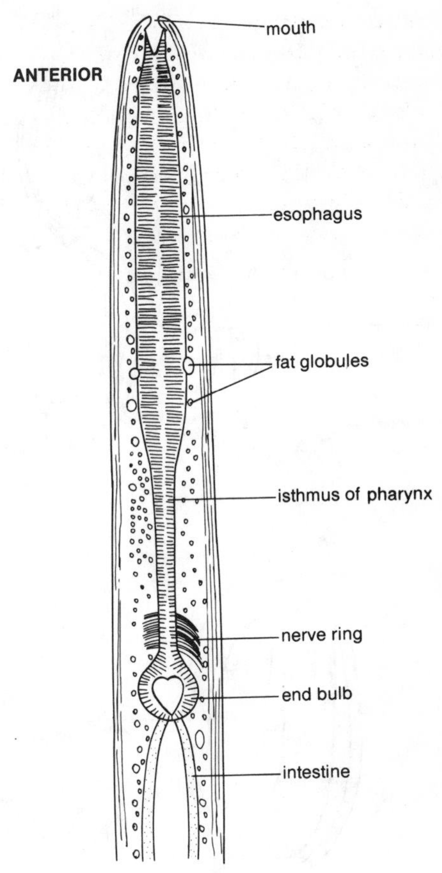

Figure 9.2. Anterior end of the vinegar eel, *Turbatrix aceti*, a free-living nematode.

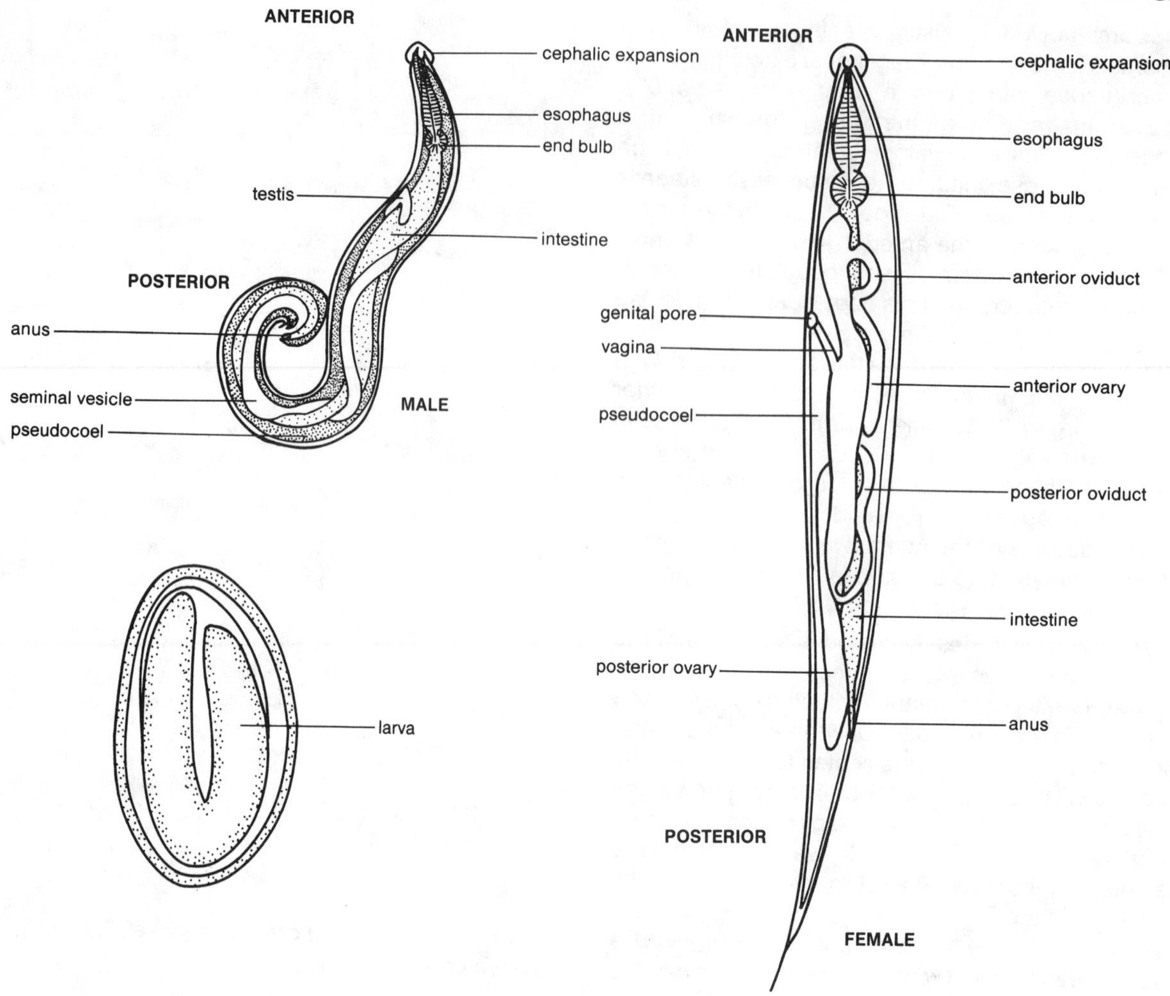

Figure 9.3. Male, female, and egg of *Enterobius vermicularis* (human pinworm).

Examine a whole-mount slide of a hookworm and note the structures in the mouth region, including a large buccal cavity with **cutting plates** or teeth. Examine a longitudinal section of a hookworm in situ and note how the worm is attached to the mucosa. In a whole-mount slide of a male, note the **copulatory bursa** and **spicules**. Determine the size of these intestinal worms. Compare the size and general morphology of *A. duodenale* rhabditiform (free-living) and filariform (infective) larvae.

Filarial Worms (Fig. 9.5). Adults of the human filarial worm, *Wuchereria bancrofti*, live in the lymph nodes of humans, where ovoviviparous females release thousands of motile embryos known as microfilaria. These make their way to the peripheral blood of humans, where female mosquitoes pick them up while taking a blood meal. The larvae migrate to thoracic and leg muscles of the mosquito, where they continue to grow. Upon reaching a certain stage of growth (usually 2 to 5 weeks) they migrate to the mosquito's mouthparts and enter the human body when the mosquito takes another blood meal. In humans, the microfilaria find their way to the lymphatic system and attain maturity in about a year. Lymphoid tissues increase in size in response to microfilaria parasitization. If, after repeated attacks, sufficient worms are present in the human host, an enlargement of certain body appendages (e.g., lower limbs, scrotum) occurs, producing the disease inappropriately called **elephantiasis**. Other diseases produced by filarial worms include River blindness (*Onchocerca volvulus*) and dog heartworm (*Dirofilaria immitis*), which is common in North America.

Observe a preserved specimen of *Dirofilaria immitis*, noting its long, thin form. How can such a long worm develop in a dog's heart? Examine slides showing blood smears of *D. immitis* and/or *W. bancrofti* microfilaria. Calculate the sizes of the microfilaria and red blood cells. Examine a cross section of a nodule produced by another filarial worm, *Onchocerca volvulus*. Note the presence of the worm's body in the nodule.

Trichina Worm (Fig. 9.6). Trichinosis is acquired by eating insufficently cooked meat contaminated with the encysted stage of *Trichinella spiralis*. Digestive juices in the stomach liberate the larvae from the meat by removing the cyst wall. Worms achieve adulthood in about 3 days. After mating the male worms die and females penetrate the intestinal mucosa, where they release juvenile worms. One female may liberate as many as 2000 larvae into the host's bloodstream before she dies. The larvae are transported throughout the body and become lodged in skeletal muscle and other muscles with a rich blood supply, such as the diaphragm and tongue. Soon after the muscles are invaded, the worms enlarge and encyst, causing harmful tissue degeneration. In the past the usual source of human infection was contaminated pig meat. However, bear meat is now a common source of infection, but any carnivore, especially rats, may be infected.

Examine a section of muscle with encysted larvae. Note the size of the cysts and their general frequency in the tissue. How might a heavy infection affect host muscle function? Examine a slide of adult *Trichinella* and compare their general anatomy with the other nematodes you have observed.

Whipworm (Fig. 9.7). The human whipworm, *Trichuris trichiura*, is slightly different in its construction from other nematodes viewed thus far. The anterior part of the body in both sexes is very elongate and threadlike, but the posterior region is swollen or saccate. The overall impression one gets when looking at the worm is that of a buggy whip, hence the name whipworm. Their life cycle is direct. Eggs, ingested via fecal contamination, hatch in the small intestine, and in about 2 weeks the juvenile worms migrate to the colon, where they attach permanently to the mucosa and feed on blood cells. Whipworms reach sexual maturity in about 3 months. It has been estimated that more than 350 million people worldwide are infected.

Examine a whole-mount slide of a whipworm. Note the slender, elongate anterior and swollen posterior ends. The esophagus is a long, thin-walled, multiglandular tube called the **stichosome**. There are no lips. The buccal cavity is armed with a minute **stylet**, which may be seen under high magnification. Determine the size of these worms. Examine a whole-mount slide of whipworm eggs. Note their oval shape with the characteristic bipolar plugs (Fig. 9.7).

Earthworm Parasite (Fig. 9.8). Adults of the species *Rhabditis maupasi* are free-living soil nematodes, but earthworms play host to juveniles. When the earthworm dies the larvae mature by feeding on bacteria that grow in the decaying tissues. Larvae are believed to enter the earthworm through the nephridiopore or genital openings. If time permits, complete the following exercise.

1. Dissect an anesthetized earthworm. Make a longitudinal cut in the earthworm's muscular body wall and pin back its edges to the dissection pan.

2. Carefully remove the digestive tract from the posterior two-thirds of the body.

3. Remove the viscera anteriorly; cut it into small sections about 1 cm in length. Place these in separate petri dishes containing 3 percent sterile agar. Label the plates with the date and the approximate location of the worm section.

4. Cut similar sections from the muscular body wall and prepare similar agar cultures with these.

5. Set the cultures aside at room temperature. In about 3 days examine the cultures using a dissecting microscope and estimate the number of nematodes per section. Which part of the worm housed the greatest number of nematodes? Transfer some specimens to a microscope slide; add a drop of water, a coverslip, and observe under higher magnification.

Intestinal Roundworm (Figs. 9.9 to 9.12). From a morphological point of view, *Ascaris lumbricoides* is not a typical nematode. Nevertheless, it is important to examine more closely because ascariasis remains a public health problem in the United States. Recent estimates suggest a childhood prevalence of about 15 percent in certain regions (Darby and Westphal 1972). The size of *Ascaris* makes it suitable for dissection, and its reproductive system is often

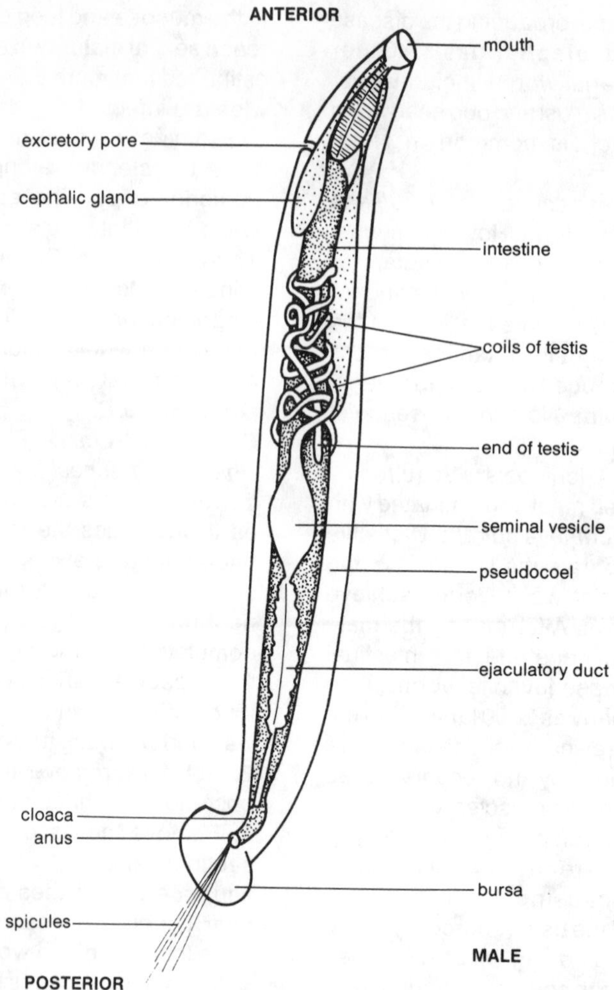

used to demonstrate mitosis and gametogenesis. Specimens used in this study are the pig strain or species *A. suum*. This species is similar morphologically to the strain found in humans. It differs only in the minute details of the lips and in aspects of its physiology.

Ascaris has a direct life cycle. When embryonated eggs are swallowed they hatch in the small intestine, where the young worms penetrate the wall and pass into the bloodstream. Upon entering the lung the worms break out into the alveoli, where they develop to the fourth stage. Then they are carried up the trachea to the pharnyx, where they are swallowed. The worms mature in the small intestine, where they mate and begin releasing eggs in the host's feces. Two related species, *Toxocara canis* and *Toxocara cati* (parasitizing dogs and cats, respectively), can cause a serious disease of humans known as **visceral larval migrans.**

General Morphology

1. Note the four thin longitudinal strips of lighter-pigmented cuticle along the length of the worm: two strips are lateral, one dorsal, and one ventral. The lateral stripes are broader and seen more easily than the other two. These are external mainfestations of internal structures, the hypodermal cords.

2. Examine the cuticular covering under a dissecting microscope and note that it appears to be finely segmented. This is not true segmentation, but superficial striations of the cuticle.

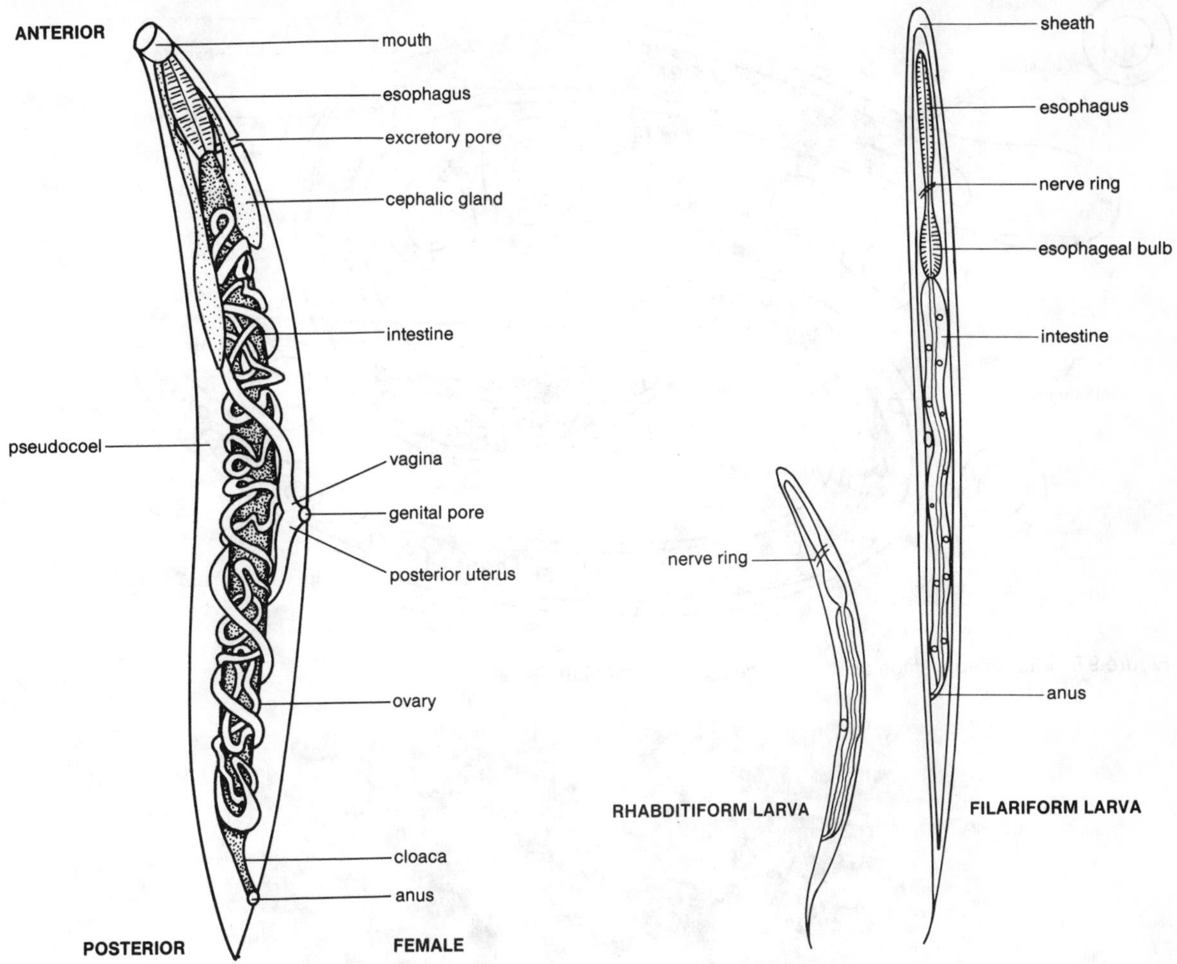

Figure 9.4. Male, female, and larval stages (rhabditiform, filariform) of *Necator americanus* (American hookworm).

3. The mouth is terminal and the anus is subterminal and ventral.

4. Observe the anterior end in a head-on position under the dissecting microscope to see the single dorsal lip and the two ventral lips surrounding the mouth. Note the somewhat dentate condition of the inner surface of the lips.

5. At the periphery of the dorsal lip are the two cephalic papillae, which house special sensory organs.

6. *Ascaris* is sexually dimorphic. In males the last 5 mm of the posterior end is curved ventrally. Do any of the specimens have a posterior hook? Occasionally, a pair of **penial spicules** (copulatory structures) can be seen protruding from the male's cloaca. Females are usually larger than the males.

7. Examine the cuticle surface of a female under the dissecting microscope and find her genital pore, which is located ventrally about one-third of the body length from her mouth.

Female Reproductive System (Figs. 9.9 and 9.10)

1. Place a female specimen ventral surface down in a dissection pan and orient the specimen and pan so that you may easily observe the worm through a dissecting microscope while dissecting it. Pin the anterior and posterior ends of the worm to the pan, cover the specimen with water, and dissect the worm as follows.

2. Make a shallow incision in the outer body wall the full length of the worm. Take care not to disturb the internal viscera. After an incision of some 6 or 8

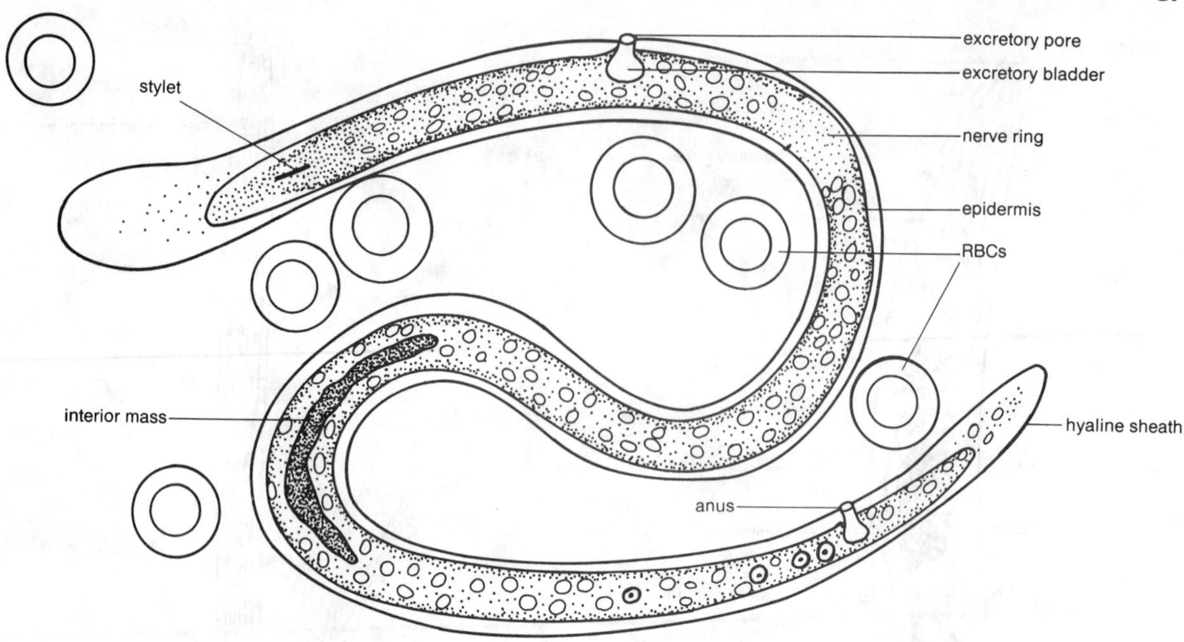

Figure 9.5. Microfilarial stage of *Wuchereria bancrofti* (human filarial worm).

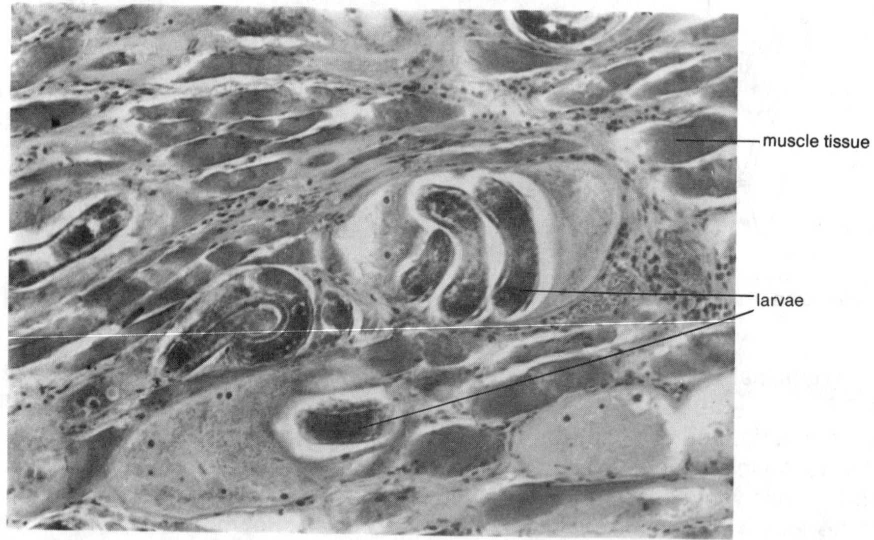

Figure 9.6. Photomicrograph of the juvenile stage of *Trichinella spiralis* (trichina worm) encysted in muscle tissue.

cm has been made, carefully pin the body wall to the pan. Continue the dissection and pinning until the entire length of the worm has been exposed (Fig. 9.9).

3. Observe the outer body wall just cut. It surrounds a central digestive tract and other organs located in the spacious pseudocoel.

4. .Note the highly convoluted pair of tubules wound about the intestine in the posterior part of the body cavity. The tube varies in dimension and constitutes the female reproductive system (Fig. 9.9).

5. Locate the most anterior part of this structure, which appears to penetrate the body wall; this is the

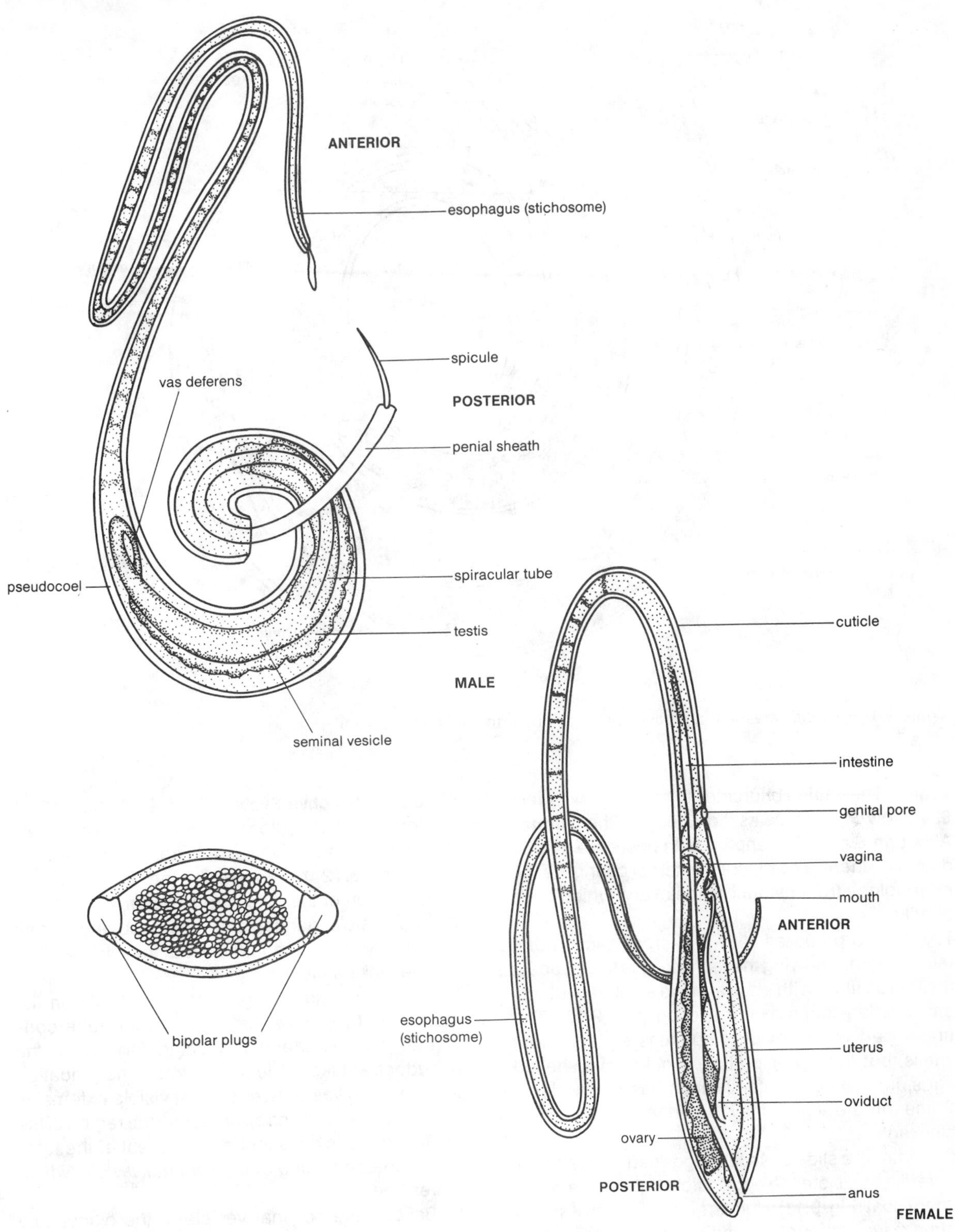

Figure 9.7. The human whipworm *Trichuris trichiura* (male, female, and bipolar eggs).

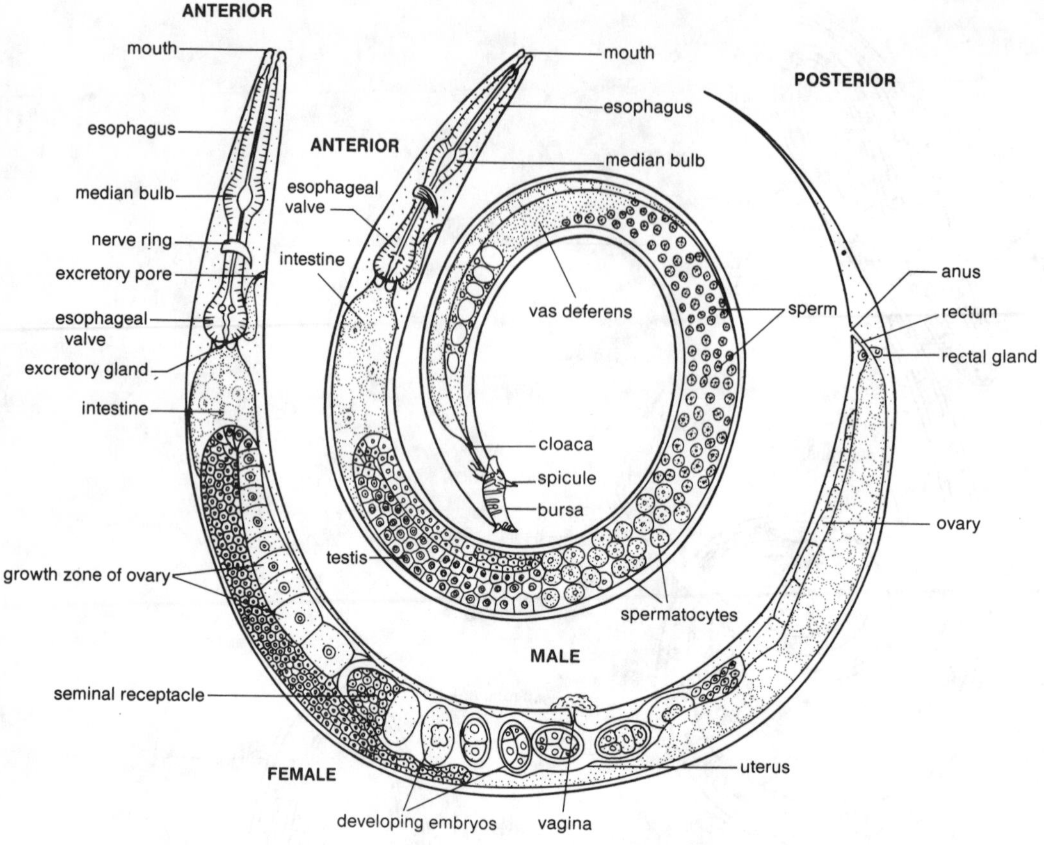

Figure 9.8. *Rhabditis maupasi*, a parasite of the earthworm (male and female).

vagina. The vagina bifurcates forming the two **uteri**.

6. Follow one uterus as it grades into the **oviduct**. Although there is an appreciable decrease in tube diameter, there is no obvious external change in the morphology from oviduct to the ovary, which ends blindly.

7. Obtain a prepared slide of a cross section of a female worm, showing the reproductive tract. Locate the uterus filled with eggs and the smaller oviducts and ovarian loops in the section (Fig. 9.10). The upper part of the oviduct contains eggs without shells, but the lower portion should have shelled eggs, indicating that fertilization has taken place. Some of the eggs may be in early stages of cleavage.

8. Examine a slide of *Ascaris* eggs (fertilized and/or unfertilized). Note the egg's thick, bumpy (mammillated) coat (Fig. 9.11). Humans infected with *Ascaris* often have only unfertilized eggs present in their stools. What simple fact would account for this situation?

Male Reproductive System (Figs. 9.12 and 9.13)

1. Complete a dorsolongitudinal incision of a male *Ascaris* in the same way as was done for the female. Use Fig. 9.12 to assist in the location and identification of the following structures.

2. The male's genital and anal openings open posteriorly into a region called the cloaca. The female lacks a cloaca.

3. Locate the single highly contorted tube in the male. The threadlike distal part of the tube constitutes the **testis** and is located forward in the pseudocoel. Like the female oviduct, the gradation from testis to **vas deferens** is not visible externally.

4. As the vas deferens approaches the region of the cloaca, there is a distinct enlargement of the tube. This is the beginning of the **seminal vesicle**, which stores sperm.

5. Beyond the seminal vesicle is the ejaculatory duct. It connects the reproductive tract with the cloaca.

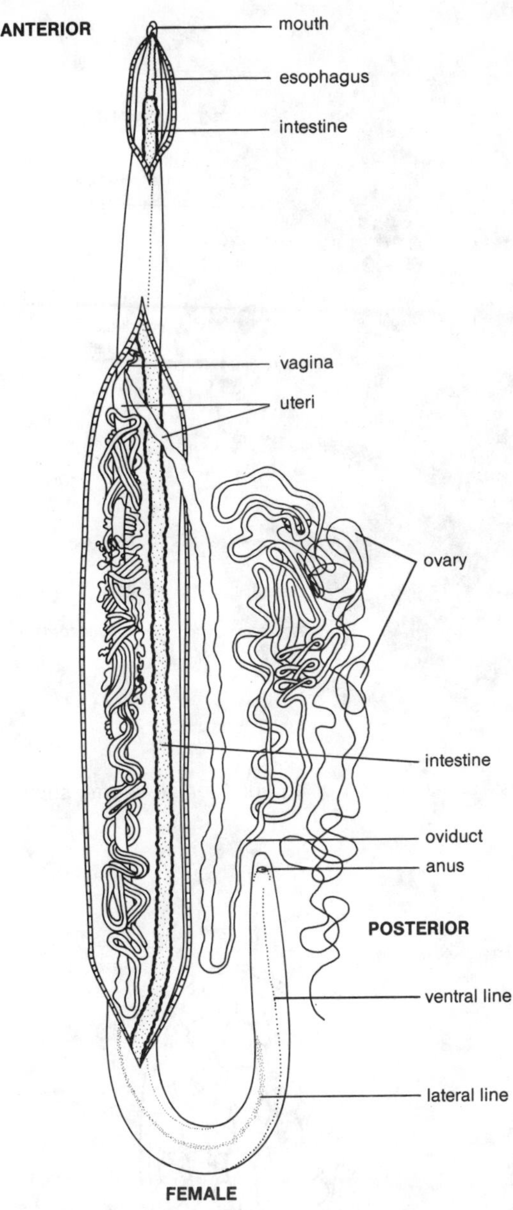

Figure 9.9. *Ascaris lumbricoides*: internal longitudinal view of an adult female.

6. Obtain a cross section of a male worm showing the reproductive tract (Fig. 9.13). Locate the testis, vas deferens, and seminal vesicle. The seminal vesicle is much larger in diameter than either the testis or vas deferens. It may be possible to find mature sperm cells in slide preparations that show portions of the ejaculatory duct and/or cloaca.

Digestive System and Other Internal Structures of *Ascaris* (Fig. 9.14)

1. In your dissected specimen, examine the digestive system: esophagus, intestine, rectum, cloaca (male), and anus. It may be necessary to move parts of the reproductive system to expose the digestive tract.

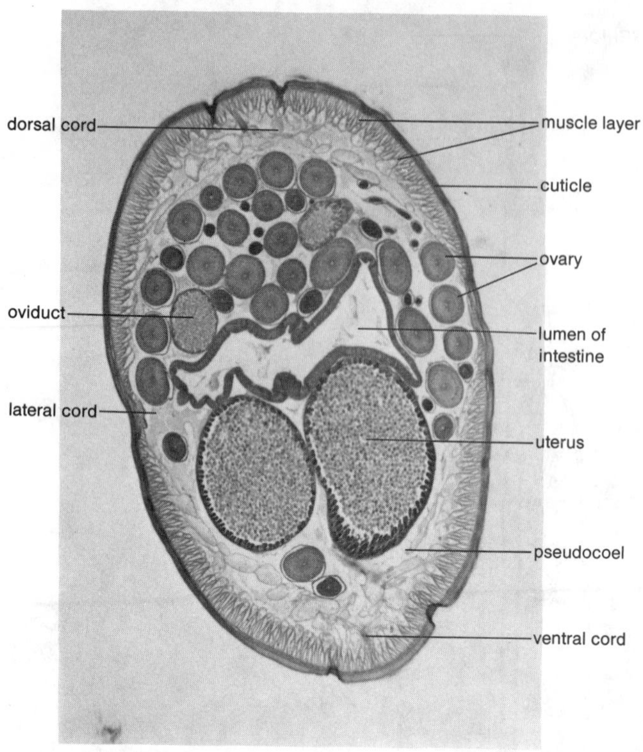

Figure 9.10. *Ascaris lumbricoides*: cross section of an adult female worm, showing parts of the reproductive and digestive systems.

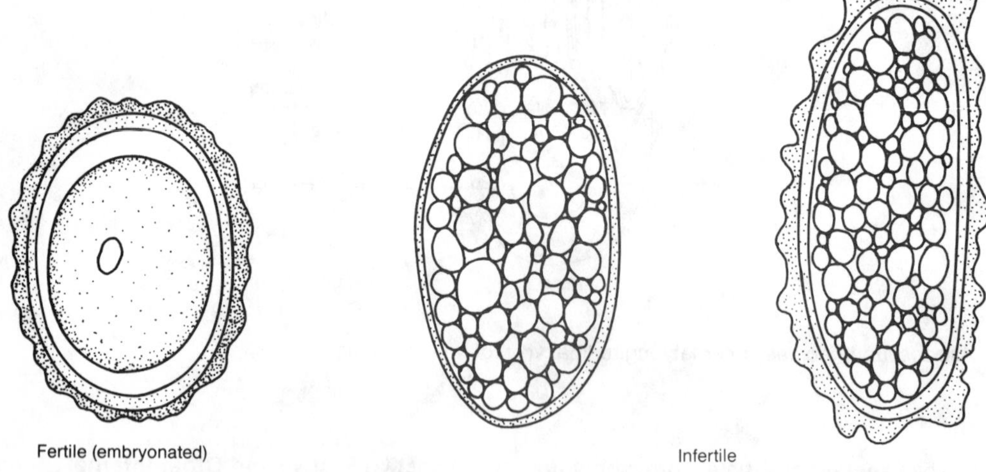

Figure 9.11. *Ascaris lumbricoides*: fertile and infertile eggs. Eggs are about 60 μm long and 40 μm wide.

2. Observe a prepared slide showing a cross section of the body of ***Ascaris*** made through the region of the esophagus (Fig. 9.14). Note the three powerful muscle units, which produce the sucking action when food is obtained from the host. Note the glandular bodies in the muscle tissue of the esophagus.

3. Strip a section of cuticle from the body and mount it on a slide with a drop of water; add a coverslip. Observe the superficial striation of the

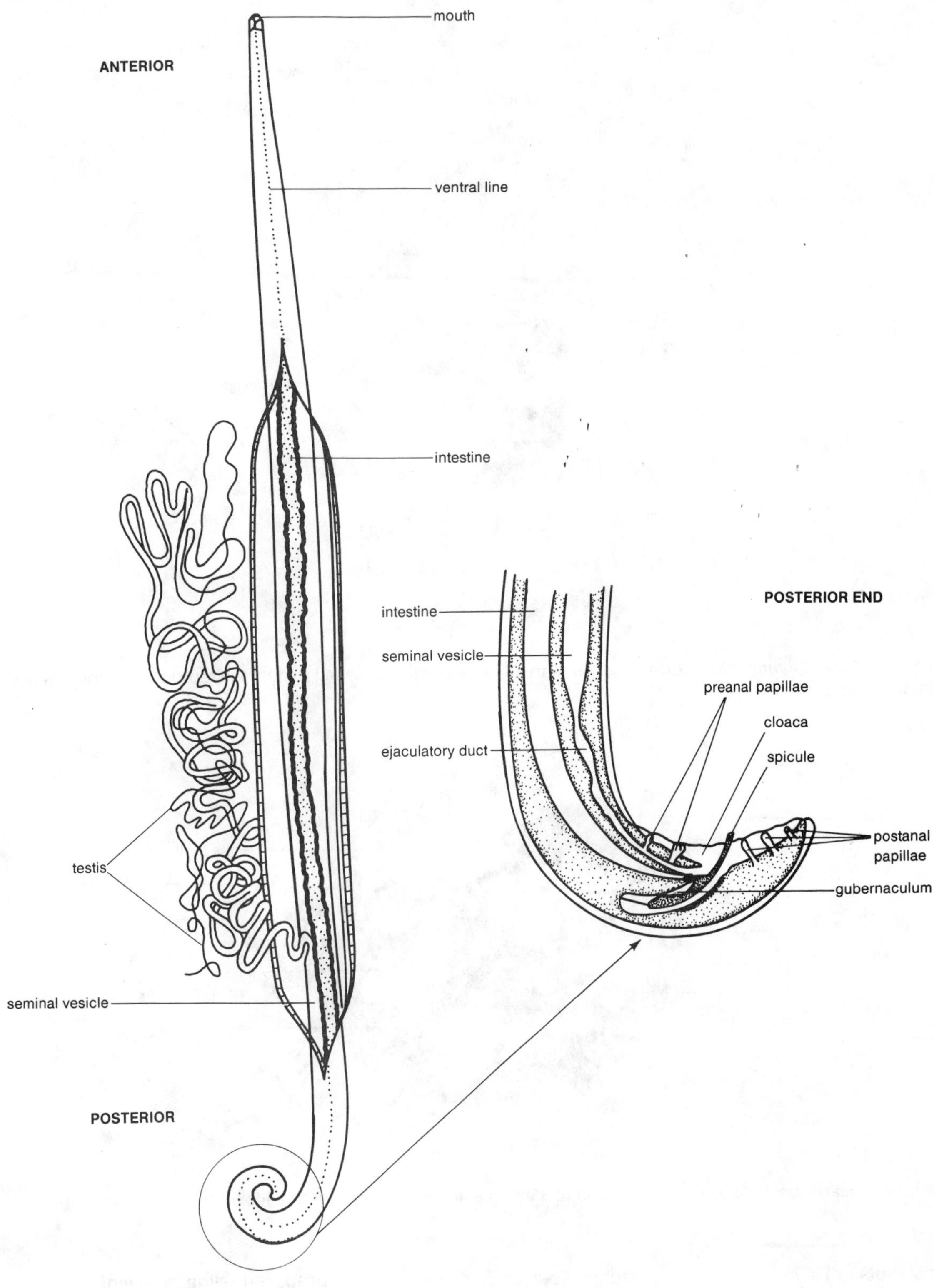

Figure 9.12. *Ascaris lumbricoides*: internal longitudinal view of an adult male.

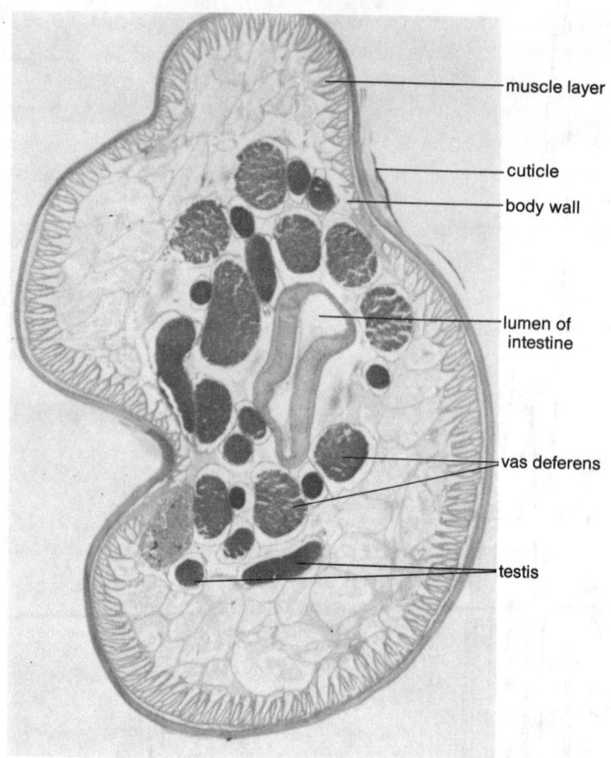

Figure 9.13. *Ascaris lumbricoides*: cross section of an adult male worm, showing parts of the reproductive and digestive systems.

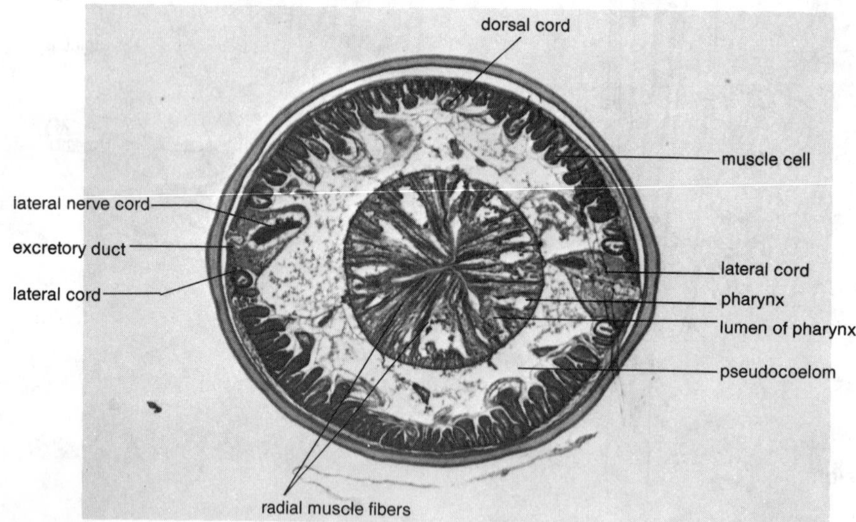

Figure 9.14. *Ascaris lumbricoides*: cross section of a worm at the level of the esophagus.

cortical (outside) layer and the crosshatched fibers of the fibrous layers. Estimate the angles the fibers make with the longitudinal axis of the worm.

4. Observe a prepared slide showing a cross section of the body wall. Use either sex. Observe the close association between the muscle cell bases and the **syncytial hypodermis**. The muscles at the inner body wall are in four quadrants, divided by the **four cords** (see below). Note that in the cross section being examined, the muscle cells have been cut transversely, which makes it somewhat difficult to visualize their proper arrangement. There are about 150 cells per quadrant and each cell is a very elongate body with a narrow, spindle form.

5. Locate the dorsal, ventral, and lateral cords. The dorsal and ventral cords contain the nerve trunks. Just inside the lateral cords you should be able to see the lateral excretory canals.

6. Internal to the muscle cells is the spacious pseudocoel.

7. Examine the intestine. The outermost perimeter of the intestine is called the basement membrane. Just inward from the basement membrane is a layer of columnar epithelial cells. These epithelial cells are elongate and bear distinct nuclei at their bases. On the side opposite the nuclei, a fuzzy coat may be seen on these cells. This coat is actually made up of many microvilli, which extend into the lumen of the intestine.

If you are to work with live ***Ascaris*** eggs, your instructor will give detailed instructions to avoid accidental contamination.

Supplemental Reading

Ambros, V., and H. R. Horvitz. 1984. Heterochronic mutants of the nematode *Caenorhabditis elegans*. Science 226:409-416.

Chitwood, M. B. 1970. Nematodes of medical significance found in market fish. Am. J. Trop. Med. Hyg. 19:599-602.

Crichton, V. F. J., and M. Beverley-Burton. 1975. Migration, growth, and morphogenesis of *Dracunculus insignis* (Nematoda: Dracunculidea). Can. J. Zool. 53:105-113.

Croll, N. A. (ed.). 1976. The organization of nematodes. Academic Press, New York.

Darby, C. P., and M. Westphal. 1972. The morbidity of human ascariasis. J. S. C. Med. Assoc. 68:104-108.

Frey, G. F., and J. G. Moore. 1969. *Enterobius vermicularis*: 10,000-year-old human infection. Science 166:1620.

Gould, S. D. (ed.). 1970. Trichinosis in man and animals. Charles C Thomas, Springfield, IL.

Hawking, F. 1975. Circadian and other rhythms of parasites. Adv. Parasitol. 13:123-182.

Kirchner, T. B., R. V. Anderson, and R. E. Ingham. 1980. Natural selection and the distribution of nematode sizes. Ecology 61(2):232-237.

Maggenti, A. 1981. General nematology. Springer-Verlag, New York.

Most, H. 1978. Current concepts in parasitology: trichinosis: preventable yet still with us. N. Engl. J. Med. 298:1178-1180.

Pramer, D. 1964. Nematode-trapping fungi. Science 144:382-388.

Rogers, W. P. 1960. The physiology of infective stages: the stimulus from the host. Proc. R. Soc. Lond. B 152:367-386.

Schad, G. A., and L. E. Rozeboom. 1976. Integrated control of helminths in human populations. Annu. Rev. Ecol. Syst. 7:393-420.

Sprent, J. F. A. 1952. Anatomical distinction between human and pig strains of *Ascaris*. Nature (Lond.) 170:627-628.

Sprent, J. F. A. 1969. Nematode *larva migrans*. N.Z. Vet. J. 17:39-48.

Tripathy, K., F. Gonzalez, H. Lotero, and O. Bolanos. 1971. Effects of *Ascaris* infections on human nutrition. Am. J. Trop. Med. Hyg. 22:212-218.

Webster, J. M. 1975. Aspects of the host-parasite relationship of plant parasitic nematodes. Adv. Parasitol. 13:225-250.

Zuckerman, B. M., W. F. Mai, and R. A. Rhode (eds.). 1971. Plant parasitic nematodes. Vol I. Morphology, anatomy, taxonomy, and ecology. Vol II. Cytogenetics, host-parasite interactions, and physiology. Academic Press, New York.

EXERCISE 10

Phylum Nematomorpha

Nematomorpha (NE-mat-o-MOR-fa; G., *nema*, thread + G., *morpha*, form) comprises about 230 species of threadlike pseudocoelomate worms that lack a functional gut. They range in length from 10 to 100 cm, but are only one to a few millimeters in diameter. As juveniles they parasitize arthropods, while the adults are free living. The common name, horsehair worms, comes from myths that said these arose from horse hairs that had fallen into water. The body wall of nematomorphs consists of a (1) thick cuticle often containing plates called **areoles**, (2) single-layered cellular **hypodermis**, and (3) muscle layer whose fibers are oriented longitudinally as in nematodes. Emergence of adults from the host that housed the juvenile stage takes place in or near water and mating begins soon thereafter.

Classification

Two classes are recognized, Nectonematoida (marine nematomorphs) and the freshwater Gordioida (most nematomorphs) based on external morphology and aspects of the cuticle.

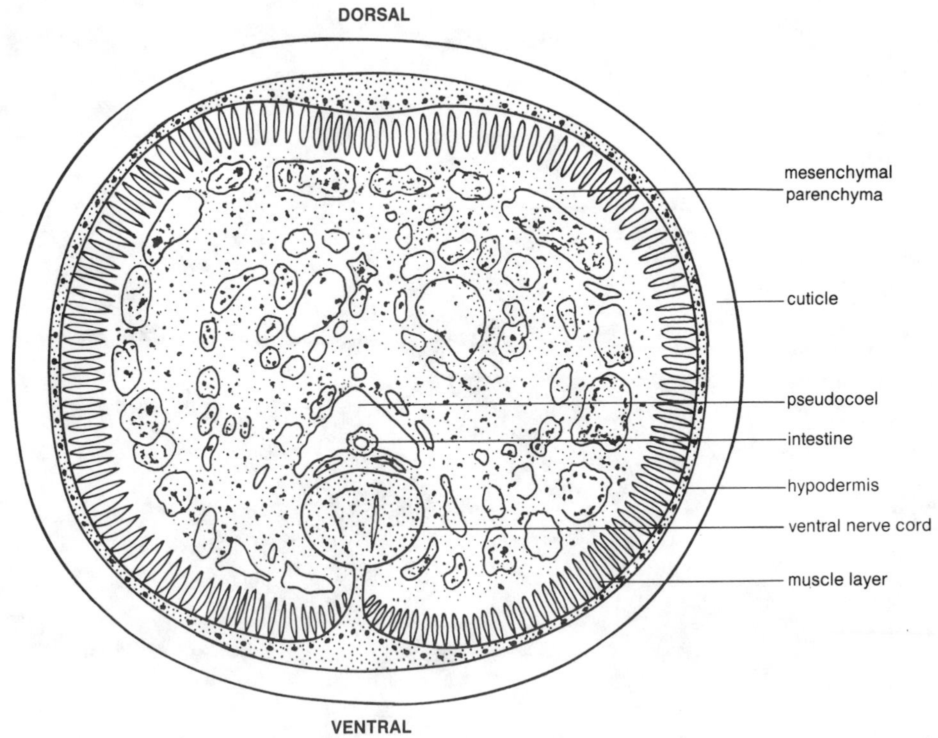

Figure 10.1. Cross section of a horsehair worm.

Observational Procedure

Examine a whole worm while keeping it immersed in a shallow dish of water. Determine the length and diameter of the specimen. Locate the anterior and posterior ends. The mouth will be located either terminally or subterminally. The head is called the **calotte** and usually has a darker area below it. The posterior end may be unlobed, bilobed, or trilobed, with a subterminal, ventral anus (cloacal aperature).

Observe a slide of a cross section of a horsehair worm and locate the structures indicated in Fig. 10.1.

Supplemental Reading

Arvy, L. 1963. Données sur le parasitisme protélien de *Nectonema* (Nématomorphe), chez les Crustacés. Ann. Parasitol. (Paris) 38:887-892.

Carvalho, J. C. M. 1942. Studies on some Gordiacea of North and South America. J. Parasitol. 28:213-222.

May, H. G. 1919. Contributions to the life histories of *Gordius robustus* Leidy and *Paragordius varius* (Leidy). Ill. Biol. Monogr. 5:1-118.

Zapotosky, J. E. 1974. Fine structure of the larval stage of *Paragordius various* (Leidy, 1851) (Gordiodea: Paragordidae). I. The preseptum. Proc. Helminthol. Soc. Wash. 41:209-221.

EXERCISE **11**

Phylum Acanthocephala

The endoparasitic worms comprising phylum Acanthocephala (a-kan-tho-SEF-a-la; G., *acantho*, spine or thorn + G., *cephala*, head) are distinguished by a protrusible **proboscis** covered with recurved spines (Fig. 11.1). The proboscis anchors adult acanthocephalans to the mucosal lining of their vertebrate host (Fig. 11.2). These worms range from about 1 mm to more than 1 m in length, depending on species; most are about 25 mm long. Like cestodes, acanthocephalans absorb nutrients through the porous body wall or **tegument**, a complex syncytial layer containing fluid-filled canals called **lacunae**. Because of their parasitic nature, acanthocephalans are of economic importance, as they affect domestic stock and game animals. Occasionally, there are reports of human infections.

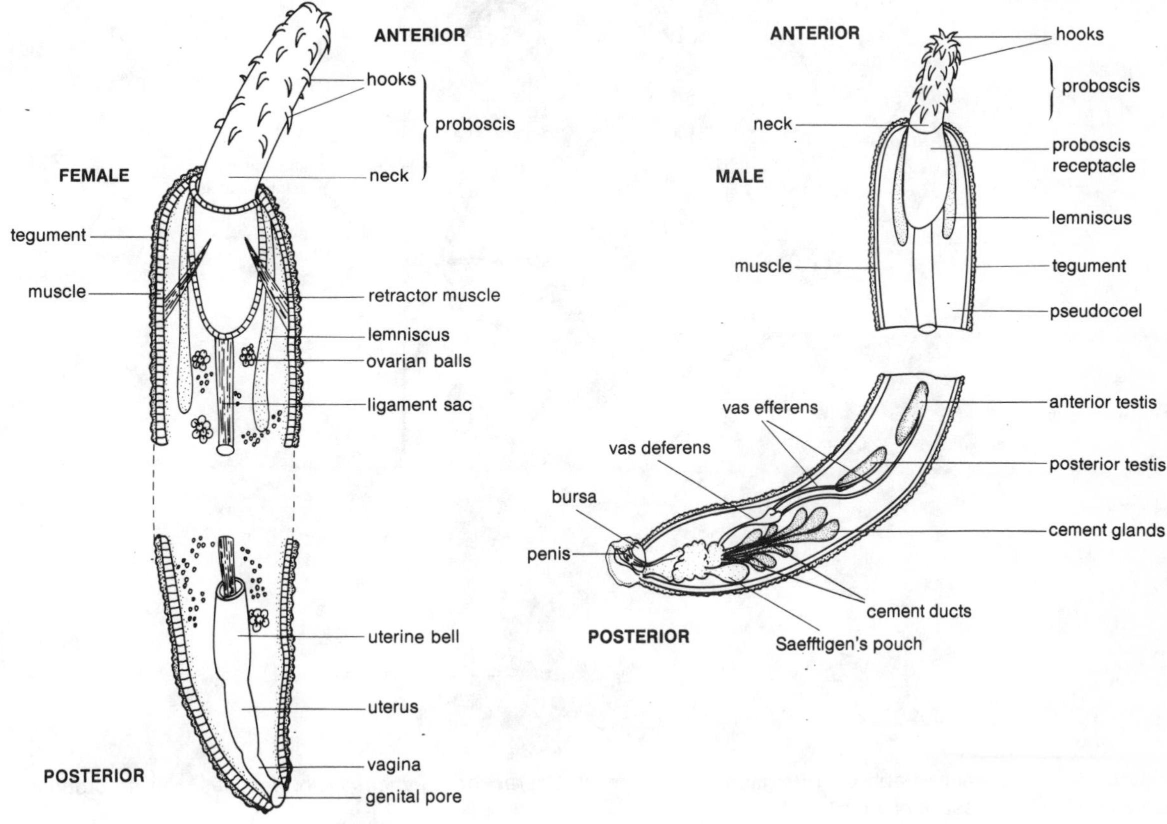

Figure 11.1. Schematic illustration of some external and internal features of male and female acanthocephalans.

Classification

Three classes are recognized in this small phylum of only about 700 species: class Archiacanthocephala (parasites of birds and mammals; insects and myriapods serve as intermediate hosts; examples include *Mediorhynchus*, a parasite of birds, *Moniliformis*, a parasite of rats, and *Macracanthorhynchus*, a parasite of pigs); class Palaeacanthocephala (parasites of all vertebrate classes; an example is *Plagiorhynchus*, a parasite of passerine birds); class Eoacanthocephala (parasites of fish and reptiles; an example is *Neoechinorhynchus*, a parasite of fish).

Observational Procedure: *Macracanthorhynchus*

General Morphology and the Proboscis. Observe a slide (whole mount) or preserved specimen of *Macracanthorhynchus hirudinaceus*, a parasite of pigs. Note the following: (1) recurved hooks on the proboscis, (2) neck, (3) genital region, and (4) trunk. Examine a cross or longitudinal section of *M. hirudinaceus* with the proboscis embedded in the intestine of a pig (Fig. 11.2). Are any spines visible in this section? Do you see any evidence of host tissue damage? How much tissue damage would you expect to find if many of these parasites were attached to the intestinal wall?

Internal Anatomy and Reproduction. Place a specimen in a dissecting pan and make an anterior-posterior incision from the region of the genital pore to the neck. Then turn the cut edges laterally and pin them to the base of the pan using fine insect pins. Using Fig. 11.1 as an aid, locate the following structures: *Male*—(1) lemnisci, (2) reproductive organs, and (3) pseudocoelom; *Female*—(1) proboscis receptacle, (2) lemnisci, (3) ligamentous sac (which posteriorly communicates with the uterine bell), (4) uterus and vagina, and (5) genital pore.

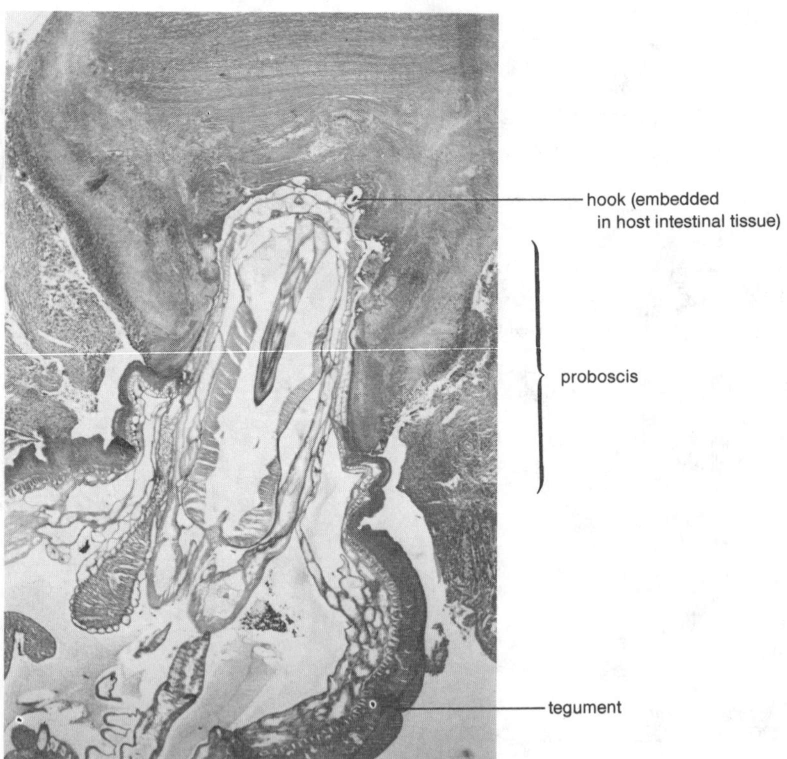

Figure 11.2. Photomicrograph of a longitudinal section of *Macracanthorhynchus hirudinaceus* with its proboscis embedded in the intestine of a pig.

Observe the muscular complex at the anterior part of the trunk. Cut away a narrow strip of the body wall and examine it under a dissecting microscope. It should appear porous. The spaces seen are the lacunar parts of the canal system in the body wall. The canal system also will be evident in prepared slides of cross and longitudinal sections of *M. hirudinaceus*.

In female specimens examine the ligament sac surface. If patches or clumps of cells are present, these probably will be egg clusters (**ovarian balls**). Examine the pseudocoel for eggs and/or bring pressure on the uterine bell to force out some eggs for collection. Place some of them on a slide with a drop of water, add a coverslip, and examine. Note the stage of development of several eggs. If available, observe the development of acanthocephalan eggs on a prepared slide (Fig. 11.3).

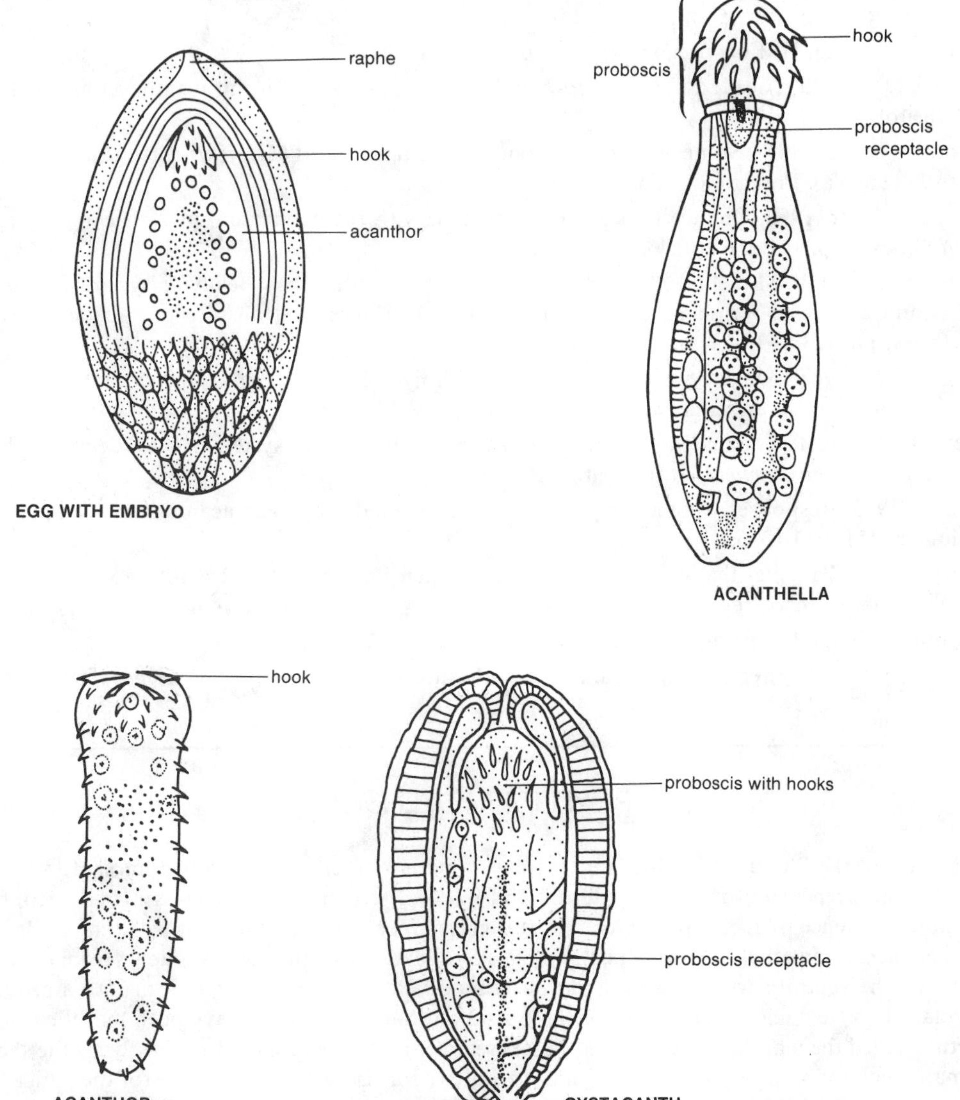

Figure 11.3. Larval stages of the acanthocephalan *Macracanthorhynchus hirudinaceus*. (After Kates from Olsen.)

Supplemental Reading

Bethel, W. M., and J. C. Holmes. 1977. Increased vulnerability of amphipods to predation owing to altered behavior by larval acanthocephalans. Can. J. Zool. 55:110-115.

Boyd, E. M. 1951. A survey of parasitism of the starling *Sturnus vulgaris* L. in North America. J. Parasitol. 37:56-84.

Brennan, B. M., and T. C. Cheng. 1975. Resistance of *Moniliformis dubius* to the defense reactions of the American cockroach, *Periplaneta americana*. J. Invertebr. Pathol. 26:65-73.

Bullock, W. L. 1963. Intestinal histology of some salmonid fishes with particular reference to the histopathology of acanthocephalan infections. J. Morphol. 112:23-44.

Byram, J. E., and F. M. Fisher, Jr. 1973. The absorptive surface of *Moniliformis dubius* (Acanthocephala). I. Fine structure. Tissue & Cell 5:553-579.

Byram, J. E., and F. M. Fisher, Jr. 1974. The absorptive surface of *Moniliformis dubius* (Acanthocephala). II. Functional aspects. Tissue & Cell 6:21-42.

DeGiusti, D. L. 1949. The life cycle of *Leptorhynchoides thecatus* (Linton), an acanthocephalan of fish. J. Parasitol. 35:437-460.

Edmonds, S. J. 1965. Some experiments on the nutrition of *Moniliformis dubius* Meyer (Acanthocephala). Parasitology 55:337-344.

Fisher, F. M., Jr. 1960. On Acanthocephala of turtles, with the description of *Neoechinorhynchus emyditoides* n. sp. J. Parasitol. 46:257-266.

Holmes, J. C., and W. M. Bethal. 1972. Modification of intermediate host behavior by parasites. *In*: E. U. Canning and C. A. Wright (eds.). Behavioural aspects of parasite transmission. Academic Press, New York. pp. 123-149.

Kates, K. C. 1943. Development of the swine thorn-headed worm, *Macracanthorhynchus hirudinaceus*, in its intermediate host. Am. J. Vet. Res. 4:173-181.

Miller, D. M., and T. T. Dunagan. 1978. Organization of the lacunar system in the Acanthocephala, *Oligacanthorhychus tortuosa*. J. Parasitol. 64:436-439.

Moore, J. 1983. Responses of an avian predator and its isopod prey to an acanthocephalan parasite. Ecology 64 (5):1000-1015.

Moore, J. 1984. Parasites that change the behavior of their host. Sci. Am. 250:108-115.

Whitfield, P. J. 1970. The egg-sorting function of the uterine bell of *Polymorphus minutes* (Acanthocephala). Parasitology 61:111-126.

Yamaguti, S. 1963. Systema Helminthum. Vol. 5. Acanthocephala. Wiley-Interscience, New York.

PHYLUM MOLLUSCA

Members of the Phylum Mollusca are **schizocoelomate**, nonsegmented **protostomes**. However, to some workers, repetition of certain external and internal structures in the class Monoplacophora constitutes evidence of segmentation or that this class arose from the line that led to more definite metamerism (i.e., annelid-arthropod line). Two distinct features set the mollusks apart from the other phyla. (1) The ventral surface has been variously modified into a muscular foot or into a group of muscular arms (tentacles). (2) Tissues of the dorsal and lateral surfaces have been modified into a structure called the **mantle**, which in most classes secretes an external shell. In some forms the shell is internal, much reduced, or absent. In the most primitive forms there is a tendency for the pericardial cavity, the metanephridia, and the gonads to be somewhat continuous. In more advanced groups these systems, especially the reproductive, become more or less separated and independent anatomical units. However, they all represent parts of the embryonic coelom, which originates in a schizocoelic manner. The schizocoelom does not function as a hydraulic skeleton.

EXERCISE **12**

Phylum Mollusca

About 50,000 described species of unsegmented schizocoelomate protostomes constitute phylum Mollusca (mo-LUS-ka; L, *molluscum,* soft). Mollusks range in size from less than 0.5 mm to nearly 15 m long and possess two distinct features that set them apart from other invertebrates. (1) Their organs are enclosed in a visceral mass and dorsal **mantle** (*pallium,* L., a mantle or cloaked) which usually secretes an external calcareous shell. (2) They possess a ventral muscular foot. In each class the foot is variously modified for locomotion and food procurement. Most mollusks have an open circulatory system, and in all but one class (Bivalvia) a long, chitinous, rasplike feeding device called the **radula** is present in the buccal cavity. The nervous system is moderately to highly cephalized, primitively with paired dorsal (visceral) and ventral (pedal) nerve cords. Well-developed excretory, respiratory, and reproductive systems are present. Mollusks date to the **Cambrian** (Fig. 12.1) and include about 45,000 fossil species. There is a rich diversity of species within most of the classes.

Classification

Seven classes of mollusks are recognized based on morphology of the foot, and presence or absence and type of shell.

1. **Class Monoplacophora,** with fewer than 20 species, is a relic class of small (less than 3 cm), nontorted, marine mollusks with a single, dorsal, caplike shell. Many fossil species are represented in this class. Of several extant genera, *Neopilina* is the best known.
2. **Class Polyplacophora** comprises mollusks that have a creeping foot and a shell divided into eight plates. Examples include *Katharina* and *Cryptochiton.*
3. **Class Aplacophora** are small (less than 1 cm to 5 cm), wormlike, marine mollusks, with calcareous spicules in the mantle. These shell-less animals crawl on, or burrow in sediments, or are epizoic on

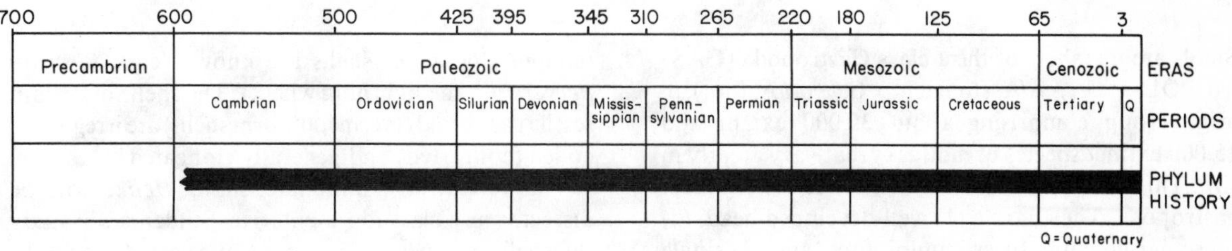

Figure 12.1. Geologic history of the phylum Mollusca.

cnidarians in deep marine waters. They are commonly called **solenogasters** (so-LEN-o-GAS-ters; G., *solen*, channel + G., *gaster*, stomach) because some species have a ventral longitudinal groove that runs nearly the entire length of the body. Recently a new class, **Caudofoveata** (KAW-do-fo-ve-AT-a; L., *cauda*, tail + L., *fovea*, pit), has been erected comprising animals formerly assigned to class Aplacophora.

4. **Class Gastropoda** is a class of snail-like mollusks possessing a well-defined head with tentacles and relatively simple eyes. Most possess a single-coiled shell. The foot is adapted for crawling, but some species can swim. Three subclasses are recognized.
   a. **Subclass Prosobranchia**: shelled marine gastropods with the mantle cavity anterior; examples include *Busycon* (whelks), *Conus* (cones), *Haliotis* (abalone), *Murex* (drills), and *Strombus* (conchs).
   b. **Subclass Opisthobranchia**: mostly marine gastropods with much reduced shell or shell lacking; mantle cavity lacking or is on the right side, or may be partly posterior; examples include *Aplysia* (sea hare) and *Aeolidia* (nudibranch).
   c. **Subclass Pulmonata**: mostly terrestrial shelled gastropods; mantle cavity modified as a lung; examples include *Lymnaea* (freshwater snail) and *Limax* (land slug).
5. **Class Bivalvia** (= Pelecypoda) includes mollusks having two shells (valves) that are hinged dorsally. The head is absent. Large curtainlike gills found in the mantle cavity are used in filter feeding. Six subclasses of freshwater and marine species are recognized; examples include *Cardium* (cockle), *Mytilus* (common mussel), *Unio* and *Anodonta* (freshwater clam and mussel, respectively), *Ostrea* (oyster), *Pecten* (scallop), *Mya* (soft-shelled clam), and *Mercenaria* (Venus clam or quahog).
6. **Class Scaphopoda** comprises some 350 species of burrowing marine mollusks with a single, dorsoventrally elongate, slightly curved, tubular shell (2 to 6 cm long) which is open at both ends. The shell resembles an elephant's tusk, so they are known commonly as tusk or tooth shells. A modified foot and set of delicate tentacles extend from the ventral orifice of the shell. An example is *Dentalium*.
7. **Class Cephalopoda** comprises elongated, free-swimming marine mollusks that are highly cephalized, possessing a well-developed nervous system and a foot modified into several highly developed tentacles. Shell external, internal, or absent. Three subclasses are recognized.
   a. **Subclass Nautiloidea**: multichambered, external shell; many arms; mostly extinct species, but with one extant genus, *Nautilus*.
   b. **Subclass Ammonoidea**: coiled, multichambered, external shells; all extinct.
   c. **Subclass Coleoidea**: internal shell or shell lacking; four orders with extant species and one wholly extinct order; well-known genera include *Loligo* (squid), *Octopus* (octopus), and *Sepia* (cuttlefish).

Of the seven molluscan classes, two (Aplacophora and Monoplacophora) will not be covered in this manual, owing to the lack of available commercial suppliers of specimens. Shells of representative species of the class Scaphopoda are discussed briefly at the end of Exercise 12.

A. Class Gastropoda

Snails are members of the a class Gastropoda (GAS-tro-POD-a; G., *gastro* stomach + G., *podos*, foot), a large group comprising about 35,000 extant and 15,000 extinct species of mollusks that vary greatly in size. Three important characteristics are seen in most gastropods. They have a (1) well-developed head, (2) muscular creeping or swimming foot, and (3) single asymmetrical shell of calcium carbonate and organic matter. The shell may be absent. Snail shells are usually coiled to the right (clockwise), but left-handed (counterclockwise) shells are known (e.g., *Busycon perversum*, the lightning whelk). The shells of certain sessile gastropods (vermetid worm shells) are irregularly coiled (contorted) and generally elongated.

The terrestrial pulmonate snail, *Helix*, will be dissected as a class representative. Students will study the shells and whole specimens of selected species to gain appreciation of the diversity found in Gastropoda.

Observational Procedure: Anatomy of *Helix*

1. Obtain a freshly killed or expanded, preserved specimen of *Helix* and examine the shell, locating the following major features: apex, spire, suture line, and aperture.

2. Identify the head and foot of the animal (Fig. 12.2). The head bears two pairs of tentacles, the anterior ones being shorter than the posterior. Both pairs can be retracted. The posterior tentacles bear an eye at their distal tip. A small opening to the **common genital duct** may be found below and slightly behind the base of the right posterior tentacle. Locate the mouth with its two lateral and one ventral lips. On the right side of the body just below the shell, locate the **pneumostome**, the opening into the mantle cavity that functions as a lung (Fig. 12.2).

3. The shell of *Helix* is thin compared to that of many gastropods. Does *Helix* possess an **operculum**? Carefully break the shell away from the animal. It is not necessary to remove all the shell material; the central portion of the shell (**columella**) may be left intact.

4. Note that the mantle is thicker along the anterior region just behind the head. This is the **collar**, which secretes shell material at the lip of the shell aperture. Carefully make a shallow cut into the lung through the pneumostome. Continue to cut the mantle cavity until the entire lung surface is exposed. Reflect the roof of the mantle cavity to one side to expose the visceral hump. Note that the floor of the mantle cavity is so thin that the internal organs may be seen underneath.

5. Carefully cut into the **pericardial cavity** and expose the heart, identifying the thin-walled auricle and the thicker-walled ventricle. The kidney is located in the roof of the mantle cavity posterior to the heart.

6. Pin your specimen to the dissection tray and immerse it in water. Cut through the floor of the mantle cavity, being careful not to damage the internal organs. Continue this cut anteriorly through the collar, extending it middorsally to the front of the head (Fig. 12.3). With the aid of Fig. 12.3, locate the following structures.

7. Identify the muscular buccal mass at the anterior end and posterior to it, the nerve ring. A slender esophagus passes from the buccal mass through the ring into the expanded crop region of the gut.

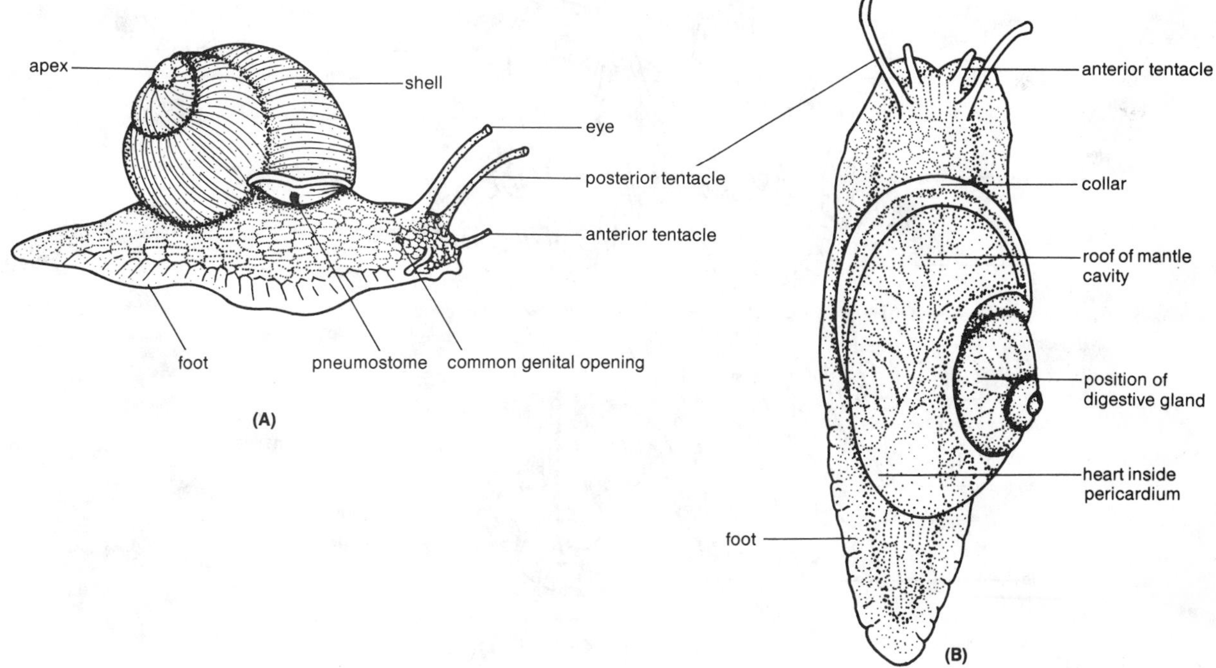

Figure 12.2. **(A)** Lateral (with shell) and **(B)** dorsal (without shell) view of *Helix, a pulmonate gastropod.*

Right and left salivary glands will be found on the lateral surfaces of the crop. A duct from each gland extends anteriorly through the nerve ring to the buccal mass. The crop leads into the large stomach, which is located in the second turn of the visceral hump. Left and right digestive glands can be observed. The left gland is larger and lies alongside the stomach; the smaller right gland occupies the apex of the visceral hump and last whorls of the shell. The intestine leads from the stomach and forms a S-shaped curve in the stomach region. *Helix* has a complex reproductive system. If you are to study it, proceed through steps 8 to 10 with care.

8. *Helix* is hermaphroditic. Carefully tease apart the reproductive organs from the gut (Fig. 12.3) and, if possible, locate the protandrous gonad (**ovotestis**) on the inner coiled face of the right digestive gland. Trace the thin, convoluted duct (**hermaphroditic duct**) that leads from the gonad to the base of an **albumen gland** near the stomach region. Where the duct enters the albumen gland is a small fertilization chamber (or **talon**); here ova are fertilized prior to receiving albumen.

9. Locate the broad **spermoviduct**, which has separate channels for both male and female reproductive products. It leads anteriorly from the albumen gland to a region where the duct divides into the sperm duct (**vas deferens**) and **oviduct**. The sperm duct leads to the **penis sac** and protrusible penis. Near the junction of sperm duct and penis sac is a long, thin, blind tube called the **flagellum**.

10. The oviduct leads to the vagina and a pair of branched mucus glands enter at about the same point. Nearby, a large bulbous **dart sac** enters too. The muscular dart sac secretes a calcareous needlelike dart that is plunged into the side of the partner before copulation. Remove the dart sac by cutting its connection to the vagina and observe its muscular structure. A long spermathecal duct joins the vagina at its posterior end. This duct ends in a spherical spermatheca that stores sperm received from another snail during copulation.

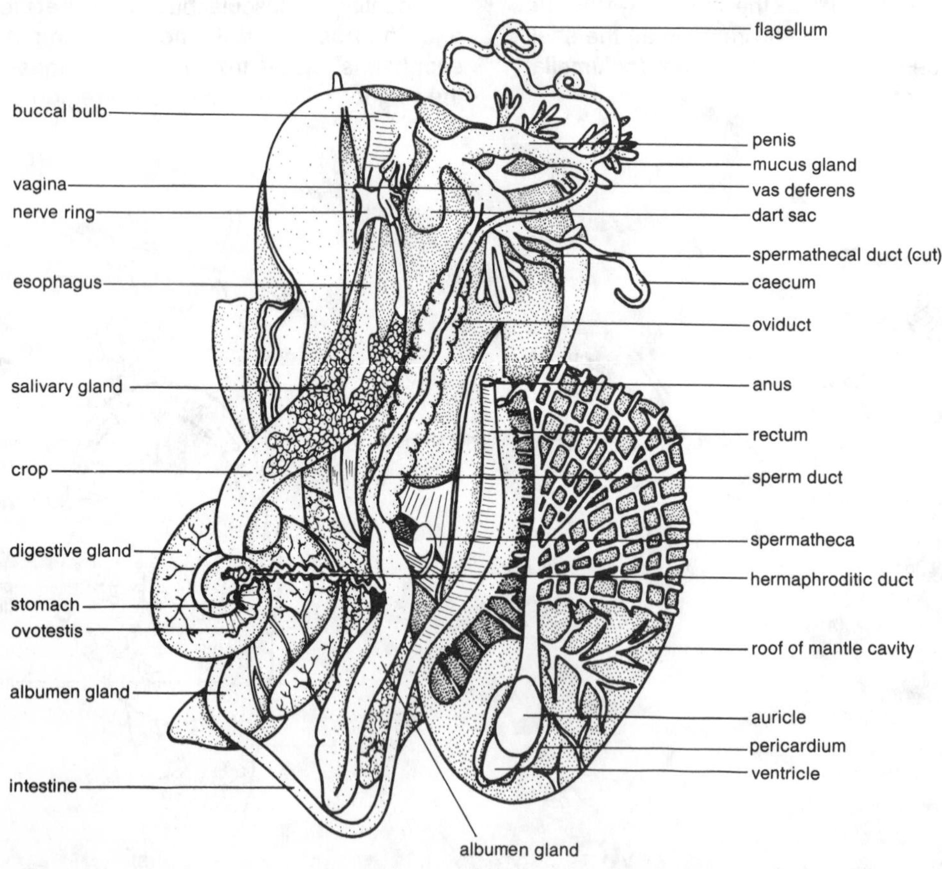

Figure 12.3. Internal anatomy of *Helix*. (After several sources.)

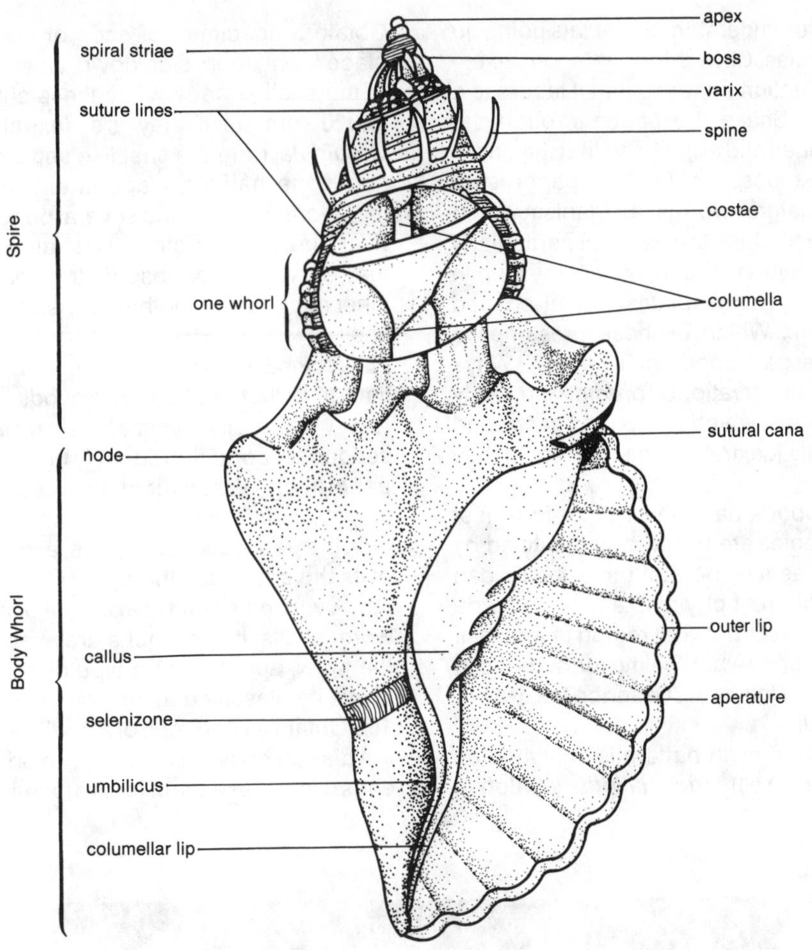

Figure 12.4. Morphology of a generalized gastropod shell. (After several sources.)

11. After you have completed your dissection, remove the buccal mass, make a longitudinal incision down its length, and with the aid of a dissecting microscope, locate the radula. Remove it from the buccal mass and examine the teeth under a microscope.

Diversity of Gastropods. There is a rich diversity of shelled forms in gastropods.

1. Set several different shells on the bench with their apices pointing away from you. What common features do these shells share (Fig. 12.4)? Are the spires of similar shape? Compare the relative sizes of the body whorls. Are there differences? Note the differences in shell apertures, size, and shape.

2. On which side of the shell is the aperture found? Is this consistent in all the specimens present? Shells with the aperture on the right side are right-handed or **dextral** (L., *dexter*, right). Those with the aperture on the left side are known as left-handed or **sinistral** (L., *sinister*, left). When viewed from the apex, dextral shells coil (grow) clockwise and sinistral shells coil counterclockwise.

3. Observe that the outer shell surface may be ornamented with **lamellae**, **striations**, **tubercles**, **spines**, and **ribs**, or may be very smooth (Figs. 12.4 and 12.5).What might be the adaptive significance of these surface features? What types of colors and color patterns are found on the outer shell surface? Are these colors found through the shell? How might these color patterns be produced?

4. Examine a shell that has been cut or broken in half longitudinally. The central column is called the columella. Is the shell divided into compartments by the columella? In life, how is the snail positioned in the shell?

5. To gain a better understanding of this point, do the following exercise. Cut a thin strip (about 20 cm by 1 cm) of construction paper so that it tapers to a point at one end. Snake the paper through the whorls of the cut shell and adjust it so that the paper achieves the best possible fit. The paper strip simulates how a bilateral, wormlike organism would fit into a helical shell. Does the paper fit particularly well? Repeat this demonstration, but cut the strip in the shape of a spiral. This simulates a spiral-shaped wormlike organism. Which of these presumptive organisms fits the gastropod shell better? Why? Based on these observations, predict the body shape of hermit crabs, which inhabit old gastropod shells. Which shells (dextral or sinistral) would they fit into best?

6. Not all gastropods have the typical form just studied. Many species are mistakenly identified by novice collectors as members of other molluscan classes or even different phyla. Examine a cowrie shell. Is the spire present? What region of the shell represents the last body whorl? Note the superficial bilateral symmetry of this shell. Describe the growth pattern of this shell.

7. A very different growth pattern is exhibited by *Crepidula* (slipper shell) and *Haliotis* (abalone). Obtain a specimen of each of these species and place them open-side down. Note the growth lines in the shell surface. View these shells from above and determine if they are dextral or sinistral. In *Crepidula*, note the shelflike **septum** that occupies about one-half of the aperture. There is no septum in *Haliotis*, but note the several perforations that run near the outer margin of the shell. What function do they serve? Also, observe the pearly luster of the inner nacreous or mother-of-pearl layer of a *Haliotis* shell.

8. Compare the shell form of *Diodora* (keyhole limpet) to that of other gastropods. Limpets appear to be bilaterally symmetrical; however, this is a secondary condition. Identify the shell aperture and perforation. What function does the perforation serve? Some limpets (e.g., *Acmaea*, true limpet) lack the perforations. In these, excurrent water flow exits the shell from the right side.

9. A very different type of shell occurs in vermetid worm shells. If specimens are available, determine the apex, aperture, and type of coiling. Can these shells be classified as dextral or sinistral?

10. Obtain a specimen of a nudibranch and place it in a dissection pan. Determine anterior and posterior ends. Note that except for the location of the genital

Figure 12.5 Diversity of shell form in class Gastropoda [including from left to right, top row: *Mitra* (papal miter), *Conus* (cone), *Vermicularia*, *Melongena*, *Busycon*; middle row: *Haliotis* (abalone), *Murex*, limpet; bottom row: *Polinices* (moon snail), *Architectonica*, sundial, *Clanculus* (strawberry top shell), cowrie, keyhole limpets].

aperture, the animal is bilaterally symmetrical, a secondary condition that probably arose with loss of the shell. Some nudibranchs have many fingerlike projections from their dorsal surfaces, called **cerata**. The cerata apparently function in gaseous exchange by increasing the surface area. Nudibranchs that lack cerata often have other secondary outgrowths that function as gills. Living nudibranchs are often brillantly colored. This condition is generally considered to be an example of warning coloration to potential predators. As a defense against predators many species produce obnoxious chemicals or use nematocysts obtained from their cnidarian prey.

11. Examine the sea hare, *Aplysia*. Unless instructed otherwise, only the external features will be examined. Determine the anterior and posterior ends of *Aplysia*. Note the large fleshy foot and winglike **parapodia** used in swimming. At the head, locate the mouth, anterior tentacles, eyespots, and **rhinophores** (olfactory tentacles). On the right side of the head between the right tentacle and rhinophore, locate the small penis. Extending posteriorly and dorsally from the penis is the ciliated **sperm groove** that ends at the gonopore. At the posterior end, locate the **anal siphon** and anus. Dorsally, find the opening in the mantle (shell aperture) under which the small, thin shell is located. Does the shell offer much protection? *Aplysia* secretes a purple dye when handled and this is said to help protect it from predators.

12. Observe other gastropods (e.g., *Busycon* and *Limax*) that are available for study. Relate the morphology and anatomy of these specimens to the ones you have just studied.

Gastropod Larvae. Examine a prepared slide of a gastropod **veliger larva** from *Crepidula*. Although the internal anatomy of most specimens will be difficult to interpret, attempt to locate the following structures: the double-lobed, ciliated **velum**, visceral mass, foot, operculum, and shell (Fig. 12.6).

The genus *Busycon* (whelks) produces young that complete their larval stages within the egg case. If available, examine a string of egg capsules of a whelk. Some whelks may produce strings of more than 175 parchment-like capsules with 100 to 200 eggs per capsule. How many capsules are present on your string? Note how the capsules are attached to the central cord. Are all the capsules the

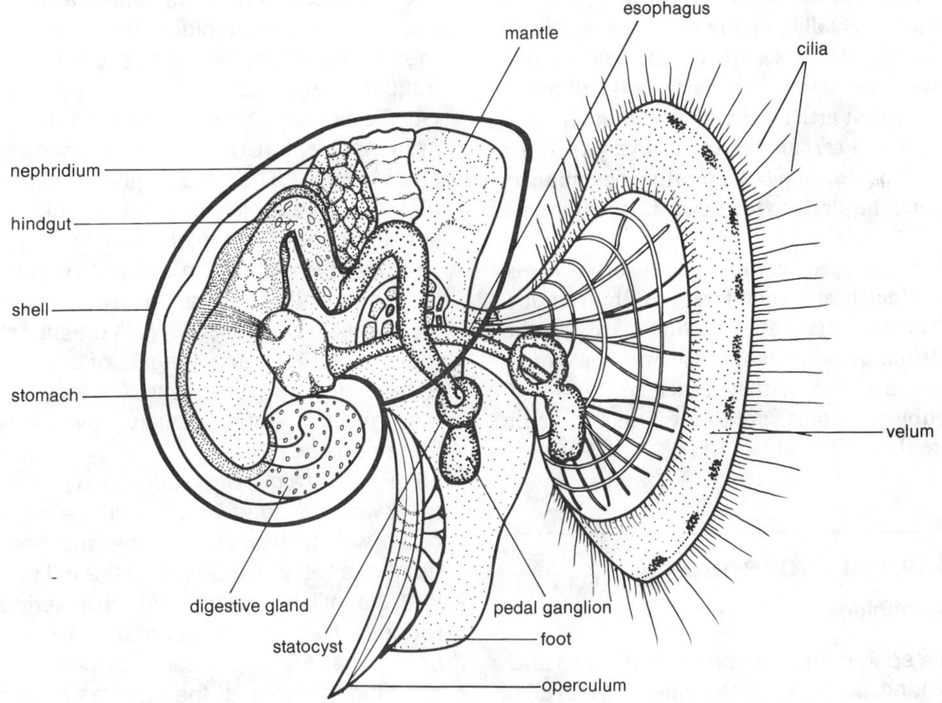

Figure 12.6. Typical veliger larvae of a gastropod. (After several sources.)

same size? The last several capsules of the string are often smaller than the rest and contain no eggs. Examine closely several capsules and determine how young snails emerge from the capsule. If permitted, remove one or more capsules from a string and dissect them. How many young snails are present in each capsule? How big are they? In which direction do the shells of these young snails coil?

Living Gastropods. If live gastropods are available, make the following observations. Observe locomotion in a living snail from the side and through a glass plate from below. How fast does the animal move? How does the foot appear to move the animal? Describe the movement of the foot when viewed from below. What roles do light and gravity play in movement?

Observe feeding in a living snail as it scrapes food from a glass surface. Note how the radula is applied to the glass. Is all the food removed by the animal? What marks are left by the radula after the animal passes by?

If living pulmonate specimens are available, observe functioning of the pneumostome. Does it remain open, or does it periodically open and close? What is the significance of this cycle?

B. Class Polyplacophora

Commonly known as **chitons**, members of the class Polyplacophora (POL-e-pla-KOF-o-ra; G., *poly*, many + G., *placo* plate + G., *phora*, carry) are a small group (about 500 species) of dorsoventrally flattened marine mollusks, usually ranging in size from 1 to 10 cm long. The ventral foot is a broad, flat, creeping sole and the shell is divided into eight partially overlapping plates or valves that articulate with one another, thus permiting chitons a certain degree of flexibility along the anterioposterior axis. Laterally, the mantle extends beyond the plates to form a girdle surrounding the animal.

Chitons have reduced nervous and sensory systems with an indistinct head. They lack eyes, but unique photosensitive structures called **aesthetes** are present dorsally. Chitons are abundant in the intertidal zones along rocky coasts, where they attach to rocks and other hard objects. Some species are found at great depths (more than 5000 m).

Observational Procedure

External Morphology

1. Obtain a specimen of the chiton *Katherina* and observe its general shape and symmetry (Fig. 12.7). Identify the anterior and posterior ends.

2. Place the specimen dorsal side up. Note that the shell consists of eight similar plates that overlap each other anteriorly. Portions of the plates are embedded in the fleshy mantle. The lateral, thickened part of the mantle is known as the **girdle**.

3. Observe the dorsal surface using a dissecting microscope. In young individuals, aesthetes may be seen in the plates. In older specimens of *Katherina*, these and other markings on the plates are frequently eroded.

4. Place your specimen ventral side up. The most prominent feature of the ventral surface is the broad muscular foot. The head region, anterior to the foot (Fig. 12.7), bears the mouth but lacks any special sense organs (e.g., tentacles and eyes).

5. Note the distinct groove between the foot and the girdle. This is the mantle cavity or **pallial groove**. It contains a single row of gills on either side of the foot (Fig. 12.7). Trace the gills of one side from the anterior to the posterior end. Are they the same size throughout? Approximately how many gills are there per side?

6. Remove a gill and place it in a depression slide filled with a drop of water. Hold the base of the gill down with a dissection probe and gently run the point of another probe along the gill surface. Note that the gill is constructed of a series of small, rounded filaments positioned like the pages of a book. What is the significance of the serial filaments?

7. Find the anus at the extreme posterior end of the pallial groove. It is located just behind the foot. With a dissecting microscope, locate the paired **nephridiopores** and **gonopores** that empty into the pallial groove on either side of the foot. The

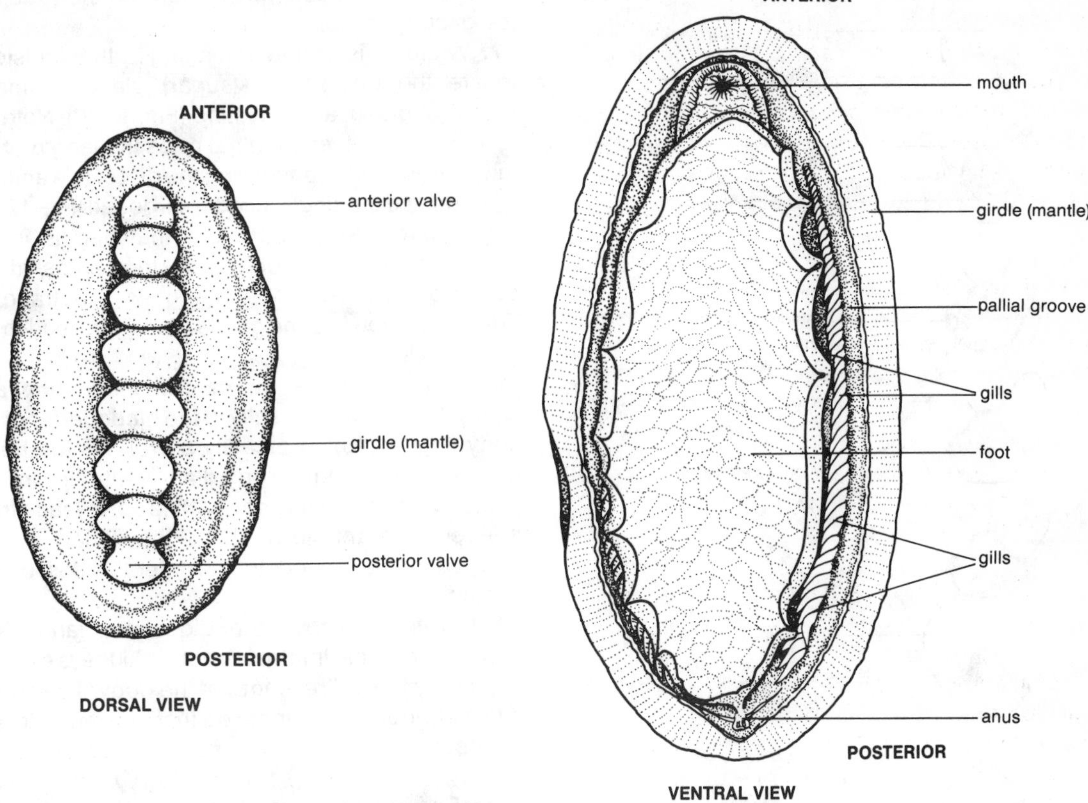

Figure 12.7. Dorsal and ventral views of *Katherina*.

nephridiopores are just anterior to the anus. Locate these pores by scanning the region between the gills and the foot, folding the gills aside with a probe as you proceed anteriorly with the search. In chitons other than *Katherina*, the nephridiopores are found on either side of the anus. The gonopore is located near and slightly anterior to each nephridiopore.

Internal Anatomy

1. The plates and internal anatomy of a chiton may be examined. Begin by carefully cutting away the girdle. Do not cut into the mantle between the plates. Cut around the plates laterally until they are nearly exposed. With a sharp scalpel, carefully remove the anterior plate by freeing it from its muscular and connective tissue attachments. When a plate is loose, pry it up using a blunt probe. Working posteriorly, remove the remaining seven plates, placing each aside in order.

2. Examine the plates in their proper order (Fig. 12.8). The medial, pigmented layer of the plate is known as the **tegumentum**, and the lateral, non-pigmented layer is called the **articulamentum**. The latter is inserted into the muscular mantle of the chiton. Note that the middle six plates are nearly identical in size and shape, but that the first (head) and last (tail) plates differ.

3. Examine the dorsal surface of the body and observe the thin mantle (Fig. 12.9). A pair of longitudinal muscles are located at midline dorsally. What is their function? Locate the oblique and transverse muscles. What are their functions?

4. Pin the specimen to the dissection pan through the anterior and posterior ends and carefully remove the dorsal mantle without damaging the viscera beneath. Submerge the specimen in water and observe it under the dissecting microscope.

5. Note the large bundles of muscles near the buccal bulb (Fig. 12.10). They attach to and operate the **odontophore** and **radula**. The salivary glands are located lateral to the muscular bands. Their ducts enter the buccal cavity posteroventrally.

6. Make a short incision through the dorsal surface of the buccal cavity, extending the cut laterally so as not to cut the odontophore muscles. Expose the

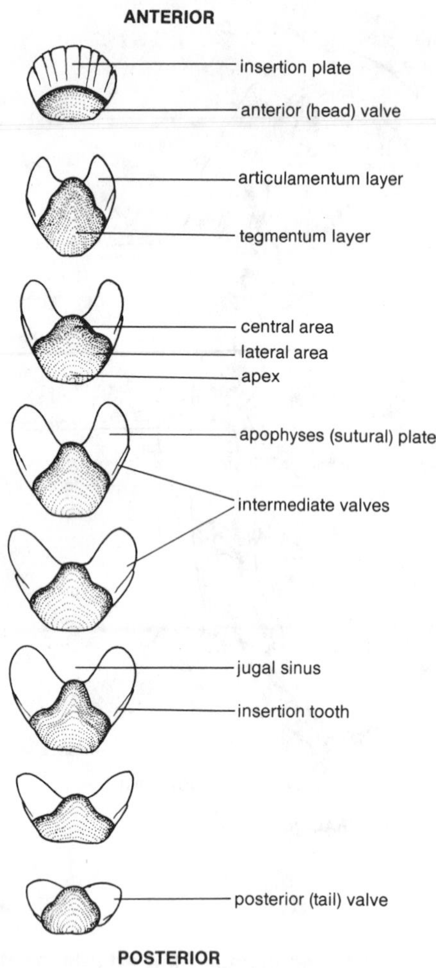

Figure 12.8. Dorsal view of isolated plates of *Katherina*.

radula and radular sac (Fig. 12.10); remove the radula intact and observe under the compound microscope. Are the teeth of uniform structure the entire length of the radula? Why? Find the cartilaginous odontophore over which the radula glides. Determine the action of this complex mechanism and compare its form to the radula of the gastropod *Helix*.

7. Without disturbing or making further incisions, locate the pharyngeal (sugar) gland, stomach, digestive gland, and intestine (Fig. 12.10). Note the length of the intestine. What does the length of the intestine indicate concerning the diet of this animal?

8. Located the single, middorsal gonad (Fig. 12.10). It is a large, lobed structure that begins in the second third of the body and terminates at the posterior end. Start at the anterior end and trace the gonad until you find the genital ducts that lead to the gonopores.

9. Next locate the **pericardial membrane** and cavity. It is located dorsally in the hindquarter of the body (Fig. 12.10). The cavity surrounds the heart. Locate the triangular median ventricle in the posterior region of this cavity. The aorta leads from the ventricle anteriorly above the gonad. Paired auricles are attached to the lateral sides of the ventricle.

10. Carefully remove the digestive glands and gonad. Locate the thin, many-lobed kidneys extending nearly the entire length of the body. They drain into nephridiopores located just posterior to the gonopores.

Other Species. Observe a specimen of *Cryptochiton* and compare its external morphology to that of *Katherina*. Note the large size and apparent lack of plates. Where are the plates located? As your instructor directs, inspect other chitons. Look for differences in the size, shape, and ornamentation of the plates and girdle (e.g., granules, scales, spines). Are there any calcareous spicules on the surface of these chitons? What might their function(s) be?

If live specimens are available, observe their general movements and feeding and homing behavior. How easy is it to remove a chiton from its substrate? If time permits, your instructor will provide instructions for observing the flow of respiratory water through the chiton's mantle cavity.

C. Class Bivalvia

Bivalvia (bi-VAL-ve-a; L., *bi*, two + L., *valva*, folding door) is a large class comprising more than 25,000 extant and about 20,000 extinct species of shelled mollusks. They differ from other mollusks in having a laterally compressed shell of two valves hinged along their dorsal margin. Many bivalves have a hatchet-shaped foot used in digging. This feature is responsible for the alternative class name of **Pelecypoda** (pel-e-SIP-o-da; G., *pelecy*, hatchet + G., *poda*, foot). Bivalves lack a recognizable head and possess neither radula nor odontophore. Individual species are known commonly by many names, including clams, mussels, oysters, cockles, and scallops. Many are economically important food sources and are used in jewelry and, in certain cultures, for money. The wood-boring bivalves, called shipworms and piddocks, do considerable damage to pilings, docks, ships, and other wooden structures in marine environments.

Students will study shells of a few selected species to gain an appreciation of the diversity of forms that comprise this class. The marine clam *Mercenaria* (= *Venus*) *mercenaria* will be dissected as a class representative.

Observational Procedure

Shell Morphologies

1. As your instructor directs, obtain the shells of several different species of bivalves (e.g., *Anodonta*, *Arca*, *Crassostrea*, *Donax*, *Mercenaria*, *Mya*, *Mytilus*, and *Pecten*). Whenever possible, study specimens with both valves. Note that these shells differ widely from one another in their size and shape, but that they have several features in common. List several of these common features and also several ways in which the shells differ. Do any have more than one type of symmetry? How would you describe shell shape in oysters?

2. Study more closely the shells of *Mercenaria* and the freshwater clam *Anodonta*. Note the thin, dark organic **periostracum** covering the outer surface of *Anodonta*. Observe the concentric lines on the external surface that recede from an elevated point near the hinged margin. These are laid down as growth proceeds. The elevation is the **umbo**, the oldest part of the shell. Umbos of opposite valves are directed toward each other and the approximating points are called **beaks**. In these clams the umbo is directed toward the anterior.

3. Orient both specimens so that the dorsal surface (hinge) is up and the umbo points away from you. Identify the right and left valves (Fig. 12.11). In other species this arrangement is not as clear.

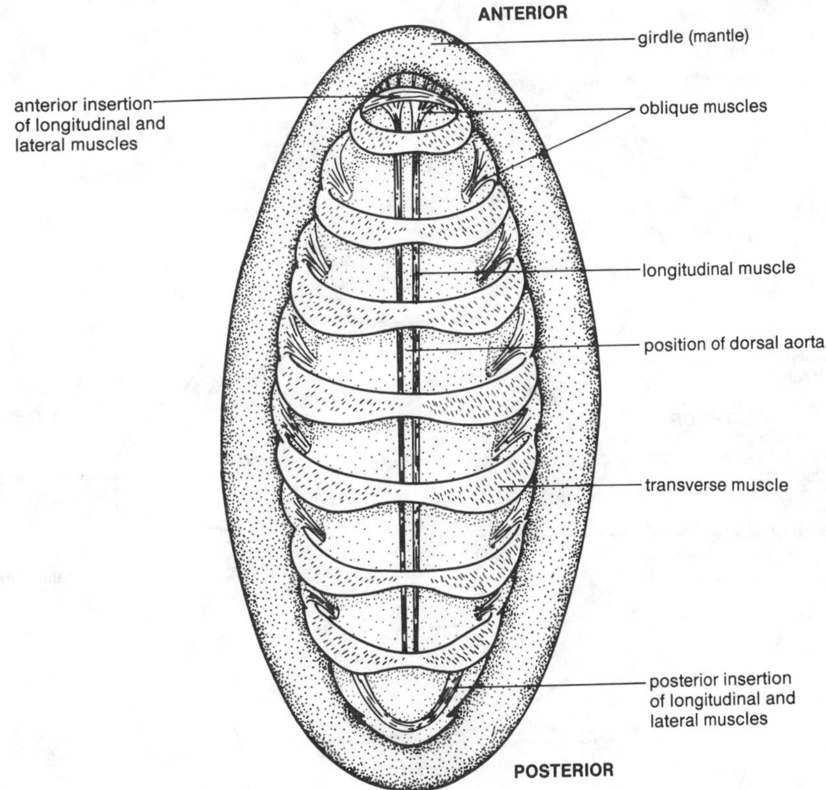

Figure 12.9. Dorsal view of *Katherina* with plates removed.

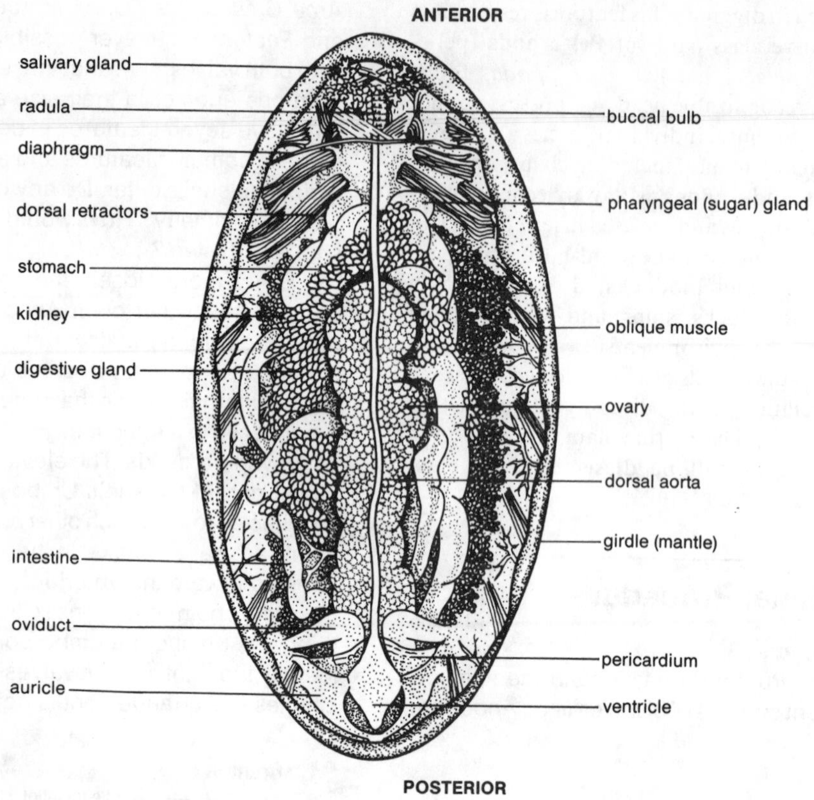

Figure 12.10. Dorsal internal view of the viscera of *Katherina*.

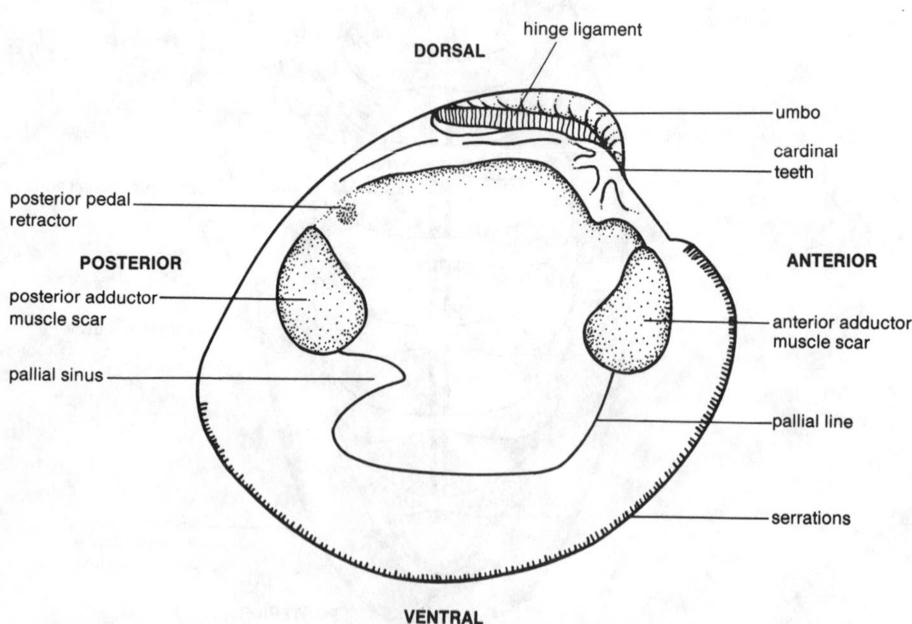

Figure 12.11. Inner surface of the left valve of *Mercenaria*. (After several sources.)

4. Observe the external surface of all specimens you have assembled. Are there any external ridges, spines, or folds on these species? What purpose might these features serve? Examine the surface of the valves closely. Is there evidence of predatory attack (e.g., drill holes or chips along the ventral margin of the shell)? Do any of your specimens show damage caused by the boring sponge *Cliona*?

5. Turn your attention to the inside of the shells and note any color patterns present. In some species such as *Anodonta* and *Mytilus* the inner surface is made up of **nacre** (mother-of-pearl). It is secreted by the entire outer surface of the mantle. Examine a prepared microscope slide of a section of mantle margin and locate the inner, middle, and outer mantle folds, pallial retractor muscle, and mantle epithelium (Fig. 12.12).

The shell is composed of a number of layers: inner (sometimes nacreous), middle, and outer. The entire shell is covered by periostracum. The middle and outer shell layers and the periostracum are formed at the perimeter of the mantle.

6. Using a hammer and chisel, break the shell, taking care to protect your eyes from shell fragments. View the broken edge of the shell fragment in cross section using a hand lens or dissecting microscope and locate the three layers. What are their relative thicknesses? Compare the similarities and differences in shell structure of various species.

7. Observe the inner dorsal margins of several species and note the presence of any prominent plates or ridges (Fig. 12.11). These make up the **dentition** or teeth of the valves. Those that run longitudinally along the hinge line are **lateral teeth**. Those grouped near the umbo are **cardinal teeth**. Carefully move the valves, noting how the teeth on opposing valves articulate. Is the fit between valves tight in all species? Note the serrated, inner ventral margin of the valves in *Mercenaria*. How do these serrations fit when the valves close? Suggest ideas that could explain the functional significance of these serrations (e.g., avoidance of predation). Do other species have similar serrations?

8. Observe the anterior and posterior muscle scars and the pallial line on the inner surface of the valves (Fig. 12.11). The pallial line represents the point of attachment of the pallial retractor muscle to the shell.

Internal Anatomy

1. Obtain a preserved specimen of *Mercenaria* in which the valves are held apart by a wooden peg. Using a heavy scalpel, carefully sever the **hinge ligament** that connects the valves along their dorsal margins. Use extreme **CAUTION** when doing this procedure; cut away from your body and fingers! Live specimens also may be used for this procedure, in which case breaking the edge of the shell with a hammer will provide sufficient space to insert a scapel. Be careful to protect your eyes from shell fragments when breaking the shell.

2. With the hinge weakened, carefully insert the scalpel inside the shell at the ventral, posterior margin; separate the delicate mantle from the right valve, progressing anteriorly. Be careful not to tear the mantle. Detach the anterior and posterior adductor muscles where they insert at the surface of the right valve. Lift the right valve off the clam, carefully separating any remaining mantle tissues or muscles from the shell. After completing the separation at the hinge, the right valve should come free while the left contains the intact animal.

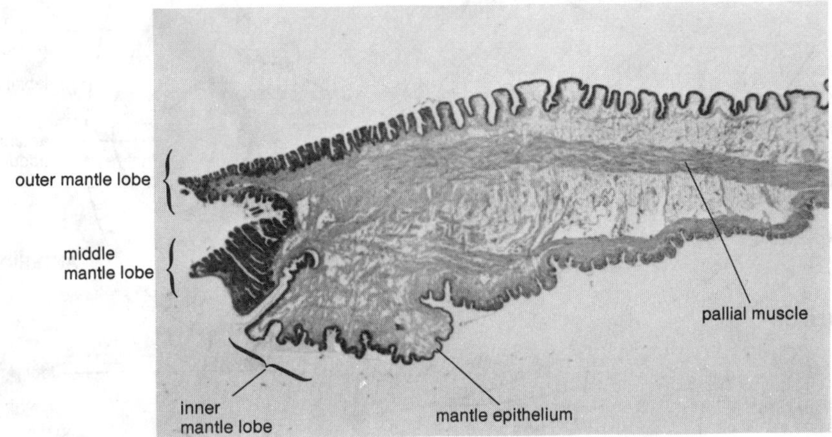

Figure 12.12. Cross section of the mantle margin of a clam (shell removed).

3. Identify the exposed muscle bundles. The large anterior muscle is the anterior adductor, important in keeping the valves closed. Above, and in close proximity, is the smaller anterior pedal retractor muscle, which withdraws the foot. The large posterior muscle is the posterior adductor, used also in closing the valves. Above it is the posterior pedal retractor.

4. Carefully cut away the right mantle lobe. This will expose the mantle cavity and the organs it contains. Locate the following structures: foot, **ctenidia** (gills), visceral mass, palps, and exhalant (excurrent) and inhalant (incurrent) siphons (Fig. 12.13). Note the arrangement of the siphonal openings. The inhalant siphon bears the aperture through which water enters the mantle cavity. The exhalant siphon is the exit channel for water that has passed through the ctenidia, for urine expelled by the kidney, for feces from the adjacent anus, and for gametes. The visceral mass houses the digestive, excretory, and reproductive organs.

5. Examine the arrangement of the ctenidia and their connection to the foot and to the mantle. Observe how they lie in an angular position in the mantle cavity on each side of the body. Note that each ctenidium (right and left sides of the mantle cavity) is composed of two **demibranchs** (inner and outer), the former typically larger than the latter. Each demibranch is composed of two **lamellae**, ascending and decending.

6. Examine a prepared microscope slide of a section of a demibranch (Fig. 12.14) and locate the numerous **ostia** in the lamellae through which water enters. Water is transported dorsally within the ctenidia through numerous **water tubes**. A complex arrangement of cilia on the vertically aligned filaments that make up the surface of the lamellae creates the flow of water. All tubes empty into the dorsal **suprabranchial cavity**, which in turn empties into the exhalant siphon. Blood vessels are profusely distributed throughout the ctenidia. Note the minute chitinous rods present on the inside of the ciliated surface. What is the significance of these rods? Water entering the mantle cavity also carries food particles, that are retained on the surface of the ctenidia. Ciliary action moves material to food grooves located on the ventral margin of each demibranch. Within the groove, food materials are moved toward the labial palps and then into the mouth.

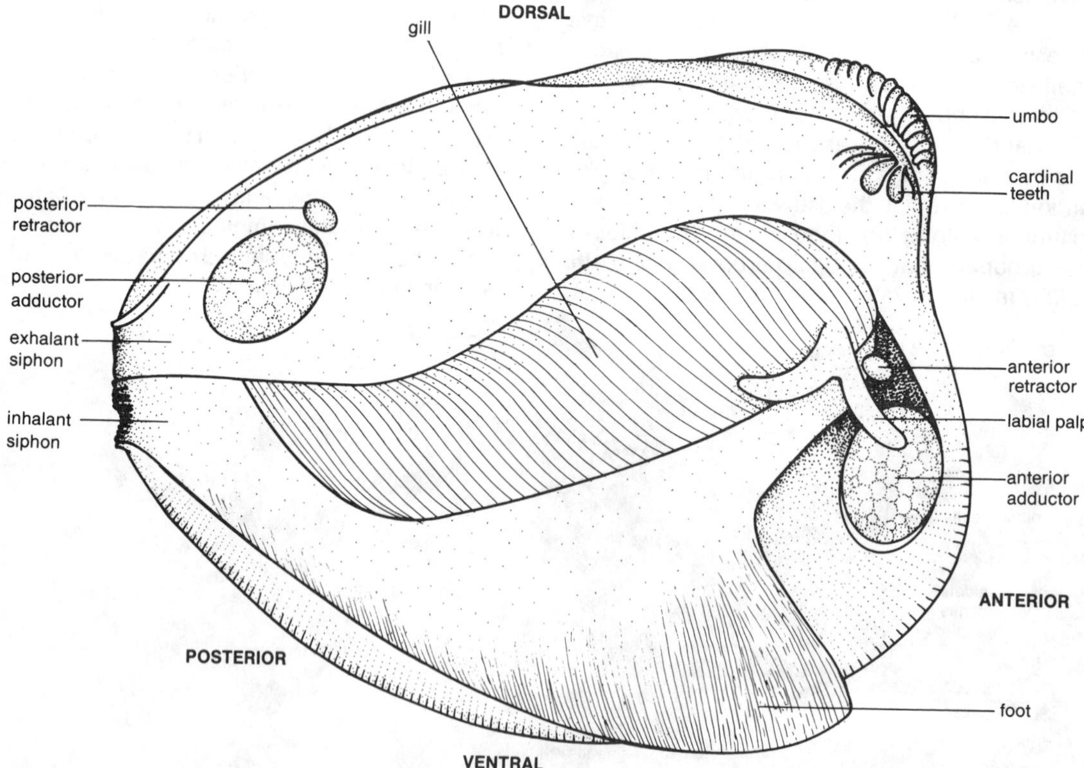

Figure 12.13. Anatomy of *Mercenaria* with right valve and mantle removed. (After several sources.)

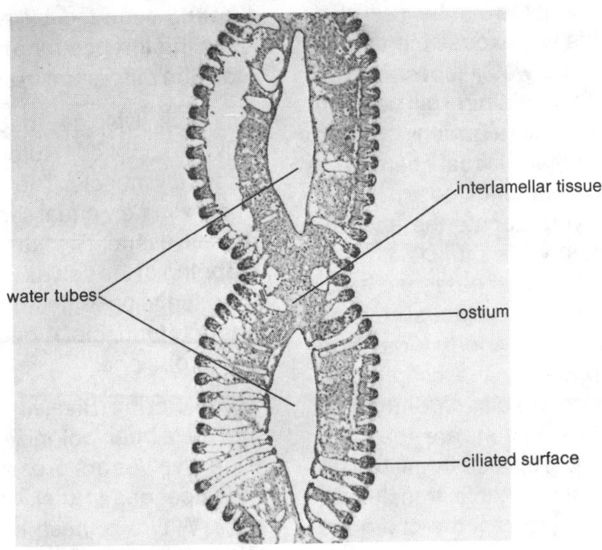

Figure 12.14. Photomicrograph of a section of a bivalve gill.

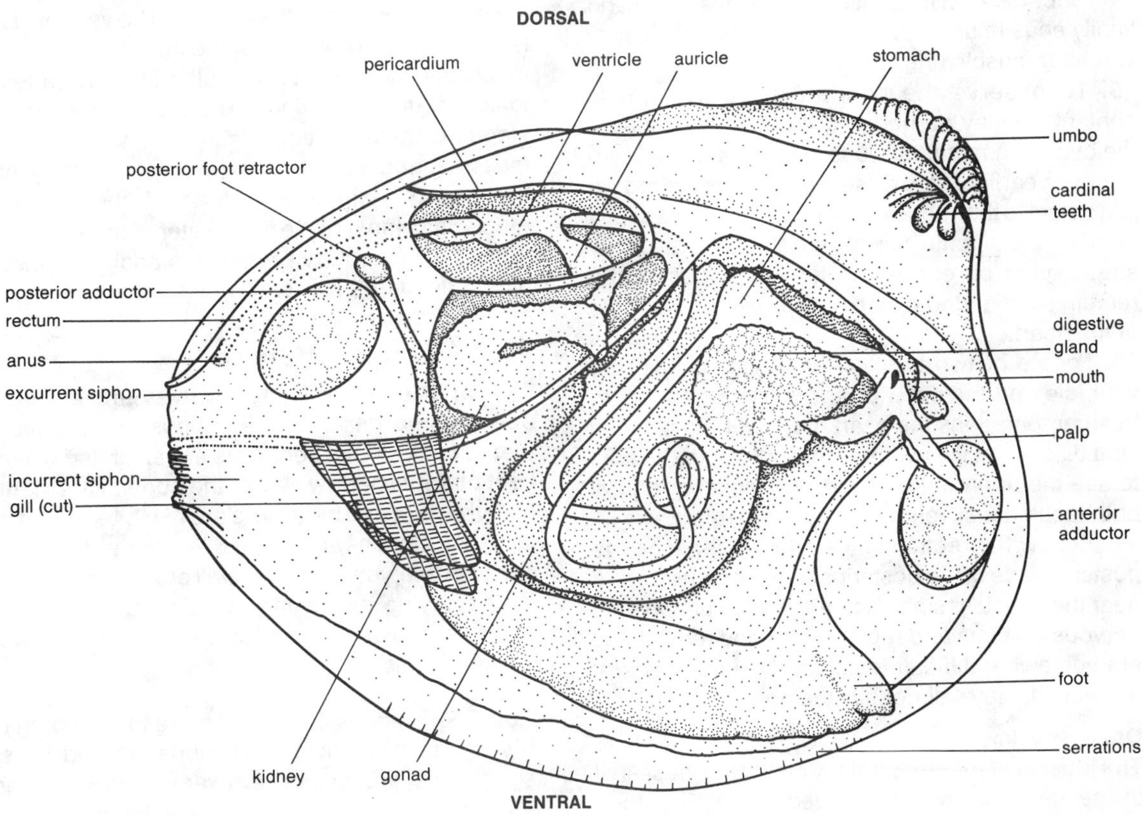

Figure 12.15. Partially dissected *Mercenaria*, depicting internal anatomy. (After several sources.)

7. With a pair of scissors cut the mantle free from the outer demibranch. This will expose the suprabranchial cavity; some of the water tubes may be visible. Carefully remove the remaining gill parts on the right. Posteriorly there is a union of both chambers, forming a **branchial-cloacal** chamber.

8. Hold the animal with the anteriormost point of the foot directed toward you. Locate the anterior adductor muscle. Just below the muscle are the labial palps and slitlike mouth with outer and inner lips (Fig. 12.15). With scissors make a lateral incision and follow the mouth cavity posteriorly for a short distance. The cavity exposed is the esophagus, which opens up into an enlarged stomach. Surrounding and opening into the stomach, are the dark **digestive diverticula** (Fig. 12.15). Also opening into it from a cecum may be a long, flexible, translucent **crystalline style** that rotates against the chitinous gastric shield of the dorsal stomach wall, releasing enzymes used in extracellular digestion. The intestine is a long, coiled tube within the visceral mass. The latter part of the intestine ascends dorsally, where it enters the anterior end of the pericardial cavity. From here the rectum passes posteriorly through the pericardial cavity, passes through the heart, and finally ends in an anus on the face of the posterior adductor muscle (Fig. 12.15).

9. To observe the pericardial cavity and its contents, it may be necessary to remove some of the overlying mantle tissues, if they have not been removed earlier in the dissection. Locate the region just dorsal to the suprabranchial cavity and carefully remove the mantle tissue; the pericardial cavity should then be exposed (Fig. 12.15). Locate the rectum surrounded by the single, median ventricle of the heart.

10. Locate the two auricles laterally attached to the ventricle and the anterior and posterior aortae. In fresh preparations the heart may still be beating. If your dissection was done properly, you may be able to see the nephrostome and the genital aperture, both located just ventral to the pericardial cavity.

11. Follow the rectum to the anus, located on the posterior face of the posterior adductor muscle and near the exhalant siphon. Since the circulatory and nervous systems and reproductive and renal ducts are somewhat difficult to differentiate in a general dissection, they will not be studied here.

Other Specimens

The Mussel *Mytilus*. A partially dissected mussel of the genus *Mytilus* will be provided. Note the reduced foot and cluster of thread-like filaments arising from the heel of the foot anteriorly. These are **byssal threads** secreted by a byssal gland. The mussel uses the threads for attachment to hard substrata. Note the difference in size of the adductor muscles.

Scallops. Display specimens of opened scallops will be supplied. Note the single, large, posterior adductor muscle. The adductor is divided into two approximately equal sections of smooth and striated muscle tissue. The latter is responsible for the rapid clapping of the valves when the scallop swims. Note the orientation of these muscle fibers in the cylindrical adductor muscle. Along the edge of the mantle are numerous ocelli.

Wood-Boring Bivalves. Examine pieces of wood that have been colonized by wood-boring bivalves. If the bivalves are present, note the size, shape, and modifications to their shells from the basic bivalve plan. Why are these bivalves called shipworms? What sort of device(s) do they possess that would allow them to accomplish burrowing? Observe the lining of the burrow.

Larval Stages. Examine specimens or prepared slides of bivalve larval stages. The first larval stage is a ciliated, somewhat barrel-shaped **trochophore**. Note the position of ciliary bands. The **veliger** stage is more complex. Although the internal anatomy of most specimens will be difficult to interpret, attempt to locate the following structures of the veliger stage: ciliated **velum**, visceral mass, adductor muscles, and valves. Compare the anatomy of a free-living veliger with that of the parasitic **glochidium** larva of the freshwater clam, *Anodonta*. Attempt to locate the visceral mass, adductor muscle, hinge ligament, valves (with hooks possessing small spines), sensory cilia, and attachment thread (Fig. 12.16).

Live Specimens. If living bivalves are available, observe and describe the various ways some of them burrow. What behaviors are exhibited when a light or dark shadow strikes the edge of the mantle or siphons? What behaviors are exhibited when the mantle or siphons are touched? If time and materials are available, your instructor will provide information for observing swimming movements in scallops and flow of water in and out of siphons in filter-feeding clams.

Fossil Specimens. Bivalves have a long geological history, dating back to the Cambrian Period. If fossil specimens are available, examine their general shell morphology and compare them to extant species.

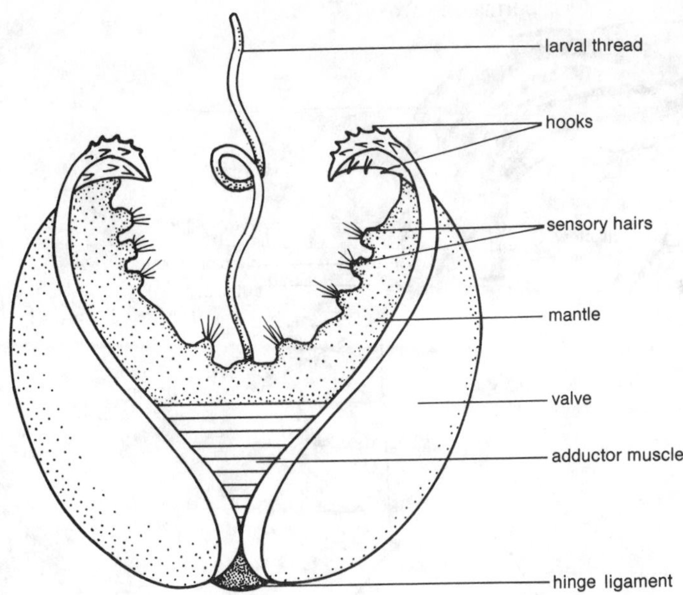

Figure 12.16. Glochidium, parasitic veliger larval stage of freshwater clams. (After several sources.)

D. Class Cephalopoda

Cephalopoda (SEF-a-lo-POAD-a or sef-a-LOP-o-da; G., *cephalo*, head + G., *poda*, foot) comprises over 1000 extant and 8000 extinct species of marine mollusks. Most cephalopods are less than 1 m long, but some very large forms are known. In general, cephalopods are adapted for swimming, although many species are bottom dwellers, crawling about on their arms. The shell is generally reduced or even absent (e.g., octopuses) in living cephalopods, but well-developed shells were possessed by members of the extinct subclasses Ammonoidea and Nautiloidea (represented today by members of the genus *Nautilus*).

Cephalopods have four major features. (1) All possess a well-developed brain and complex nervous system. (2) The anterior portion of the foot is modified into a circle of prehensile muscular **arms** and **tentacles** that surround the mouth. Except in *Nautilus*, these appendages bear numerous muscular suckers that aid in prey capture. (3) Cephalopods have a muscular buccal apparatus containing beaklike **jaws** and a radula. (4) Locomotion in this group is unique. Water taken into the mantle cavity can be expelled in a forceful contraction of the mantle walls. This exhalation furnishes a type of jet propulsion.

The common Atlantic coast squid, *Loligo pealeii*, will be studied in detail as the class representative. Given the diversity within the class, *Loligo* should not be considered to be a typical cephalopod.

Observational Procedure

External Anatomy

1. Orient the squid in the dissection pan so that the arms and tentacles are directed away from you and the broad flat surface of the fins is directed upward. The posterior end of the animal is now nearest you, the anterior end farthest away, and the dorsal surface is upward (Fig. 12.17). The anterior end contains the mouth, surrounded by arms and tentacles. At the direction of your instructor, compare this orientation to that of other molluscan representatives.

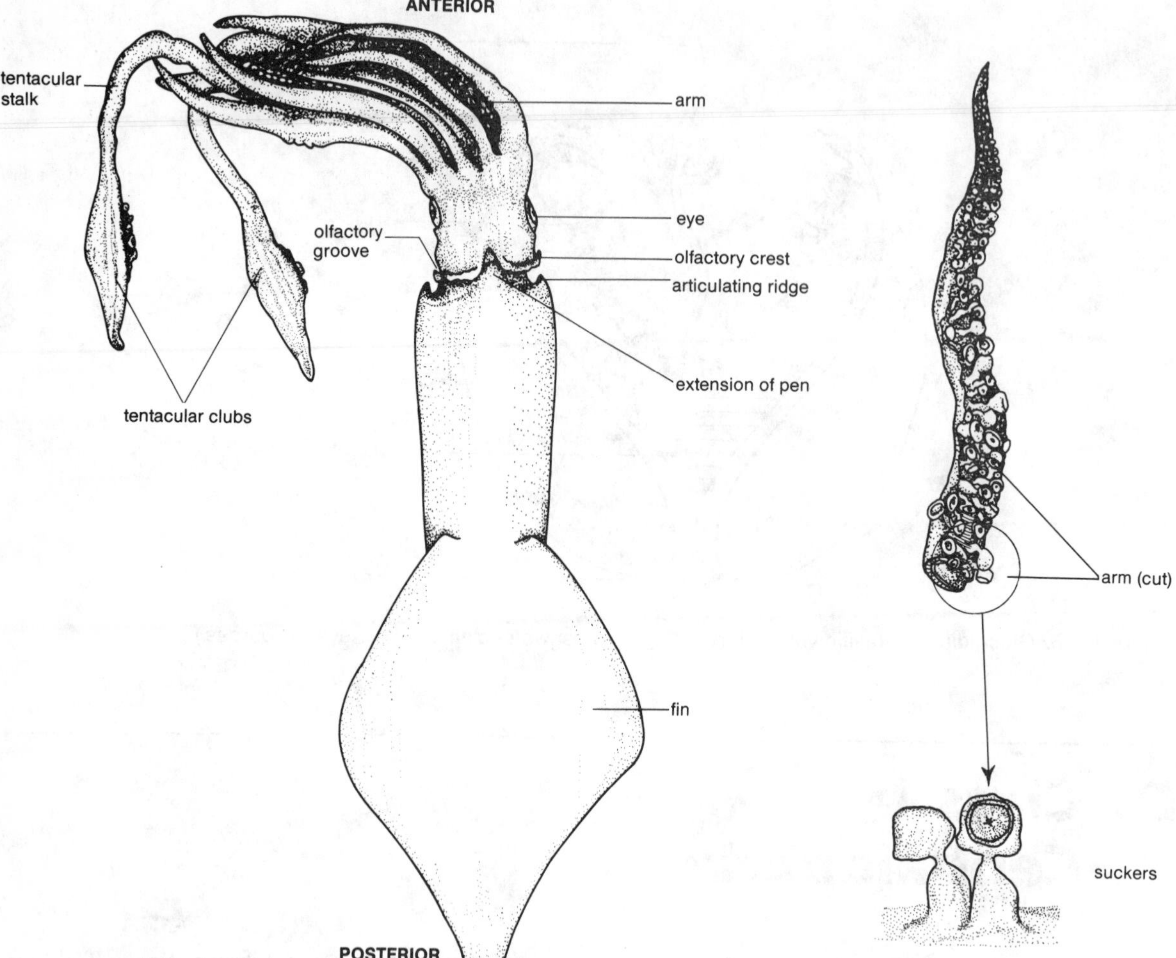

Figure 12.17. Dorsal view of the squid *Loligo pealeii.*

2. Note the finely striated appearance of the surface of the fins. These striations are muscle bundles that control fin movement. The spotted condition of the body surface is due to the presence of pigmented cells called **chromatophores**.

3. Gently scrape the surface of the body, removing the transparent outer cuticle. Each chromatophore is a small membranous sac to which tiny muscle fibers are attached. Contraction of muscle fibers enlarges the chromatophore and makes the pigment visible.

4. Arrange the squid with the ventral side up and the head nearest you. The body form is elongate and arrow-shaped (Fig. 12.18). The head region appears as though it is inserted in a central tube. This is the tough, muscular mantle.

5. Pick up the squid and view the animal from the oral aspect. While applying lateral pressure to the tube, look into the mantle cavity. The blunt lateral projections are extensions of the **mantle locking ridges**. Cartilaginous locking grooves are located on the inside surface and will be observed later. The pointed extension on the dorsal surface of the mantle marks the location of the internal shell (**gladius**) or **pen** (Fig. 12.17). The cone-shaped structure extending beyond the mantle on the ventral surface is the **funnel**.

6. Locate the eyes on either side of the head in the region of the funnel. They will be considered in detail later.

7. Two kinds of appendages are located at the anterior end around the mouth; shorter ones are arms and the longer ones with club-shaped ends are tentacles. How many arms and how many tentacles are present?

8. Examine the arms and tentacles. How do they differ as to size and shape? Note any differences in size and position of sucker cups on the arms and tentacles (Fig. 12.17).

9. Remove several suckers and determine how they function. What do they resemble?

10. Locate the **aquiferous pore** at the anterior edge of the eye (Fig. 12.19). This pore is probably used for equalizing pressure in the eye chamber.

11. Observe the articulation of the mantle ridges with the funnel-locking cartilages on each side of the funnel. Near the eye, the integument is folded and thickened as the **olfactory crest**, under which is found the **olfactory groove** (Figs. 12.17 and 12.19).

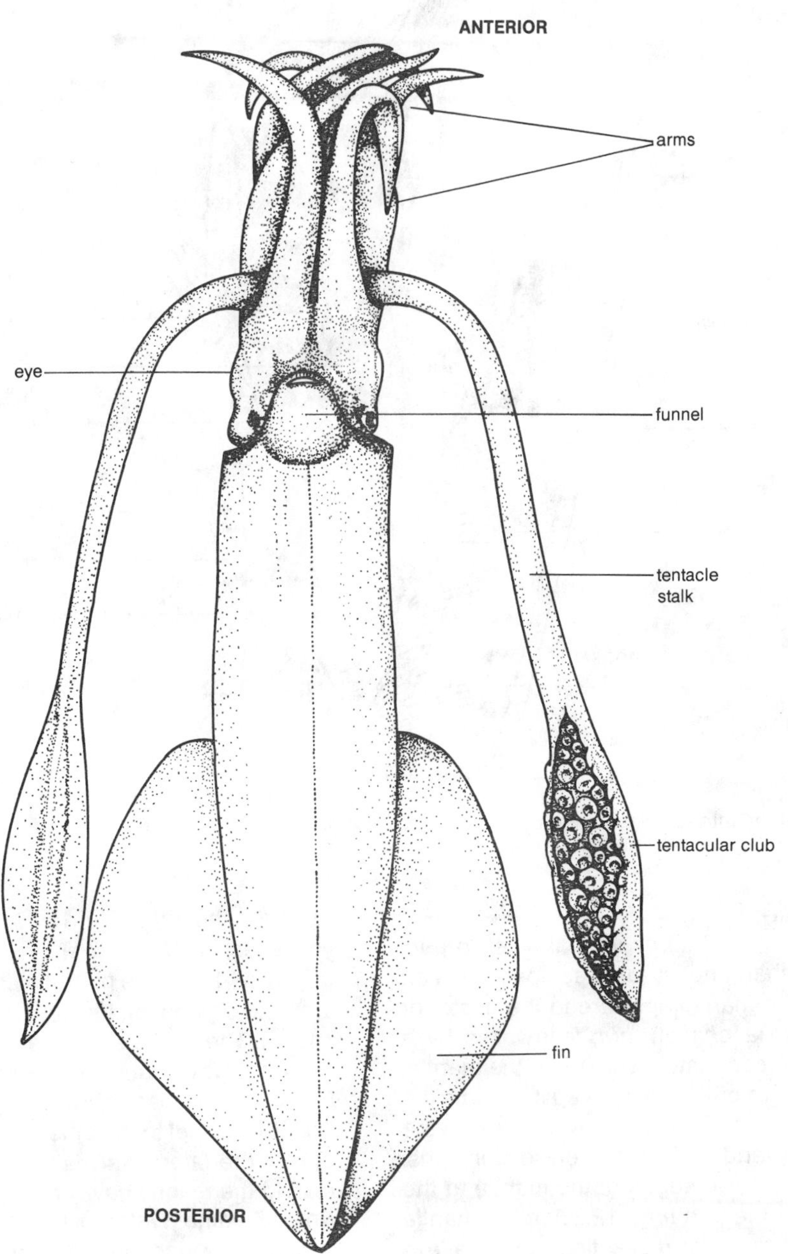

Figure 12.18. Ventral view of the squid *Loligo pealeii*.

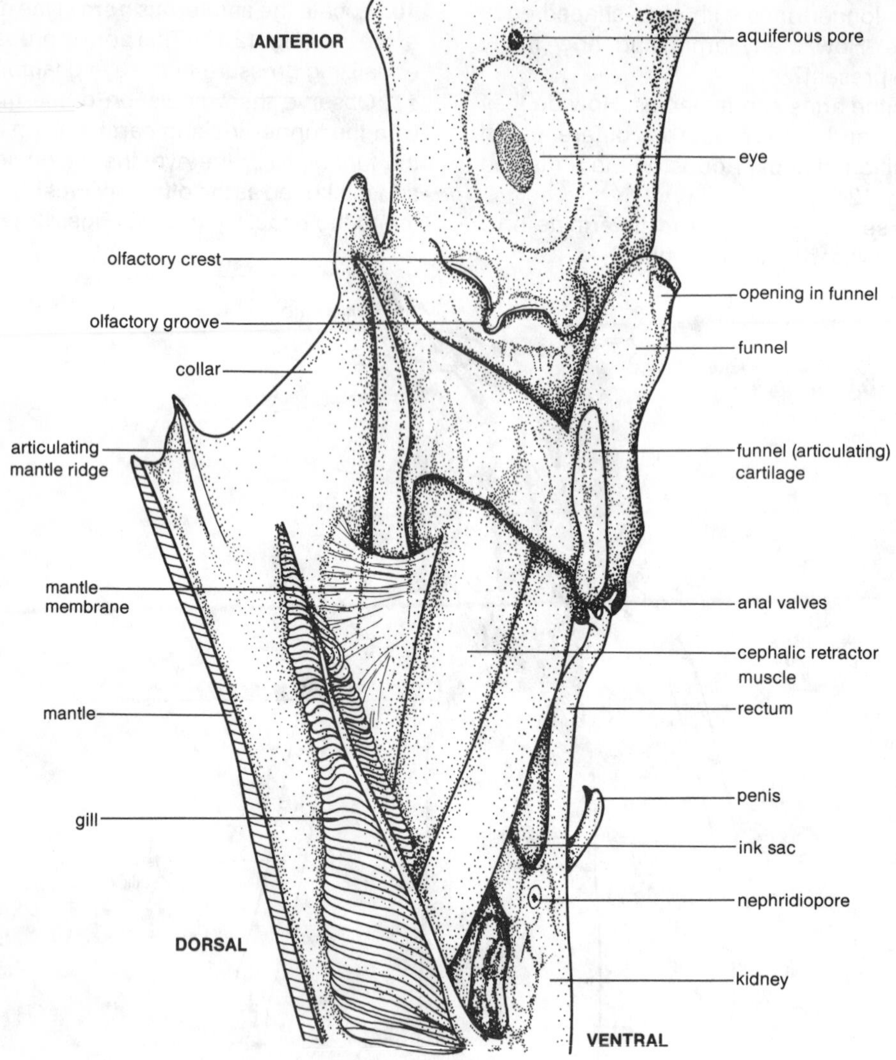

Figure 12.19. Lateral internal view of the trunk-head region of a male squid (*Loligo pealeii.*)

Internal Anatomy

1. Orient the squid with its ventral side up (Fig. 12.18). With a sharp, heavy-duty scalpel, make an incision from the mantle aperture to the posterior end. First make a *shallow cut*; follow this by a second cut that completely separates the mantle. Make sure that this cut is not so deep it disturbs the internal organs. Gently fold the two sides of the mantle laterally and, with large dissection pins, firmly attach the outer edges of the mantle to the paraffin base of the pan (Fig. 12.20). If the mantle cannot be secured with dissection pins, make a shallow longitudinal cut the length of the mantle on either side.

2. With the aid of Fig. 12.20, determine the sex of your specimen. Sever the funnel lengthwise, exposing its cavity. The large paired glands inside the funnel are the **organs of Verrill,** whose function is not known. The liplike structure at the forward edge of the funnel is the **funnel valve**, which regulates water flow out of the mantle cavity. Note the fold extending from the funnel to the other side of the head. This is one of the **lateral valves** that prevents water from exiting the mantle cavity along the same path from which it came. Water exits only through the funnel. Describe the functional morphology of the structures involved with circulation of water through the mantle cavity.

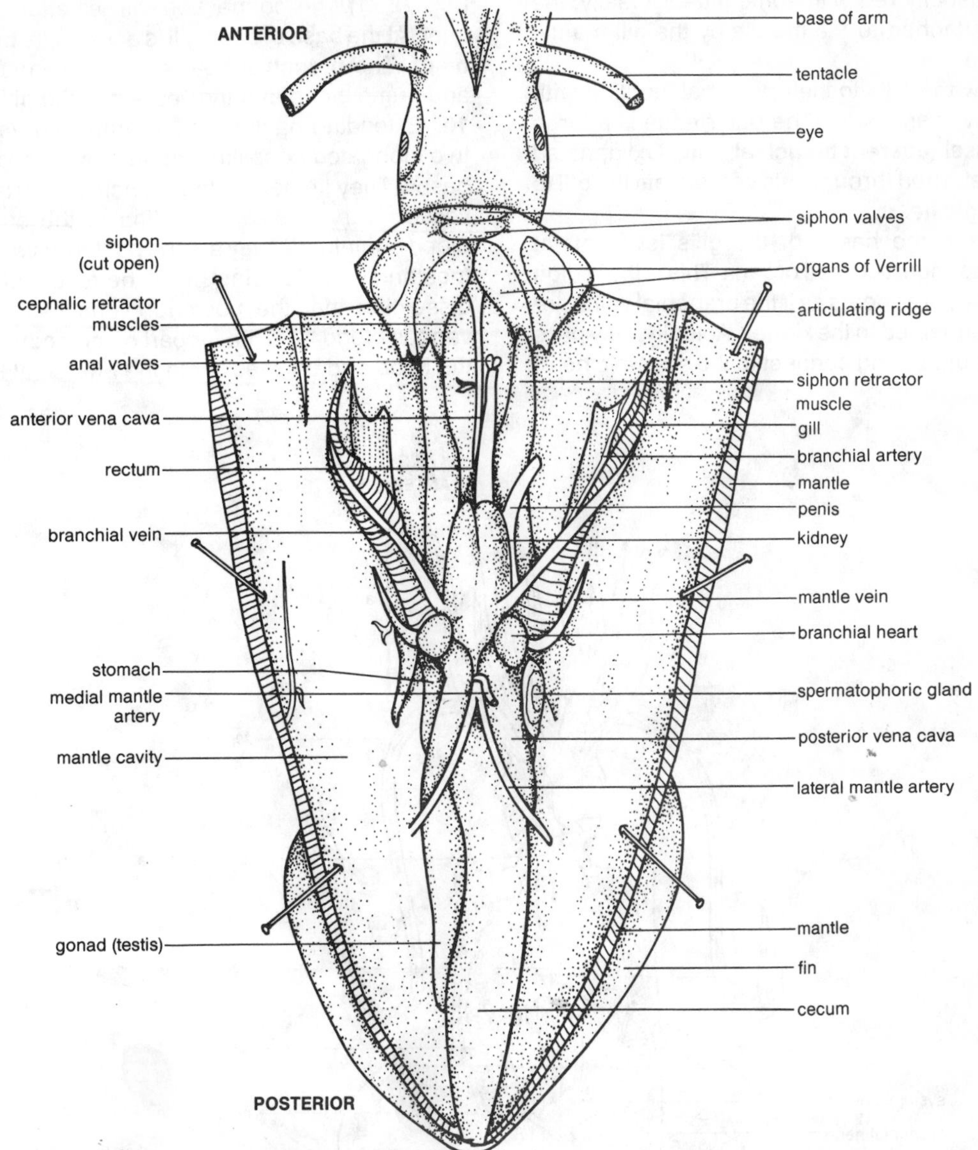

Figure 12.20. Ventral view of the anatomy in the mantle cavity of a male squid (*Loligo pealeii*).

3. Locate the **rectum** and the anal opening at its end. The anus bears two flaplike valves. How do they function?

4. Below the rectum is the **ink sac** (Fig. 12.20), which is filled with a black pigment called **melanin.** Observe the membranous tissue binding the ink sac to the rectum. Using sharp, fine-pointed scissors and a scalpel, carefully separate and remove the ink sac from the rectum.

5. Open the sac in a small dish. Remove some of the thick, dark ink with a pipette or syringe and place a drop into some water. Note how deeply the ink colors the water.

6. Two pairs of large muscles are found in the area of the ink sac. Locate the **funnel retractor muscles**, which extend dorsally from under the funnel and dip strongly under the central organs. Beneath these muscles and still larger in size are the paired **cephalic retractor muscles** (Fig. 12.20).

7. The paired large pinnate structures in the mantle cavity are the gills (Figs. 12.19 to 12.21). In specimens injected with latex, the gills will be colored, usually red with some blue. Dorsally, the gills are attached to the mantle by the gill mantle membrane.

8. Follow the gills to their proximal junction with the other viscera. Along the outer edge is a large blood vessel (**efferent branchial vein**). Oxygenated blood is returned through this vessel from the gill to the **systemic heart**.

9. Between the bases of the gills is a pair of elongate sponge-textured organs. This is the region of the kidney. It appears as if the branchial veins are directly connected to the kidney. Actually, they pass under this organ and connect to the systemic heart (Fig. 12.21). This connection will be better observed later. At the base of each gill is a bulb-like **branchial heart**. The branchial hearts collect venous return and pump blood into the vessels of the gills.

10. Extending back from the branchial hearts are two elongated conelike structures with broadened bases. They lie next to the branchial hearts. These vessels vary in size according to the amount of blood contained. They are the posterior **vena cavae** (posterior mantle veins). It will be revealed by later dissection that the posterior vena cava gathers venous blood from the far part of the body opposite the head and empties it into the branchial heart.

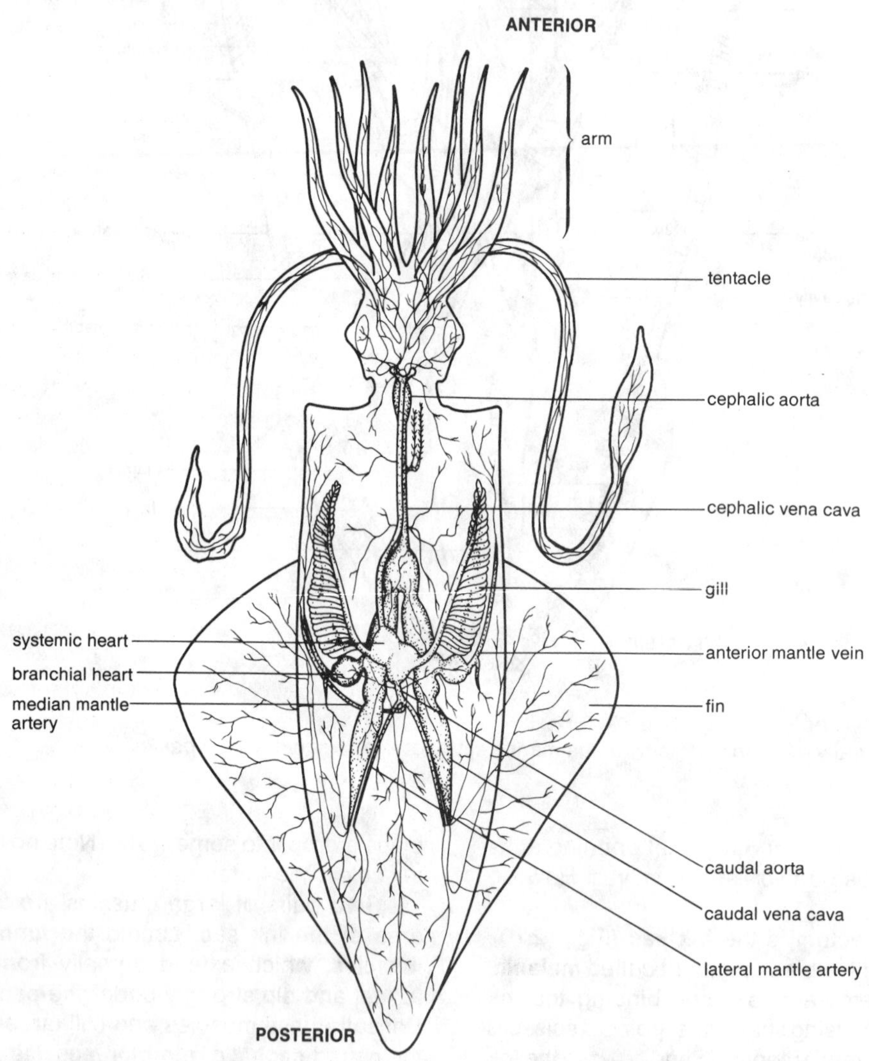

Figure 12.21. Circulatory system in *Loligo.* (Modified from Williams.)

11. Coming from the opposite area of the body is another venous vessel, which forks in the region of, and courses through, the kidney. Because of the close association of the forks, the vessels are hard to see. They are described as the nephridial portions of the anterior vena cava. Carrying venous blood, it too empties into the branchial heart. With blood from both the anterior and posterior vena cavae, the branchial heart functions as an accelerating pump and forces blood into the gills for aeration.
12. In addition to the larger vessels that empty into the branchial heart, there are three smaller vessels that bring blood directly from the mantle (**anterior** and **medial mantle veins**) and from the gonad (**gonadal vein**). The latter vein will not be seen at this time. The anterior and medial mantle veins may be seen lateral to the branchial hearts.
13. Arrange the squid so that the head is nearest you. Take the head in your hand and spread the arms and tentacles. Observe a ring of small fingerlike projections, **buccal lappets**, surrounding the mouth. Note their arrangement and the suckers they bear. The buccal lappets are united with the buccal membrane which they support. Inward to and below the buccal lappets surrounding the mouth is another membrane, the **peristomial membrane**. The margin of this membrane forms a rugose collar.
14. Grasping the free edge of the funnel with a pair of forceps, lift it and sever the two small funnel protractor muscles extending between the head and funnel. Also cut the lateral funnel valves.
15. Separate the pair of arms immediately below the funnel by making a medial lengthwise incision at the base of the arms. Again note the buccal and peristomial membranes. As the arms are separated, notice their arrangement around the buccal mass. The general muscular organization of the specialized foot is complex and will not be studied.
16. Apply slight pressure to this region of the body to flatten the arms and tentacles, revealing the **buccal mass** (Fig. 12.22).
17. Examine the buccal mass.
18. Cut the peristomial and peripheral buccal membranes, exposing a compact body with the end displaying dark chitinous jaws (mandibles) that are similar to an inverted parrot's beak. The upper jaw is overlain by the lower jaw (Fig. 12.22). What function do these mandibles serve?
19. Using a sharp scalpel, make a midsagittal cut through the buccal mass. Locate the jaws and, in the floor of the mouth, a structure called the **tongue**. A duct from the median posterior salivary gland opens at the tip of the tongue.
20. With forceps, grasp the distal end of the tongue and bend it backwards until the radular area is exposed. The **radula** is surrounded by the upper jaw and is easily identified by having numerous teeth. Note the flaplike plates at the side of the radula. The lower (ventral) edges of the paired flaps lead to the esophagus. With a dissecting microscope, study the dentition of the radula. The space enclosed between the jaws is the buccal cavity; it communicates with the esophagus.
21. To locate the esophagus, it will be necessary to make a deep, medial incision on the surface of the head at the base of the arms. The incision must be exact, as the esophagus is a relatively small tube. As you dissect, a hardened structure, the **cephalic cartilage**, will be encountered at about the level of the eyes. Also exposed are whitish pulpy masses on either side of the tubular esophagus. These two tissue masses are the **cerebral** and **pedal ganglia**. The pedal ganglion is on the same side of the esophagus as the funnel.
22. Follow the esophagus from the buccal mass attachment to the place where it penetrates the cream-colored **digestive gland** (Fig. 12.23).
23. Remove the digestive gland and expose the esophagus, which lies in a groove between the digestive gland and the body wall. The digestive gland also surrounds the elongated median posterior **salivary gland**. Anterior salivary glands are embedded in the buccal mass and may be found with careful dissection. The anterior glands produce a lubricating secretion and the posterior salivary gland a poison and **hyaluronidase**. What are the functions of the poison gland and the enzyme hyaluronidase? The small vessel located alongside the esophagus and connected with the systemic heart is the **cephalic aorta** (Fig. 12.21).

Study of the digestive system will be discontinued temporarily and attention given to the circulatory and reproductive systems. It is difficult to dissect the digestive canal at this point and still leave the circulatory and reproductive systems intact.

Two types of hearts (**branchial** and **systemic**) have been mentioned (Fig. 12.21). The branchial hearts are venous and the systemic heart is arterial. The systemic heart receives aerated blood from both gills. Blood moves from the systemic heart to the body via the anterior (cephalic) aorta, posterior aorta, and genital artery. Valves in each vessel prevent blood from flowing back into the heart. Blood returns to the central circulatory system through a complex system of veins. The anterior (cephalic) vena cava and posterior vena cava are

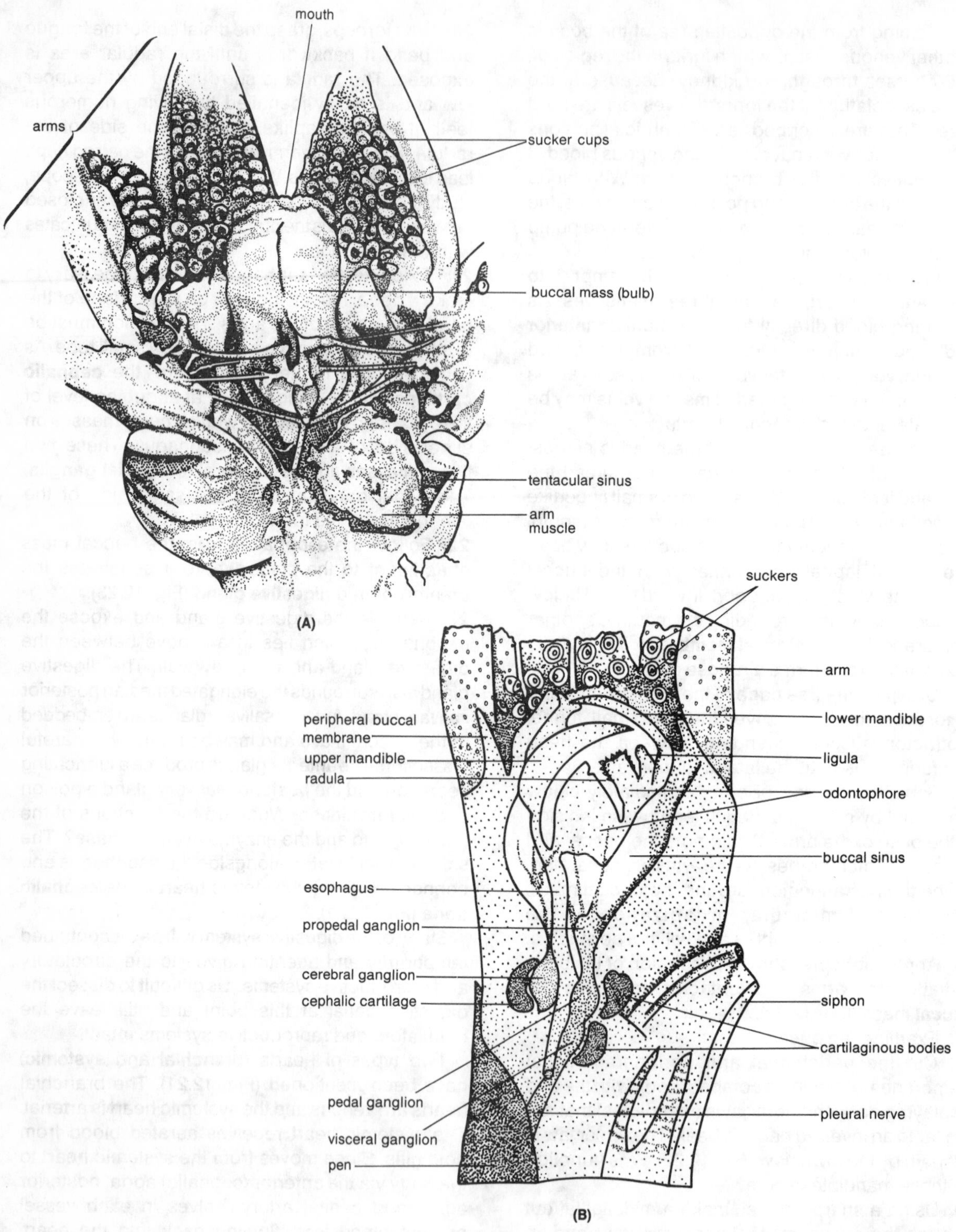

Figure 12.22. **(A)** Base of arms separated to expose the buccal mass and **(B)** midsagittal diagrammatic section of the head region of *Loligo*.

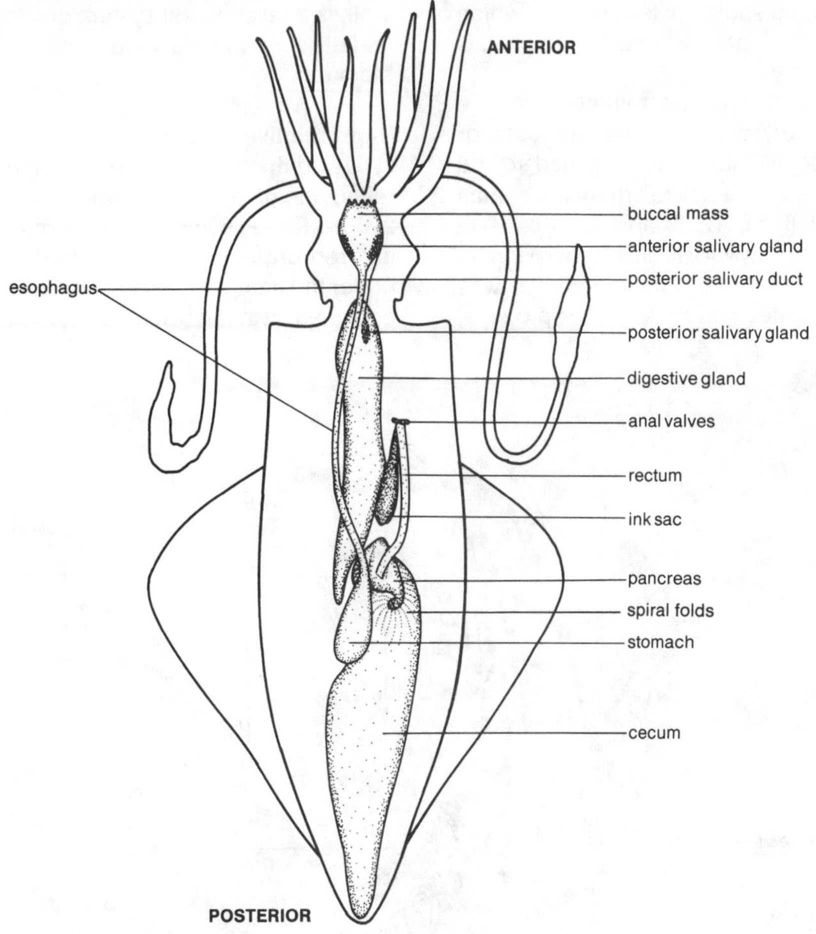

Figure 12.23. Schematic view of the digestive organs of *Loligo.*

enlarged portions of this system. Blood in the vena cavae passes to the gills via the branchial hearts. Blood vessels leaving the brachial hearts possess valves that prevent return of the blood.

24. Locate the small vessel that extends between and slightly posterior to the branchial hearts. This is the posterior aorta, which communicates with the systemic heart. It is three-pronged, with two lateral vessels and one medial vessel. The central one is the median mantle (pallial) artery. The other two forks are the lateral mantle (pallial) arteries. That portion of each posterior vena cava in close association with the kidney before entering the branchial heart is known as the nephridial portion of the posterior vena caval vein. The right and left lateral mantle arteries run along the inner margins of each posterior vena cava (Fig. 12.21).

25. Using a broad-bladed scalpel, gently scrape away the nephridial tissue between the bases of the gills. A somewhat convoluted mass of tissues beneath the kidney is the digestive duct appendages. When additional pancreatic tissue has been removed the quadrate-shaped systemic heart can be seen (Fig. 12.21). The forwardmost corner of the systemic heart is the part communicating with the anterior aorta, supplying the head and foot. Large lateral vessels running to the inside of the gills are the efferent branchial veins. Smaller blood vessels may be found attached to the systemic heart. They are very small and only careful dissection will ensure their being located. These vessels carry arterial blood to the branchial hearts.

26. Locate the genital artery arising from the systemic heart. The dorsal corner of the heart leads into the conspicuous short posterior aorta, which subsequently branches into the three mantle arteries mentioned previously.

27. Turn your attention to the remainder of the digestive system and trace the esophagus posterior to the large, muscular stomach. Attached to the stomach is a thin-walled, saclike organ, the **cecum** or **gastric pouch** (Figs. 12.23 and 12.24). Both stomach and cecum may vary in size, depending on the amount of food eaten before the animal was preserved. Inside the cecum, located on the proximal wall, is a large spiral system of folds. These ciliary leaflets sort minute food particles sent from the stomach.

Reproductive System: Male

1. Using the handle of your scalpel, gently lift and reflect the gastric pouch laterally.

2. The large, somewhat firm but flattened, light-colored organ in the dorsal part of the cavity and under the gastric pouch is the **testis** (Fig. 12.25). It is enclosed by the perivisceral (peritoneal) membrane.

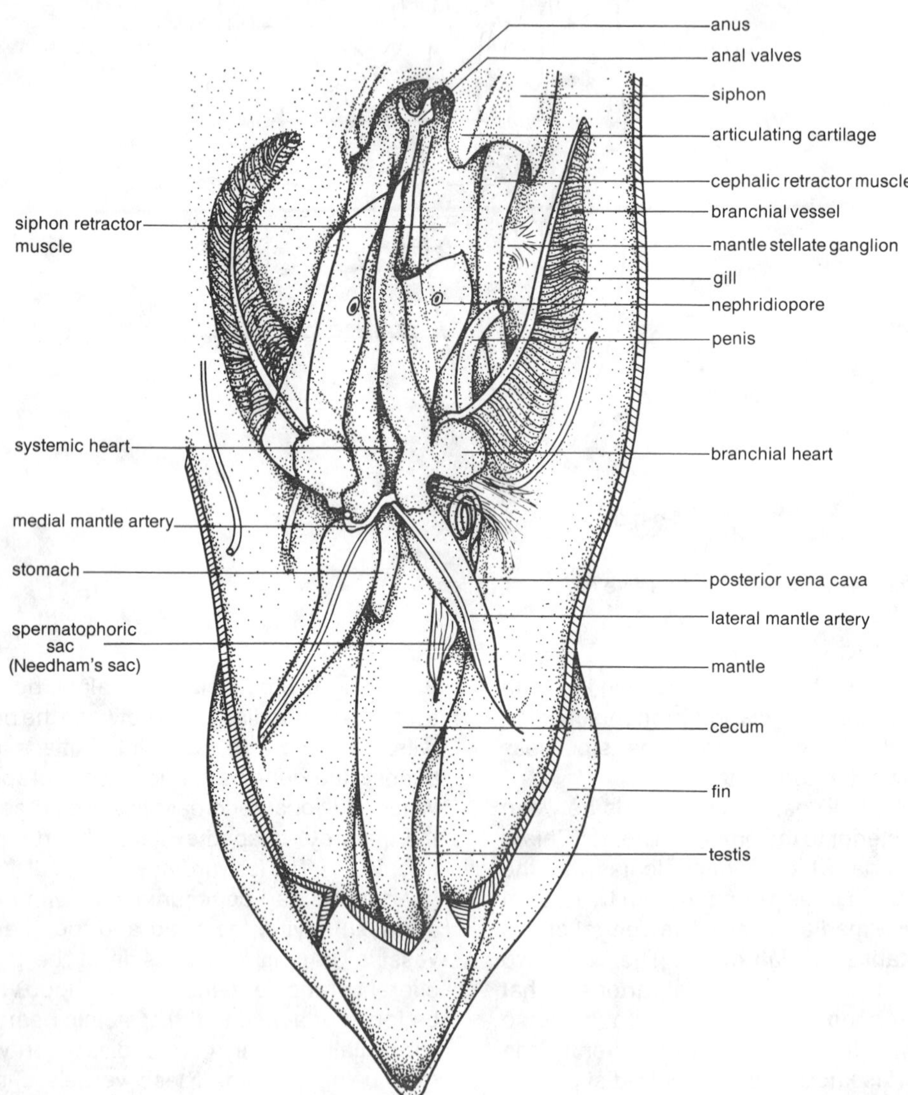

Figure 12.24. Central visceral mass with perivisceral membrane turned back to expose nephridial and circulatory organs of a male *Loligo*.

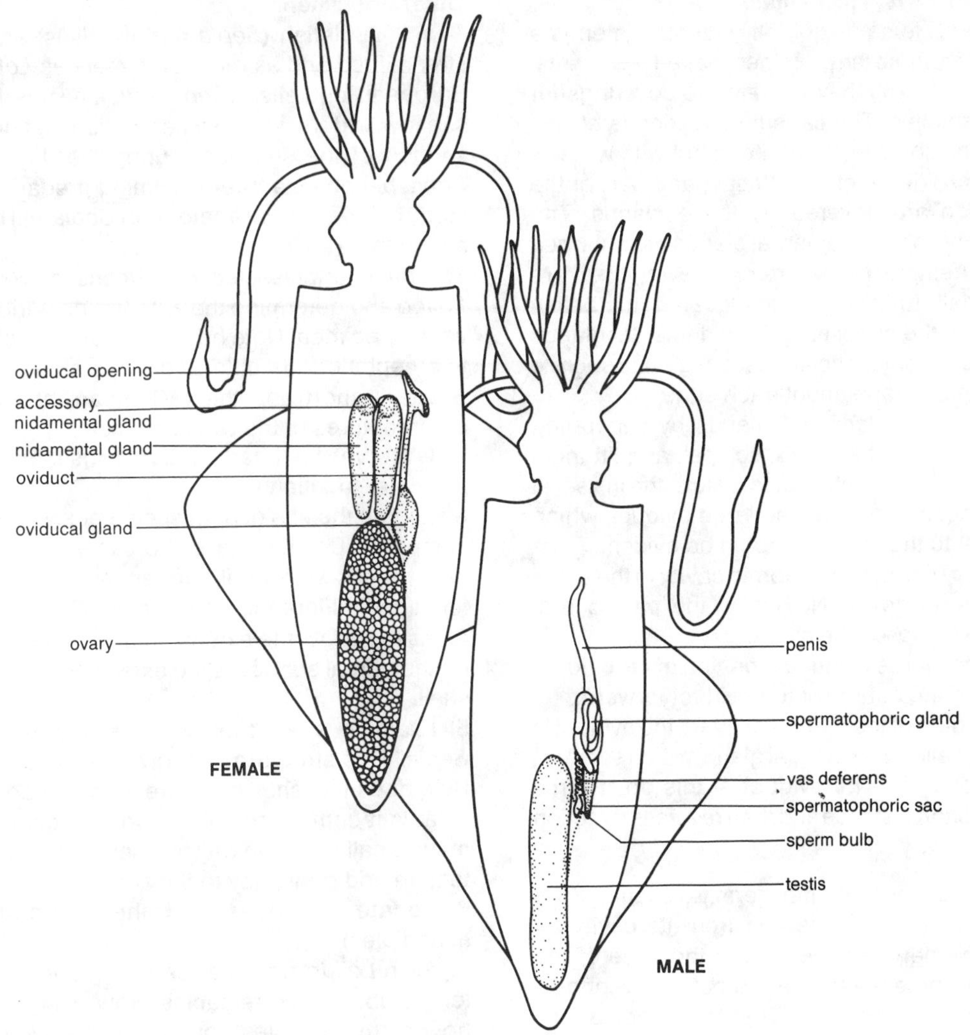

Figure 12.25. Female and male reproductive systems of *Loligo pealeii*.

Sperm are passed freely into the coelom surrounding the testis. The sperm exit from the testis through a slitlike opening on its surface.

3. Locate the spindle-shaped organ, the **spermatophoric sac** (Fig. 12.25). Alongside the sac is the coiled **vas deferens**. The free end of the vas deferens is enlarged as the **sperm bulb** (ciliated funnel). Sperm are swept by ciliary currents into the bulb. After entering the bulb they pass into the convoluted vas deferens.

4. The vas deferens connects to an enlarged convoluted tubule, the **spermatophoric glands**. These glands provide secretions that enclose the sperm in spermatophores (Fig. 12.25). Spermatophores exit the spermatophoric glands by way of the spermatophoric duct to the spermatophoric sac or **Needham's sac**, where they are stored. Exclusive of the testis, Needham's sac, when filled with spermatophores, is the largest structure of the male reproductive system. The spermatophores exit to the mantle cavity through the muscular penis that extends anteriorly from Needham's sac.

5. Remove some of the spermatophores from the sac or from the penis and examine them under the low power of the compound microscope. Study the shape and size of the spermatophores.

Reproductive System: Female

1. Examine a female squid. If your specimen is a mature female, note the prominent paired **nidamental glands** (Fig. 12.25).They secrete the coverings for the egg capsules. The capsule covering is elastic and hardens somewhat upon contact with water.

2. The nephridia, rectum, ink sac, and part of the heart region are covered by these glands. The anterior accessory nidamental glands are enlarged when the female is mature. However, they are usually small, rounded bodies located just below (as you view the specimen) the nidamental glands on the anterior side. The accessory glands secrete an elastic membrane about each egg.

3. Remove the nidamental glands by separating them with the blunt end of a scalpel, and lift them from the body. Be careful not to disturb the ink sac if it has not been removed. The large oviduct, which lies parallel to the rectum, should be evident.

4. The ovary lies in the coelomic cavity in the same position as the testis. Note how the perivisceral membrane encloses the ovary.

5. Locate the large oviducal opening of the oviduct by following the course of the oviduct forward.

6. In the region of the branchial heart, the oviduct is dilated and called the oviducal gland, which secretes the egg covering. The oviduct at this point turns down for a short distance and then reverses direction toward the ovary.

The Eye and Shell. If time permits, examine the squid's eye. Remove one eye from its cavity by severing the nerve connection at the base of the cavity. With forceps remove the outer covering or false cornea of the eye. The true cornea lies underneath the false one. Remove the cornea. The opaque ball-like structure is the lens. Surrounding the lens is the iris. Under the iris the surface of the eye is dark in color. Remove this membrane. The ciliary muscles controlling the lens may be seen as a concentric ring around the lens. By holding the lens and its band of ciliary muscles to the light, one will discover that what appears as a single band is really made up of several colored bands. Remove the inner lining or retina of the eye and examine it with the aid of a dissecting microscope. Your instructor will provide additional information if you are to compare the eye of a squid to that of a vertebrate.

Push the viscera to one side and carefully dissect free the translucent, amber-colored gladius or pen (internal shell) embedded within the mantle.

Other Specimens

Sepia. Cuttlefish (*Sepia* and its allies) are not as streamlined and as rapid swimmers as *Loligo*. The European cuttlefish (*Sepia officinalis*) is found in shallow waters, where it lies buried just under the surface of the sediment during the day. At night *Sepia* becomes a free-swimming predator. A brief study of the external anatomy of *Sepia* will be made and compared to that of *Loligo*.

1. Obtain undissected specimens of *Sepia* and *Loligo* and determine the anterior-posterior axis of each specimen. Note how the body shape of these two cephalopods differ. *Loligo* is elongate, while *Sepia* is short and thick. Compare and contrast other features of the two specimens.

2. Observe the head of *Sepia* inserted into the muscular mantle tube.

3. Locate the fins of both specimens. How do they compare? Do they have the same shape? How might these structural differences be manifest as functional differences in swimming?

4. As in *Loligo*, the extension of the mantle margin on the dorsal surface is the extension of an internal shell.

5. Examine a specimen of the internal shell of *Sepia*. This structure is known as the **cuttlebone**. How does this shell compare to that of *Loligo*? Of what is it composed? Cuttlebone is comprised of many small camerae with thin septa. How could this feature add bouyancy to the shell?

6. Locate the eyes and the chitinous jaws (mandibles).

7. As in *Loligo*, note that *Sepia* has short arms and long club-shaped tentacles. How many arms and how many tentacles does *Sepia* have? Examine closely the arms and tentacles.

Octopus. Octopuses are very different types of cephalopods from those studied so far. Although capable of swimming by jet propulsion, octopuses more often crawl along the bottom in search of prey or hide, waiting for prey. *Octopus* is capable of learning simple tasks and may be one of the most intelligent of all invertebrates. A brief comparison will be made of the general morphology of *Octopus* to that of the other cephalopods studied.

Obtain undissected specimens of *Octopus*, *Sepia*, and *Loligo*, and compare their general external anatomy. Identify the anterior and posterior ends of *Octopus*. Locate the funnel of the specimen. Where is water taken into the mantle cavity in this organism? How is it expelled? Is there any noticeable internal shell as in *Loligo*? (N.B.: *Octopus* has paired internal shell vestiges called **stylets**.) How many arms does *Octopus* have? Compare the suckers to those of *Loligo* and *Sepia*. How are they different? Locate the eyes and funnel on *Octopus*.

If available, examine a female specimen of the pelagic octopod, *Argonauta*. Note the paired dorsal arms with their broadly expanded membranes. The membranes secrete the shell, which functions as a bouyancy organ and also as an egg case. Examine the shell of *Argonauta*. Note that it has a general spiral shape.

Nautilus. *Nautilus* is the only extant cephalopod with a well-developed external shell. This species lives in the tropics of the western Pacific ocean, ranging from the surface to depths of about 500 m. *Nautilus* is an active diurnal hunter, grasping prey with its 38 suckerless tentacles.

1. Obtain shells of *Nautilus* and a few gastropods that have been cut in a medial longitudinal section. Compare the external morphology of the *Nautilus* to that of the gastropods. Is shell coiling the same in both of these shells? What type of coiling is found in *Nautilus*? Note that each successive whorl is in the same plane as previous ones.

2. Examine the outer surface of *Nautilus* and note any external color patterns that may be present. Also note the central depression called the **umbilicus**. This represents the axis of coiling that develops because the body whorls do not fully reach the axis. Is there external evidence of internal septa?

3. Examine the inner cut face of the *Nautilus* shell (Fig. 12.26) and locate the aperture, septa, **camerae** (chambers), nacreous layer, and connecting rings of the **siphuncle**.

4. What do the different-size camerae represent? Where was the major portion of the body of the living organism located?

5. Where is the perforation of the siphuncle located on the septal face?

6. In the axis of coiling, locate the umbilical perforation. This is the internal manifestation of the umbilicus noted above.

7. Also locate the **protoconch**, a small camera that is the initial chamber of the shell.

8. *Nautilus* regulates its bouyancy by altering the relative volumes of liquid and gas in the camerae through action of the siphuncle. Given this information, describe how *Nautilus* is oriented in the water. In which direction does it swim?

9. If available, examine a preserved *Nautilus* and observe the position of hood, eyes, tentacles, funnel, and shell.

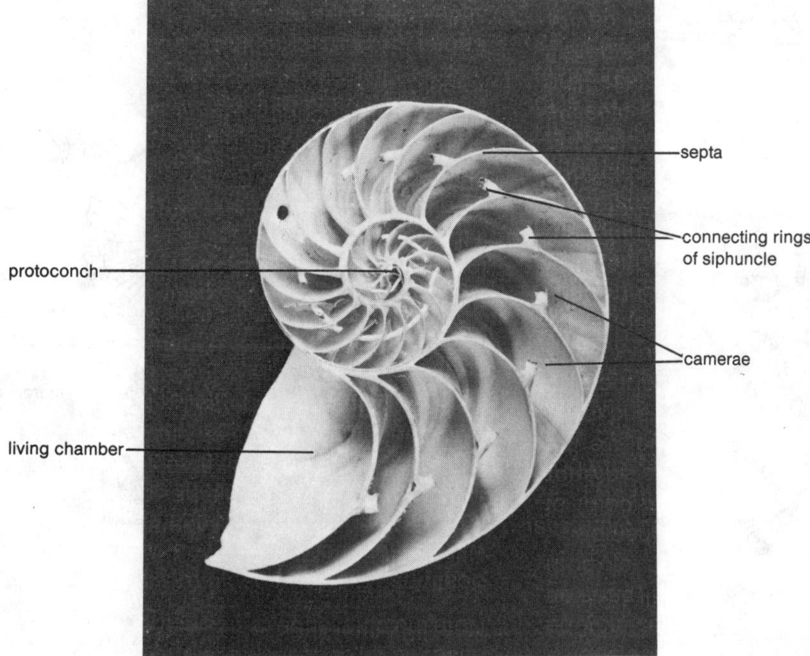

Figure 12.26. Photograph of the internal structure of the shell of *Nautilus*.

10. Examine the surface of the tentacles, noting the lack of suckers.
11. Determine the thickness of the hood. What is the function of this structure?
12. Compare the general anatomy of *Nautilus* to that of the three other cephalopods studied. Note that the funnel in *Nautilus* is not a closed tube; the edges only overlap.

Fossil Forms

Examine the fossilized cephalopods on display. In coiled forms, what type of coiling is exhibited? In ammonoids note the suture pattern formed by the septa where they meet the outer margin of the shell. How would you describe these patterns? In a sectioned specimen locate the camerae. Compare shells of ammonoids with that of *Nautilus*. Are there any similarities or differences? Do any of the fossil specimens exhibit evidence of the siphuncle (i.e., septal neck)? The internal anatomy of these forms is believed to be similar to that of *Nautilus*.

Not all fossil cephalopods had coiled shells; some were loosely coiled, arc-shaped, or even had straight shells (Fig. 12.27). Examine a fossil species that had a straight shell (e.g., *Geisenoceras*). Where were the camerae located with respect to the living animal in these forms? How did straight-shelled species maintain a balanced position in the water? Is there any evidence of cameral or siphonate deposits that could help counterbalance the animal?

If specimens are available, compare shells of fossil and extant scaphopods (e.g., *Dentalium*) to the cephalopods you have just examined. Determine the anterior-posterior axes, and the dorsal (concave) and ventral (convex) sides (Fig. 12.28). In complete specimens, locate the foot, captacula, mouth, and excurrent and incurrent openings to the mantle cavity.

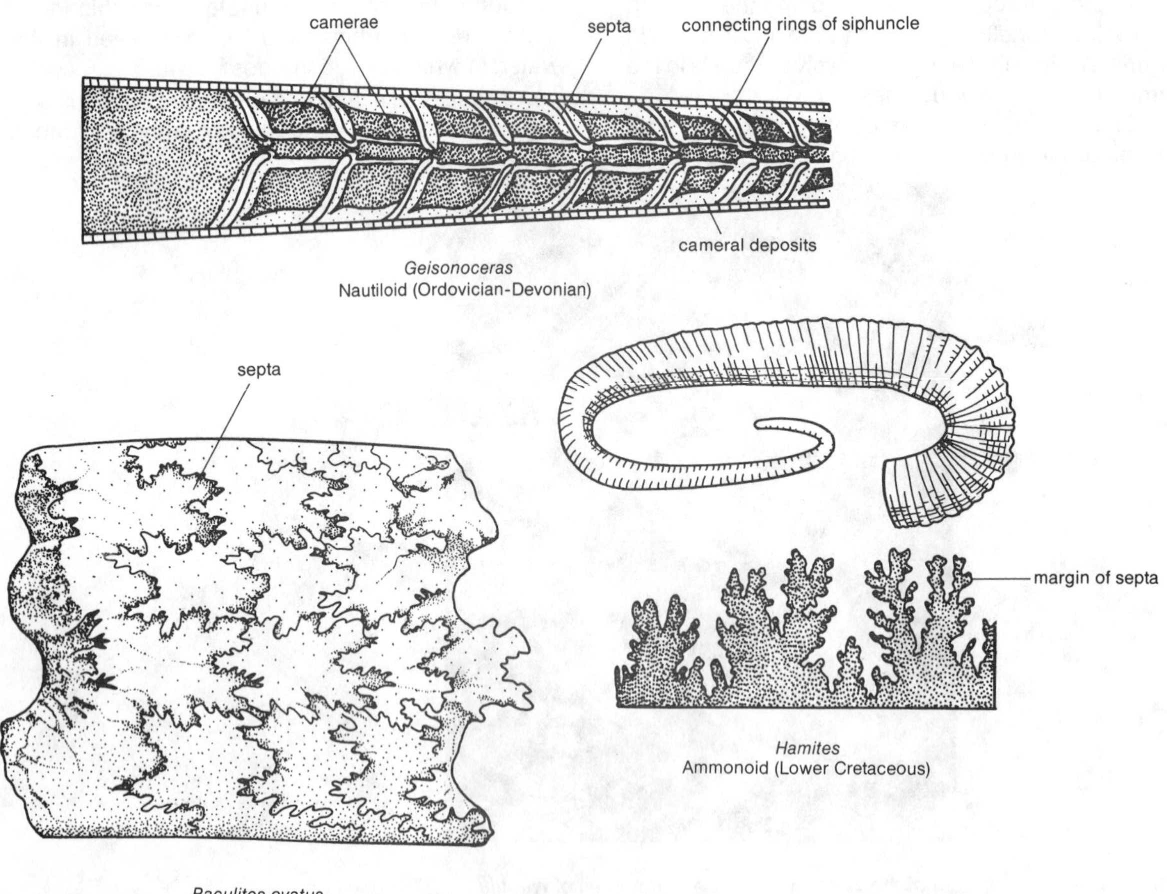

Geisonoceras
Nautiloid (Ordovician-Devonian)

Hamites
Ammonoid (Lower Cretaceous)

Baculites ovatus
Ammonoid (Upper Cretaceous)

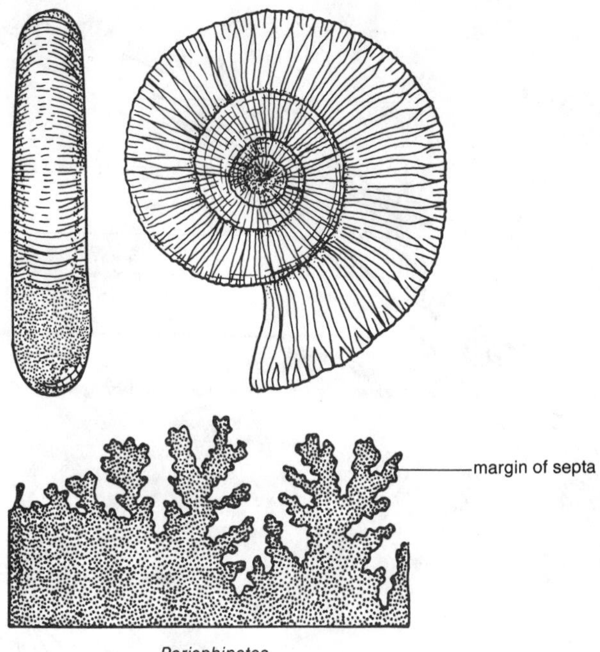

Perisphinctes
Ammonoid (Upper Jurassic)

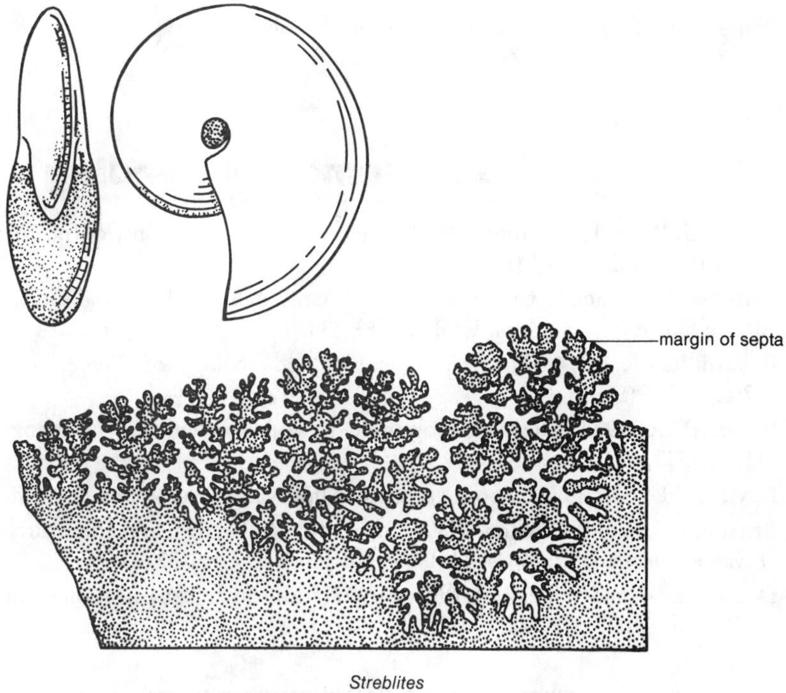

Streblites
Ammonoid (Upper Jurassic - Lower Cretaceous)

Figure 12.27. Example of fossil cephalopods. (After several sources.)

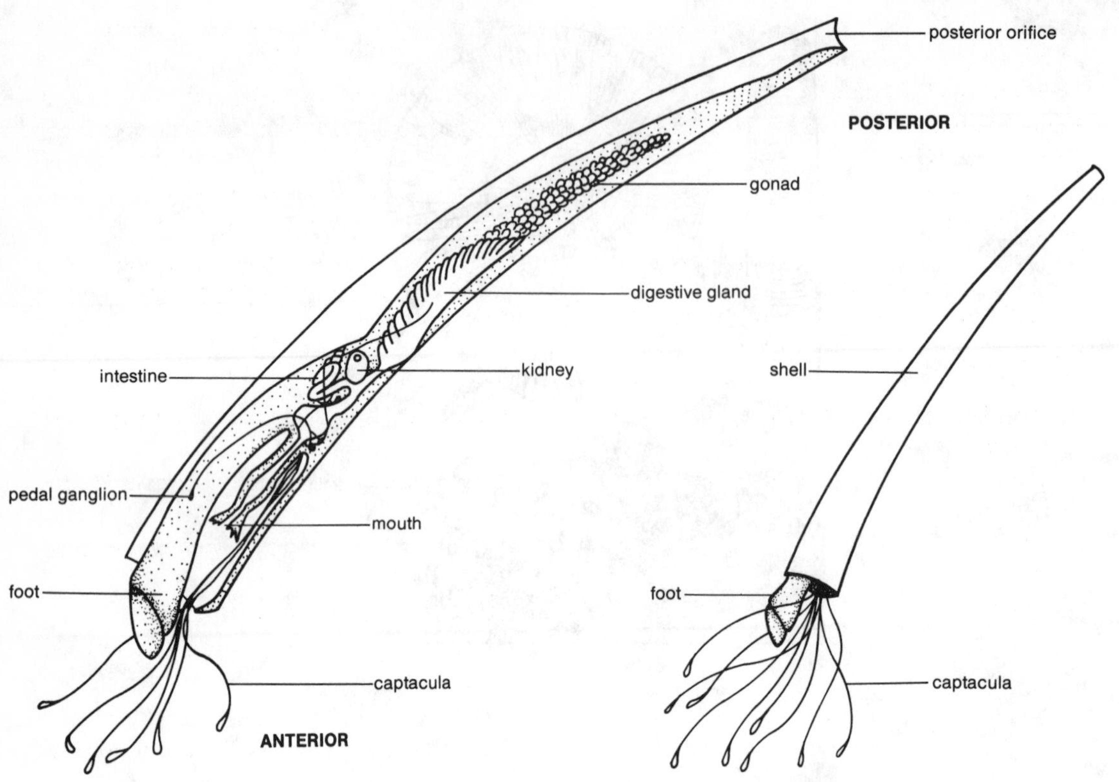

Figure 12.28. External and internal views of a scaphopod.

Supplemental Reading*

Barnes, J. R., and J. J. Gonor. 1973. The larval settling response of the lined chiton *Tonicella lineata*. Mar. Biol. 20:259-264. (P)

Bertness, M. D., and C. Cunningham. 1981. Crab shell-crushing predation and gastropod architectural defense. J. Exp. Mar. Biol. Ecol. 50:213-230. (GA)

Bilyard, G. R. 1974. The feeding habits and ecology of *Dentalium entale stimpsoni*. Veliger 17:126-138. (S)

Boyle, P. R. 1977. The physiology and behavior of chitons. Oceanogr. Mar. Biol. Annu. Rev. 15:461-509. (P)

Boyle, P. R. (ed.). 1983, 1986. Cephalopod life cycles. Vols. 1 and 2. Academic Press, New York. (C)

Britton, J.C., and B. Morton. 1982. A dissection guide, field and laboratory manual for the introduced bivalve *Corbicula fluminea*. Malacol. Rev. 3 (Suppl.) :1-82. (B)

Denton, E. J., and J. B. Gilpin-Brown. 1973. Flotation mechanisms in modern and fossil cephalopods. Adv. Mar. Biol. 11:197-268. (C)

*The supplemental literature listed here is coded to indicate the general topic covered by the paper: GN, general Mollusca; A, Aplacophora; B, Bivalvia; C, Cephalopoda; GA, Gastropoda; M, Monoplacophora; P, Polyplacophora; S, Scaphopoda.

Emerson, W. K. 1976. The American Museum of Natural History guide to shells. Alfred A. Knopf, New York. (GN)

Feder, H. M. 1972. Escape responses in marine invertebrates. Sci. Am. 227:92-100 (GN)

Fretter, V. (ed.). 1968. Studies in the structure, physiology and ecology of mollusks. Academic Press, New York. (GN)

Gilles, R. 1972. Osmoregulation in three molluscs: *Acanthochitona discrepens* Brown, *Glycymeris glycymeris* L. and *Mytilus edulis* L. Biol. Bull. 142:25-35. (P, B)

Hadfield, M. G. 1978. Metamorphosis in marine molluscan larvae: an analysis of stimulus and response. *In*: F.-S. Chia and M. E. Rice (eds.). Settlement and metamorphosis of marine invertebrate larvae. Elsevier, New York, pp. 165-175. (GN)

Harman, W. N. 1972. Benthic substrates: their effect on freshwater Mollusca. Ecology 53:271-277. (GN)

Hughes, R. N. 1986. A functional biology of marine gastropods. Johns Hopkins University Press, Baltimore. (GA)

Lane, F.W. 1960. Kingdom of the octopus: the life history of the Cephalopoda. Sheridan House, Dobbs Ferry, NY. (C)

Lemche, H. 1957. A new living deep-sea mollusk of the Cambro-Devonian class Monoplacophora. Nature (Lond.) 179:413-416 (M)

Linsley, R. M. 1978. Shell form and the evolution of gastropods. Am. Sci. 66:432-441. (GA)

Lutz, R. A., D. Jablonski, and R. D. Turner. 1984. Larval development and dispersal at deep-sea hydrothermal vents. Science 226:1451-1454. (GA)

Morris, P. A. 1975. A field guide to shells. 3rd ed. Houghton Mifflin, Boston. (GN)

Palmer, A. R. 1977. Function of shell sculpture in marine gastropods: hydrodynamic destabilization in *Ceratostoma foliatum*. Science 197:1293-1295. (GA)

Palmer, A. R. 1979. Fish predation and the evolution of gastropod shell sculpture: experimental and geographical evidence. Evolution 33:697-713. (GA)

Purchon, R. D. 1977. The biology of Mollusca. 2nd ed. Pergamon Press, Elmsford, NY. (GN)

Rhoads, D. C., R. A. Lutz, E. C. Revelas, and R. M. Cerrato. 1981. Growth of bivalves at deep-sea hydrothermal vents along the Galápagos Rift. Science 214:911-913. (B)

Runnegar, B., and J. Pojeta, 1974. Molluscan phylogeny: the paleontological viewpoint. Science 186:311-317. (GN)

Sabelli, B. 1979. Simon and Schuster's guide to shells. Simon and Schuster, New York. (GN)

Saunders, W. B. 1984. *Nautilus* growth and longevity: evidence from marked and recaptured animals. Science 224:990-992. (C)

Scheltema, A. H. 1978. Position of the class Aplacophora in the phylum Mollusca. Malacologia 17:99-109. (A)

Stanley, S. M. 1975. Why clams have the shape they have: an experimental analysis of burrowing. Paleobiology 1:48. (B)

Trueman, E. R. 1968. The burrowing activities of bivalves. Symp. Zool. Soc. Lond. 22:167-186. (B)

Van der Spoel, S., A. C. van Bruggen, and J. Lever (eds.). 1979. Pathways in malacology. Dr. W. Junk, The Hague. (GN)

Vermeij, G. J. 1975. Evolution and distribution of left-handed and planispiral coiling in snails. Nature (Lond.) 254:419-420. (GA)

Vermeij, G. J. 1978. Biogeography and adaptation. Harvard University Press, Cambridge, MA. (GN)

Ward, D. V. 1972. Locomotory function of squid mantle. J. Zool. (Lond.) 167:437-449. (C)

Wilbur, K. M. 1983-1988. The Mollusca. Vols. 1-12. Academic Press, New York. (GN)

PHYLUM ANNELIDA

Members of the phylum Annelida are schizocoelomate, segmented protostome worms. The most striking feature of annelids is their body construction, consisting of a series of similar compartments or segments. This feature is called **metamerism** (me-TAM-er-ism; G., *meta*, after + G., *mere*, a part). Annelida is the first phylum to be covered here that exhibits this characteristic; the repetition of body units in cestodes is not metamerism. Metamerism is probably the greatest advancement displayed by annelids. In two of the three classes (Polychaeta and Oligochaeta), the external rings generally correspond to an internal division of the body. In the third class (Hirudinea or leeches), external annulation does not correspond to internal segmentation. Most major organ systems are affected by metamerism (e.g., body wall musculature, nervous, excretory, and vascular). Metamerism is important in burrowing activities of annelids. The schizocoelom acts as a hydostatic skeleton when circular and longitudinal muscles of the body wall contract alternately. This is because the volume of each segment remains constant despite variations in their linear proportions.

EXERCISE 13

Phylum Annelida

Phylum Annelida (an-NEL-i-da; L., *annellus*, ring) is a diverse group of over 15,000 marine, freshwater, and terrestrial worms having a body that is composed of a longitudinal series of similar cylindrical compartments called **segments** or **metameres**. There is a spacious schizocoelom that is compartmentalized by septa in polychaetes and oligochaetes. In the class Hirudinea (leeches) there are many external rings called **annuli**. They correspond to internal segmentation, but not in a very simple fashion. Also, in the leeches the schizocoelom is filled with connective tissue and has been reduced to a complex system of sinuses. Many members of the class Polychaeta have well-developed lateral locomotory appendages called **parapodia**, which exhibit a high degree of neuromuscular organization. Parapodia are lacking in the other classes. In the polychaetes and oligochaetes, bristlelike **setae** usually are present on many segments; setae are absent in most leeches.

Classification

Three classes of annelids are commonly recognized (Polychaeta, Oligochaeta, and Hirudinea), although in some classification systems the latter two are grouped together (e.g., subphylum **Clitellata**) because of certain similarities (Table 13.1). The polychaetes are then placed in the subphylum **Aclitellata**. Myzostomaria is a fourth group of highly specialized ectoparasites of crinoids (phylum Echinodermata). They commonly are separated from the polychaetes. A fifth unusual group, the Aeolosomatida, is placed with the oligochaetes. The Archiannelida was a catch-all group comprised of primitive members from several families. Because affinities of its members to other groups are recognized, the taxon is no longer necessary.

1. Class Polychaeta. Mainly marine and a few freshwater worms, consisting of 25 orders and more than 85 families with 8000 species distinguished by either paired, lateral appendages called parapodia bearing numerous setae, or numerous anterior tentacles. Most with distinct head with eyes, palps, and cirri; some possess antennae. Examples include the sea mouse, *Aphrodita*, the scaleworm, *Lepidonotus*, the sand or clam worm, *Nereis* (= *Neanthes*), the Samoan palolo worm, *Palolo* (= *Eunice*), the parchment tube worm, *Chaetopterus*, the lugworm, *Arenicola*, and the fanworm or feather-duster, *Sabella*.

2. Class Oligochaeta. Terrestrial and aquatic annelids (more than 6000 species), with segmentation both external and internal. Parapodia are absent, but setae are present; a distinct, bandlike clitellum is formed on several anterior segments; three orders are recognized (Lumbriculida, Moniligastrida, Haplotaxida). Most oligochaetes are found in the latter order of 24 families. Examples include both aquatic tubificids (*Tubifex*) and terrestrial forms (the night crawler, *Lumbricus terrestris).*

3. Class Hirudinea. Predominantly fresh water leeches (about 700 species) with some terrestrial and marine forms having a conspicuous, superficial external annulation which is not reflected internally as segmentation in any simple fashion. Parapodia lacking; setae are usually absent, but a distinct clitellum is present. Three subclasses are recognized (Branchiobdellida, Acanthobdellida, Hirudinoidea). Examples include *Hirudo medicinalis* (which is still used under special circumstances for medicinal blood letting) and *Glossiphonia complanata* (which parasitizes snails).

Fossil Forms

Because they have few hard parts, annelids do not fossilize well. Most putative fossils consist of trails and burrows. However, several good specimens of the class Polychaeta are represented in the **Burgess Shale Fauna**. Therefore, this class dates at least to mid-**Cambrian** and probably earlier (Fig. 13.1).

Table 13.1 Comparative Annelid Characteristics

Characteristics	Polychaeta	Oligochaeta	Hirudinea
Metameric segments	Many and variable in number	Numerous to constant	Few (ca. 33) and constant in number; annuli vary in number
Parapodia	Well developed in mobile species, less developed in sedentary and sessile species	Absent	Absent
Setae	Many and varying in size and length	Usually, few, often reduced in size or absent, occasionally numerous	Generally absent
Jaws	Well developed or absent	Generally absent	Present in some
Sex	Mostly dioecious, few asexual and hermaphroditic forms	Mostly hermaphroditic by cross-fertiliz-ation; self-fertile and parthenogenic forms occur; a few via asexual fission	Hermaphroditic; no asexual reproduction
Gonads	Peritoneal lining seasonally proliferates gametes	Typically, four or fewer pairs of gonads in the anterior segments	A few to many paired testes; single, paired ovary
Fertilization	Generally external	Generally external in a cocoon (internal in one family	Internal, but embryos are deposited in a cocoon
Regenerative abilities	Very strong	Strong	Absent
Trochophore larva	Present	Absent; direct development in the cocoon (= capsule)	Absent; direct development in the cocoon (= capsule)

A. Class Polychaeta

The largest class of annelids is Polychaeta (POL-e-KE-ta; G., *poly*, many + G., *chaeta*, hair or bristle), comprising more than 8000 species that range in length from a few millimeters to more than 3 m. Two general groups were previously recognized in this diverse class: free-moving forms, the **Errantia** (er-RAN-she-a; L.,

TIME (IN MILLIONS OF YEARS)

	700–600	600–500	500–425	425–395	395–345	345–310	310–265	265–220	220–180	180–125	125–65	65–3	3–
ERAS	Precambrian	Paleozoic						Mesozoic				Cenozoic	
PERIODS		Cambrian	Ordovician	Silurian	Devonian	Missis-sippian	Penn-sylvanian	Permian	Triassic	Jurassic	Cretaceous	Tertiary	Q
PHYLUM HISTORY													

Q=Quaternary

Figure 13.1. Geologic history of the phylum Annelida.

erran, wander), and sedentary forms, the **Sedentaria** (sed-en-TA-re-a; L., *sedent*, sit), although the distinction is not completely clear-cut. Modes of food procurement also show great diversity and include raptorial predators, herbivores, scavengers, and sediment and filter feeders. Few polychaetes are parasitic.

The body comprises a series of similar cylindrical segments, each having a pair of lateral appendages called parapodia, which are important in locomotion and gaseous exchange (Fig. 13.2). The schizocoelom is compartmentalized by intersegmental septa, which are perforated to allow the coelomic fluid to pass from one segment to another. Many of the same body functions are performed in each segment. Not all polychaetes are uniformly segmented and the parapodia may be grouped into regions that differ in size, shape, and function. These specialized, functional regions are called **tagmata** (TAG-ma-ta; G., *tagma*, a division). Tagmatization is often quite pronounced in some forms.

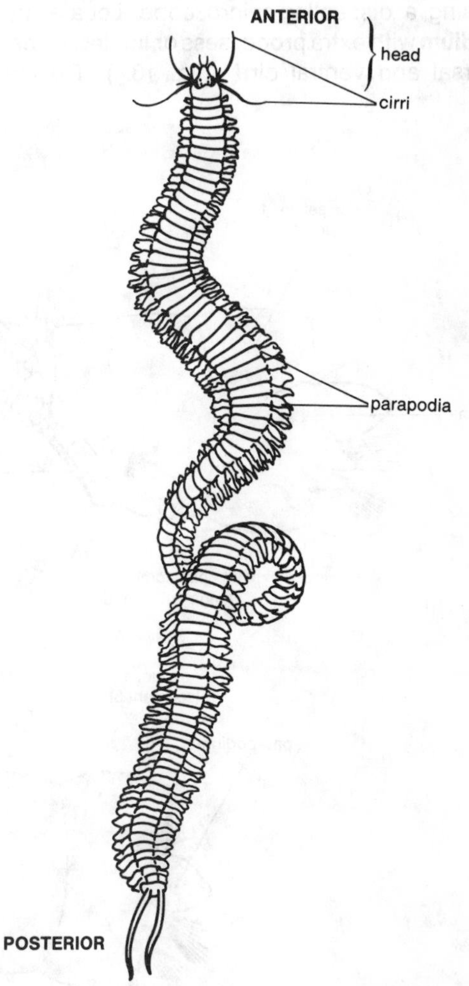

Figure 13.2. Dorsal view of *Nereis*.

Observational Procedure

***Nereis*.** Members of the genus *Nereis*, commonly known as rag, sand, or clam worms, are found in shallow marine and estuarine waters worldwide. These worms are often predatory on other invertebrates, but they may be scavengers or herbivores. In turn, they are eaten by bottom-feeding fish. One of several different species will be available for study.

External Morphology

1. Obtain a specimen of *Nereis* and examine both surfaces. There is a slight, midventral depression which runs the length of the worm. Observe the distinct, uniform segmentation with paired, lateral parapodia, and the appendages on the head region (Fig. 13.3). Examine the anterior region and locate the **prostomium**, eyes, **palps**, antennae, **peristomium** (segments 1 and 2), and peristomial **cirri**. Note the terminal position of the anus.

2. Is the pharynx extended or retracted? How can a living worm extend the pharynx? Examine an extended pharynx and locate the jaws and pharyngeal paragnaths.
3. Observe the worm under a dissecting microscope. Do the segments vary in size or morphology along the length of the worm? The first parapodia-bearing segment (number 3) is called the first **setiger**.
4. Using a scalpel, remove several parapodia from different places along the worm's body. Make sure they are cut close to the body. Place the parapodia in a watch glass, cover with water, and examine them using a dissecting microscope. Locate the **notopodium** with extra processess or **ligules**, setae, and dorsal and ventral cirri (Fig. 13.4). Do the various parapodia differ greatly from one another? Is there much evidence for tagmatization in *Nereis*?
5. Compare a whole mount of a parapodium with your specimen (Fig. 13.4). Species may be differentiated on the basis of morphological details of their parapodia.
6. Note that both the notopodium and neuropodium are supported internally by stiff, chitinous rods called the **acicula.**
7. Remove a few setae from several parapodia, place them in a drop of water on a slide, add a coverslip, and observe under a microscope. Note that the setae are not just uniform rod-shaped structures. How do the setae function in locomotion?

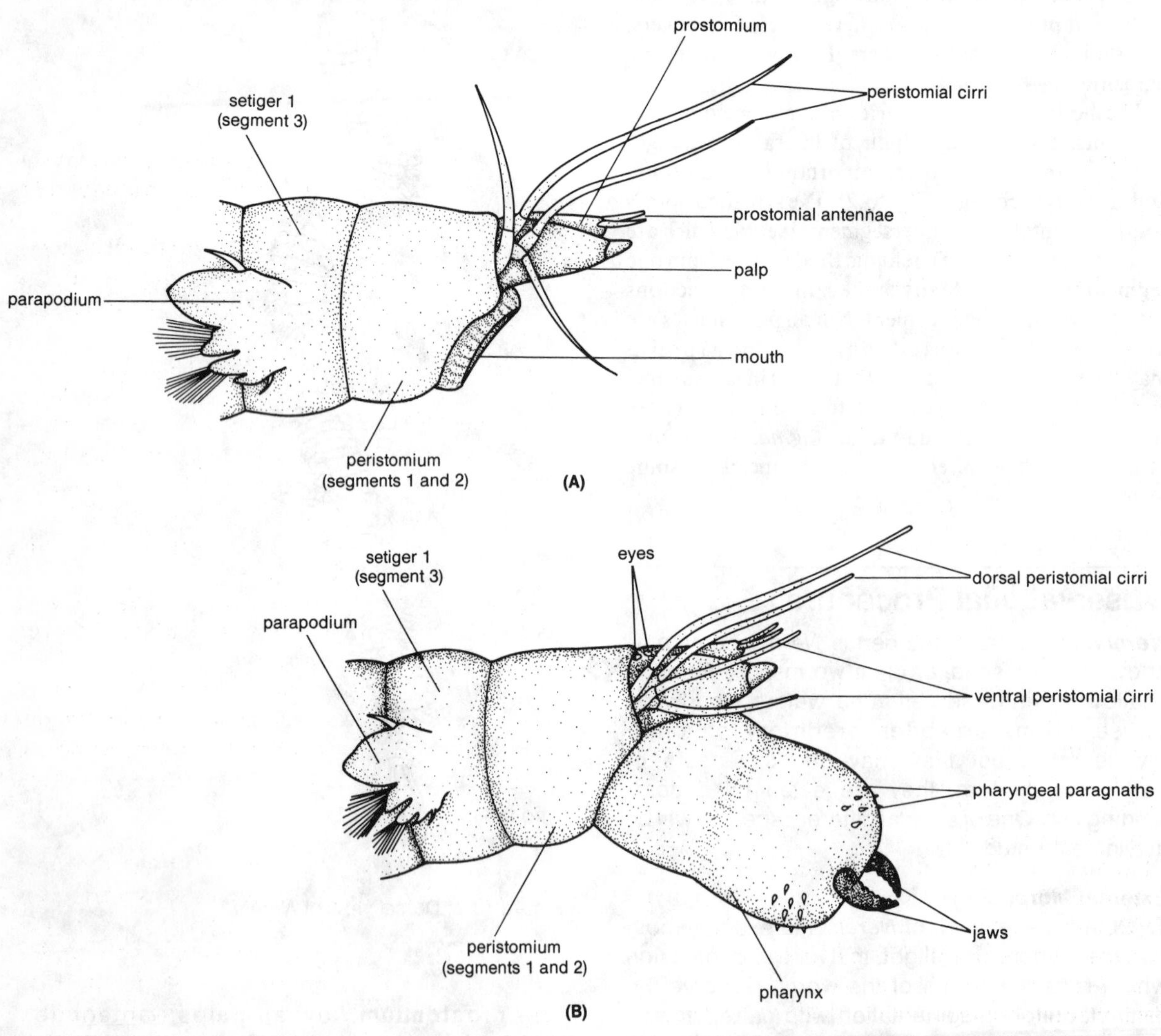

Figure 13.3. Lateral anterior view of the anterior end of *Nereis*. **(A)** Pharynx retracted. **(B)** Pharynx extended.

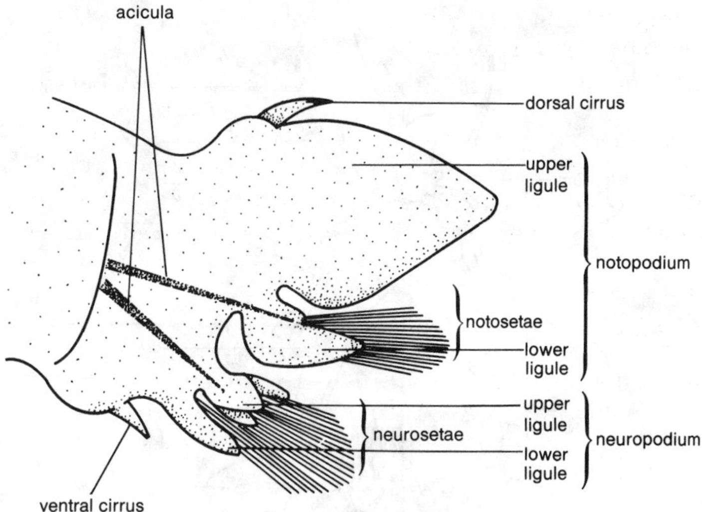

Figure 13.4. Typical *Nereis* parapodia.

Internal Anatomy

1. With a scalpel carefully make a parasagittal, shallow incision from the dorsal side, about one-third of the way from the anterior end.

2. With a small water dropper, siphon some liquid from the coelomic compartments of several segments. Examine this fluid by placing several drops on a glass slide, cover with a coverslip, and view under low power of a compound microscope. Gametes may be present, depending on the season the worm was collected. Abundant corpuscles should be seen. Following your examination of the coelomic fluid, cover the worm with water and proceed with the dissection.

3. Continue the dissection forward until the peristomium (segments 1 and 2) is reached. Pin the body wall to the wax pan as the incision is made.

4. Observe the numerous **septa**. These may have been torn during the parasagittal incision.

5. Count the segments posteriorly from the peristomium to the seventh segment (fifth setiger). Note the enlarged pharyngeal region (segments 3-6, setigers 1-5) and the absence of septa there. Locate the dorsal blood vessel, which runs longitudinally over the gut (Fig. 13.5). Coming off the vessel are paired lateral vessels that communicate with the parapodia and body wall. The lateral vessels may be difficult to observe in your specimen.

6. Trace the gut toward the posterior end. The esophagus occurs in segments 7 to 11 (setigers 5 to 9). In the region of the esophagus, paired esophageal cecae are attached laterally to the gut. Posterior to the esophagus the gut is designated as the stomach-intestine. In each segment are paired nephridia located ventrolaterally (Fig. 13.5).

7. Sever the gut between the fifteenth and twentieth segments. Carefully remove the gut up to the pharynx and sever again. Make sure that you do not damage the ventral nerve cord.

8. Once the pharyngeal region is reached, continue to remove the gut and note the position of the nerve cord. Just above the nerve cord the ventral blood vessel may be found. Locate the **circumpharyngeal connectives**, which encircle the pharynx and communicate with the dorsal ganglia or brain.

9. Find the nerves that extend anterior from the brain to the prostomium.

10. Observe the nerve cord along its length and note the arrangement of ganglia and lateral nerves that innervate each segment.

11. Obtain a prepared slide of a cross section of *Nereis* and locate the gut, dorsal and ventral gut suspensors, dorsal and ventral blood vessels, dorsal and ventral longitudinal muscle masses, and ventral nerve cord (Fig. 13.6).

Aphrodita. Observe a specimen of the sea mouse, *Aphrodita*. The dorsal side appears hairy because of fine setae arising from the notopodia. Turn the specimen over and identify the anterior and posterior ends. Locate the terminal anus. Note that the parapodia are similar along the length of the worm. As in *Nereis*, tagmatization is lacking in *Aphrodita*.

Arenicola. Observe a specimen of *Arenicola*. This

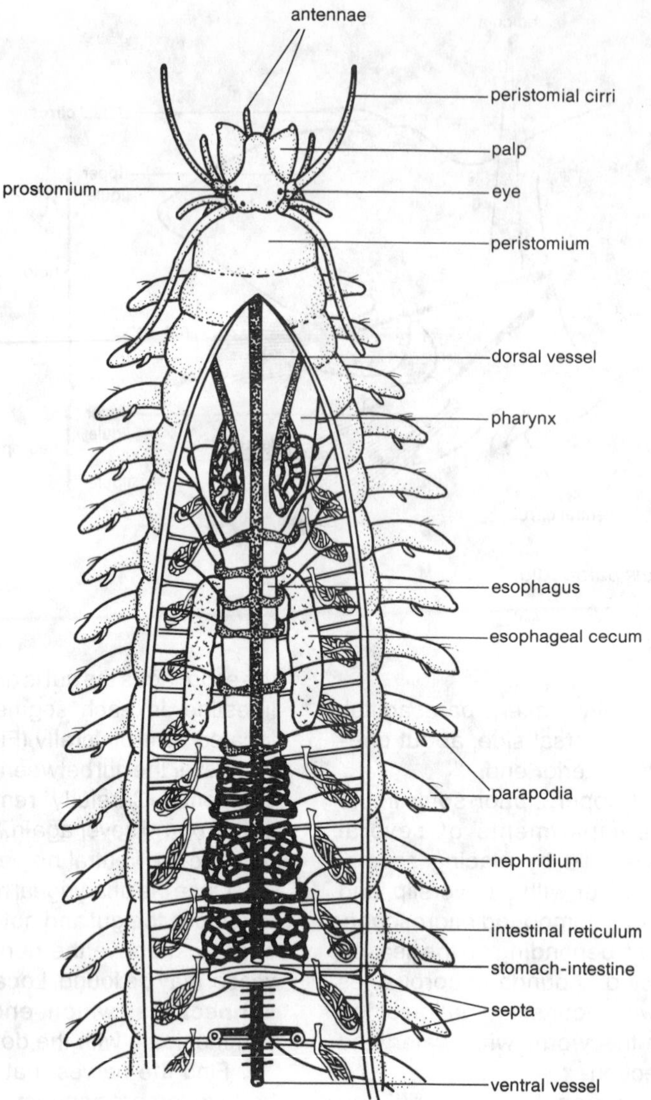

Figure 13.5. Internal dorsal view of *Nereis.*

worm, commonly known as the lugworm, lives on both U.S. coasts at about the low-tide mark, where it burrows in sandy to muddy sediments.

1. There is a shallow, narrow, linear depression running along the entire ventral length of *Arenicola*. Locate the terminal anus.

2. Note that the parapodia are modified and reduced in size in comparison with the other worms just studied (Fig. 13.7).

3. *Arenicola* also has distinct **annulation** between segments. Body segments extend from the first annulation behind a parapodium to the first annulation behind the next parapodium.

4. At the anterior end, a fleshy pharynx and buccal mass, covered with small papillae, may be extended from the mouth (Fig. 13.7). Just behind the pharynx is the minute prostomium that lacks appendages. Behind the prostomium, the peristomium may be seen; it is the first annulus.

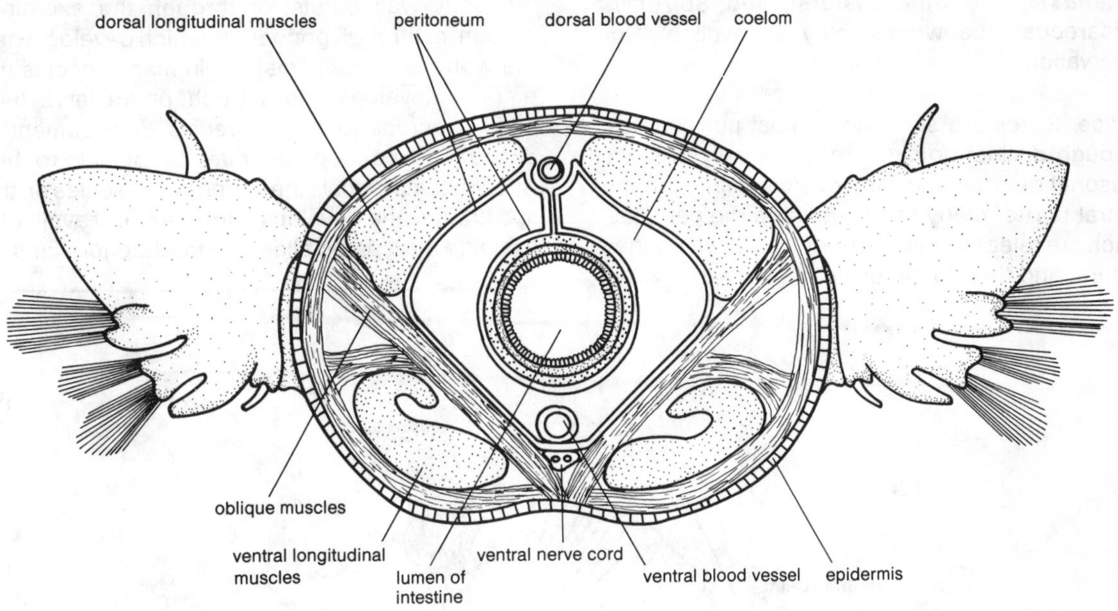

Figure 13.6. Cross section of *Nereis*.

5. Note that the body appears to be divided into three distinct tagmata: (a) anterior, prebranchial, (b) branchial, and (c) posterior, postbranchial (Fig. 13.7). The first tagmata past the peristomium may be identified by the presence of reduced notopodia. They increase slightly in size toward the posterior end. The neuropodia are very difficult to see in this region. Segments in the next tagmata have biramous parapodia and possess gills, branched outgrowths of the body wall. In this region the neuropodia appear as slightly raised welts with slits. Appendages are lacking beyond the last gill-bearing segment.

Chaetopterus. Obtain a specimen of *Chaetopterus* which has been removed from its tube.

1. Locate the spatulate anterior end. There is a small, dorsal, ciliated groove at this end. The outer lip of the mouth is the **peristomal collar**.

2. Beginning anteriorly, observe that the segments and parapodia are not of a uniform morphology. Rather, there is a differentiation of the animal into three regions: (a) anterior with uniramous parapodia, (b) middle with greatly modified parapodia, and (c) posterior with biramous parapodia (Fig. 13.8). These worms live in U-shaped tubes opened at both ends. Water containing food and oxygen is drawn into the tube at the worm's anterior end and forced out the posterior vent. This movement is accomplished by special modified parapodia called fans (segments 14, 15, and 16). Notopodia of segment 12 collect food. They are long and winglike (**aliform**) and contain mucus glands which produce a film of mucus that stretches between the notopodia, forming a baglike structure.

3. At segment 13 is a cup-shaped structure where food is concentrated and formed into small balls. These balls are passed anteriorly along the small, dorsal ciliated groove that extends to the mouth, a shallow, funnel-like depression.

4. Note that parapodia posterior to the fans are well developed and similar in shape, although they decrease in size at the posterior end. The anus is terminal.

Other Polychaetes. Diversity in the Polychaeta is very great. One should not have the impression that the few specimens reviewed here give a complete view of the diversity of this class. Other polychaetes may be available for your study, and you should examine additional specimens as your instructor directs. For example, free-moving forms such as *Glycera* (beakthrower or blood-worm) and *Lepidonotus* (scale worm) and sedentary species such as *Pectinaria* (= *Cistenides*) (trumpet worm), *Diopatra* (plume worm) and *Sabella* (fanworm),

Sabellastarte (featherdusters), and *Spirorbis* (calcareous tube worms) may be available for observation.

Larvae. Sexes are separate in most polychaetes, although distinct gonads are not usually present. Seasonal outgrowths of the coelomic lining in the ventral part of many segments form the gametes, which are released into the body cavity, where they mature, and then to the environment, either when the body wall bursts, or through the excretory system, or through gonoducts which develop when the worm is sexually mature. In many species the embryo develops into a trochophore larva (Fig. 13.9), whereas in other species development is direct. The trochophore may be planktotrophic. After metamorphosis the young worm settles on the substratum and develops into an adult. If available, examine a prepared slide of a trochophore larva.

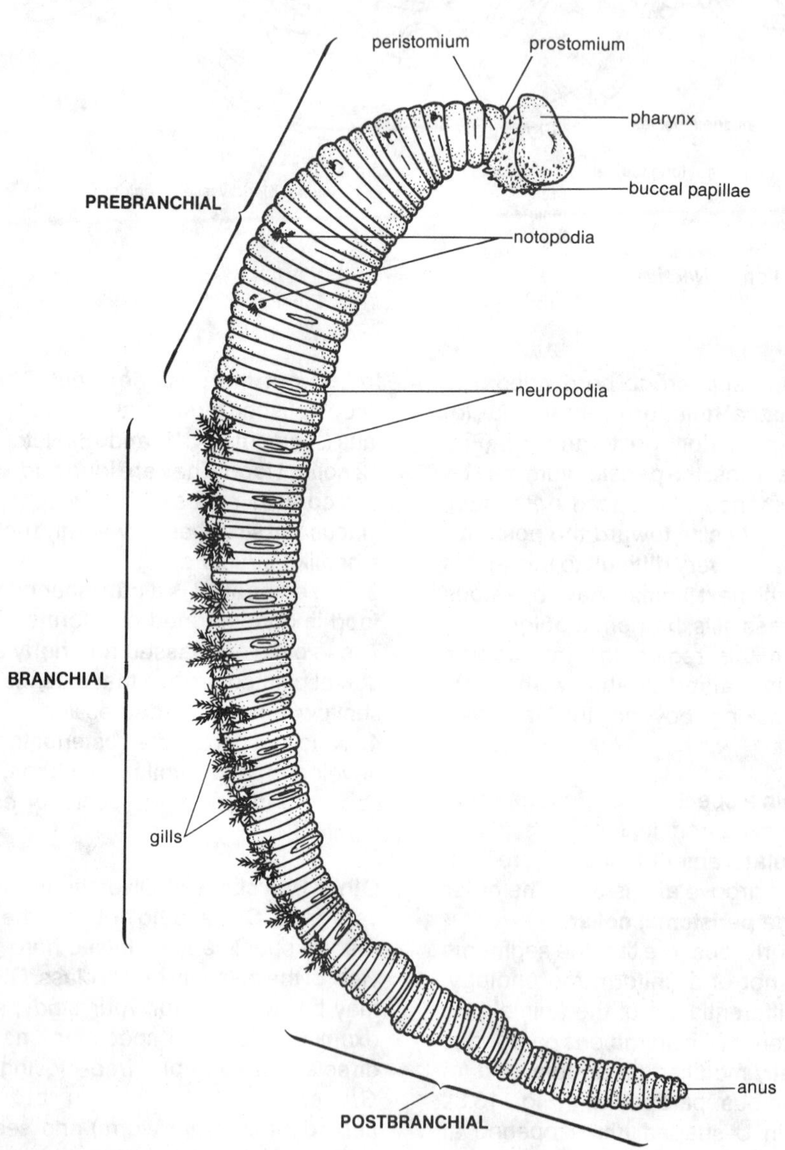

Figure 13.7. Lateral view of *Arenicola* with pharynx extended.

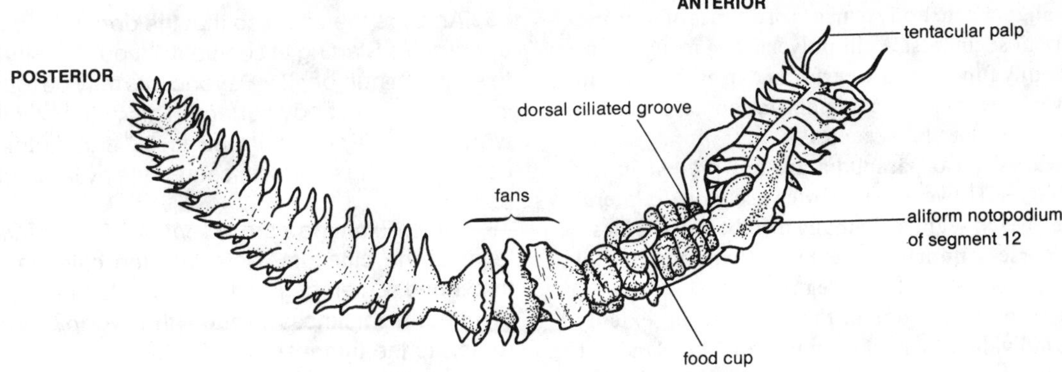

Figure 13.8. Dorsolateral view of *Chaetopterus*.

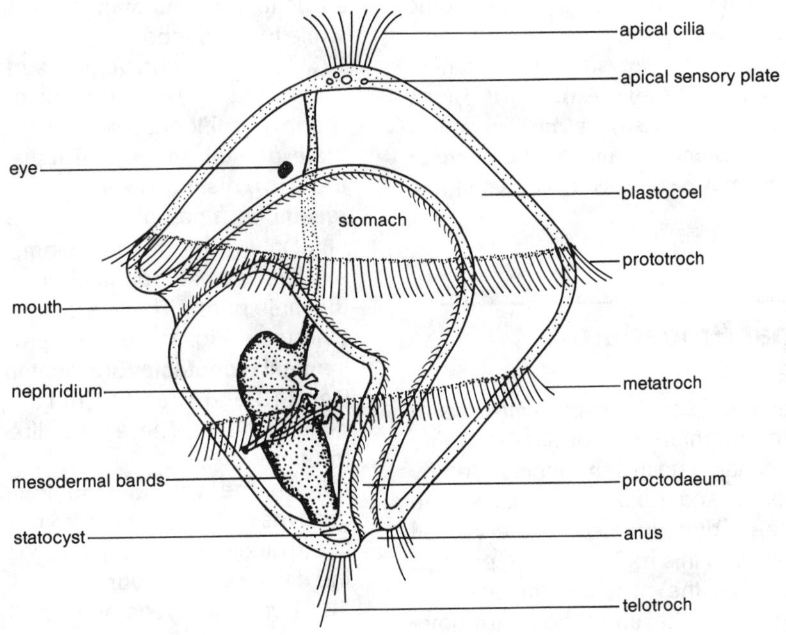

Figure 13.9. Typical polychaete trochophore larva.

B. Class Oligochaeta

The Oligochaeta (OL-e-go-KE-ta; G., *oligo*, few + G., *chaeta*, long hair or bristle) is the second largest class of annelids, containing more than 6000 species. Although oligochaetes are commonly called earthworms, the term is technically incorrect because many species are aquatic. Most inhabit fresh waters, but there are many microscopic marine species too. Most feed on dead and decaying materials, but earthworms ingest soil to feed on the microfauna and flora. The night-crawler represents a group that browses on grass and leaves on the surface of the soil. Oligochaetes range in length from <0.5 mm to more than 7 m.

The oligochaete body consists of a series of uniform cylindrical segments. As in polychaetes, many of the same bodily functions are performed in each segment. However, the extensive tagmatization of segments present in some polychaetes is not found in the oligochaetes. Also parapodia are not present in the oligochaetes. However, setae often occur on the lateral surface of each segment in many members of the class. Some possess numerous setae, perhaps up to 200 around the equator of each segment.

One prominent region of the epidermis in sexually mature individuals is the **clitellum**, a bandlike, glandular structure which may be one to many cells thick. The size of the clitellum varies depending on the species and the time of the year the specimen was collected. The clitellum produces mucus during copulation and later secretes a **cocoon** or capsule, where embryonic development takes place.

In this exercise *Lumbricus terrestris* will be studied in detail as the class representative. Although it is not a typical species of oligochaete, it is a convenient organism with which to work. Some sections of this exercise (e.g., viewing slides) may be omitted to conserve time for other activities.

Observational Procedure

External Anatomy

1. Obtain a preserved specimen of *Lumbricus* or one recently killed in chloretone or similar agent. Note the lack of any tagmata in this species. Identify the worm's anterior and posterior ends. At the anterior end is the mouth, located on the ventral portion of segment 1. In this region are two parts: a middorsal projection, the prostomium, and the peristomium (segment 1) that surrounds the mouth (Fig. 13.10).

2. At the posterior end, locate the **pygidium** bearing the anus.

3. Gently pull the animal between your fingers to feel the setae. Eight setae in four pairs are found on every segment except the first and last. One pair is located at each ventrolateral edge, and one pair is found on each lateral surface. Setae function in locomotion, copulation, and as sensory bristles.

4. *Slide*: Select a cross section of *Lumbricus* which shows setae. Note the position of the setae and the muscles attached to them (Fig. 13.11). Setae are highly birefringent and may be easily identified in the section by examining it for birefringent structures.

5. Arrange the worm so that the dorsal surface is uppermost. Starting at segment 1, count posteriorly to segment number 31. Beyond this may be found a swelling on the body surface. This is the clitellum, which covers segments 32 to 37 and forms the cocoon (or capsule) from secretions by large gland cells.

6. *Slide*: Examine a cross section of *Lumbricus* at level of the clitellum. Note that the clitellum is a thickened region of glandular cells. Is the clitellum of uniform thickness around the worm? Which region is the thinnest? Under high magnification, observe the large, globular secretory cells that make up the bulk of the clitellum. The dark-staining cells closest to the surface secrete mucus and cocoon material; those cells with less stain secrete albumin for nourishing the developing embryos while in the cocoon.

7. Turn the worm ventral surface uppermost and count the segments from the peristomium until you reach the fifteenth segment; observe the paired **male genital pores**. The male pores are tiny, but there is a distinct swelling around the pores which resembles a pair of lips.

8. On the fourteenth segment locate a pair of extremely small openings in the same position as the male pores. These are the external pores of the oviducts (Fig. 13.10). The paired openings of the **seminal receptacles** are located between segments 9 and 10, and 10 and 11, but they are very small and difficult to find. There is no direct transfer of sperm in *Lumbricus*.

9. On the ventral side locate the two sperm grooves and trace them from each male gonopore to the clitellum. Review the process of copulation in the earthworm in your textbook.

10. Two other types of pores that may be seen on the surface of *Lumbricus* are the **nephridiopores** and **coelomic pores**. Nephridiopores are paired and located by the segmental furrows just anterior and slightly above the ventrolateral setae.

11. Coelomic pores are single and release fluid which helps keep the worm's surface moist. They are located middorsally in the intersegmental furrows in the middle and posterior body.

Body Wall. The outermost layer of the body wall is the noncellular **cuticle**. This is secreted by special cells in the epidermis. Microscopic pores present in the cuticle allow epidermal secretions to pass to the outside of the body.

1. Scrape some cuticle from the epidermis. Place the sample on a glass slide, add a drop of water, and

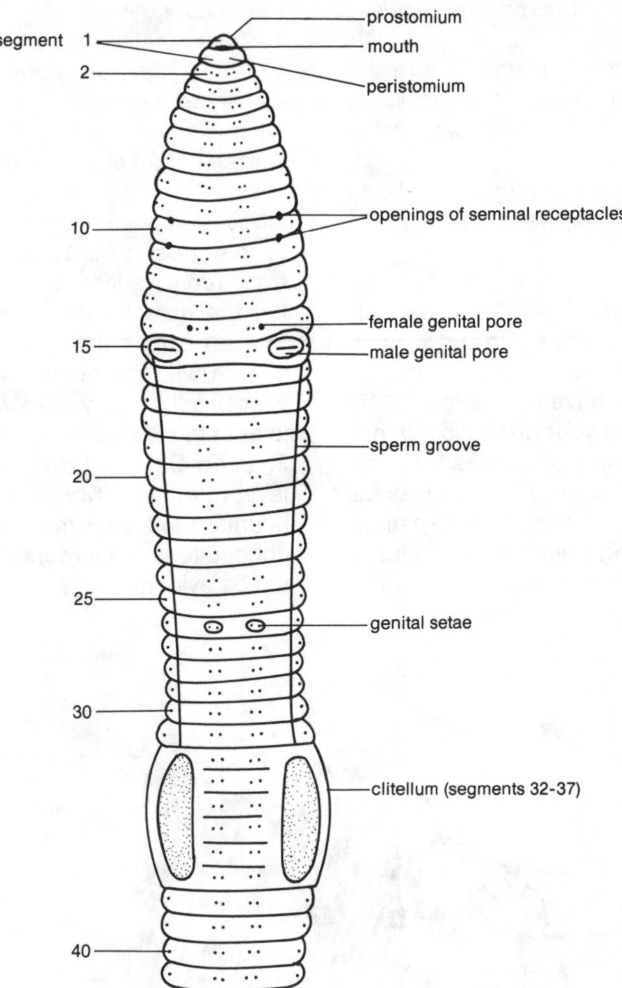

Figure 13.10. Ventral anterior view of *Lumbricus terrestris.*

apply a coverglass. Observe under low power of a compound microscope. What do you observe? The smooth texture of the cuticle plus mucoid secretions give the characteristic slippery feel to the worm's surface. Immediately below the cuticle is the epidermis, composed of several kinds of columnar epithelial cells.

2. *Slide*: Obtain a median longitudinal section of *Lumbricus* and examine the body wall using high magnification. Note the different cell types found in the epidermis. Inward from the epidermis are two muscle-tissue layers. The outermost of these is the circular muscle layer. Scattered among the fibrils are brownish pigment granules.

3. Note that the circular muscles in any segment are not continuous with the muscles in other segments.

4. The innermost muscle layer consists of longitudinal muscle cells that are continuous from segment to segment.

5. Just below the longitudinal muscles is the peritoneum, which lines the schizocoelom and which is continuous with the septa.

6. Between each pair of segments, locate a septum.

Internal Anatomy

1. Extend the worm full length with the ventral surface down. Fasten the specimen to the bottom of a dissection pan by placing pins through the prostomium and the pygidium.

2. Using a razor blade or very fine scissors, make a *shallow* midsagittal incision through the anterior third of the body. Be *very careful* not to disturb the underlying viscera. *Work slowly and patiently*; a *rushed dissection can render the specimen useless for detailed anatomical study.*

3. About halfway along the cut, start to turn the body wall over laterally and pin it to the pan. Work posteriorly, continuing to pin the body wall to the pan as you cut.

4. Note the septa which connect the gut to the body wall. The coelom is found between each pair of septa.

5. Add water to the pan so that the worm is covered completely. All observation should be made with the worm submerged to prevent drying and to allow structures to float freely for better observation.

6. Several organ systems have been exposed by your dorsal incision. Study your dissection with a dissecting microscope. Some structures, such as the seminal funnels and the vas deferens, may be difficult to locate. However, students who are patient will be successful at locating them. These structures also should be observed in prepared microscopic sections.

Circulatory System (Fig. 13.12).

1. The dorsal blood vessel is located on the upper surface of the gut. In preserved worms this vessel usually appears as a light brown line. Close examination will reveal the small vessels joining the main vessel at right angles in each segment. The dorsal vessel is a collecting tube and pulsates, moving the blood from posterior to anterior.

2. In segments 7 to 11 locate five pairs of hearts that extend dorsoventrally around the esophagus, connecting the dorsal and ventral vessels (Fig. 13.12). These hearts pulsate, contracting in coordinated rhythm with the dorsal vessel. The ventral blood vessel will be observed when the gut is removed later in the dissection.

3. *Slide*: Obtain a cross section of *Lumbricus* at the level of a heart. Note the position and size of the heart(s) and the thin muscular wall. In most sections, connections to the dorsal and ventral vessels will not be evident. Why?

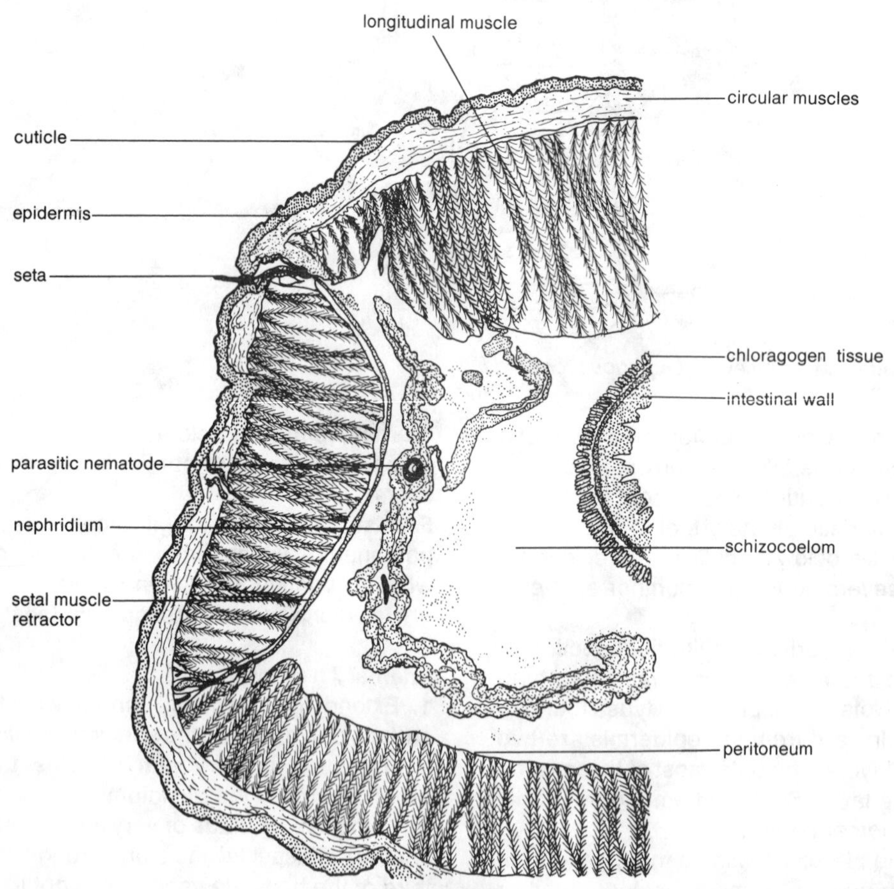

Figure 13.11. Cross section of *Lumbricus terrestris* showing the setae and other internal structures.

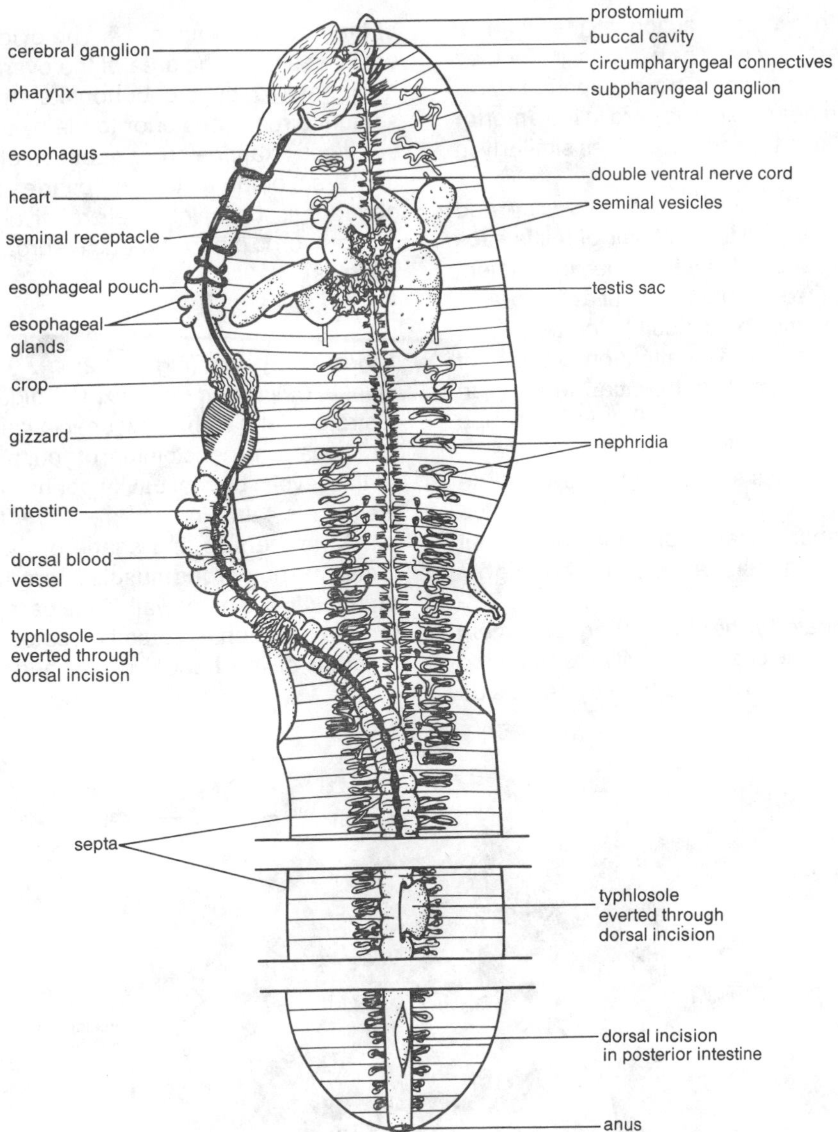

Figure 13.12. Internal dorsal view of *Lumbricus terrestris.*

Reproductive System (Fig. 13.12). Posterior to the second pair of hearts in segment 8 are three pairs of large light-colored structures that almost fill the whole area of the worm in this region. These are the **seminal vesicles.** Their bases are attached ventrally, but they expand dorsally to occupy space in segments 9 to 12, bulging septum 12/13.

1. Using a scalpel, remove a portion of one seminal vesicle. What is the general internal texture of this structure?

2. *Slide*: Obtain a cross section of *Lumbricus* at the level of the seminal vesicles and observe the large size and general construction of these structures. Note that they are not just large sacs. What is their function?

3. Scan the seminal vesicles for sporocysts of the gregarine protozoan parasite *Monocystis.* These structures resemble a small sphere filled with several minute spindle-shaped structures (Fig. 13.13).

Under the lobes of the two anterior pairs of seminal vesicles (segments 9, 10) are the paired spherical seminal receptacles (Fig. 13.12), where sperm are stored.

4. *Slide*: Obtain a cross section of the **seminal receptacles (spermatheca)**. The vesicles are located ventrolaterally and may contain many spermatozoa. Testes are found near the nerve cord in the anterior of segments 10 and 11, and the ovaries similarly in segment 13.

5. Near the bases of the middle seminal vesicle and located midventrally are the first pair of **testis sacs** (segment 10). The second pair is located in segment 11. The **sperm funnels** are buried under a membrane, the testis sac, and may be difficult to locate.

6. If this is the case, turn your attention to segment 15 and locate the vasa deferentia, which run anteriorly along the body wall. Follow that tube anteriorly as it leads to a funnel.

7. *Slide*: View a cross section at the level of the sperm funnels. Sperm funnels may be located easily, as the sperm stain dark. In longitudinal sections, sperm funnels appear to be highly convoluted.

8. Ovaries are located in the anterior part of segment 13. Eggs set free in the coelomic cavity are collected by the **ovarian funnels**. Oviducts open to the outside of the body in segment 14. The oviduct will not be easily seen, but the area of the ovarian funnel may be identified by the buttonlike opening on the septum directly posterior to the ovary.

9. *Slide*: Examine cross sections of *Lumbricus* at the levels of the testes and ovaries and locate the gonadal tissues. Note their relatively small size in comparison to other reproductive organs (i.e., seminal vesicles).

Digestive System (Fig. 13.12)

Buccal Cavity and Pharynx. The digestive system is a straight tube with a few specialized regions.

1. Locate the peristomium of your preserved worm. At this level a buccal cavity opens into the pharynx (segments 4 and 5).

2. Note the size and shape of the pharynx, and locate the numerous muscle fibers that connect the pharynx to the body wall. Take care not to damage the brain, which is dorsally located at the anterior junction of the buccal cavity and pharynx (Fig. 13.12).

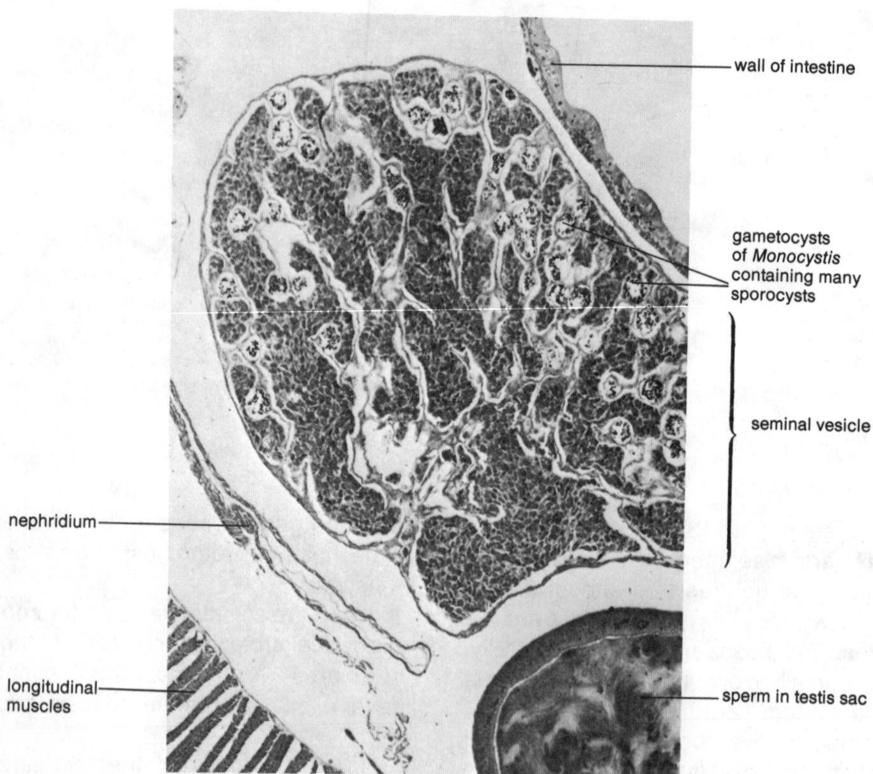

Figure 13.13. Cross section of *Lumbricus terrestris* at the level of the seminal vesicles.

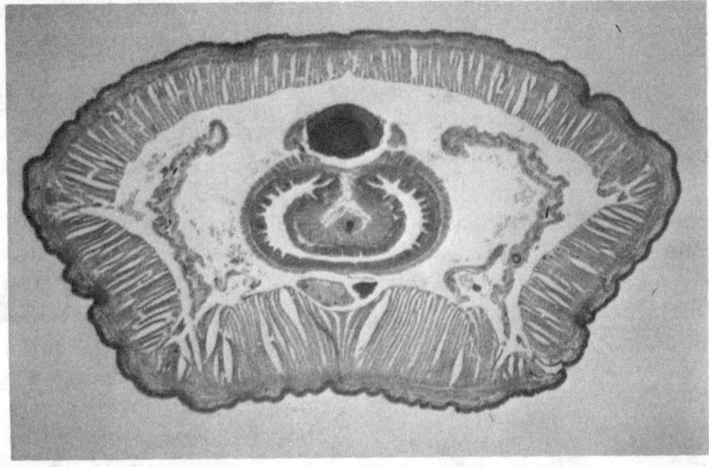

Figure 13.14. Cross section of *Lumbricus terrestris* at the level of the intestine.

3. Using your finger, gently press on the pharynx and note the general texture of this organ.

4. *Slide*: Examine a cross section of the pharynx and note the distribution of the powerful pharyngeal muscle. What is the function of this powerful muscle?

5. Observe a medial section through the anterior of the worm and locate the pharynx.

6. In this section note the pharyngeal dorsal diverticulum. What function might this structure perform?

7. Note the muscle fibers that connect the pharynx dorsally to the body wall.

Esophagus. The esophagus is a long tube (segments 6 to 13) that carries food from the pharynx to the crop (segments 14 and 15).

8. Follow the pharynx posteriorly where it opens into the esophagus (Fig. 13.12).

9. On either side of the esophagus are two pairs of bulbous evaginations (segments 11 and 12) called **calciferous glands.** One of their functions is regulation of calcium concentration in bodily fluids, and consequently ionic and pH balance.

10. *Slide*: Locate these glands in a cross section of the esophagus. Note the cavities between elongate cells of the gland.

Crop and Gizzard. From the esophagus food passes into a thin-walled storage organ called the **crop** (segments 14 and 15) and then into a muscular, cuticle-lined **gizzard** (segments 16 to 19), which is a food-grinding organ (Fig. 13.12).

11. Locate these organs in your dissected specimen. Using your finger, gently press on the crop and gizzard and compare the general texture of these organs.

12. *Slide*: With an anterior medial section of *Lumbricus*, locate the crop and gizzard; compare the thickness of their walls.

13. Also observe these two organs in cross section. Are muscles present in the walls of the crop or gizzard? How are the muscles distributed in the crop?

14. Observe the inner lining of the crop and gizzard. Why would the gizzard have a thick cuticular lining?

Intestine

15. Continue the midsagittal incision of the body wall for the full length of the worm, pinning the body wall to the dissection pan. Trace the intestine, which extends from the gizzard to the anus (Fig. 13.12).

16. Observe the external morphology of the intestine along its length. Note that externally there is only minor regionalization for the remainder of the gut.

17. Remove 1-cm sections from anterior, mid, and rectal portions of the intestine. Cut these sections open lengthwise, place them in small glass dishes, and observe the lining with the aid of a dissecting microscope.

18. In the anterior section, note the presence of the **typhlosole**. Is the typhlosole found in the other sections of the intestine? What does it look like in each section? What is the function of the typhlosole?

19. *Slide*: Examine a cross section of the intestine. The intestine is just below the dorsal blood vessel. Both are surrounded by a special tissue called **chloragogen**. Suspended from the dorsal side of the intestine is the typhlosole (Fig. 13.14). From what region of the intestine was your section made? How can you tell?

20. Note that chloragogen tissue is found inside the typhlosole also. Locate the thin layer of cells between the intestinal epithelium (gastrodermis) and the chloragogen tissue. This layer is sometimes called the submucosa, but it is actually a layer of circular muscle.

Excretory System (Fig. 13.12). A pair of **metanephridia**, one on each side, is located in the lateral area of each body segment, with the exception of the first two segments and pygidium.

1. In the region of the gizzard and posterior to it, examine the inner body wall surface of each segment and locate the metanephridia.

2. Using a dissecting microscope, examine a metanephridium in situ. The nephrostome and nephridiopore will be difficult to locate.

3. *Slide*: Compare this structure to a whole-mount slide of a metanephridium. Locate the convoluted tubules, bladder, and the vessels of the circulatory system.

4. Locate the nephridiopore in a prepared cross section. What other structures of the metanephridium do you observe? *Lumbricus* is a host to a parasitic nematode, *Rhabditis maupasi*, which is believed to enter the worm via nephridiopore or genital openings (see phylum Nematoda).

Nervous System (Fig. 13.12).

1. Dorsal to the buccal cavity, at a point where the buccal cavity and the pharynx meet, locate the brain or **suprapharyngeal ganglia**. Projecting anteriorly into the prostomial region are the prostomial nerve fibers. Extending laterally from each cerebral ganglion are the **circumpharyngeal connectives** that lead to the **subpharyngeal ganglia**.

2. Using scissors, carefully cut through the pharynx and esophagus and remove that section to expose the circumpharyngeal connectives and the subpharyngeal ganglia.

3. Observe the size difference between the supra- and subpharyngeal ganglia. Note the ring form made by the circumpharyngeal connectives.

4. *Slide*: Locate the suprapharyngeal and subpharyngeal ganglia in cross sections made at the appropriate levels.

5. Trace in your dissection the ventral nerve cord as it extends posteriorly from the subpharyngeal ganglia. Just above the nerve cord is the ventral blood vessel.

6. In each body segment locate a ganglionic enlargement of the nerve cord.

7. Find the three pairs of lateral nerves that arise from the nerve cord in each segment. Lateral nerves are more or less independent in all their sensory and motor functions, allowing localized action to one segment or several widely separated or even contiguous segments.

8. *Slide*: Examine several cross sections of *Lumbricus* below the level of the subpharyngeal ganglion and locate the ventral nerve cord.

9. Just above the nerve cord, locate the ventral blood vessel and within the nerve cord, find the **midventral subneural blood vessel** and two **lateral neural blood vessels**.

10. In the dorsal area of the nerve cord note a centrally placed tubelike body with two smaller ones subequal in size to either side. These are **giant fibers**. Giant fibers are large neurons that transmit impulses unimpeded the full length of the worm, thereby permitting a sudden withdrawal of the worm down its burrow.

11. *Slide*: Examine the structure of the supra- and subpharyngeal ganglia and the ventral nerve cord using cross sections. Below the giant fibers are two circular structures which are composed of motor and sensory neurons. In some cross sections you may be able to locate lateral nerves which leave the ventral nerve cord in each segment.

12. *Slide*: In medial sections the ventral nerve cord may be traced posteriorly for some distance. Occasionally, segmental ganglia and lateral nerves may be seen.

Live Specimens. Obtain a live *Lumbricus* and gently hold opposite ends of the worm in each hand, noting the strength the worm exhibits on contraction. The dissection of *Lumbricus* described herein also may be done with an anesthetized worm. Some anatomical details are best seen with a live worm, including other stages of the parasite, *Monocystis*. An excellent view of the functioning of the dorsal vessel and hearts also is possible. Worms anesthetized in an ice-water bath or a solution of 1 part magnesium sulfate (saturated solution) to 3 parts water for about

1 hour will be provided by your instructor. As with the preserved worm, the animal should be kept submerged during the dissection. Razor blades do not work well on a living worm; very fine scissors should be used. If time permits, your instructor will provide directions for observing earthworm locomotion.

Other Species. *Lumbricus terrestris* is not a typical oligochaete, but it is a fairly easy one with which to work. If time permits, your instructor may provide specimens of other oligochaetes, and observational instructions for their study (e.g., *Aeolosoma, Enchytraeus, Dero, Stylaria, Tubifex*).

C. Class Hirudinea

Members of the class Hirudinea (or Hirudinoidea) (hi-ru-DIN-e-a; L., *hirudin*, leech), commonly known as leeches, are a small group (ca. 500 species) of highly specialized annelids. They are mainly freshwater, but terrestrial and marine forms are known. Most leeches range in length from less than 1 mm to over 5 cm. The very largest may reach more than 0.25 m. Leeches are popularly identified as blood-sucking parasites. The large leeches, *Hirudo medicinalis* (Europe) and *Macrobdella decora* (native to the United States), were used in bloodletting during the eighteenth and nineteenth centuries. Under some rather unusual circumstances they still are used today. Heavy infestations of leeches are not uncommon and may cause anemia or even substantial mortality to fish in hatcheries and in natural habitats. Most leeches are temporary parasites, abandoning their host after becoming engorged with blood. However, as a class, leeches are not strictly ectoparasitic; many are predaceous or scavengers.

Leeches have fewer segments than do other annelids. What appear to be segments in leeches are called annuli and are actually surface subdivisions of individual segments. This is similar to what was observed in the polychaete *Arenicola*. Unlike other annelids, the number of segments in leeches is fixed at 34 (some accounts, which do not count the prostomium, report 33). The body is rounded to flattened and may be divided into five regions: (1) head (segments I to VI), (2) preclitellum (segments VII to IX), (3) clitellum (segments X to XII), (4) trunk (segments XIII to XXIV) and (5) a terminal region (segments XXV to XXXIV) with an anus (segments XXV to XXVII) and posterior sucker (segments XXVIII to XXXIV).

Observational Procedure

External Morphology

1. Obtain one or two preserved leeches, such as *Hirudo medicinalis* and *Haemopis grandis*. Observe the elongated body with the numerous annuli. Note the body pigmentation.

2. Place the leech on a dissection tray and identify the dorsal and ventral surfaces (Fig. 13.15). The dorsal surface is usually the more pigmented. Why might this be advantagous to the animal in its natural habitat?

3. Locate the anterior and posterior **suckers**. The anterior sucker is the smallest and surrounds the mouth. Note the general shape of the posterior sucker.

4. Several minute **eyespots** are located on the dorsal surface at the anterior end just above the mouth. Observe the eyespots with the dissecting microscope or hand lens.

5. On the middorsal surface of the worm, just before the posterior sucker, is the anus. It is often helpful to dry the surface of the animal with a paper towel to see the anus. Why is the anus located here and not at the very posterior part of the animal?

6. Turn the worm over on its dorsal side. About one-fifth to one-fourth of the way from the anterior end in the ventral midline are two small **gonopores**; the male gonopore is anterior to the female gonopore. In some specimens, the penis may be everted from the male pore. It is easier to locate these pores if the surface is dry.

7. Although sensory papillae are difficult to see on preserved worms, attempt to locate them. They

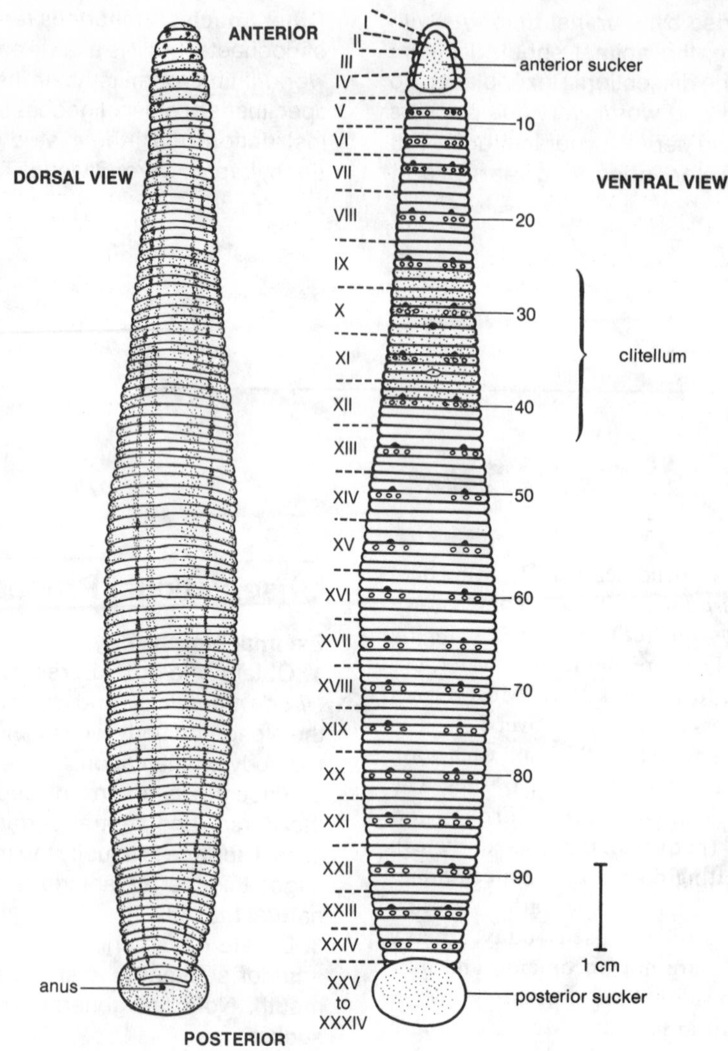

Figure 13.15. Dorsal and ventral views of *Hirudo medicinalis.*

occur on the dorsal and ventral sides of the body. The annulus on which sensory papillae are found marks the middle of a segment that extends through several annuli (Fig. 13.15).

8. On the second annulus of each segment, almost in the depression between the annuli, are the paired nephridiopore openings (Fig. 13.15). There are five annuli per segment in the midsection of *H. medicinalis.*

Internal Anatomy

Glossiphonia complanata (Fig. 13.16). Stained prepared microslide mounts of *G. complanata* will now be observed. Under both low and high powers of a dissecting microscope, locate the following anatomical features: annuli, eyespots, proboscis in its proboscis cavity, salivary gland, crop, gastric ceca, intestine, intestinal ceca, rectum, anus, posterior sucker, ovary, and testes. Note the paired eyespots and the highly muscular proboscis. Although the various anatomical parts shown in the illustration are not exactly the same for all leech species, the basic structure is similar.

***Hirudo medicinalis*.** Place a specimen of *H. medicinalis* in a dissection pan on its dorsal surface. Firmly pin the leech to the pan by inserting a dissection pin through the tip of the anterior sucker and another pin through the tip of the posterior sucker. Cover the specimen with water.

1. Carefully cut away the lower lip to expose the three jaws, arranged as a triangle; two jaws are lateral and one is dorsal (Fig. 13.17). Observe the position of the jaws and the numerous dilator muscles connected to the pharynx. Carefully remove one jaw with a fine-pointed scalpel. Place the jaw in a drop of water in a depression slide and examine under a compound microscope. Note the numerous minute serrations along the cutting edge of the jaw.

Set aside your specimen and obtain a cross section of a leech at the level of the pharynx and observe the very muscular, triangular lumen of the pharynx. What is the functional significance of a triangularly shaped pharynx?

2. Return to your dissection and turn it over on its ventral side. Pin the specimen firmly to the dissection pan. Make a shallow medial incision from about the level of the anterior sucker posteriorly for about 25 annuli. At the posterior and anterior limits of this incision, make two lateral cuts. Using a sharp scalpel, cut the body wall from the fibrous internal mass of tissue. Pin the flap to the dissection pan. Note that the complete internal septa and spacious body cavity observed in the other annelids are absent. Throughout the coelom is a loose mesenchymal mass called **botryoidal tissue.** This is a highly vascularized mass of tissue whose main component is pigmented cells. Spaces, called

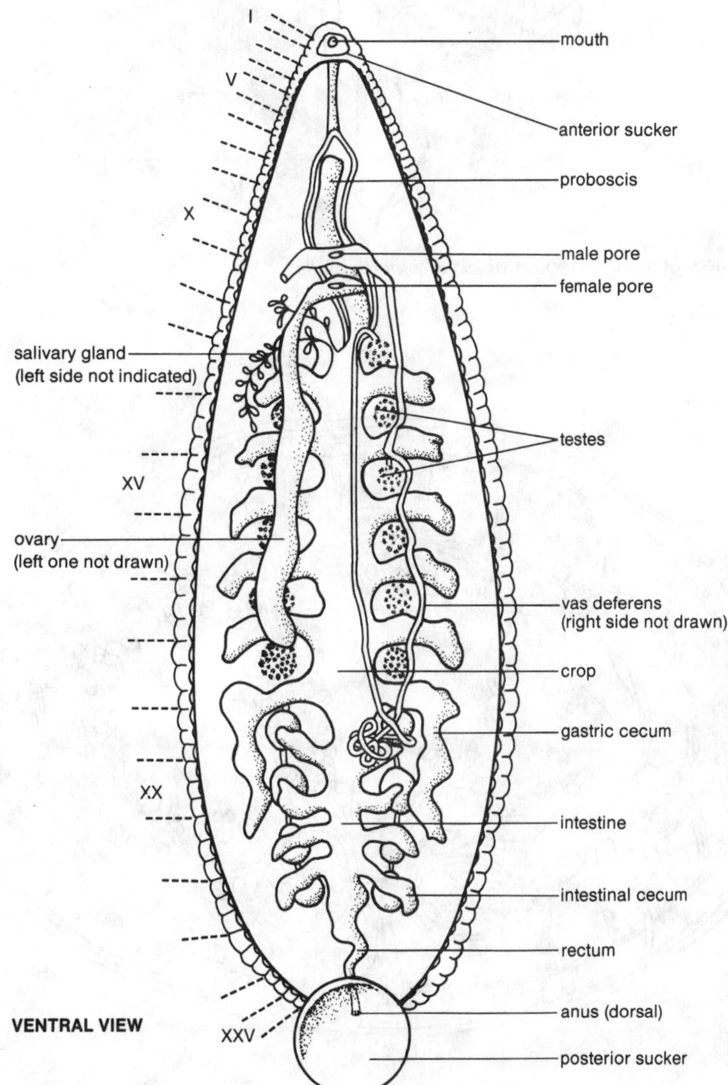

Figure 13.16. Ventral view of *Glossiphonia*, showing the relative positions of the digestive and reproductive systems.

coelomic sinuses, between the loose-connected mass of fibers are the remains of the coelom.

3. Obtain a cross section of a leech at the midbody level and note the position of the botryoidal tissue. Also note the relative positions of the gut and sinuses (Fig. 13.18).

Live Specimens. If live specimens are available, your instructor will provide instructions to make the appropriate observations.

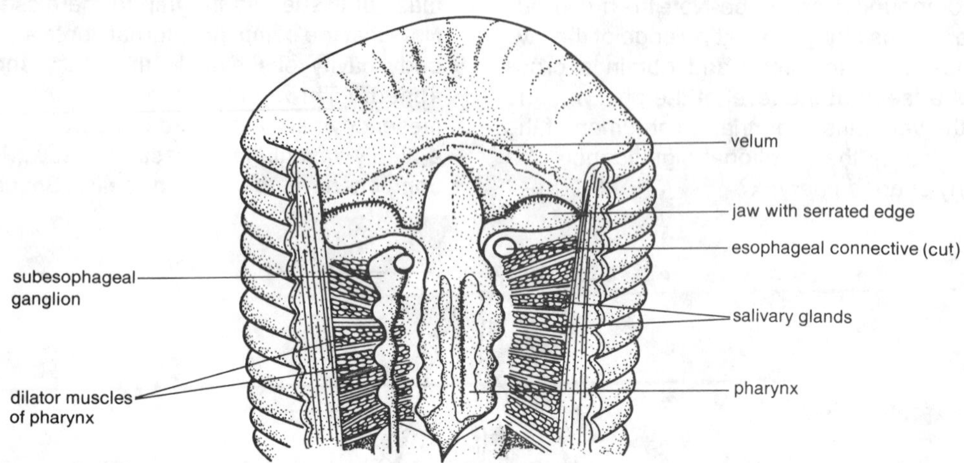

Figure 13.17. Interior view of the oral region of *Hirudo medicinalis.*

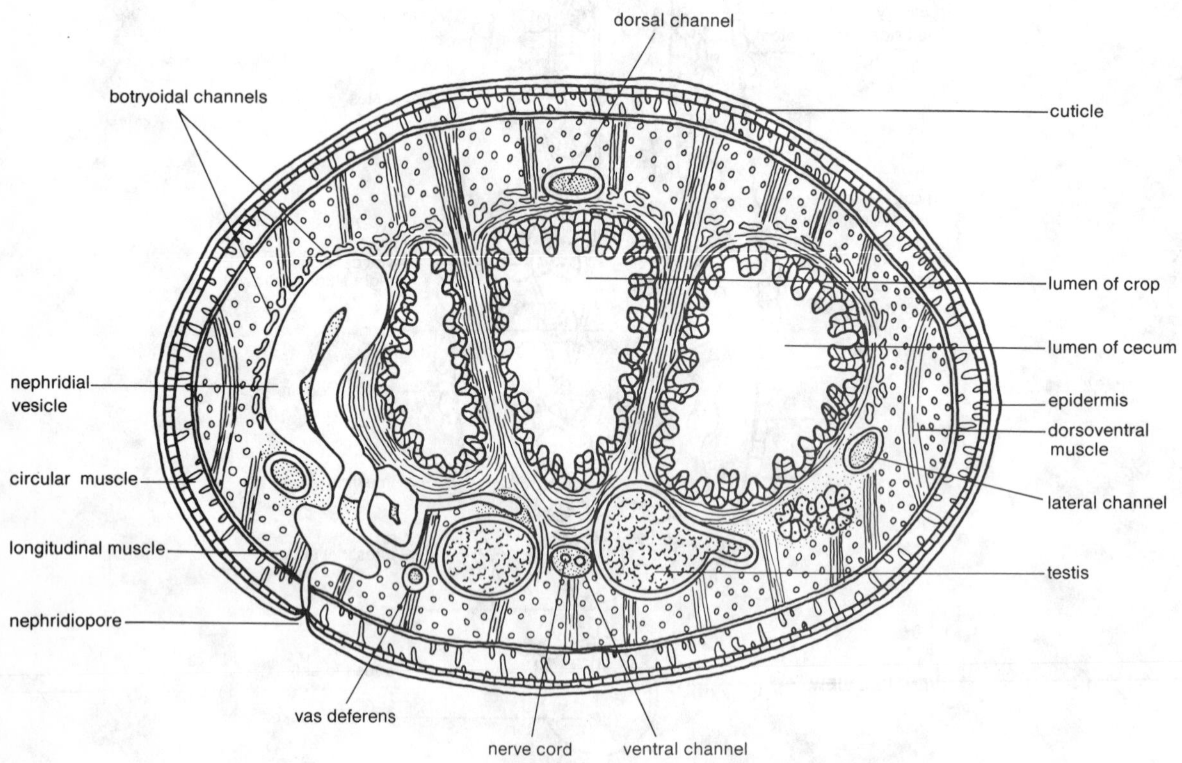

Figure 13.18. Cross section of a leech at the midbody level.

Supplemental Reading*

Aston, R. J. 1973. Tubificids and water quality; a review. Environ. Pollut. 5:1-10. (O)

Barnes, R. D. 1965. Tube-building and feeding in chaetopterid polychaetes. Biol. Bull. 129:217-233. (P)

Bonomi, G., and C. Erséus (eds.). 1984. Aquatic Oligochaeta. Developments in hydrobiology 24. Dr. W. Junk, The Hague. (O)

Brinkhurst, R. O. 1982. Oligochaeta. *In*: S. P. Parker (ed.). Synopsis and classification of living organisms. Vol. 2. McGraw-Hill, New York, pp. 50-61. (O)

Brinkhurst, R. 1984. Comments on the evolution of the Annelida. Hydrobiologia 109:189-191. (O)

Brinkhurst, R. O. 1984. The position of the Haplotaxidae in the evolution of oligochaete annelids. Hydrobiologia 115:25-36. (O)

Chapman, G. 1958. The hydrostatic skeleton in the invertebrates. Biol. Rev. Camb. Philos. Soc. 33:338-371. (G)

Chapman, P. M., and R. O. Brinkhurst. 1984. Lethal and sublethal tolerances of aquatic oligochaetes with reference to their use as a biotic index of polution. Hydrobiologia 115:139-144. (O)

Clark, R. B. 1962. Structure and function of polychaete septa. Proc. Zool. Soc. Lond. 138:543-578. (P)

Darwin, C. 1881. The formation of vegetable mould, through the action of worms, with observations on their habits. John Murray, London. (O)

Eakin, R. M., G. G. Martin, and C. T. Reed. 1977. Evolutionary significance of fine structure of archiannelid eyes. Zoomorphologie 88:1-18. (G)

Fauchald, K. 1975. Polychaete phylogeny: a problem in protostome evolution. Syst. Zool. 23:493-506. (P)

Foster, N. 1972. Freshwater polychaetes (Annelida) of North America. Biota Freshwater Ecosyst. Identif. Man. 4:1-15. (P)

Fransen, M. E. 1977. Ultrastructure of spermatids and spermatozoa in Archiannelida. Zoon 5:97-105. (G)

Gerard, B. M. 1967. Factors affecting earthworms in pastures. J. Anim. Ecol. 36:235-252. (O)

Glaessner, M. F., and M. Wade. 1966. The late Precambrian fossils from Ediacara, South Australia. Palaeontology 9:599-628. (G)

Greene, K. L. 1974. Experiments and observations on the feeding behavior of the freshwater leech *Erpodbella octoculata* (L.) (Hirudinea: Eropobdellidae). Arch. Hydrobiol. 74:87-99. (H)

Jouin, C. 1971. Status of the knowledge of the systematics and ecology of Archiannelida. Smithson. Contrib. Zool. 76:47-56. (G)

Klemm, D. J. (ed.). 1985. A guide to the freshwater Annelida (Polychaeta, Naidid and Tubificid Oligochaeta, and Hirudinea) of North America. Kendall/Hunt, Dubuque, IA. (G)

Knight-Jones, E. W., J. E. Bailey, and M. J. Isaac. 1971. Choice of algae by larvae of *Spirorbis*, particularly of *Spirorbis spirorbis*. *In*: D. J. Crisp (ed.). Fourth European marine biology symposium. Cambridge University Press. Cambridge, pp. 89-104. (P)

Mikulic, D. G., D. E. G. Briggs, and J. Kluessendorf. 1985. A Silurian soft-bodied biota. Science 228:715-717. (H)

Nicholls, J. G., and D. Van Essen. 1974. The nervous system of the leech. Sci. Am. 230:38-48. (H)

Pettibone, M. H. 1982. Polychaeta. *In*: S. P. Parker (ed.). Synopsis and classification of living organisms. Vol. 2. McGraw-Hill, New York, pp. 3-42. (P)

Rupp, R. W., and M. C. Meyer. 1954. Mortality among brook trout, *Salvelinus fontinalis*, resulting from attacks of freshwater leeches. Copeia 1954:294-295. (H)

*The supplemental literature listed here is coded to indicate the general topic covered by the paper: G, general Annelida; P, Polychaeta; O, Oligochaeta; H, Hirudinea.

Sawyer, R. T. 1984. Arthropodization in the Hirudinea: evidence for a phylogenetic link with insects and other Uniramia. Zool. J. Linn. Soc. 80:303-322. (H)

Schmidt, G. D., and L. S. Roberts. 1985. Foundations of parasitology. 3rd ed. C. V. Mosby. St. Louis, MO. (G)

Sims, R. W., and B. M. Gerard. 1985. Earthworms. Synopses of the British fauna 31. Brill, Leiden, The Netherlands. (O)

Wallwork, J. A. 1983. Earthworm biology. Studies in biology 161. Edward Arnold, London. (O)

Woodin, S. A. 1974. Polychaete abundance patterns in a marine soft-sediment environment: importance of biological interactions. Ecol. Monog. 44:171-187. (P)

ARTHROPODOUS PHYLA

Members of the phylum Arthropoda are schizocoelomate, metameric (segmented) protostomes. Annelids also exhibit these characteristics, but arthropods possess a unique **chitinous exoskeleton** that is often strengthened with calcium deposits. The **exoskeleton** is not equally thickened in all regions of the body. Separate plates are associated with each body segment or **somite** (G., *soma*, body) and these are connected to one another by flexible, thinner regions called **articulating membranes**. Periodically, the arthropod forms a new exoskeleton underneath the old, which is then shed in a **molt**. The new exoskeleton is then stretched to a larger size before hardening. In primitive arthropods, each segment possesses a pair of appendages, but this condition does not hold for most species. There is a strong tendency for reduction in number of appendages and segments, and for tagmatization of segments. The schizocoelom is much reduced and does not function as a hydraulic skeleton as occurs in annelids. The large body cavity in arthopods is actually a hemocoel (a modified blastocoel).

Some workers believe that the Arthropoda may have arisen several different times (i.e., polyphyletic arthropodization) from the annelid-arthropod line (e.g., Manton 1973, 1977). This makes the Arthropoda a superphylum comprised of as many as four phyla: Trilobita, Chelicerata (these two may be closely related), Crustacea, and Uniramia (including Insecta and the four myriapodous classes). However, Boudreaux (1979), among others, argues just the opposite (i.e., that the phylum is monophyletic).

Phylum Onychophora is considered in this section of the manual because their anatomy and morphology reflects both annelidan and arthropodian characteristics.

Boudreaux, H. B. 1979. Arthropod phylogeny with special reference to insects. Wiley, New York.

Manton, S. M. 1973. Arthropod phylogeny—a modern synthesis. J. Zool. (Lond.) 171:111-130.

Manton, S. M. 1977. The Arthropoda: habits, functional morphology, and evolution. Clarendon Press, Oxford.

EXERCISE 14

Phylum Onychophora

The approximate 70 living, wormlike species of Onychophora (o-NEE-kof-or-a; G., *onychos*, claw + G., *phora*, to bear) have features similar to both arthropods and annelids. Because of this, some authors regard onychophorans to be a "missing link" between Arthropoda and Annelida. Arthropod characteristics include an open circulatory system with a dorsal tubular heart bearing ostia, a large hemocoel, a reduced schizocoelom that is limited to small sacs associated with the gonads and nephridia, a respiratory system of tracheae and spiracles, and a pair of modified appendages (e.g., **mandibles**) for feeding. The presence of ciliated **nephridia**, lack of jointed appendages, and structural configuration of the body wall are annelidan in character. Although the body is externally and internally unsegmented, the nervous system and nephridia are segmented and the position of the legs indicates external segmentation.

Distribution of onychophorans is discontinuous in both tropical and southern temperate regions. The animals are nocturnal and most are tropical, living in moist areas under leaves, rocks, and logs. A fossil specimen from the **Burgess Shale fauna** of the **Cambrian Period** indicates the phylum is ancient (Fig. 14.1).

Classification

Onychophora is divided into two families. The Peripatidae contains members, such as *Peripatus*, having primarily an equatorial distribution, whereas the Peripatopsidae, exemplified by *Peripatopsis*, is limited to the southern hemisphere. Present distribution of the phylum can be explained by movements of the continents.

Observational Procedure: *Peripatus*

Observe one or more specimens of *Peripatus* under the dissecting microscope. The elongate wormlike body has two indistinct regions: (1) a head and (2) a trunk of 14 to 43 fused segments. Note the short, stumpy, unsegmented paired legs on each trunk segment (Fig. 14.2). How many pairs of legs are on your specimen? The number varies with the species and sex. Examine closely the pair of curved terminal claws and transverse pads on each leg.

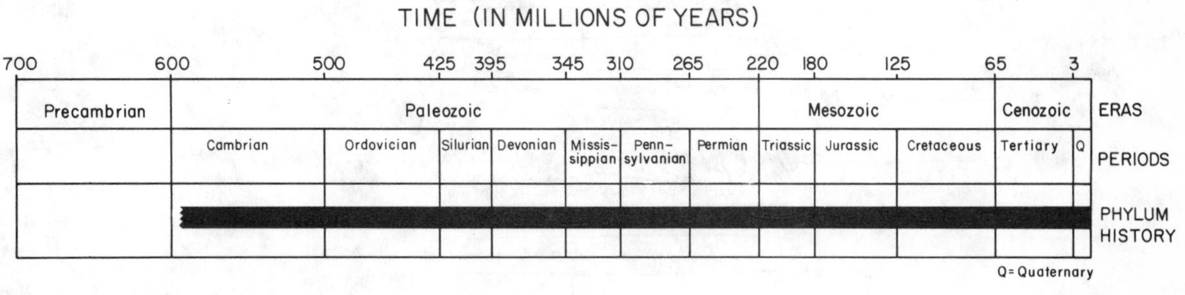

TIME (IN MILLIONS OF YEARS)

700 600 500 425 395 345 310 265 220 180 125 65 3

ERAS	Precambrian	Paleozoic							Mesozoic			Cenozoic	
PERIODS		Cambrian	Ordovician	Silurian	Devonian	Missis-sippian	Penn-sylvanian	Permian	Triassic	Jurassic	Cretaceous	Tertiary	Q
PHYLUM HISTORY		■	■	■	■	■	■	■	■	■	■	■	■

Q = Quaternary

Figure 14.1. Geologic history of the phylum Onychophora.

Note the **tubercles** and their arrangement on the legs and body. Attempt to see the **nephridiopore** on the ventral surface near the inner base of each leg.

The body wall is essentially annelidan. A thin, flexible chitinous **cuticle** covers the body surface. The cuticle is composed of the same layers as that of arthropods and is molted. Below the cuticle are the following layers in order of position: epidermis, dermis, and circular, diagonal, and longitudinal muscles (Fig. 14.3). Onychophorans lack a rigid exoskeleton.

Locate the two annulated **antennae** on the head (Fig. 14.2). Sensory bristles cover the antennae. A small ocellus may be seen near the base of each antenna. Posterior to each antenna and on the ventral surface are two **oral papillae**. Glands in each papilla discharge a secretion that hardens, forming a net of adhesive threads that catch prey of worms, insects, and other small invertebrates.

Between the oral papillae is the mouth, containing paired mandibles. Note the peribuccal lobes or lips surrounding the mouth. Food entering the mouth

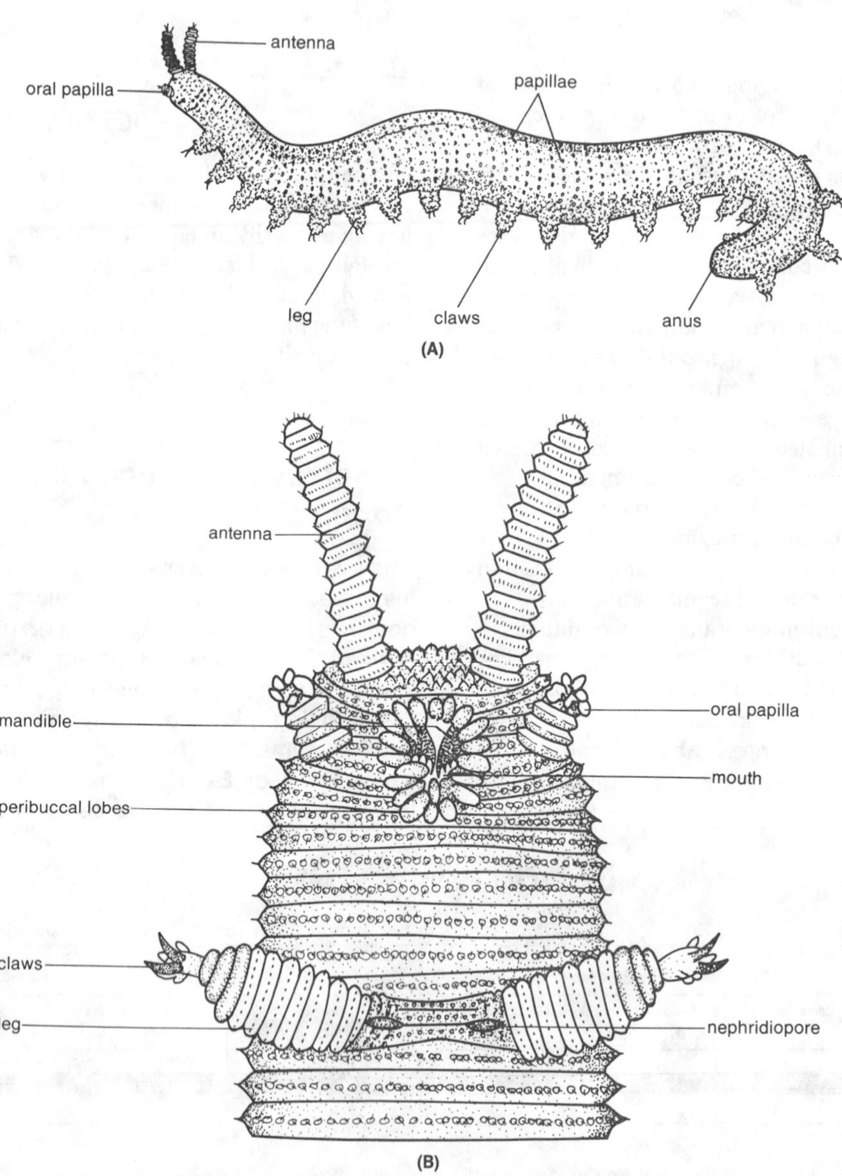

Figure 14.2. **(A)** *Peripatus*, and **(B)** anteroventral area of *Peripatus*.

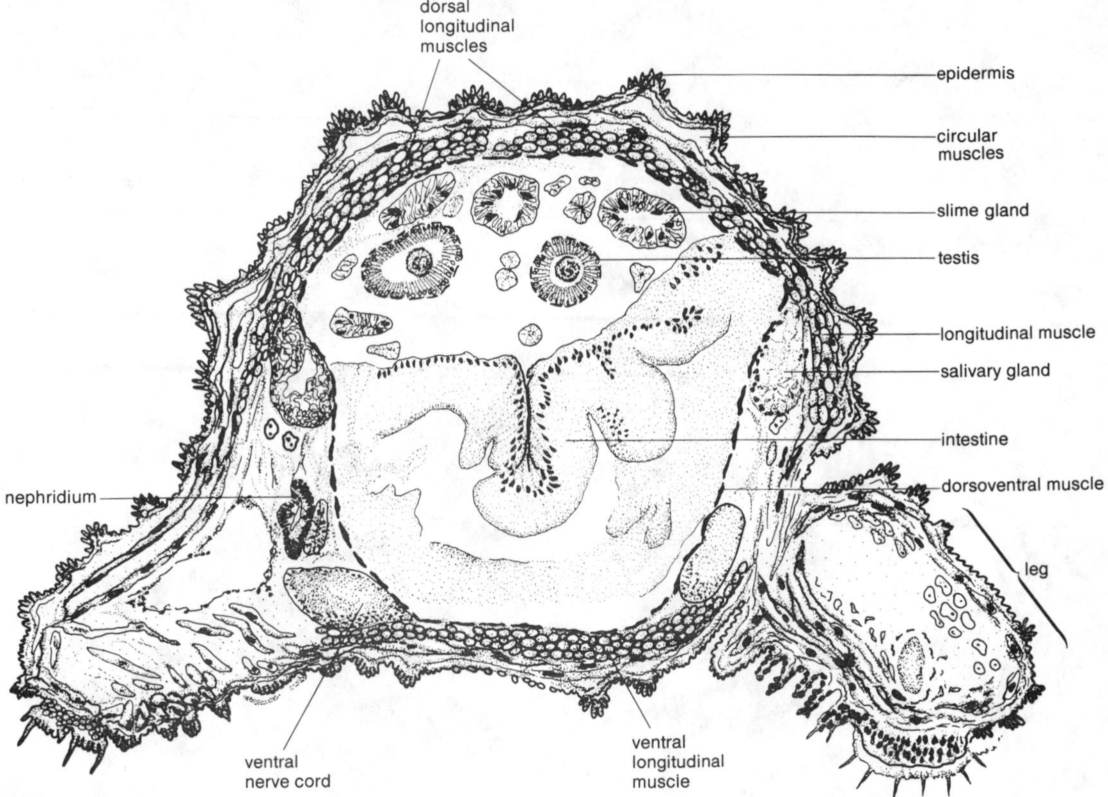

Figure 14.3. Cross section of *Peripatus* through the testes.

passes into the pharynx and then into an esophagus. From there, food enters the straight intestine composed of columnar cells (Fig. 14.3). The intestine leads to the rectum, which opens to the anus on the ventral side of the posterior end.

The nervous system consists of a **brain**, two **circumpharyngeal connectives**, and two ventrolateral **nerve cords**. Observe the latter in the cross section (Fig. 14.3).

The sexes of *Peripatus* and all onychophorans are separate; a larva is absent. Males are smaller and have fewer legs than do females. The paired ovaries or testes may be seen in a cross section (Fig. 14.3). Attempt to see the gonopore just anterior to the anus on the preserved material.

Supplemental Reading

Boudreaux, H. B. 1979. Arthropod phylogeny with special reference to insects. Wiley, New York.

Manton, S. M. 1937. Studies on the Onychophora. II. The feeding, digestion, excretion, and food storage of *Peripatopsis*. Philos. Trans. R. Soc. Lond. B 227:411-464.

Manton, S. M. 1949. Studies on the Onychophora. VII. The early embryonic stages of *Peripatopsis*, and some general considerations concerning the morphology and phylogeny of Arthropoda. Philos. Trans. R. Soc. Lond. B 223:483-580.

Manton, S. M. 1973. The evolution of arthropodan locomotory mechanisms. II. Habits, morphology and evolution of the Uniramia (Onychophora, Myriapoda, Hexapoda) and comparisons with Arachnida, together with a functional review of uniramian musculature. J. Linn. Soc. Lond. Zool. 53:275-375.

Manton, S. M. 1977. The Arthropoda: habits, functional morphology, and evolution. Clarendon Press, Oxford.

EXERCISE 15

Phylum Arthropoda

Arthropoda (ar-THROP-o-da or ar-thro-PO-da; G., *arthro*, joint + G., *pod*, foot) is an enormous phylum of segmented, metameric protostomes possessing an exoskeleton with jointed appendages. Arthropods range in size from microscopic to several meters in length. Estimates of the number of described species range from 750,000 to 1 million. However, it is predicted that up to 10 million species of insects are extant, and perhaps one species per day becomes extinct due to clear-cutting the Brazilian rain forest. Arthropods are found from the ocean depths to mountain peaks, and even several thousand meters in the air. They have enormous medical, veterinary, ecological, and economic importance. Some of the most virulent diseases to plague humans are carried by arthropod vectors: the mosquito *Aedes aegypti* transmits the virus causing yellow fever, the flea *Xenopsylla cheopis* carries the bacterium causing black death, and the mosquito *Anopheles* spp. carries the protozoan causing malaria. The immense size of this phylum precludes anything but a brief overview to be presented in the following exercises.

Arthropods possess four important features. (1) A unique **exoskeleton** or **cuticle** made of **chitin** and **scleroprotein** completely covers the body. The exoskeleton is often strengthened by deposits of calcium salts. (2) Primitively, arthropods possess one pair of jointed appendages per body **segment (somite)**, but as segments have been lost or fused in the evolution of advanced species this condition is lost. (3) The arthropod body can usually be divided into two or three distinct tagmata. (4) Cilia are nearly absent.

Additional features of the phylum include (1) presence of a reduced coelom, (2) presence of a modified blastocoel (a **hemocoel**) that functions as an open circulatory system, (3) absence of a nephridial excretory system, (4) a high degree of cephalization, and (5) sexual reproduction with development characterized

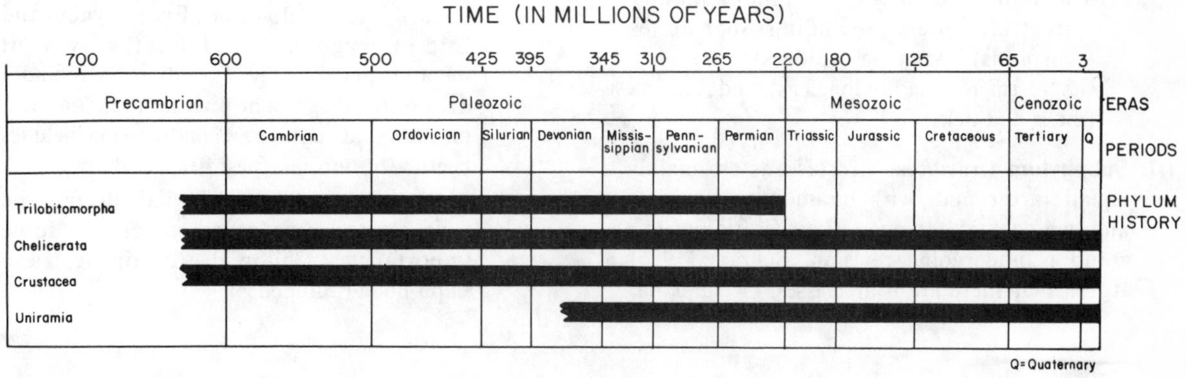

TIME (IN MILLIONS OF YEARS)

700 600 500 425 395 345 310 265 220 180 125 65 3

ERAS	Precambrian	Paleozoic							Mesozoic			Cenozoic	
PERIODS		Cambrian	Ordovician	Silurian	Devonian	Missis-sippian	Penn-sylvanian	Permian	Triassic	Jurassic	Cretaceous	Tertiary	Q

PHYLUM HISTORY: Trilobitomorpha, Chelicerata, Crustacea, Uniramia

Q = Quaternary

Figure 15.1. Geologic history of the four subphyla in phylum Arthropoda.

by distinct morphological stages. Arthropods have a geologic history dating from the **Cambrian Period** (Fig. 15.1).

Classification

Arthropoda is comprised of four subphyla: Trilobita, Chelicerata, Crustacea, and Uniramia (cf. Manton 1973, 1977; Schram 1986).

I. Subphylum Trilobita. Extinct arthropods possessing three longitudinal lobes (two pleural and one axial) and three tagmata (head, thorax, and pygidium). Trilobite systematics is based on external features alone (Moore 1959, but compare Bergström 1973).

II. Subphylum Chelicerata. Chelicerates possess two body regions (cephalothorax, or prosoma, and abdomen, or opisthosoma) and no antennae. The first pair of appendages are called chelicerae. This subphylum is divided into three classes: Merostomata, Pycnogonida, and Arachnida.

A. Class Merostomata. Two subclasses of mostly extinct aquatic chelicerates having book gills and a posterior telson: Xiphosura (containing the horseshoe or king crab, *Limulus polyphemus*); Eurypterida (entirely extinct group of aquatic chelicerates known as giant water scorpions).

B. Class Pycnogonida (Pantapoda). The Pycnogonida comprises about 600 species of spiderlike marine arthropods with four to six long legs. Examples include *Achelia*, *Pycnogonum*, and *Nymphon*.

C. Class Arachnida. Most chelicerates are arachnids. There are 12, predominantly terrestrial orders, including Scorpiones (scorpions), Araneae (spiders), Opiliones (harvestmen or daddy longlegs), and Acarina (mites and ticks).

III. Subphylum Crustacea. Crustaceans are mainly aquatic arthropods with biramous appendages and two pairs of antennae. Classification of this group is undergoing revision. Schram (1986) argues that there are four classes: Remipedia, Phyllopoda, Maxillopoda, and Malacostraca. An older classification divided the crustaceans into seven classes, sometimes referred to as the lower crustaceans or **entomostracans** (en-to-MOS-tra-kans; G., *entom*, insect + G., *ostracum*, shell) and a single class, Malacostraca (mal-a-KOS-tra-ka; G., *malac*, soft), referred to as the higher crustaceans.

Entomostracans include four major classes: Branchiopoda, fairy (*Branchinecta*) and brine shrimp (*Artemia*), and water fleas (*Bosmina, Daphnia*); Ostracoda, ostracods (*Cypris*); Copepoda (*Calanus*); Cirripedia, the barnacles (*Balanus* and *Lepas*).

Malacostraca comprises two subclasses with about 15 orders, including four important ones: Isopoda, pill bugs (*Oniscus*) and wood lice (*Limnoria*); Amphipoda, freshwater scuds (*Gammarus*) and beach fleas (*Orchestoidea*); Euphausiacea, krill (*Euphausia*); Decapoda, some shrimps (*Penaeus*), crabs (*Callinectes*, *Cancer*, *Paralithodes*, *Pagurus*, *Uca*), crayfish (*Cambarus*), and lobsters (*Homarus*).

IV. Subphylum Uniramia. Arthropods with appendages of one branch and single pair of antennae. The subphylum contains three major classes (Insecta, Chilopoda, and Diplopoda) and two minor ones, Symphyla and Pauropoda.

A. Class Insecta (Hexapoda). Arthropods having three body regions (head, thorax, and abdomen), a single pair of antennae (rarely absent), and wingless or with one or two pairs of wings borne on the thorax. Two subclasses are recognized: Apterygota (primitive wingless insects, including silverfish and springtails) and Pterygota (winged or wingless insects divided into two divisions, Exopterygota and Endopterygota, based on the type of metamorphosis). The first division includes dragonflies, grasshoppers, cockroaches, termites, and bugs. The second division includes beetles, butterflies, flies, fleas, and ants.

B. Myriapodous arthropods. Myriapods comprise four classes, two of which are of some importance: Chilopoda (centipedes) and Diplopoda (millipedes).

A. Subphylum Trilobita

Trilobita (tri-lo-BI-ta; L., *tri*, three + G., *lob*, lobe) is a moderately sized group of about 15,000 extinct species of primitive, dorsoventrally flattened arthropods with a chitinous exoskeleton strengthened with calcium and phosphate deposits. These marine animals were most abundant during the **Cambrian** and **Ordovician**

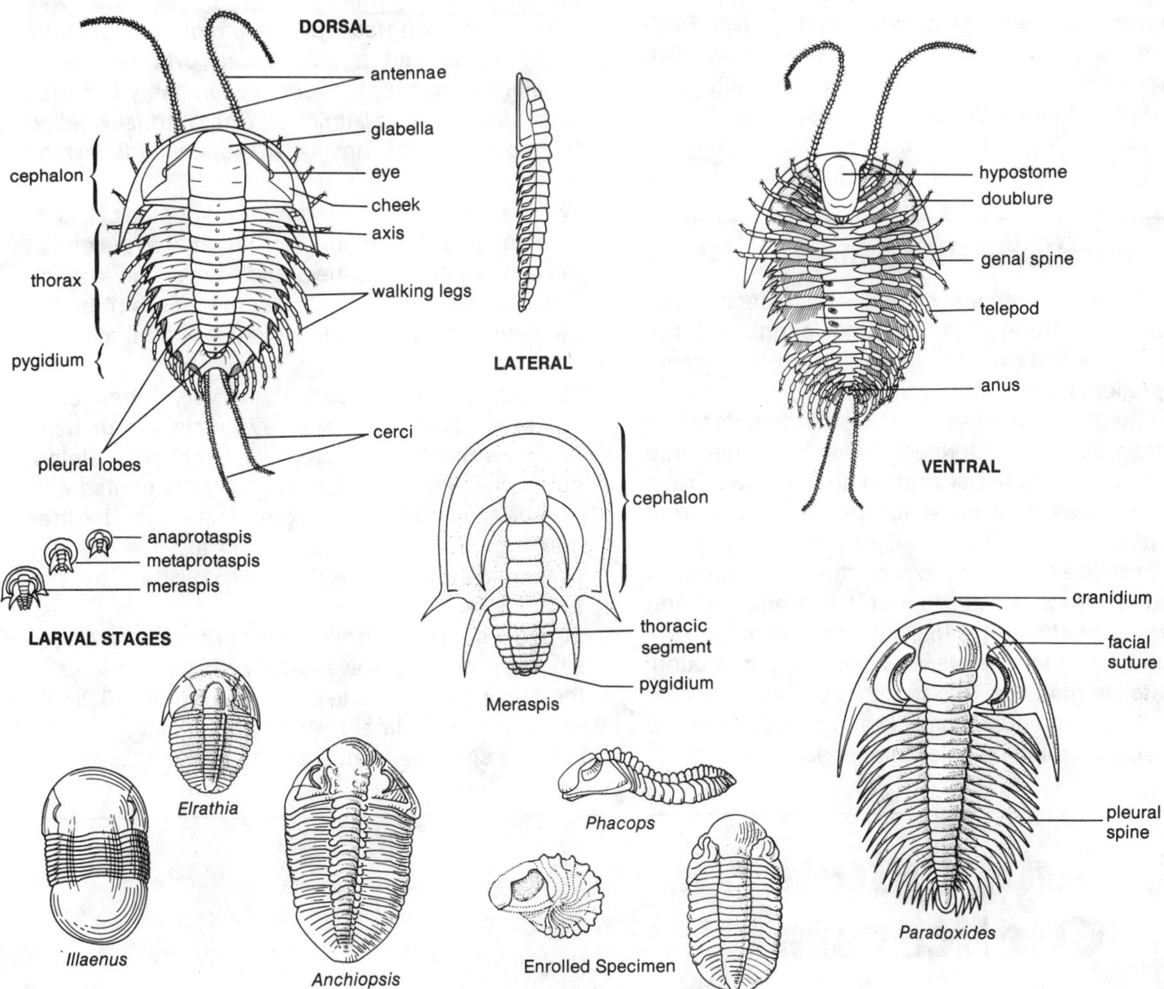

Figure 15.2. Examples of trilobites. Shown here are a schematic diagram of a generalized trilobite (dorsal, ventral, and lateral views), the larval stages (protaspis and meraspis), and several examples of adults. (Not drawn to scale. After several sources.)

Periods of the **Paleozoic Era**, but were extinct by its end (Fig. 15.1). Although a few large species are known to reach nearly 1 m, most adult trilobites were small (less than 10 cm long). The name *trilobite* refers to their oval body being longitudinally divided into three lobes: two lateral **pleural lobes** and one medial **axial lobe** (Fig. 15.2). The body consisted of three tagmata: **cephalon** (head), **thorax**, and **pygidium** (abdomen) (pi-JID-e-um; G., *pyg*, rump).

Although extinct for over 200 million years, trilobites possessed many features common to the phylum Arthropoda. As in other phyla with a rich fossil history, a complex nomenclature has been developed by paleontologists. Only the basic morphology of these remarkable arthropods will be studied.

Observational Procedure

1. Observe specimens of several species and identify the anterior and posterior ends, the three lobes (axis and two pleural), and the three tagmata (cephalon, thorax, and pygidium).

2. The cephalon of several fused segments forms a large dorsal head shield which is divided into various parts by **facial sutures** and **furrows**. Trace the central **axial furrow**, which delimits the axis from the pleural lobes through all three tagmata.

3. Examine specimens to see if their cephalons are divided into a central region, the **cranidium** and lateral **cheeks** (Fig. 15.2). In the center of the cranidium, often elevated above the cephalon, locate the **glabella**. This structure probably housed intestinal diverticula, stomach, and provided a broad surface to which muscles attached.

4. Lateral to the glabella find the compound eyes. In some species you may be able to see the functional units of the eye, the **ommatidia** (e.g., in *Phacops rana*). Unfortunately, in most fossils the ommatidia will not be visible.

5. At the posteriolateral corners of the cephalon locate the genal angles. If elongated, these are called genal spines.

6. The thorax consists of 2 to 42 unfused, articulating segments each bearing a pair of appendages similar to those on the cephalon. The thorax was capable of flexing, so that many trilobites were able to fold over like a pill bug, a process called **enrollment** or **conglobation**. Presumably this protected the vulnerable (thinner) ventral surface called the **doublure**. Do any of the specimens exhibit enrollment?

7. The pygidium was usually constructed of 2 to 30 fused segments with the anus located at the end of the last segment. Are the pygidia well defined in the specimens? In some species the axis is ribbed in the pygidium, giving a false impression that it was articulate in this tagma.

8. Although appendages are very infrequently preserved, examine your study specimens for them.

9. The entire tergal (L., *tergum*, back) exoskeleton, particularly the cephalon, may be ornamented with bumps, granules, tubercles, lines, and other markings. Using Fig. 15.2 as an aid, identify any other anatomical features exhibited by the specimens.

10. If a whole-mount slide of a "trilobite" larva from a horseshoe crab is available, compare the larva to the drawings of trilobite larval stages (Fig. 15.2) and to the adult specimens. What similarities are found in these specimens; how do they differ?

B. Subphylum Chelicerata

Members of subphylum Chelicerata (ke-LIS-e-ra-ta; G., *chela*, claw + G., *cera*, horn) are distinguished from other arthropods by having two important features. (1) Chelicerates do not possess antennae. (2) The first pair of appendages are modified distally to form opposing clawlike legs called **chelicerae**. Following the chelicerae are the **pedipalps**, and posterior to these are four pairs of walking legs. Mandibles are lacking. The body is divided into two tagmata: anterior **cephalothorax** or **prosoma**, and posterior **abdomen** or **opisthosoma**. All appendages are associated with the prosoma; the opisthosoma lacks appendages except in the Xiphosura. Most chelicerates are predaceous or parasitic on arthropods or other invertebrates.

Class Merostomata

Two subclasses comprise the class Merostomata (mer-o-STO-ma-ta; G., *mero*, part + G., *stoma*, mouth),

Xiphosura (including horseshoe crabs) and the extinct Eurypterida.

Subclass Xiphosura. Horseshoe crabs, members of the subclass Xiphosura (ZIF-o-SU-ra; G., *xiphos*, sword + G., *oura*, tail), have existed since the Ordovician, but only four species are extant. All are marine. *Limulus polyphemus* is widely distributed along the Gulf of Mexico and east coast of the United States. The other species occur along coasts in the far east Pacific.

In some ways *Limulus* resembles a trilobite. *Limulus* possesses three indistinct longitudinal lobes and has a convex dorsal shield covering the ventral appendages (Fig. 15.3). However, *Limulus* possesses only two tagmata. From above, *Limulus* resembles an upside-down shovel. Specimens may reach more than 50 cm in length, including the tail spine or **telson**. The sexes are separate and mating pairs may be observed along U. S. coasts in the spring. Up to several males may cling to a female as she comes to shore, where she deposits her eggs in the sand as the males release sperm.

Observational Procedure: *Limulus*

1. Place a specimen of *Limulus* dorsal side up with the telson to one side. Identify the anterior and

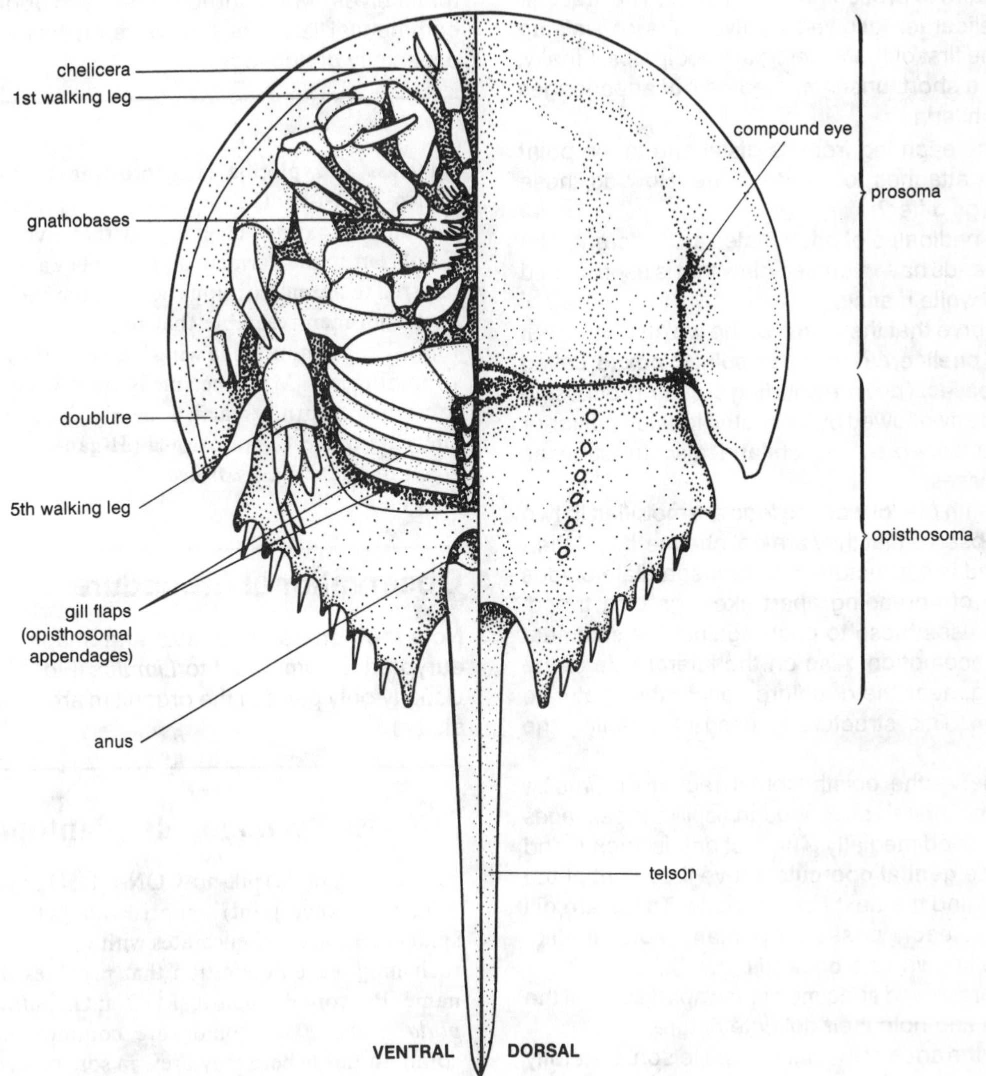

Figure 15.3. Dorsal and ventral views of *Limulus polyphemus*.

posterior ends. Note the shovel-shaped prosoma. How is *Limulus* adapted for shallow burrowing? Notice the general resemblance of *Limulus* to that of a trilobite. Are the ventral sides of the two organisms similar (cf. Figs. 15.2 and 15.3)?

2. Return the animal to its dorsal side and examine the **medial** and **compound eyes** with a hand lens. Of what type of eye does this remind you? What defensive structures does *Limulus* possess on its dorsal surface? Examine a molted exoskeleton (**exuvia**) and identify the **suture line** that breaks open when the animal molts.

3. Examine the ventral surface of a preserved specimen. As in the trilobites, the ventral extension of the carapace is called the doublure. Locate the seven pairs of prosomal appendages. The first pair are chelicerae, followed by five pairs of **walking legs**. The first of these being the pedipalps. Finally, there is a short, unsegmented pair of appendages called **chilaria**.

4. Trace each leg from its distal end to the point where it attaches to the doublure. How do these seven legs differ?

5. The pedipalps of adult males are different. The chelate ends have a curved claw that is used to hold females while mating.

6. Observe that the bases of the walking legs, but not the chelicerae, possess spined **coxae** called **gnathobases**. They are grinding organs, processing food to be swallowed by the mouth located posterior to the chelicerae. The chilaria also function as gnathobases.

7. The fifth pair of walking legs are modified in two ways. Observe that they are not chelate, but at their distal end is a structure with four spatulate blades capable of spreading apart like legs of a tripod. *Limulus* uses these to push against the sediment during locomotion. Also on the lateral side of the same leg, near the doublure, locate the spatulate **flabellum**. This structure is used in cleaning the gills.

8. Observe the opisthosomal region of *Limulus* and examine the six pairs of tough, flaplike appendages that are fused medially. The first pair is thicker and called the **genital opercula**. Move these out of the way and find the next five opercula. These are gill opercula, each possessing many soft, leaflike lamellae known as a **book gill**.

9. In a preserved specimen tease apart some of the gill folds and note their delicate nature.

10. Find the anus at the base of the telson. Carefully move the telson and determine its range of motion. From that observation do you think that the telson is a particularly good defensive structure?

11. Examine a whole-mount slide of a trilobite larval stage of *Limulus*. Note that the larva looks like a miniature adult. All the legs are present, but some of the gills are still absent and will be gained in subsequent molts. The medial and compound eyes should be visible through the thin exoskeleton. *Limulus* larvae superficially resemble larval and adult trilobites. Compare the general morpology of a *Limulus* larva to larval and adult trilobites (Fig. 15.2).

12. Your instructor will provide directions if adult specimens are to be dissected. If live animals are available for study, note how they move, right themselves when turned over, and look for the commensal flatworm, *Bdelloura*, on the ventral side, especially on the legs.

Subclass Eurypterida (Gigantostraca). Eurypterida (U-rip-TER-i-da; G., *eury*, wide + G., *pteron*, wing) is a group of extinct chelicerates that lived from the **Cambrian** to the **Permian periods**. They are sometimes referred to as sea scorpions because the chelicerae and long abdomen resemble that of true scorpions (Fig. 15.4). The etymon "wide wing" refers to the paddlelike swimming legs. Most specimens are <30 cm long, but some giant forms reached nearly 3 m, hence the alternate name **Gigantostraca** (JI-gan-tos-tra-ka; G., *gigantos*, giant + G., *ostraca*, shell).

Observational Procedure

Note the general body size and shape of a fossil eurypterid. Compare it to *Limulus* (Fig. 15.4). (N.B.: Usually only parts of the organism are available for study.)

Class Pycnogonida (Pantopoda)

The Pycnogonida (pik-no-GON-i-da; G., *pycno*, thick + G., *gony*, knee, joint) comprises about 600 species of spiderlike marine chelicerates with very long legs. It is their long-legged condition that provides their other name, **Pantopoda** (pan-to-POD-a; G., *panto*, all +G., *poda*, feet). Pycnogonids are common in benthic communities, where they prey on soft-bodied animals such as bryozoans and cnidarians by sucking the body

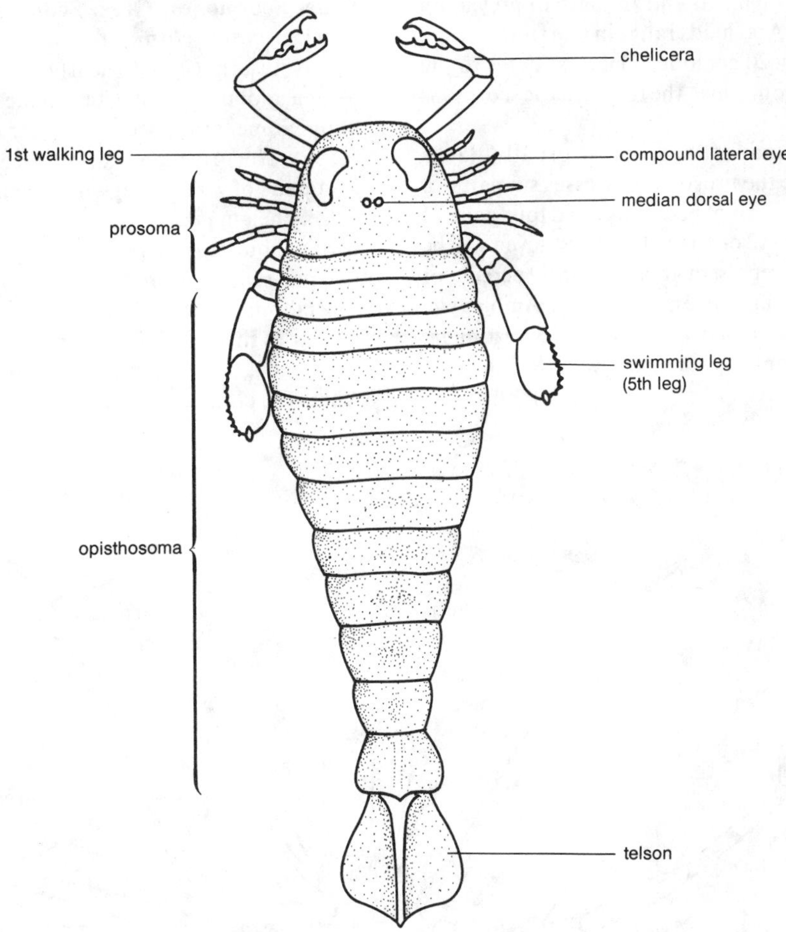

Figure 15.4. A fossil eurypterid (sea scorpion).

fluids or consuming bits of tissue. Some are not predaceous, but feed on algae accumulated on the surface of sessile organisms. Most pycnogonids are less than 10 mm long, but some have very long legs (ca. 75 cm).

Observational Procedure

Examine a preserved pycnogonid under a dissecting microscope and note the narrow body, composed of a **prosoma** with cephalon and trunk, and a small, unsegmented abdomen (**opisthosoma**) bearing a terminal anus (Fig. 15.5). Locate the proboscis at the anterior end, and on the anterior dorsal part of the cephalon, find a small tubercle that bears **ocelli**. The number of appendages differs among species; *Nymphon* has seven pairs. Locate the chelicerae that lie alongside the proboscis. Also locate the pedipalps, the second pair of appendages near the chelicerae. The third pair of appendages are the very slender **ovigerous legs** that normally are used by males to carry eggs; females of many species lack these legs. Note that the remaining four pairs of walking legs are long (Fig. 15.5).

Your instructor will assist you in making observations if live specimens are available for study.

Class Arachnida

Class Arachnida (a-RAK-ni-da; G., *arachne*, spider) comprises the vast bulk of chelicerates (more than 60,000 species) and is of great importance because some are parasitic on plants and animals, or prey upon agricultural pests. Arachnids range in size from minute (200 μm) to nearly 20 cm long. The class dates to the **Silurian Period**, from which the first aquatic scorpions are known.

The opisthosoma of arachnids usually lacks appendages, while the prosoma possesses six pairs: a pair of chelicerae, a pair of pedipalps, and four pairs of walking legs. Of the 12 orders in this class, seven will be studied here. They represent greater than 99 percent of arachnids. The remaining five minor orders, Amblypygi, Cyphophthalmi, Palpigradi, Ricinulei, and Schizomida, will not be studied unless your instructor has specimens for study.

Order Scorpiones. Order Scorpiones (skor-pi-O-nez; G., *skorpios*, scorpion) is a well-known group of relatively large (2 cm to nearly 18 cm) chelicerates with a long mobile tail that terminates with a poisonous **sting**. Some fossil species were very large (greater than 0.5 m). The prosoma bears chelicerae, pedipalps, and four pairs of walking legs. The chelicerae are small; the large, chelate pedipalps are used to hold prey and are used in courtship. The opisthosoma is elongate and of two parts, a **mesosome** (preabdomen) and a **metasome** (postabomen). The latter is a segmented tail that possesses the sting. Scorpions are viviparous and live in tropical, subtropical, and south temperate zones where the climate is warm. *Centruroides* is a common

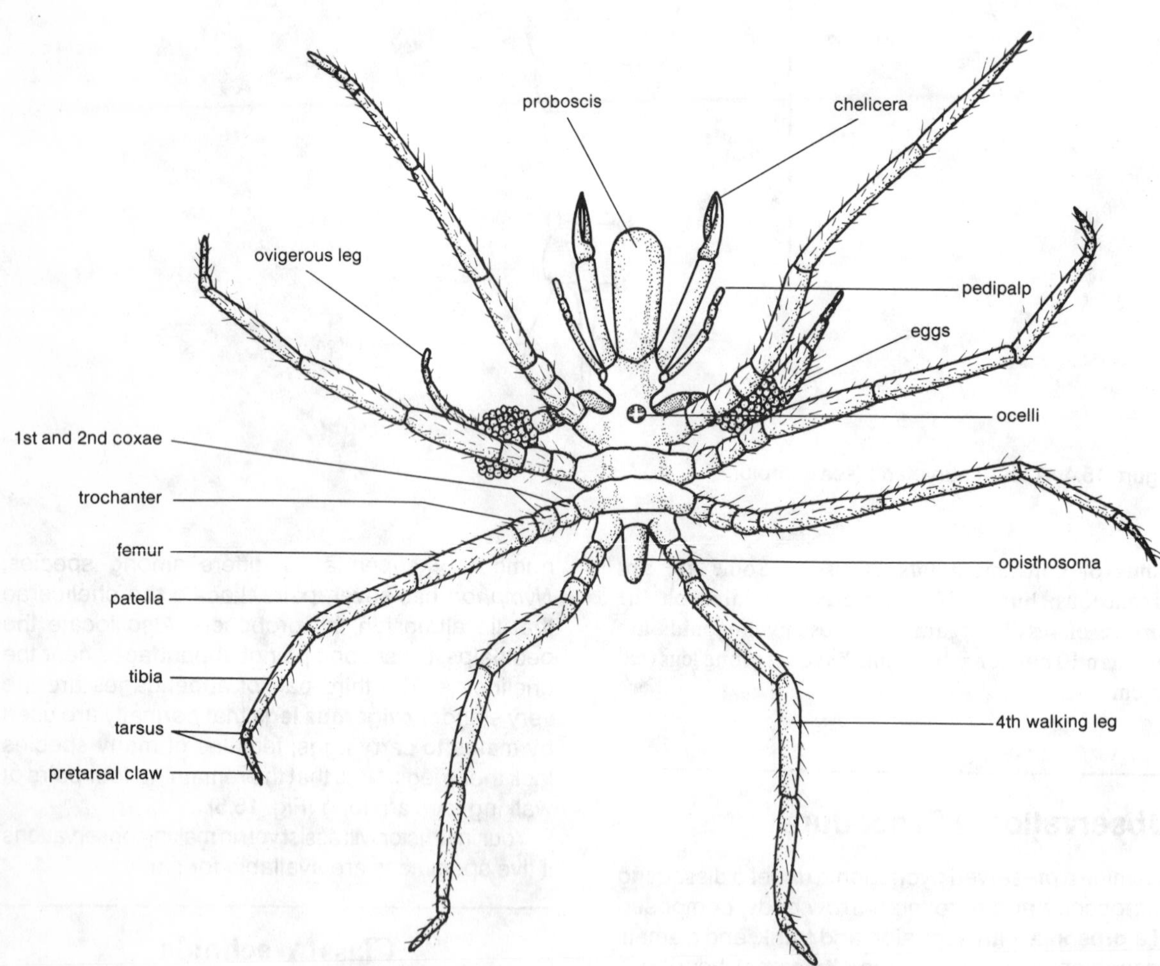

Figure 15.5. Dorsal view of the pycnogonid, *Nymphon*.

genus found in the southwestern United States. The species *C. sculpturatus* can be dangerous to humans, especially children.

Observational Procedure: *Centruroides*

1. Examine a specimen of *Centruroides* or of another genus available for study (Fig. 15.6). Locate the chelicerae, pedipalps, and walking legs at the anterior end.
2. Dorsally the prosoma is coverd by a subquadrate plate or **carapace** that bears, at the anterior end, three groups of simple eyes.
3. Immediately behind the carapace is a short segment (eighth) that begins the opisthosoma. Observe that the opisthosoma is divided into two sections. The first is a broader region called the mesosoma (preabdomen), comprising segments 8 to 14. The second part (metasoma or postabdomen) is thinner, tail-like, and comprises segments 15 to 19.
4. Locate the **basal bulb (telson)** and sting attached to segment 19. The sting is capable of injecting a neurotoxin into prey from a gland located in the basal bulb.
5. Examine the chelicerae closely. Note that the finger on the outer side has two cusps that overlap the inner finger. The legs end in two pairs of claws on the tarsus.
6. Turn the animal ventral side up. Most of the ventral prosoma is covered by the first joint (**coxae**) of the walking legs. Find the ninth body segment. It may be recognized by the presence of **pectines,** comblike, sensory structures unique to scorpions.
7. Just anterior to the pectines locate the genital operculum of the eighth segment.
8. Find segments 10 to 13 just posterior to the pectines, and at their lateral margins, locate the spiracular openings to the respiratory organ called the **book lung.**

Order Pseudoscorpiones. Order Pseudoscorpiones (SU-do-SKOR-pi-O-nez; G., *pseudo*, false) comprises about 1500 species of small (less than 10 mm) chelicerates that resemble true scorpions, less metasoma and sting. Book lungs are also missing; gas exchange takes place through a tracheal system. These animals are common occupants in nests of ground-inhabiting animals as well as in organic debris. *Chelifer* is commonly found in houses. Pseudoscorpions feed on other small arthropods by grasping them and injecting a poison from glands in the pedipalps.

Observational Procedure

Examine a whole-mount slide of a pseudoscorpion, noting their general resemblance to scorpions except for the absence of metasoma and sting. Locate the chelicerae, pedipalps, and four pairs of walking legs. The dorsal surface is covered by a **carapace** or **scutum** (Fig. 15.7). The segmented opisthosoma is broader than the prosoma and rounded at its posterior end. As in scorpions, coxae cover most of the ventral surface of the prosoma. Note that the bulbous pedipalps are chelate and bear several sensory hairs, called **trichobothria,** scattered over their surfaces. Eyes may be present at the anteriolateral regions of the carapace.

Order Araneae. Order Araneae (a-RA-ne-a; L., *arane*, spider) comprises over 30,000 species of true spiders. Spiders differ from most other chelicerates in that the prosoma and opisthosoma are connected by a short, narrow isthmus called the **pedicel**; both body regions are unsegmented in the adult. The opisthosoma, usually referred to as the abdomen, contains the majority of the viscera, including the silk-spinning organs. Spiders range from less than 1 mm to nearly 10 cm long and are found in most habitats, including the tropics, deserts, arctic and alpine regions, dorm rooms, and at several thousand meters in the air.

Observational Procedure

1. Obtain several specimens for study (e.g., *Argiope*, orb weaver; *Latrodectus*, black widow; *Tarantula*, tarantula) and examine them, locating the prosoma, opisthosoma, and pedicel (Fig. 15.8).
2. Find the four pairs of walking legs and pedipalps on the prosoma. Note the large, two-jointed chelicerae with **fangs** (Fig. 15.8). The basal joint is the larger and the second joint is composed of a daggerlike claw or fang. In males the terminal joint

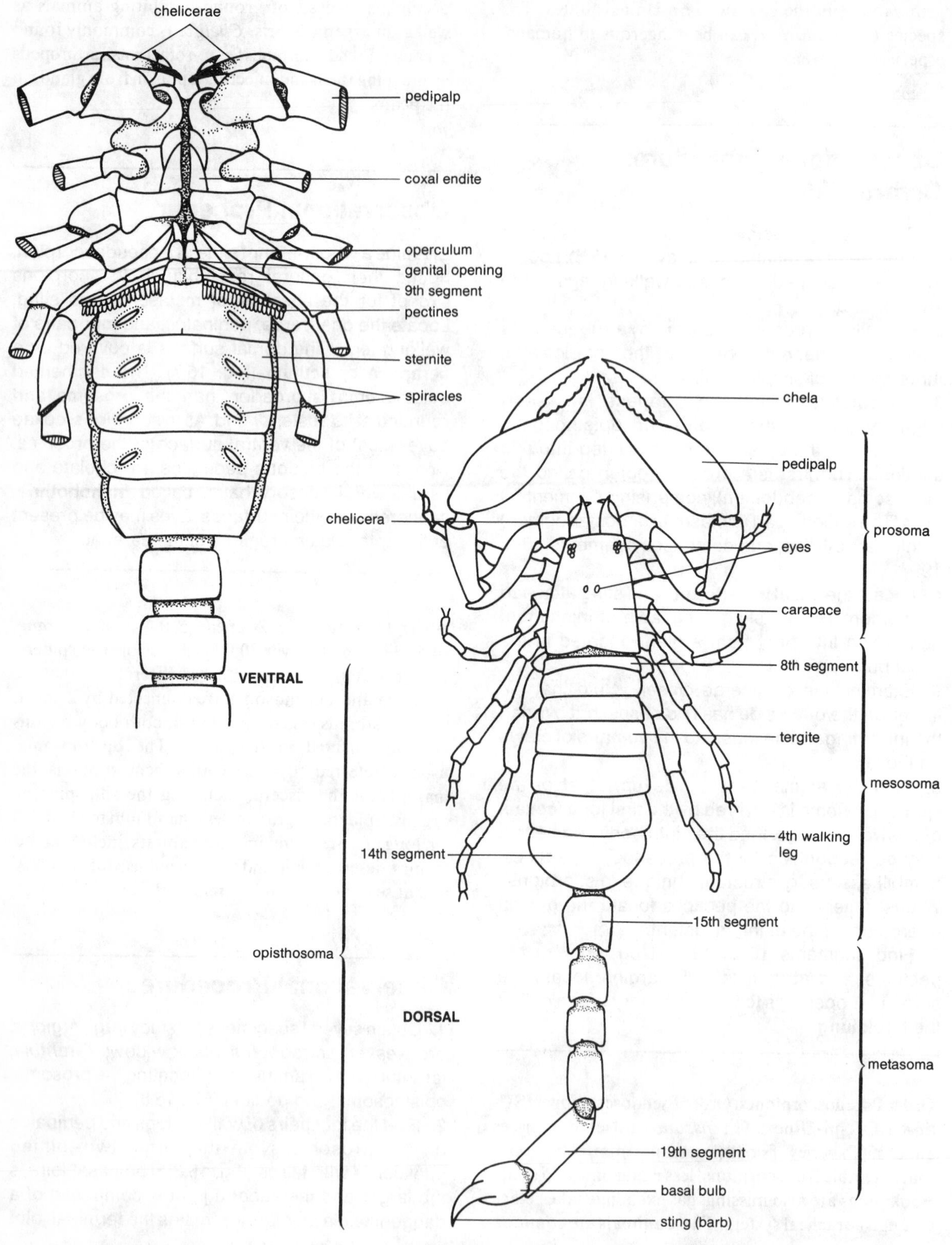

Figure 15.6. Generalized drawing depicting dorsal and ventral features of a scorpion.

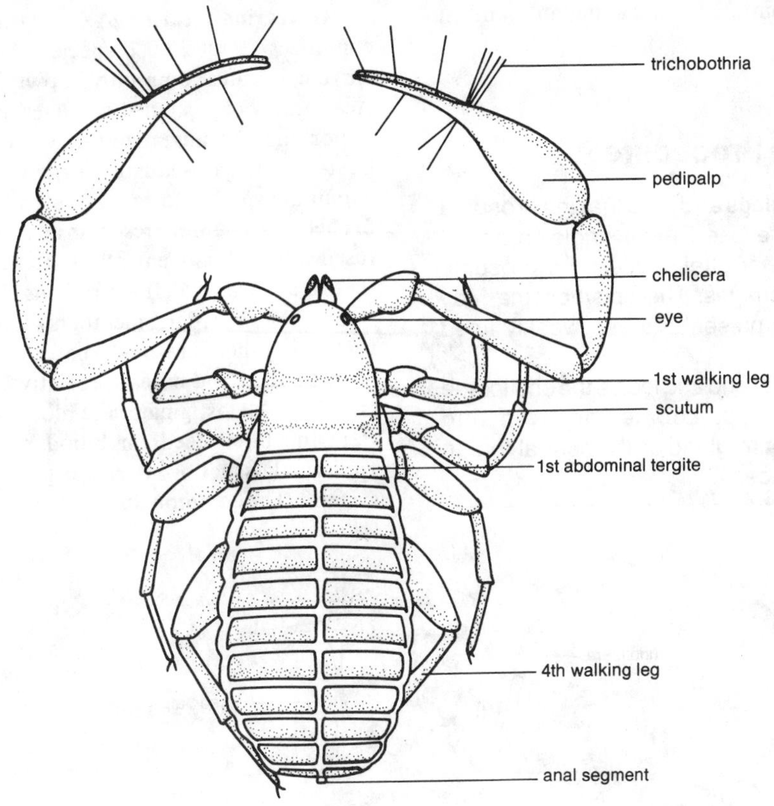

Figure 15.7. Generalized drawing depicting dorsal features of a pseudoscorpion. (After several sources.)

of the pedipalp is modified as an organ by which sperm are conveyed to the female's genital receptacle. This feature is an important taxonomic character.

3. Note the dorsal shield or carapace that covers the prosoma. Locate two rows of four eyes each at the anterior end.

4. Find the pedicel, the first segment of the opisthosoma (Fig. 15.8). Ventrally, segment 2 possesses the genitalia as well as openings to the book lungs (Fig. 15.8). The latter may be found by locating a transverse line called the **epigastric furrow** or fold. The lateral extensions comprise the openings (**spiracles**) to the book lungs.

5. Medial and slightly anterior to the furrow, find the genital opening. In the vicinity of the genital opening of most female spiders is the **epigynum**, which contains the opening to the seminal receptacles.

6. Find the **spinnerets** on the posterioventral surface of the opisthosoma near the anal lobe (Fig. 15.8). There are four to six spinnerets, depending on the species. Spinnerets are silk-producing glands and are considered appendages of the ventral body segments 10 and 11.

7. Anterior to the spinnerets, locate another one or two spiracle openings. In primitive spiders (mygalomorphs), the posterior spiracles are paired and lead to book lungs. In advanced forms the respiratory organs are tracheae, most often with a single spiracle.

8. If specimens are available, observe web-building and prey-capturing behaviors of live spiders according to directions provided by your instructor.

Order Opiliones. The familiar daddy longlegs or harvestmen comprise the order Opiliones (o-pil-e-O-nes; L., *opilio*, shepherd) or Phalangida (fa-LAN-ji-da; G., *phalange* spider), about 3000 species of arachnids noted for their long legs. Opiliones have the prosoma and opisthosoma fused into an egg-shaped body (5 to 20 mm long) that is elevated above the substratum by four pairs of very long legs. They have a cosmopolitan distribution, but are abundant in moist regions.

Observational Procedure

Obtain a daddy longlegs and note that the prosoma and opisthosoma are fused into a single carapace (Fig. 15.9). As in other chelicerates, the prosoma supports the appendages. The opisthosoma may be identified by the presence of ten weakly lined segments.

Examine the anterior end and locate the chelicerae and leglike pedipalps. Locate the eyes and anteriolaterally the stink gland on the central region of the dorsal carapace.

Order Acarina. Acarina (AK-a-RI-na; G., *akari*, mite) comprises about 25,000 species of small (less than 1 cm) chelicerates commonly known as ticks and mites. Acarines are of great medical, veterinary, and economic importance, because many are vectors or intermediate hosts of disease-causing organisms (e.g., Rocky Mountain spotted fever). The standard tagmatization of chelicerates is not present in acarines. The tagma are fused into a dorsoventrally flattened, ovoid body of several arbitrarily defined regions (Fig. 15.10).

Of seven suborders, members of five are parasitic or vectors of other parasites. Anatomical differences that separate ticks and mites are relatively minor. All ticks (suborder Metastigmata) are blood-sucking parasites. Parasitic mites feed on blood or burrow through epidermal tissues or internal organs (e. g., respiratory system). The common dust mite (*Dermatophagoides*)

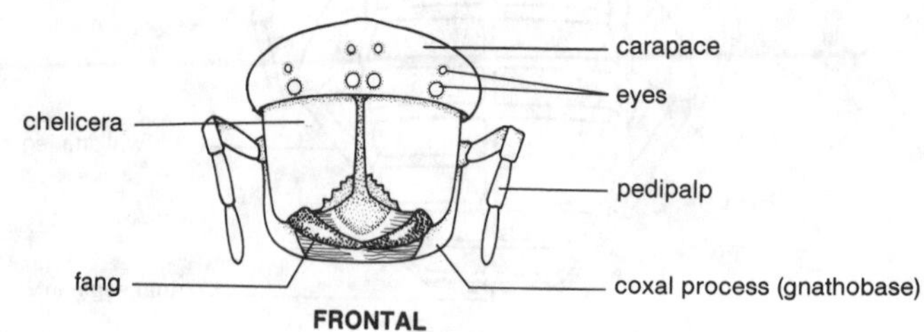

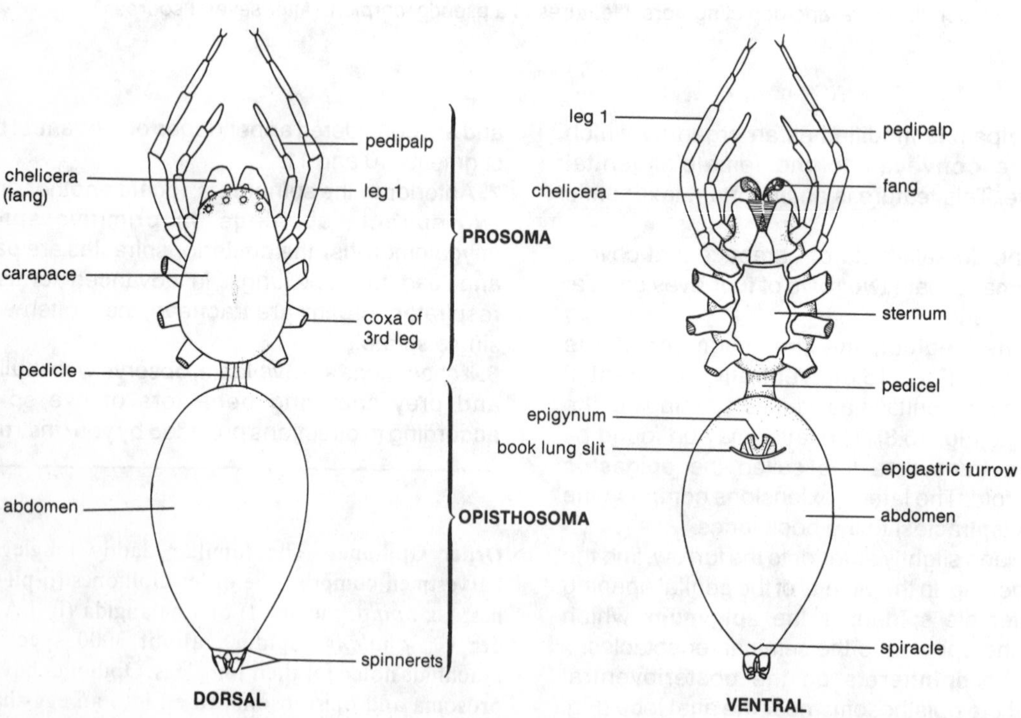

Figure 15.8. Generalized drawing showing spider morphology including frontal, dorsal, and ventral views.

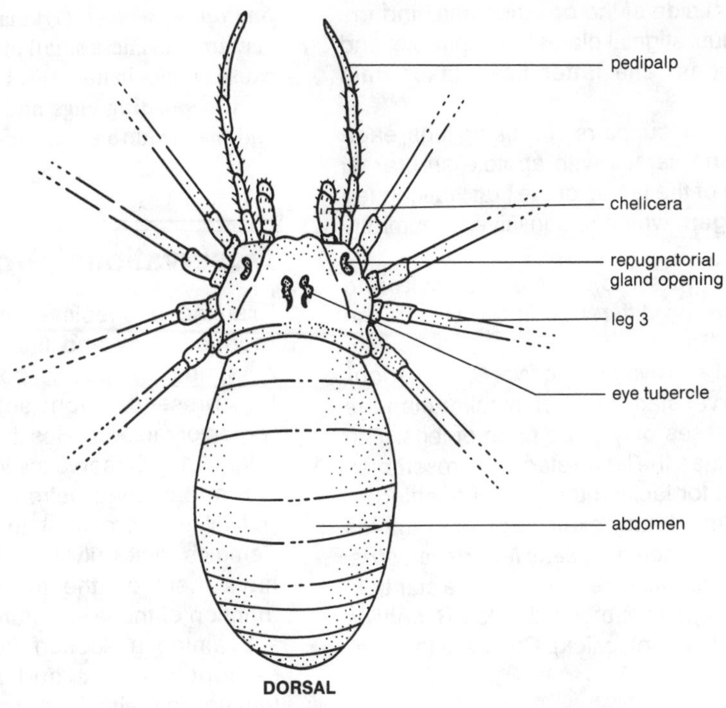

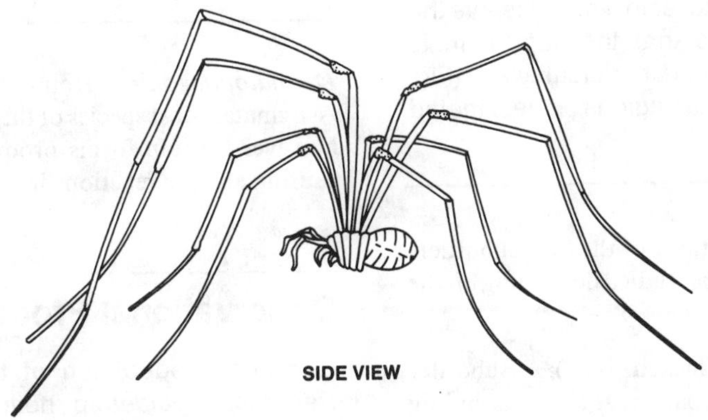

Figure 15.9. Dorsal and lateral views of a daddy longlegs. (After several sources.)

is not parasitic, but releases materials that may cause severe allergic reactions in humans.

Students will examine several examples of ticks and mites for an understanding of their diversity of form.

Observational Procedure

Ticks

Hard Ticks (family Ixodidae): *Dermacentor* spp.

1. Examine male and female specimens of either *Dermacentor andersoni* (Rocky Mountain wood tick) or *D. variabilis* (American dog tick) (Fig. 15.10). Both species are vectors of the rickettsial disease, Rocky Mountain spotted fever, and tick paralysis. Identify the two body regions: the body or **idiosoma** consists of a region bearing the legs (**podosoma**), and one in which legs are missing (**opisthosoma**).

2. Locate the anterior **capitulum (gnathosoma)**, bearing the feeding appendages, including the toothed **hypostome**, chelicerae, and pedipalps. Note that the capitulum may be seen in a dorsal view. This feature distinguishes hard ticks from soft ticks.

3. On the ventral side of the opisthosoma, find the genital pore, anus, stigmal plates with spiracle, and marginal **festoons**. The latter have taxonomic importance.
4. Also observe the four pairs of walking legs, each bearing a terminal tarsus with apotele and tarsal claws. At the tip of the tarsus of the first walking leg find **Haller's organ**, which is sensitive to humidity and odors.
5. Compare the dorsal idiosoma of a male and female and note the difference in the size of the **scutum** (Fig. 15.10).
6. Examine larval and nymph stages of *D. andersoni.* Note that the larval stage is much smaller than the adult and possesses only three pairs of legs. The nymph is larger than the larval stage and resembles the adult, except for lacking the genital opening.
7. For more advanced study, examine other examples of hard ticks (e.g., *Haemaphysalis leporispalustris,* rabbit tick; *Amblyomma americanum,* lone star tick; *Rhipicephalus sanguineus,* brown dog tick; *Boophilus annulatus,* American cattle tick). Compare these to *Dermacentor.*

Soft Ticks (family Argasidae): *Argas persicus.* Examine a specimen of *Argas persicus,* the fowl tick, a vector of relapsing fever in birds. Observe the sculptured cuticle. Note that the capitulum is obscured by the idiosoma in dorsal view (Fig. 15.10). How would the body regions be designated in this tick?

Mites. Representatives of three of the four suborders comprising the mites may be studied depending on the time available.

Dermanyssus gallinae (Chicken Mite): Suborder Mesostigmata. These mites parasitize chickens, hiding in the roosts by day and coming out to feed at night. Large infestations may be sufficient to kill the hosts from loss of blood.

Observational Procedure

Examine a specimen of *Dermanyssus gallinae* (Fig. 15.11). Locate the capitulum with unarmed hypostome, chelicerae, and pedipalps. Note that the chelicerae are needlelike; if you examine them under high magnification, you may be able to see their minute chelate ends.

Sarcoptes scabiei (Human Itch Mite): Suborder Astigmata. These small mites (less than 500 μm long) cause **scabies** in humans. Females burrow through the skin, depositing eggs and excreta that cause itching and may lead to secondary infection by bacteria.

Observational Procedure

Examine a specimen of *S. scabiei,* noting the rounded body and fine striations on the cuticle. Also, note the position of the four pairs of stubby legs, presence of long setae, and dorsally, a central region of raised scales that extend posteriolaterally (Fig. 15.12). Observe the legs of your specimen. The first and second pairs possess extensions called **pulvilli** just proximal to the tarsal leg segment. Females lack pulvilli on the rear pairs, but males have them on the fourth pair. What might the function of these structures be?

Examine a section from a mammal that has sarcoptic **mange** and attempt to locate adult females in their winding burrows. You may also be able to locate eggs, unhatched embryos, and ecdysed cuticles.

Dermatophagoides (House Dust Mite): Suborder Astigmata. Most species of this genus are not parasitic; however, certain forms produce powerful allergens, causing allergic reactions in sensitive people.

Observational Procedure

Examine a specimen of the house dust mite, *Dermatophagoides* sp., noting the ovoid body, long setae, and fine striations present on the cuticle as in *S. scabiei.* If time permits, you will be instructed on how to collect your own specimens.

Demodex folliculorum (Follicle Mite): Suborder Prostigmata. The follicle mites, *D. folliculorum* and *D. brevis,* live in human follicles and sebaceous glands, respectively, usually without causing serious damage. They are small (less than 500 μm) mites with short legs and an elongated, annulated opisthosoma.

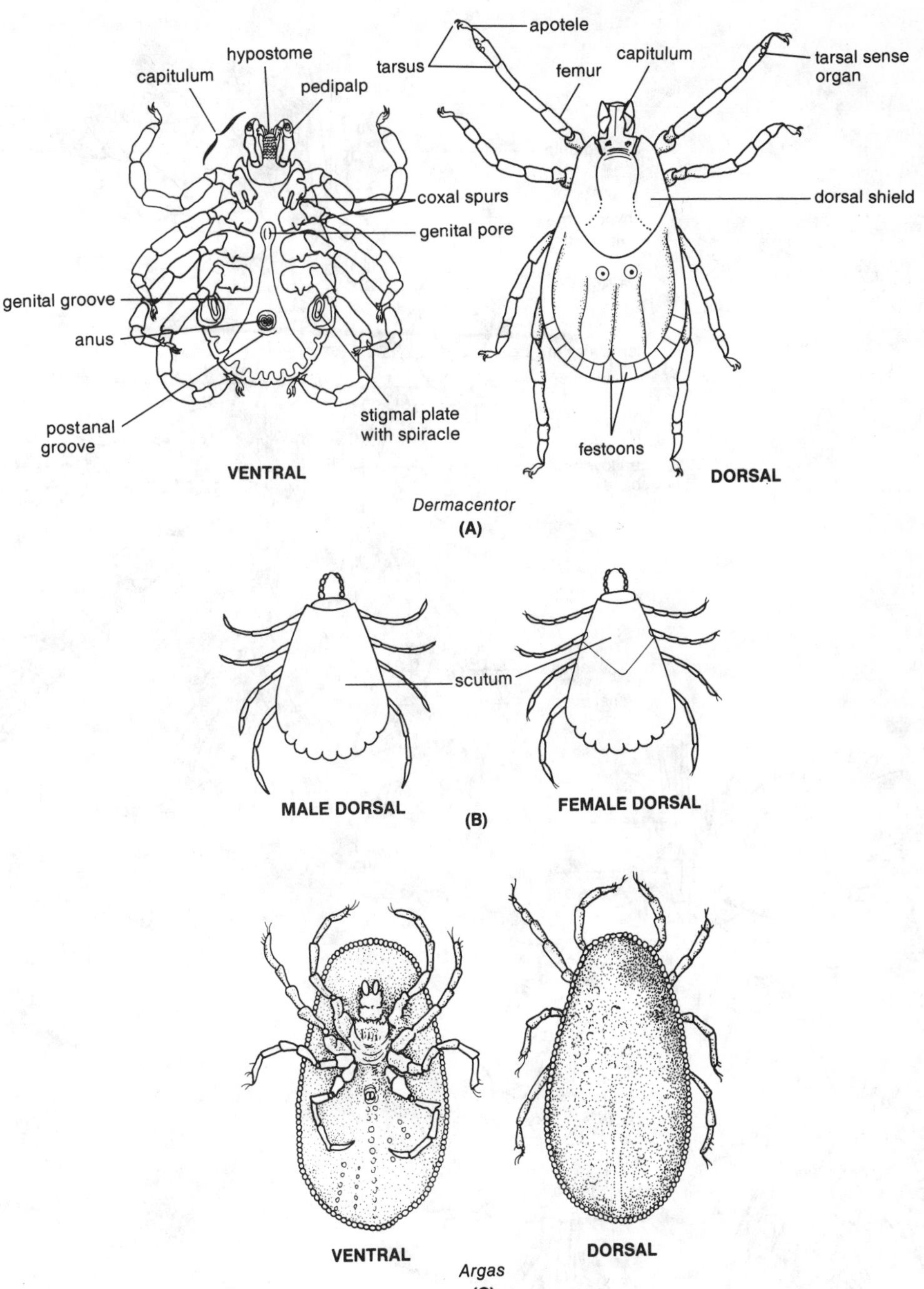

Figure 15.10. Dorsal and ventral views of hard (*Dermacentor*) and soft (*Argas*) ticks. (After several sources.)

Observational Procedure

Examine a specimen of a follicle mite, noting the four pairs of short legs with claws and the elongate, annulated opisthosoma. Also, examine a cross section of *D. folliculorum* in situ and locate the mites in the hair follicles. Do you notice anything unusual about the orientation of these organisms in the

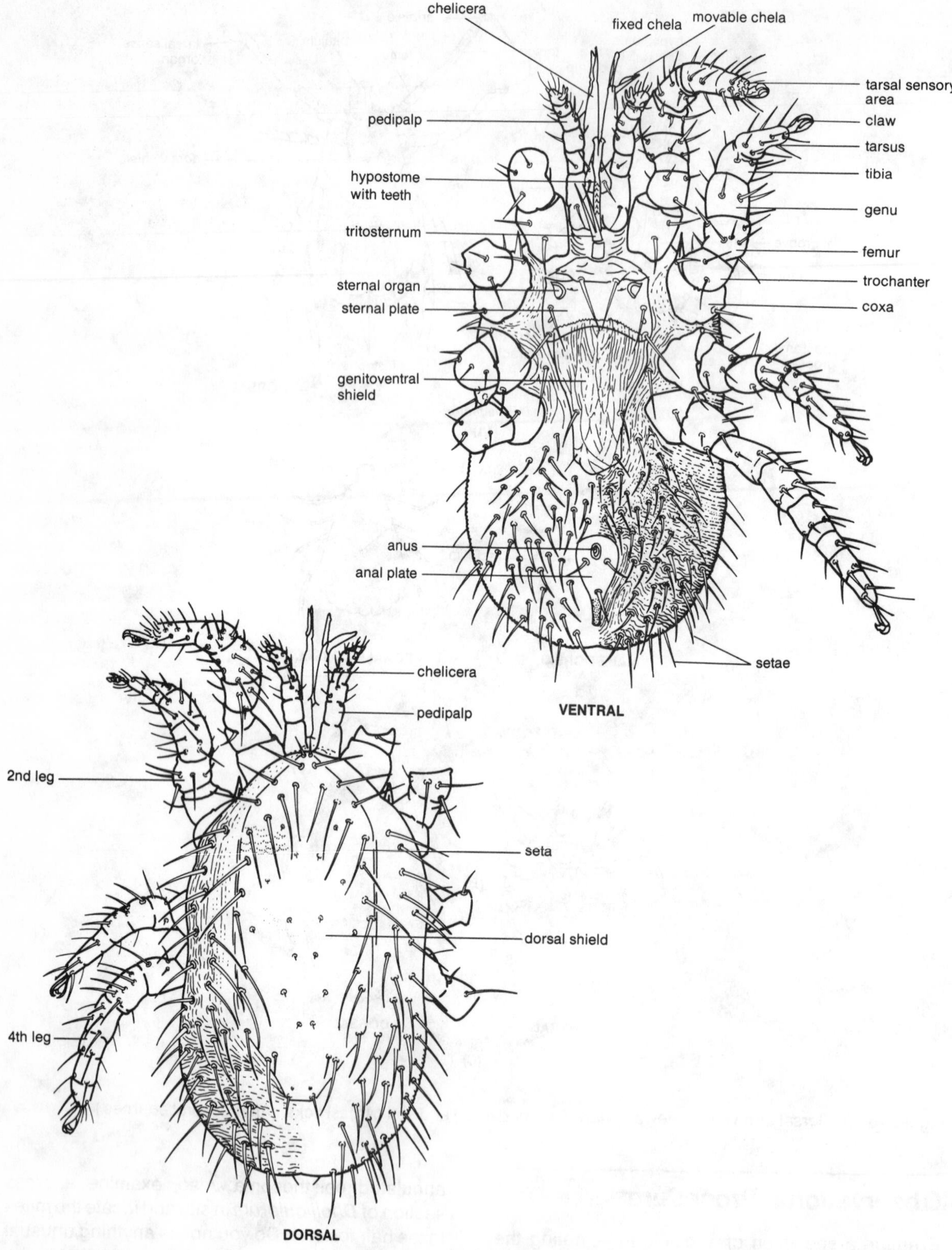

Figure 15.11. Dorsal and ventral views of *Dermanyssus gallinae*. (After T. M. Evans.)

follicle? What is the significance of the elongated body? If time permits, you will be instructed on how to collect your own specimens.

Order Solifugae. The Solifugae (so-LIF-u-je; L., *soli*, sun + L., *fug*, flee), also Solpugida (sol-PU-ji-da; L., *solpug*, venomous sider), is a group of relatively large (1 to 7 cm), aggressive chelicerates. They are commonly known as sun, camel, soldier, or wind spiders. Solifugids feed on arthropods and vertebrates, tearing prey apart with large, stocky chelicerae. They have a cosmopolitan distribution in tropical, subtropical, and warm, arid temperate regions. They are abundant in desert regions of the southwestern United States. Respiration in the entire order is accomplished by trachea.

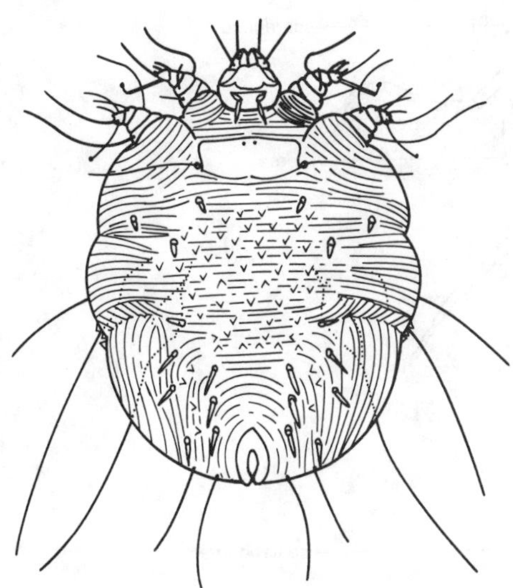

Figure 15.12. Dorsal view of a female *Sarcoptes scabiei*. (After Cheng, from Hirst.)

Observational Procedure

Examine a specimen of a solifugid (Fig 15.13). Observe that the hirsute (hairy) body appears to be divided into three divisions. Actually, the prosoma has a constriction that separates it into two regions of three somites each: head and thorax (or anterior and posterior carapaces, respectively). In non-burrowing species, the head and thorax sections articulate. Note that the thorax has dorsal transverse lines that indicate the position of the three somites. Four to six simple eyes are present on the head, depending on species. Locate the large chelicerae and determine in which direction they articulate. Note the robust nature of the pedipalps and the relatively delicate first legs. The pedipalps are not chelate but are used in prey capture. The first pair of legs are not used in locomotion but are tactile organs. On the coxae of the posterior legs are the racquet organs, whose function is unknown. The segmented opisthosoma is broadly joined to the prosoma and comprises ten segments.

Order Uropygi. The Uropygi (U-ro-PI-ge; G., *uro*, tail + G., *pyge*, rump) contains small chelicerates (a few millimeters to about 7 cm) that are flattened both in the prosoma and opisthosoma. They are commonly called whipscorpions because of the long **flagellum**, or whip, at the end of the opisthosoma. They are also called vinegaroons, due to their ability to spray a defensive fluid composed mainly of acetic acid from a pair of glands on either side of the anus. Whip-scorpions have a worldwide distribution, but tend to be common in tropical, subtropical, and warm temperate regions. One of the largest forms (ca. 6.5 cm), *Mastigoproctus giganteus*, is found in the southern parts of the United States.

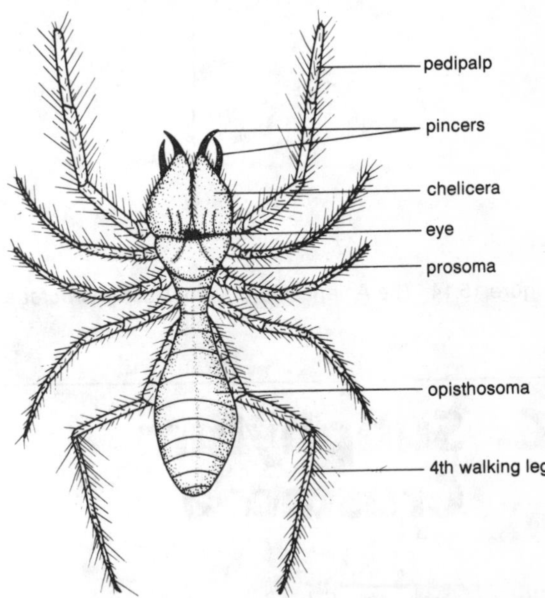

Figure 15.13. Dorsal view of a solifugid. (Total length including prosoma and opisthosoma about 2.5 cm.)

Observational Procedure

Examine a specimen of *Mastigoproctus*, noting the flattened body and long flagellum. At the anterior end distinguish the small, stout chelicerae from the larger, chelate pedipalps used in prey capture (Fig. 15.14). The first pair of legs are thinner and sensory, but the remaining three pairs are used in walking. Note that the opisthosoma is segmented and broadly attached to the prosoma. It is composed of a larger mesosoma of eight somites and a shorter metasoma of three narrow segments before the flagellum. Locate the paired openings to the defensive anal glands on the ventral side of the metasoma. Note how the metasoma can be oriented to direct the spray in many directions.

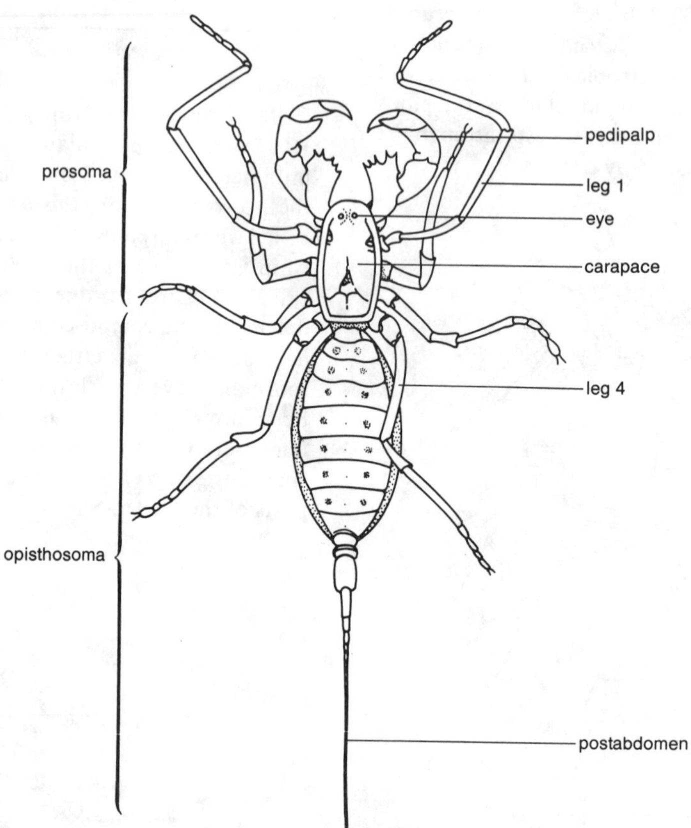

Figure 15.14. The American uropygid, *Mastigoproceteus giganteus.* (After several sources.)

C. Subphylum Crustacea

Shrimps, crabs, crayfish, lobsters, barnacles, and pill bugs are members of a large, diverse subphylum of arthropods called Crustacea (krus-TA-shi-a; L., *crusta*, crust or rind). Over 40,000 species are recognized, most of which are aquatic. Crustaceans are unique arthropods possessing two pairs of antennae, biramous appendages, and mandibles (jawlike appendages). They range in size from minute (less than 500 μm) to

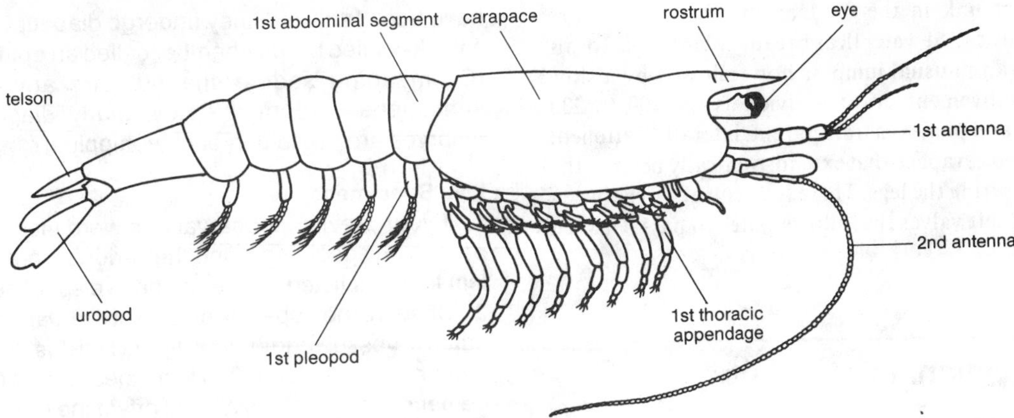

Figure 15.15. Generalized malacostracan crustacean.

extremely large; the lobster *Homarus americanus* may reach more than 0.5 m long. Many crustaceans are commercially important to the shellfish fishery (e.g., shrimps, crabs, lobsters in marine coastal waters and crayfish in freshwater). Small free-swimming crustaceans, their larvae and the larvae of larger forms, collectively called **zooplankton**, are very important ecologically, as they constitute a trophic level intermediate between algae and zooplanktivorous fish.

Primitively, the crustacean body is composed of a series of several segments each possessing a pair of biramous appendages. In advanced forms there is a strong tendency toward tagmatization, loss and fusion of segments, and loss of appendages. Three tagmata are usually recognized: head, thorax, and abdomen. The head and thorax may be fused into a single cephalic region called the **cephalothorax** (prosoma). Many crustaceans possess a shieldlike extension of the cephalothorax called the **carapace**, which may surround the animal entirely except for a small ventral gap. Most members of the class Malacostraca have a shrimplike body form with obvious tagmatization (Fig. 15.15). This arrangement is called the **caridoid facies** (L., *carid*, shrimp + L., *facies*, appearance).

Typical crustacean appendages are paired and biramous (Fig. 15.16), consisting of a proximal **protopod** (or protopodite) with two distal rami: the **endopod** (inner or medial branch) and **exopod** (outer or lateral branch). The protopod may be composed of three units (precoxa, coxa, and basis) and have other outgrowths on the lateral side (epipodites or exites) and on the medial side (endites).

Different classification schemes for the crustaceans are found in general and advanced texts, and your instructor may choose to discuss with you the merits of these various schemes. However, for purposes of laboratory study, taxonomic position is not particularly important, so the following exercises do not emphasize taxonomy. The crustaceans available for study are to be considered only as representatives of this very diverse group. Your instructor may provide additional representatives or substitute others for study.

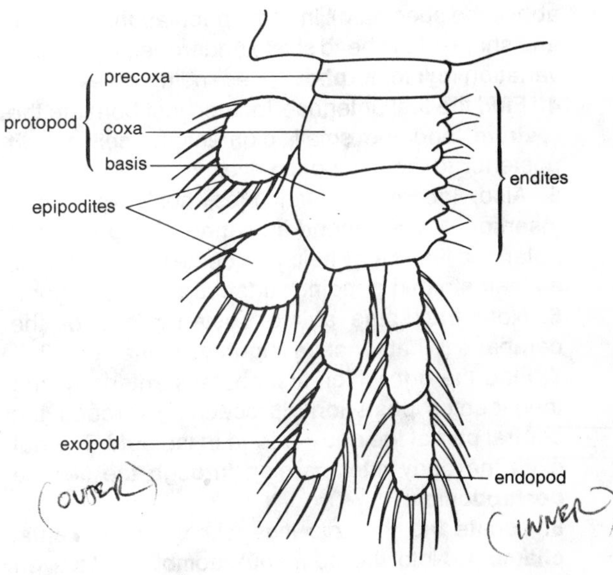

Figure 15.16. Generalized crustacean appendage. (After Schram 1986.)

Cladocerans

Cladocera (kla-DOS-e-ra; G., *cladus*, branch + G., *cera*, horn) is a small suborder of predominantly freshwater crustaceans that represent an extremely important link in the food chain. Cladocerans are commonly called water fleas because planktonic forms perform an unusual jump or hop followed by a slow sinking movement. Most individuals are 200 to 300 μm long and possess a transparent, laterally flattened, nonhinged carapace that extends ventrally beyond the body, covering the legs. There is a ventral gap between right and left valves that allows water to pass between them.

Observational Procedure

Morphology

1. Obtain a cladoceran such as *Daphnia* and observe it under a compound microscope (Fig. 15.17).

2. Examine the ovoid body and identify the anterior, posterior, dorsal, and ventral ends.

3. Observe the prominent second antennae and the single compound eye enclosed within a head shield at the anterior end. The prominant muscles, which originate on the inner surface of the head shield and insert on the inner surface of the second antennae, are birefringent and may be visualized using crossed polarizing filters (one below and one above the specimen). In some populations, the size and shape of the head shield undergoes a dramatic variation (**cyclomorphosis**) depending on season.

4. Find the first antennae, located just beneath the **rostrum**, and the small, pigmented **ocellus**, just posterior to the compound eye.

5. Also, locate the mandibles, posterior to the insertion of the second antennae. Using crossed polarized filters may help you locate the mandibles, as their striated grinding surfaces are birefringent.

6. Note the spine at the posterior end of the carapace. What function might the spine serve?

7. Find the gut, which first courses anteriorly and then curves posteriorly, proceeding through the central part of the body. Is food in the gut? The gut exits the body after passing through the flexible **postabdomen**.

8. Locate the thoracic legs within the **branchial chamber**. Note the numerous comblike setae on each leg. In life, the legs are almost constantly in motion, producing a water current that flows between the valves of the carapace, bringing food to the animal.

9. Dorsal to the gut is the **brood chamber**, which may contain developing embryos. How many embryos are present in your specimens? At what stage of development are the embryos? If eggs have been fertilized, they undergo diapause within a thick-walled brood chamber called an **ephippium** (G., meaning saddle-shaped). Are any of the specimens ephippiate? How many diapausing embryos are present in each ephippium?

Live Specimens

1. If your previous observations were made using preserved specimens, find the various morphological features listed above on the live specimens.

2. Observe the hop-sink motion of live *Daphnia* in a culture vessel under low light. How is the hop motion accomplished? How does drag and the center of gravity of the animal affect the orientation of the animal as it sinks?

3. How does *Daphnia* respond to light shined from above? From the side?

4. Place a single specimen into a drop of pond water on a shallow depression slide and cover it with a cover glass.

5. Observe the beating heart located dorsally near the midpoint of the body. Under high magnification you may be able to see blood cells coursing through the **hemocoel.**

6. Observe feeding activity of *Daphnia* by adding a small drop of a suspension of food particles. Can you see feeding currents? How does *Daphnia* process food particles? How do the mandibles work? Is food present in the food groove? Attempt to observe rejection of food by *Daphnia*. How is this accomplished? Does the postabdomen have any function in food processing? Remember that your observations do not reflect the normal feeding conditions of *Daphnia*. In your preparation the animal is confined in a small amount of water with abnormally high levels of food.

Ostracods

Ostracoda (os-tra-KO-da; G., *ostracum*, shell) is a class of small, bean-shaped crustaceans, commonly called seed shrimps. Most are less than 1 mm to 3 mm long, although some larger (greater than 2 cm) marine forms are known. Ostracods resemble bivalves in possessing a dorsally hinged carapace of right and left **valves**, with an adductor muscle for closure. Resemblance to bivalves provides another common name, mussel shrimp. The valves are reinforced with calcium carbonate, except at the hinge. Ostracods have a cosmopolitan distribution in benthic freshwater and marine environments; several terrestrial genera are known. As with the cladocerans, ostracods use their antennae for locomotion. Many ostracods are omnivorous.

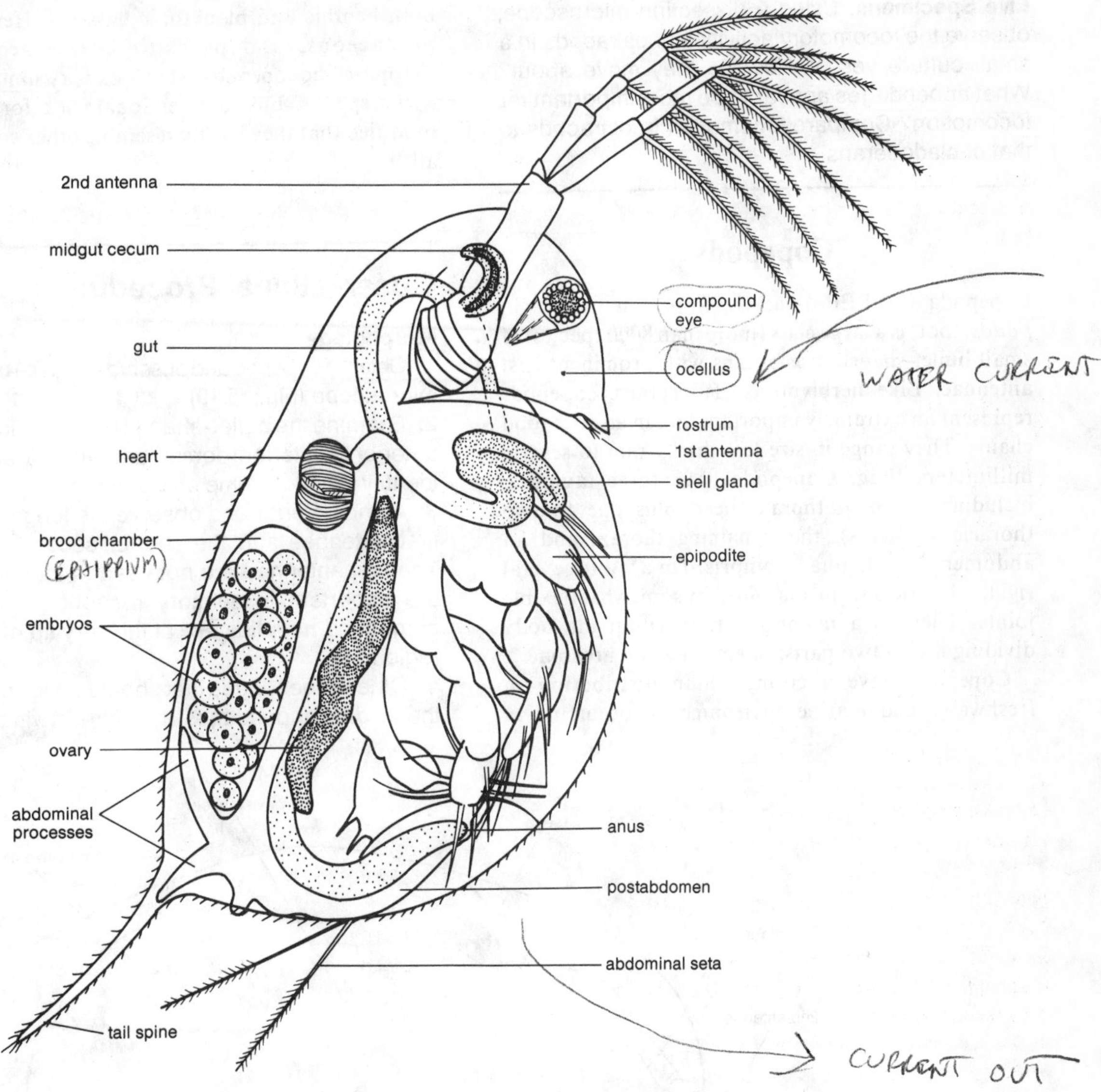

Figure 15.17. Lateral view of the freshwater cladoceran *Daphnia*. (After several sources.)

Observational Procedure

Morphology

1. Obtain an ostracod and observe it under a compound microscope (Fig. 15.18).

2. Note the ovoid body and determine the anterior, posterior, dorsal, and ventral ends.

3. Observe at the anterior end the prominent, paired first and second antennae. How do these differ from those of *Daphnia*? Note how the valves surround the body and the appendages except for the flagella of the antennae. Much of the internal anatomy of ostracods is obscured by the valves.

4. Observe the single compound eye just posterior to the insertion of the first antennae.

5. In the central region of the valve, locate the adductor muscle.

6. Attempt to locate the gut. Occasionally, portions of the gut may be identified by the presence of food material. The mandibles may been seen about one-third of the way back from the anterior end near the ventral margin of the valves.

Live Specimens. Using a dissecting microscope, observe the locomotory activities of ostracods in a small culture vessel. How do they move about? What appendages appear to be most important for locomotion? Compare locomotion in ostracods to that of cladocerans.

Copepods

Copepoda (ko-PEP-o-da; G., *kope*, an oar + G., *podos*, foot) is a large class (more than 8000 species) of small bullet-shaped crustaceans with prominent first antennae. Like herbivorous cladocerans, copepods represent an extremely important link in aquatic food chains. They range in size less than 1 mm to several millimeters long. Copepods have three tagmata, including a cephalothorax (head plus one to two thoracic segments), the remaining thorax, and an abdomen. Each tagma is comprised of a few to several rigid, cylindrical segments joined by somewhat flexible joints. There is a major junction within the body dividing it into two parts: **metasome** and **urosome**.

Copepods have a cosmopolitan distribution in freshwater and marine environments, occurring in both benthic and planktonic habitats. Herbivorous, predaceous, and parasitic forms are known. Ectoparasitic copepods often look very similar to free-living species, but some endoparasitic forms are so modified that they hardly resemble other copepods at all.

Observational Procedure

Morphology

1. Obtain a copepod and observe it under a compound microscope (Fig. 15.19).

2. Examine the bullet-shaped body and identify the anterior, posterior, dorsal, and ventral ends. Also identify the metasome and urosome.

3. At the anterior end observe the long uniramous first antennae and the shorter second antennae. The first antennae are not used in locomotion but are important in sensory reception. Locate the compound **nauplius eye** at the very tip of the head region.

4. Other appendages that should be found include the feeding appendages, the paired swimming legs

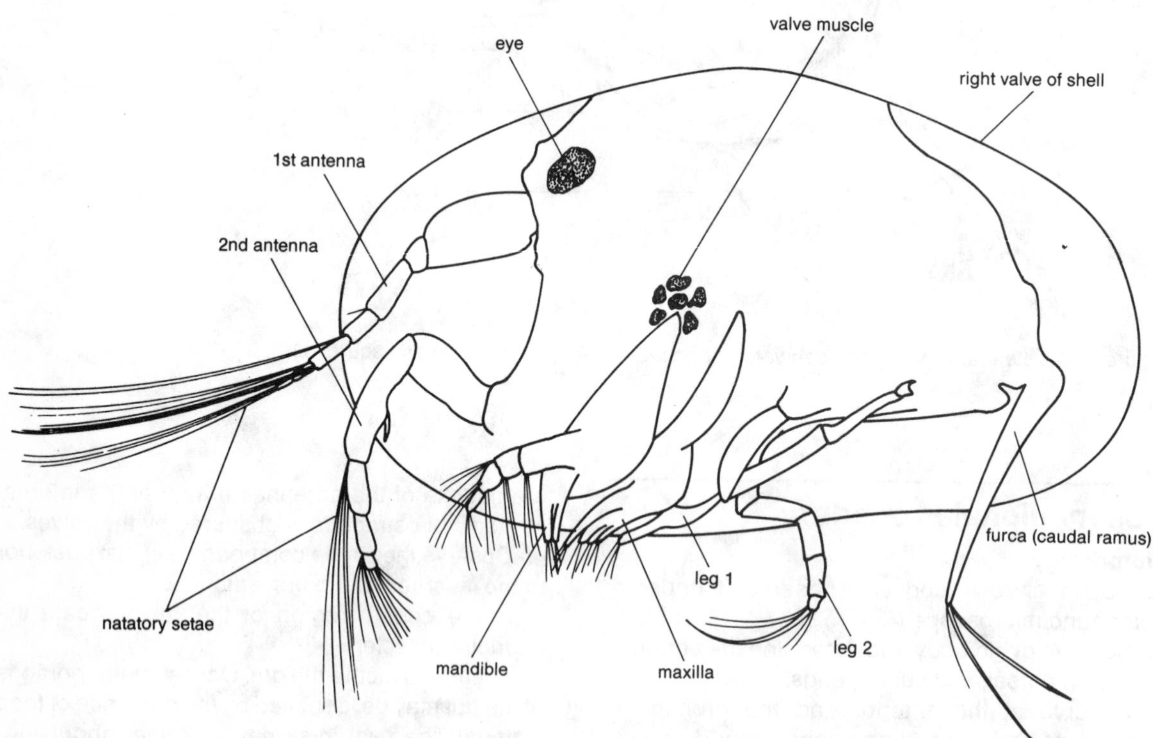

Figure 15.18. Lateral view of a generalized ostracod. (After several sources.)

(in the thorax), and a pair of setaeous **caudal rami** at the posterior end of the urosome.

5. Females may carry a pair of egg sacs near the junction of the metasome and urosome.

6. Internally, look for prominent bands of longitudinal and circular muscles.

Live Specimens

1. Observe the locomotory activities of copepods in a small culture vessel using a dissecting microscope. How do they move about? What appendages appear to be most important for locomotion? Compare locomotion in copepods to that of cladocerans and ostracods.

2. Place a single specimen into a drop of pond water on a shallow depression slide and cover it with a coverslip. The specimen should be restrained as much as possible.

3. Note that there is no heart in the organism. Are there any colored materials present outside the gut? Colored lipid drops are often found in the hemocoel of copepods (and other microcrustaceans) as forms of stored energy.

Barnacles

Cirripedia (sir-ri-PED-i-a; L., *cirrus*, curl of hair + L., *pedis*, foot) is a small class of about 1000 species of sessile, marine crustaceans commonly called barnacles. Barnacles hardly resemble crustaceans at all. In fact, Linnaeus classified barnacles in the Mollusca, where they remained until 1830, when their larvae were discovered and it was deduced that they were crustaceans. Barnacles have tremendous economic importance, chiefly by fouling ocean-going vessels.

Most people are familiar with free-living thoracican barnacles, which are separated into two groups: (1) stalked or goose-neck barnacles (e.g., *Lepas*) and (2) stalkless or sessile barnacles (e.g., *Balanus*). Parasitic barnacles, which will not be studied here, have a morphology that is very much different from the thoracicans.

Observational Procedure

Morphology

1. Obtain a specimen of *Lepas*, a stalked barnacle, and place it in a dissection pan with the muscular peduncle toward you.

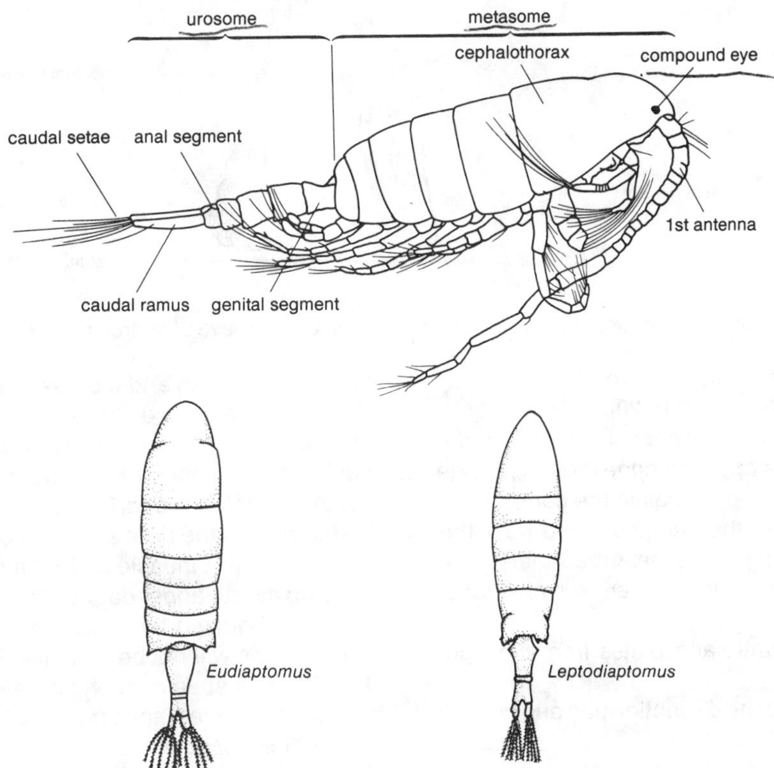

Figure 15.19. Dorsal and lateral views of typical free-living copepods. (After several sources.)

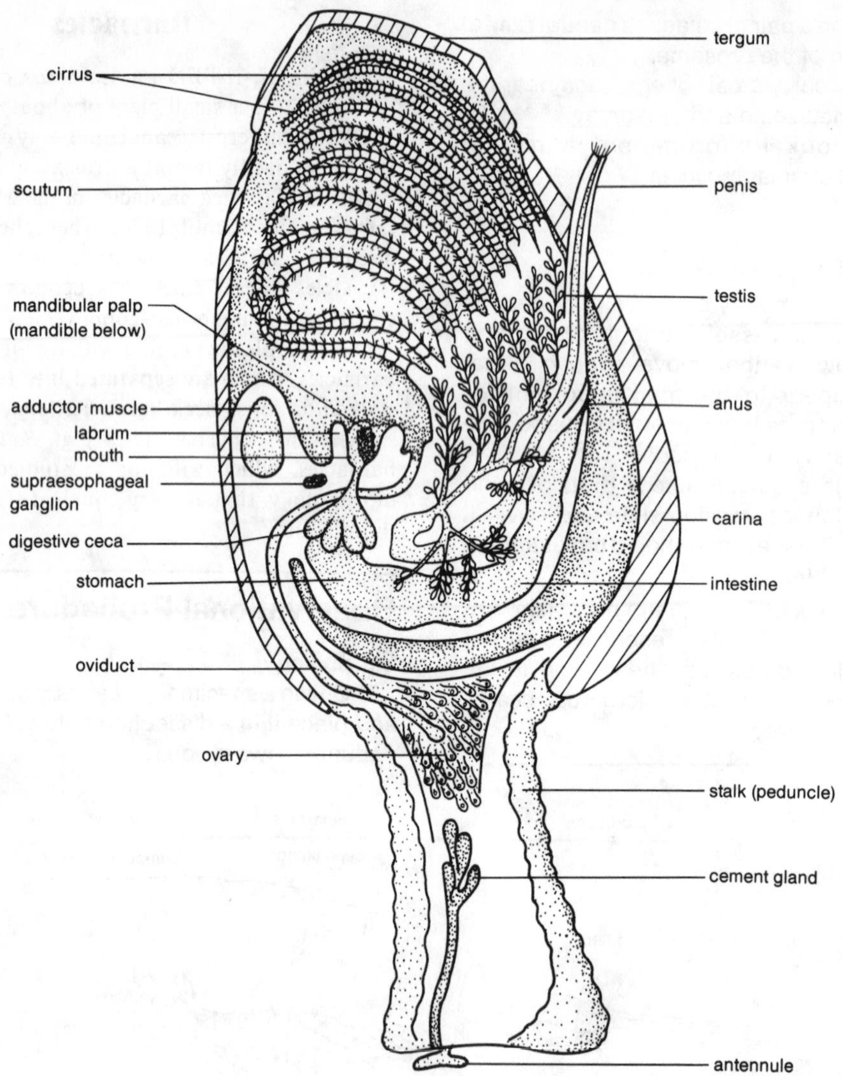

Figure 15.20. Lateral view of the stalked thoracican barnacle, *Lepas*. (After several sources.)

2. Distinguish the **peduncle** from the laterally compressed **capitulum** (Fig. 15.20).

3. On one edge of the capitulum, locate the **aperture**, and on the opposite edge (dorsal), locate an unpaired, keel-like plate called the **carina.**

4. On either side of the capitulum and near the peduncle, locate a pair of large, broad plates, the **scuta**. Above the scuta is another pair of smaller plates, the **terga.**

5. Remove the mantle and plates from the right side of the animal.

6. Pin the animal to the dissection pan and cover it with water.

7. The most conspicuous structures are six pairs of biramous **cirri** with numerous setae. The cirri curl toward the mouth and increase in size posteriorly. How is the setation distributed on the cirri?

8. Posterior to the cirri, locate the long, segmented penis and at its base, the posterior anus.

9. At the anterior end, locate the adductor muscle that connects the right and left sides of the mantle. Posterior to this muscle find the mouth and accompanying mouth appendages. The mandibles will be easily recognized by their strong dentition. Other appendages will not be identified.

10. Obtain a specimen of a sessile barnacle such as *Balanus*. Dried specimens will suffice, as they will not be dissected.

11. Determine the anterior-posterior orientation of the barnacle and the various plates (Fig. 15.21).

Note that the operculum is composed of two pairs of movable plates (terga and scuta).

12. The internal anatomy of *Balanus* is similar to that of *Lepas*. If advanced dissection is to be done, your instructor will provide additional directions.

13. Obtain a prepared slide of a **nauplius larva** of a barnacle under a compound microscope and examine the shield-shaped body.

14. Determine the anterior, posterior, dorsal, and ventral ends.

15. At the anterior end locate the first antennae, nauplius eye, and frontal horns.

16. Two prominent bulges may be seen on either side of the ventral surface. These are the second antennae, which possess cement glands.

17. Toward the posterior part of the body a metameric tail-like structure may be seen. These metameres will develop into the thoracic legs in subsequent molts.

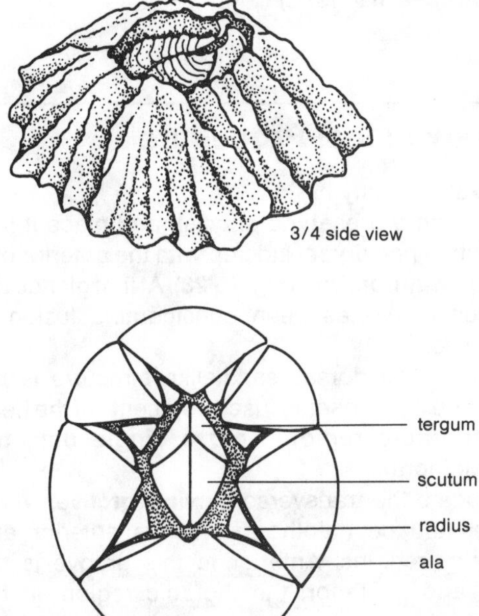

Figure 15.21. External plates of the sessile barnacle, *Balanus*. (After several sources.)

Live Specimens. Examine a living barnacle in a small dish filled with seawater and observe movements of the cirri. The rate of movement of the cirri and the way in which they are used to filter particles from the water depends on factors such as temperature, water movement, and concentration of food. Place a drop of milk near the cirri and observe the movement of milk. How do the cirri respond to the presence of the milk suspension? At the direction of your instructor, attempt feeding the barnacles other materials.

Amphipods

Amphipoda (am-FIP-o-da; G., *amphi*, double + G., *podos*, foot) is a large order (about 6000 species) of shrimplike malacostracans which range in size from less than 1 mm to nearly 30 cm in a deep-sea benthic form. Most amphipods are marine, but many freshwater species and a single terrestrial group are known. Amphipods are commonly called scuds, indicating a swift swimming movement, but benthic burrowing and crawling forms are common. The laterally flattened body is slightly rounded on the dorsal side. Three tagmata are recognized: head, thorax (preaeon), and abdomen with pleon plus urosome. Amphipods lack a carapace.

Observational Procedure

Morphology

1. Obtain an amphipod such as *Gammarus* and observe it under a compound microscope (Fig. 15.22).

2. Determine the anterior, posterior, dorsal, and ventral ends of the organism.

3. Locate the paired, prominent first and second antennae at the anterior end. Note the well-defined head with paired, sessile compound eyes.

4. Find the feeding appendages in the buccal region. They are usually small and hidden from view. Locate the two pairs of **gnathopods**, which are used in grasping food and, by males, in holding the female during copulation. Gnathopods are the second and third pairs of thoracic legs called **pereiopods** (G., *pereio*, on the other side; also peraeopods). The third through seventh pereiopods are unspecialized ambulatory limbs.

5. **Coxal gills** are difficult to locate, but may be seen as ovoid sacs on the inner surface of the coxae of legs 2 to 6.

6. In the pleon region of the abdomen, locate the three pairs of setaceous **pleopods** (G., *pleo*, swim + G., *podos*, foot) used in swimming.

7. Posterior to the pleon is the urosome of three segments, each with paired uropods.

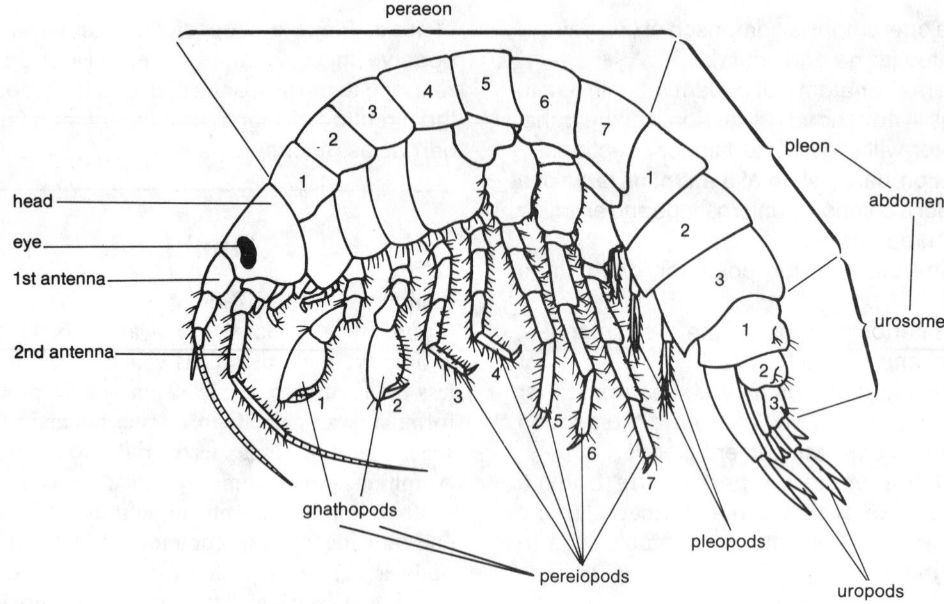

Figure 15.22. Lateral view of the freshwater amphipod, *Gammarus*. (After Pennak 1978.)

Live Specimens. Observe locomotory activities of amphipods in a small dish with the aid of a dissecting microscope. How do the animals move about? What appendages appear to be most important for locomotion? Compare locomotion in amphipods to that of the other crustaceans examined so far. If advanced study of the amphipod is to be done, your instructor will provide directions.

Crayfish

Crayfish, crawdads, or crawfish are commercially impoortant as food, especially in Louisiana. However, crayfish often are considered to be pests. In some regions they consume young crops and their burrows interfere cultivation, while in others their feeding habits deteriorate habitats previously suitable for game fish. Crayfish, crabs, and lobsters are members of the order Decapoda (G., *deca*, ten + G., *podos*, foot) by virture of having five pairs of pereiopods. We will study the crayfish as a representative decapod. The study of the blue crab, which then follows, may be substituted for that of the crayfish, or these two decapods may be compared.

Observational Procedure

General Anatomy

1. Obtain a preserved crayfish and place it in a dissection pan dorsal side up with the anterior end facing away from you (Fig. 15.23). At first glance it is obvious there has been considerable fusion of segments.

2. The large dorsal, saddlelike structure is the carapace. It represents fused segments of the head and thoracic regions into a single unit, the cephalothorax.

3. Locate the transverse **cervical groove**, which divides the cephalothorax into its anterior and posterior regions. Anterior to this groove is the head, and posterior, the thoracic region. In the middorsal region of the thoracic portion of the carapace and bounded by the **branchiocardia grooves** is the **areola**.

4. The fused terga of the carapace curve ventrolaterally to form a rooflike cover known as the **branchiostegite**. The space underneath it is the branchial chamber.

5. Turn the specimen over and reflect to one side one of the branchiostegites; observe the branchial chambers containing the gills.

6. Return the specimen to its dorsal side and locate the **rostrum**. To each side of the rostrum are located the stalked compound eyes that are set in ocular depressions (Fig. 15.23).

7. Turn the specimen ventral surface up and find the biramous **antennules** (first antennae). These are the most anterior appendages.

8. Just behind them are the elongate antennae (second antennae).

9. The next six pairs of appendages surround the mouth. These feeding structures are, from front to rear, the **mandibles**, first and second **maxillae**, and the first, second and third **maxillipeds**.

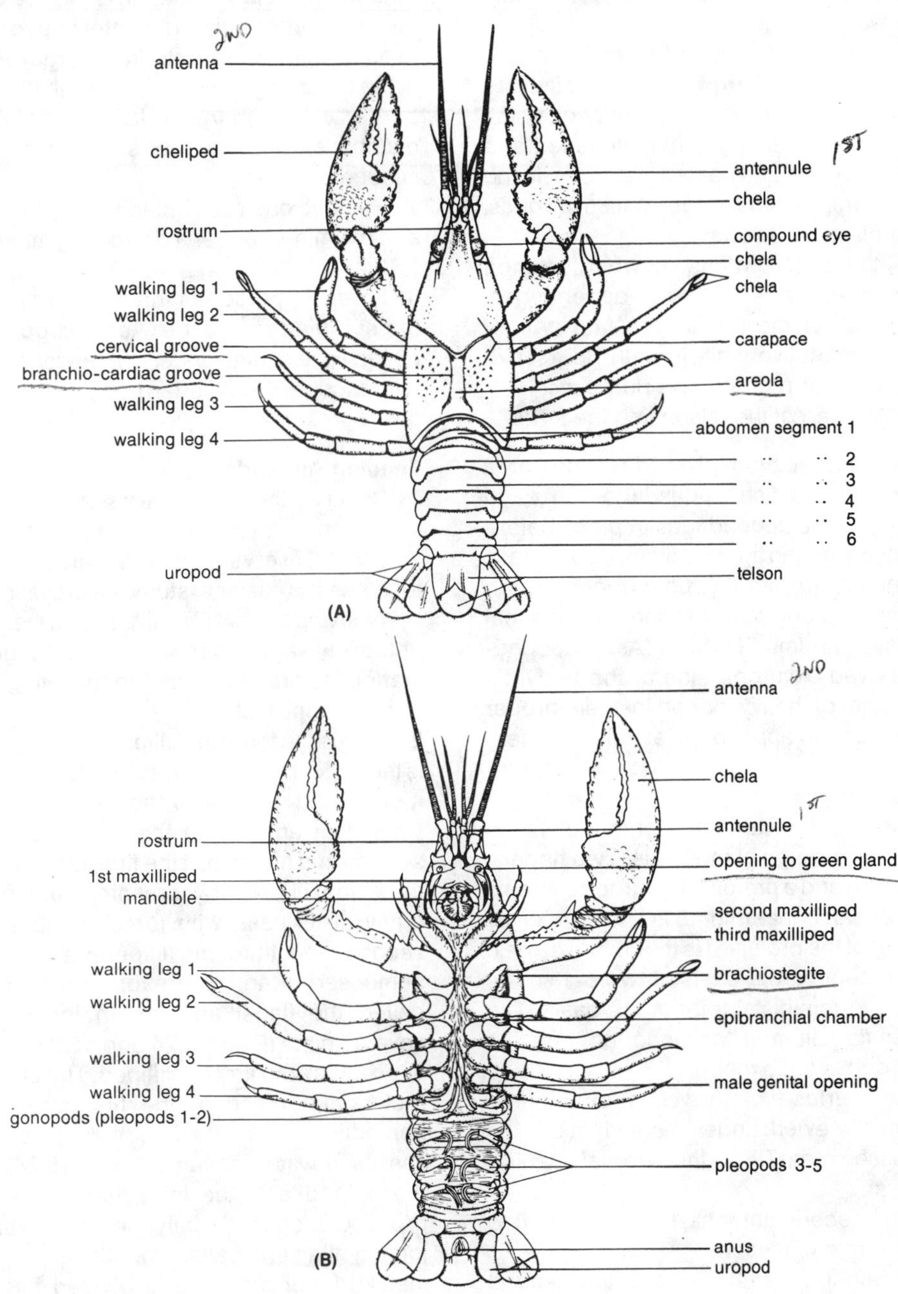

Figure 15.23. Dorsal and lateral views of a male crayfish.

10. The next five pairs of appendages are termed **pereiopods**. The first pair of pereiopods (**chelipeds**) are large and stout, distally bearing massive chelae. They are used in prey capture, defense, and aggression, and in manipulation of food.
11. The next two pairs of appendages posterior to the chelipeds are also chelate, but are much smaller. These and the following two nonchelate pairs are the walking legs.
12. Turn your attention to the abdomen. There are six abdominal segments (**tergites**). The first five are somewhat bandlike and extend laterally over the sternal plates. The sixth segment is flattened and broader than the others, and bears the tail, consisting of two fanlike biramous appendages, called **uropods**, and a central plate, the **telson.**
13. Ventrally, the abdomen is somewhat flattened and possesses six pairs of lateral appendages, called **pleopods**. In the female the first pair of appendages is diminutive, while in male the first two pairs are modified for reproductive purposes.
14. In the central part of the telson find the anus.

Appendages. This section gives direction for a detailed study of crayfish appendages. In each instance, observe the appendages in place before they are removed, noting their position, movements, and relationship to other appendages. Each appendage will be compared to the hypothetical biramous crustacean limb (Fig. 15.18). As the appendages are removed from one side of the body, fix them to a piece of heavy paper in their proper sequence; label each appendage as this is done.

Sensory Appendages

1. Manipulate the first antenna (antennule) and determine what its range of motion is. Two **flagella** of unequal length and a protopod of three segments compose the first antennae (Fig. 15.24). The proximal, single segment of the protopod is the coxa. The next two distal segments are a basis of two parts. The flagella attach to this distal joint of the basis. The shorter, inner flagellum is the endopod, and the longer, outer one is the exopod.
2. The dorsal surface of the coxa of the first antennae when viewed under magnification will show a depressed area. This is the external opening to the **statocyst**.
3. Observe the second antenna, manipulating it to determine its range of motion.
4. Using a scalpel, carefully remove the second antenna, making sure that the coxa remains attached. The large leaflike structure is the exopod, also termed the **squame** or **antennal scale** (Fig. 15.24).
5. Locate the slightly elevated structure on the coxa and note the opening at the apex. This is the excretory organ commonly called the **green gland**.
6. Anterior and lateral to the coxa there is the broad second joint of the protopod, the basis. Inward to the basis is the first joint of the endopod, called the ischiopodite. The endopod consists of three joints. First is the ischiopodite, next the meropodite, and then the carpopodite. The latter supports the many-jointed flagellum. The straight margin of the squame is the outer edge; the inner margin is clothed on its border with a fringe of long setae. Examine the distribution and kinds of setae on all parts of the second antenna.
7. Remove one eye, including the **peduncle** (stalk), and examine it under a dissecting microscope (Fig. 15.24).
8. Scrape a portion of the eye's surface and place the scraping on a slide. Add a drop of water and coverslip; examine under low power of a compound microscope. What pattern do you see? What does this represent?

Feeding Appendages

1. Observe the appendages in the region of the mouth and move them about using a probe (Fig. 15.23B). Observe how they articulate and nest together. It is easier to study the crayfish mouthparts by removing the third maxilliped and then proceeding anteriorly, appendage by appendage, until the mandibles are reached. The following procedures use this approach.
2. Locate the three maxillipeds. The third maxilliped is the largest and most posterior feeding appendage. It is located anterior to the base of the chelipeds. Lying just anterior to the third maxilliped is the second and ahead of it the first. Carefully remove all three maxillipeds by grasping and pulling each firmly at its base with forceps. Observe that the second and third maxillipeds are similar, being composed of an outer exopod with an extension called the flagellum and an inner endopod of several joints (Figs. 15.24 and 15.25).
3. Examine the first maxilliped. The elongate, basal rectangular piece of the first maxilliped is the epipodite, which fits into the gill chamber and assists in water movement (Fig. 15.24). Anterior to the epipodite is the long, narrow, and medially grooved exopod. Distally the exopod has a many-jointed filament called the flagellum. The large, leaflike inner structure is divided into two parts, endite 1 proximally and endite 2 distally. At the base of and in between endite 2 and the exopod is the relatively diminutive endopod. As with the second

maxilla, the coxa and basis are not distinctly separated. The central complex of chitinized structures represents a fusion of parts and constitutes the protopod. A large epipodite is attached to a coxa, as are the endites. The endopod and exopod are attached to the basis.

4. Locate the second maxilla. The largest part is the **scaphognathite** or **gill bailer** (Fig. 15.24). The narrower part lies next to the mandible, while the broader portion extends into the branchial chamber and assists water movement in the chamber. The lateral (inner) part of the second maxilla is divided into four narrowly foliaceous subdivisions. The anterior pair is considered to be endite 2 and the posterior endite 1. The slender, elongate appendage between endite 2 and the scaphognathite is the endopod. The basis and coxa are not distinguishable as separate components.

5. The first maxilla is leaflike (Fig. 15.24). The smaller, pointed part is the endopod. The two parts lateral to the endopod are endites; the broadest one is endite 1 and the longest is endite 2.

6. Locate the anteriormost feeding appendages, the mandibles, composed of broad, smooth, convex surfaces, with an inner sawlike edge and a blunt fingerlike process (Fig. 15.24). The basal part of the mandible is the coxa and the fingerlike structure the **palp**, comprised of three joints. The first joint of the palp is considered the basis and the outer two segments compose the endopod. The exopod is not present on the mandible.

Walking Appendages. The first, second, and third maxillipeds are considered the first appendages of the thorax. The next five to be studied are the pereiopods.

1. Locate the first pair of pereiopods, the chelipeds (Figs. 15.23 and 15.25). They are the largest of the appendages and are used to capture prey and to tear food apart. The remaining four pairs of pereiopods are the walking legs. Of these, only the first two pairs are chelate (Fig. 15.23).

2. Carefully remove all the pereiopods from one side of the body, severing them from the body at the coxa. Note that the exopod is not present in these appendages.

3. On the posteriormost part of the cheliped is a bilamellate, folded **podobranch gill** (arthrobranchia). The lamellated epipodite is membranous and folded as in the second and third maxillipeds. The upper lamella is covered with the gill filaments. Near the point where the long, threadlike coxal filaments are located, the plumose arthrobranchia are attached. Even with the greatest of care, the arthrobranchia may be difficult to remove intact.

4. Observe that the chela is made up of the propodite and dactylopodite (Fig. 15.25). The propodite is extended fingerlike at its outer edge, while the dactylopodite is movable and opposes the propodite, forming a pincerlike structure.

5. Remove a chela from one of the large chelipeds. This is best done by severing the appendage at the carpopodite joint (Fig. 15.25).

6. Remove the exoskeletal covering from one side of the chela. Note the two sets of muscles. The larger is the adductor muscle, which inserts on the inner edge of the dactylopodite and is instrumental in pulling the dactylopodite toward the propodite, thus providing the pincer action. The smaller abductor muscle reverses the movement.

7. Pereiopods 2 and 3 are chelate and other than in size are very similar to the cheliped. The female crayfish has a **genital operculum** on the inside surface of the coxa of the third pereiopod. The male genital opening is located on the inner surface of the coxa of the fifth pereiopod (Fig. 15.23).

8. Identify the sex of your specimen.

9. The fourth and fifth pereiopods are strictly ambulatory and are nonchelate. Between the coxopodites of the fifth pereiopods, the sternum in some female crayfish is modified as a **seminal receptacle**. This structure is termed the **annulus ventralis**. Sperm released from the receptacle at egg-laying time fertilize the eggs, which are then retained on the under surface of the tail region.

10. The remainder of the crayfish appendages are located on the abdominal region and are called pleopods or **swimmerettes** (Fig. 15.23). Some of them are similar in structure, but others differ considerably, depending on the sex.

11. Compare the position and morphology of the pleopods on the first and second abdominal segments in male and female crayfish.

12. In the male, the first pair of pleopods are closely associated, directed forward, and modified for sperm transport to the female during copulation. They are called **gonopods** (Fig. 15.25).

13. In the female the first pleopod is small and is composed of a single stalk (of two joints) with a flattened, distal flagellum (Fig. 15.25). The second female pleopod is the typical biramous appendage characteristic of most abdominal segments. The sternite to which the pleopod is attached is narrow and ringlike.

14. The remaining pleopods are the same in both sexes (Fig. 15.25). The coxa is a small ringlike

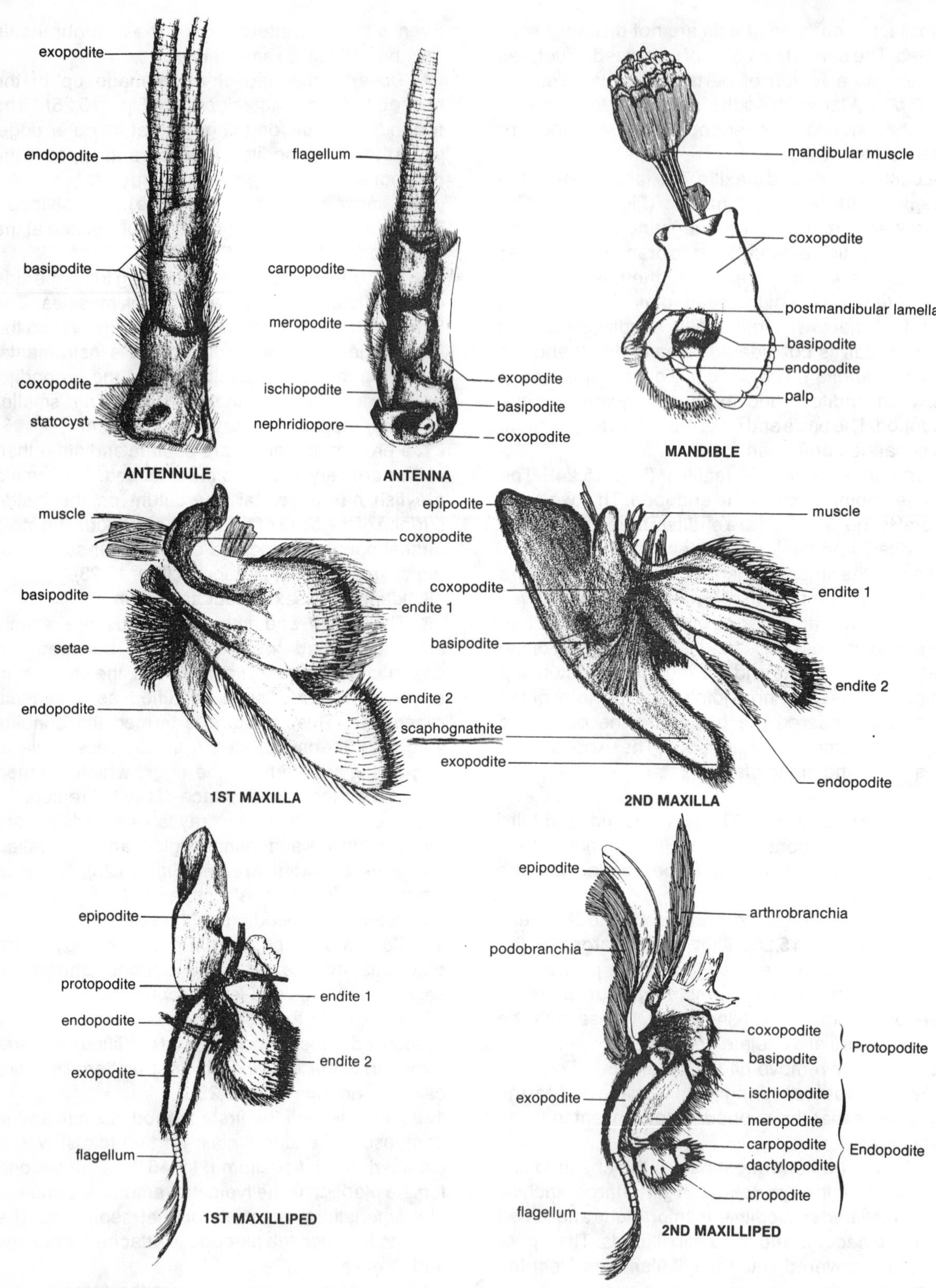

Figure 15.24. Crayfish appendages: first antennae through second maxilliped.

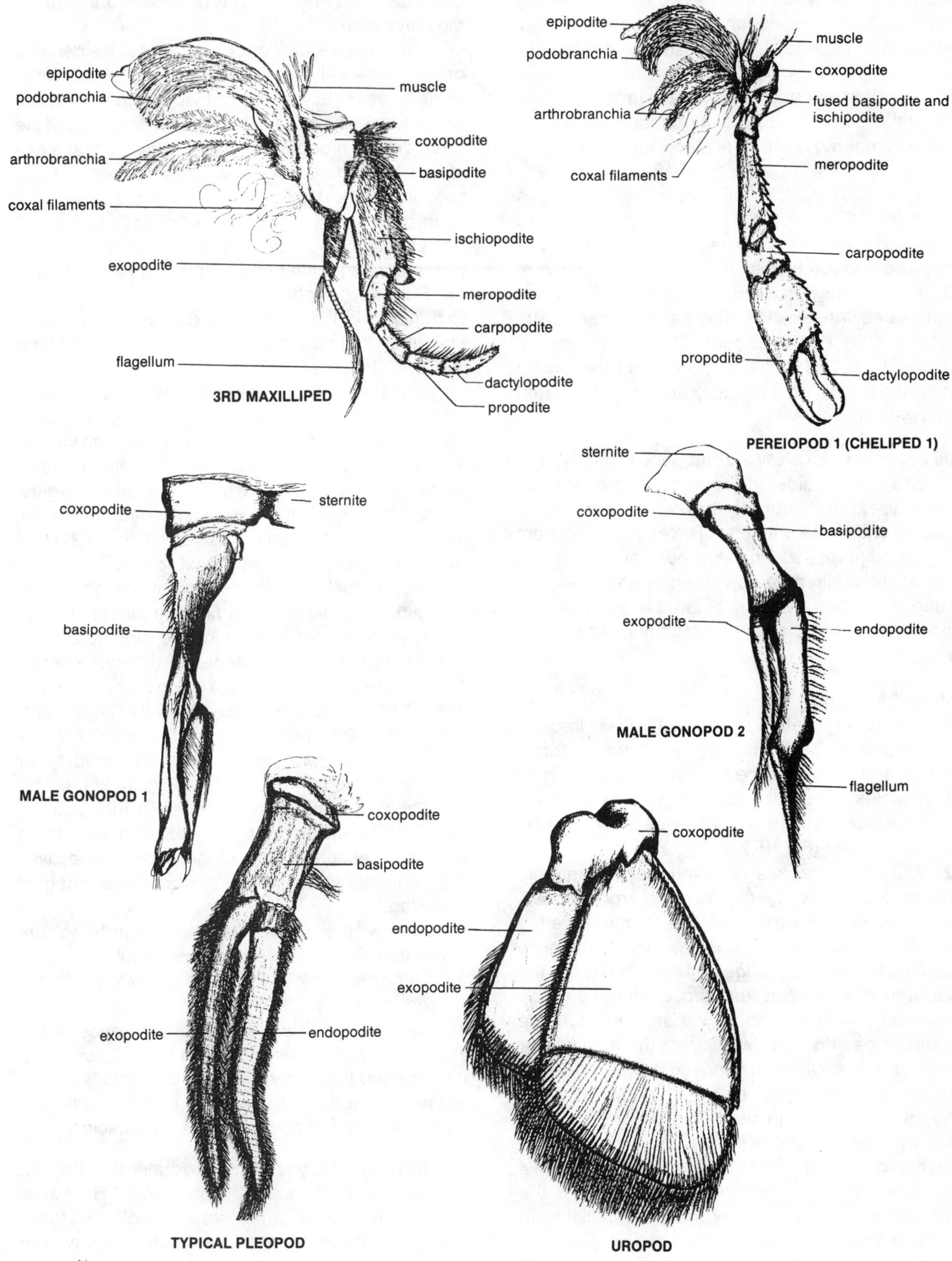

Figure 15.25. Crayfish appendages: third maxilliped through uropod.

structure and may be lost unless the limb is carefully removed. The endopods and exopods are flattened, setose, and flagellate. They arise from the stocky basis.

15. The sixth pair of pleopods is the uropods. They are similar in both sexes, but radically different in shape compared to the other pleopods. The exopod and endopod are large paddlelike flaps (Fig. 15.25). The exopod is divided into two sections along a broad joint (Fig. 15.25). The coxa and basis are fused into a large, angular segment from which the exo- and endopods arise.

16. The telson is not considered to be an appendage in the same sense as the others studied here. Note that the anus opens through the telson's ventral surface (Fig. 15.23). The uropods and the telson comprise the tail fan. This structure is used for rapid backward movement.

Gill Chamber. Cut away the carapace which covers the gills on the side where appendages are still intact. Be careful to cut only the branchiostegite, not the gills within the chamber. The entire gill system should be exposed undisturbed. Starting at the anteriormost gill, note how the gills are arranged. Using dissection needles, tease the gills apart to understand their places of attachment and serial arrangement.

Internal Anatomy

1. Make a very shallow, midsagittal incision through the carapace, beginning at the posterior dorsal margin, and extend it to the rostrum. Avoid damaging the muscles and other internal organs just beneath the carapace. Remove the carapace anterior to the cervical groove (Fig. 15.23).

2. With the carapace removed, several muscle groups and other internal organs are exposed. Extending to the rostrum from the stomach are the paired **anterior gastric muscles**. The two large muscles at the lateral edge of the stomach are the **mandibular adductors**, and those attached to the posterior dorsal margin of the stomach are the **posterior gastric muscles**. One of the mandibular muscles may have been removed previously when the appendages were being studied, but there should be one left in position. The anterior and posterior set of muscles are important in the mechanical operation of the stomach. Dissection a little later will show an intricate set of **ossicles** in the stomach that are moved by muscles assisting in grinding food that enters the stomach.

3. Gentle movement of the mandibular muscles to one side will reveal the dorsal anterior portion of the digestive gland (Fig. 15.26).

4. Cut away the carapace posterior to the cervical groove. This will expose the heart and the principal arterial vessels, reproductive bodies, and the remainder of the hepatic bodies. On each side of the heart are the elongate extensor muscles that manipulate the abdomen. Closely examine the heart and note the **ostia** (L., opening). If the specimen is a female, the ovary will be found directly in front of and slightly below the heart. If the sex is male, the testes will be located in about the same position.

5. Remove the right lateral wall of the cephalothorax to the base of the legs. The expansive liver is best shown by this dissection. Identify the **green gland** (Fig. 15.26).

6. Remove the stomach by severing the esophagus from the intestine. Place it in a dissection pan, partially filled with water. Turn the stomach ventral surface upward. Examine the organ with the dissecting microscope. Make a longitudinal midsagittal incision and then pin the stomach membrane aside. Identify the several sets of **ossicles** (Fig. 15.27) that give the stomach the name **gastric mill**.

7. Also note the unusual lining of portions of the stomach. This serves as a filtering mechanism. In the anterior, lower portion of the stomach, there may appear rounded, flattened structures on each side. These are **gastroliths**. They occur seasonally and will not be present in recently molted individuals.

Now that the stomach and other viscera in the cephalothorax have been removed, direct your attention to the nervous system (Fig. 15.26).

8. Locate the stub of the esophagus and posterior to it find the **subesophageal ganglion**. Extending around the esophagus are single nerve connectives that communicate with the **supraesophageal ganglion**.

9. Follow the nerve tract posteriorly to the abdomen. This will necessitate careful dissection as the nerve trunk is closely associated with the inner surface of the exoskeleton.

10. Remove the upper exoskeletal surface of the abdomen to the telson (Fig. 15.26). Locate the hindgut and the superior abdominal artery with its lateral branches. Note the arrangement and distribution, of the abdominal flexor muscles.

Live Specimens. Observe the live crayfish on display. How do they move about the aquarium? How does the crayfish react when threatened? As your instructor directs, observe the feeding behavior of crayfish.

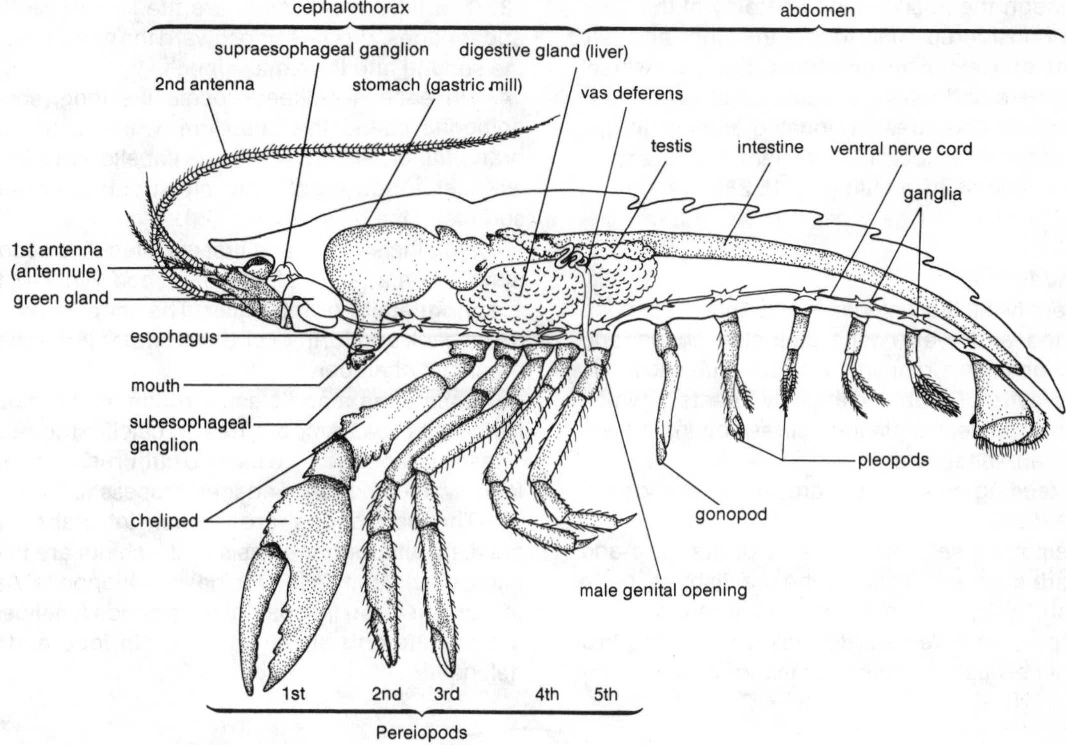

Figure 15.26. Internal anatomy of the crayfish (diagrammatic).

Blue Crab

The blue crab, *Callinectes sapidus*, is a common Atlantic and Gulf coast inhabitant of the United States, ranging from Massachusetts Bay to the northern shores of South America. It is commercially important to the fishing industry in the United States. This crab lives in waters ranging in salinity from brackish to normal seawater.

Observational Procedure

External Morphology. Obtain a specimen and compare its morphology with that of the crayfish (see Figs. 15.23 and 15.28A and B).

1. Place the crab dorsal side up in a dissection pan. Note that, unlike the crayfish, the blue crab is compressed dorsoventrally and is greatly expanded laterally. The crab is very broad and the legs are spread comparatively farther apart than in the crayfish.

2. Turn the crab ventral side up and find the abdomen, which is folded under the thorax (Fig. 15.28B). Note that the abdomen is proportionally smaller than that of the crayfish. The abdomen consists of six segments and is narrow and T-shaped in males, triangular in immature females, and broad in mature females.

3. What is the sex of your specimen? Observe the abdomen of a crab of the opposite sex. Find the tubercles on the fifth thoracic sternum of males and immature females that lock the abdomen to the thorax.

4. Locate the anus on the ventral side of the abdomen at its distal end.

5. The ventral side is divided into several regions (Fig. 15.28B). The area reflected around from the spines to the legs is called the **sub-branchial carapace**. Most of the ventral surface is composed of the thoracic sterna, separated from one another by distinct suture lines. The thoracic sterna are more-or-less covered by the abdomen, depending on the sex.

6. Although the position and anatomy of the gills will be considered later, locate the right and left **inhalant** and **exhalant apertures**, through which water enters and leaves the branchial chambers. The inhalant apertures are located anterior to the coxae of the chelipeds; the exhalant apertures are on either side of the mouth (Fig. 15.28B).

Appendages

1. Identify the first antennae (antennules), second antennae, and eyes, noting their size, shape, and relative position. Compare these structures to those of the crayfish. Determine the movements of which each is capable. Locate the depression into which the first antennae fold.

The feeding appendages are similar to those of the crayfish.

2. Remove a set from one side of the crab and compare them with those of the crayfish. As in the crayfish, it is best to locate the three pairs of maxillipeds in reverse order, followed by the two pairs of maxillae, and then the mandibles.

3. The third maxillipeds are the largest feeding appendages. Progressing toward the mouth, locate the second and third maxillipeds.

4. On each maxilliped, locate the long, setous epipodite called the **flabellum**, which enters the branchial chamber. The three flabella are fringed with setae and function to clean debris from the surface of the gills.

5. Find the second and first maxillae. The second maxilla has a large, platelike exopodite called the **scaphognathite** or **gill bailer**. The scaphognathite beats back and forth to circulate water through the branchial chamber.

6. Locate the mandibles surrounding the mouth. Note their broad, smooth, heavily calcified surfaces. These appendages are used to cut, crush, and hold food, while other appendages process it.

7. The pereiopods are similar in crabs and crayfish, although the basis and ischium are fused into a single unit called the basi-ischiopodite. As in the crayfish, the first pair of pereiopods (chelipeds) are chelate and are used to obtain food and for defense.

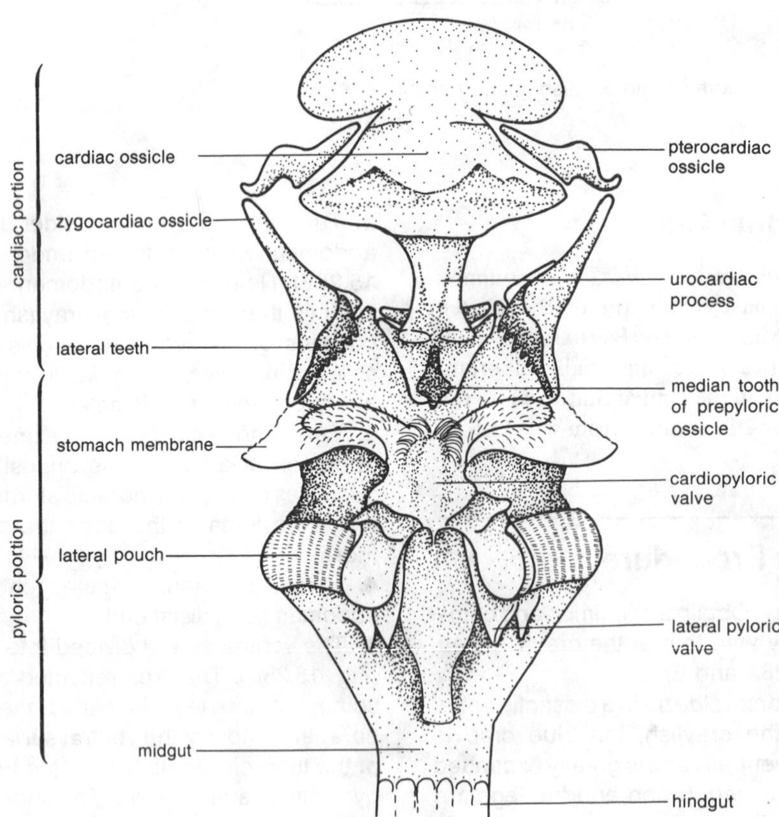

Figure 15.27. An internal, ventral view of the gastric mill.

8. Note how the articulation of this pereiopod permits it to be folded closely against the body. The next three pairs of pereiopods are not chelate and are used for walking. If a leg is caught or damaged, it is cast off or **automized** by the crab at the basi-ischiopodite and is later regenerated.

9. Find the last pair of pereiopods. These paddle-shaped legs are used in a sculling fashion for swimming.

10. On the ventral surface of the abdomen, locate the pleopods. Only the first two pairs of pleopods are present in the male. They are highly modified

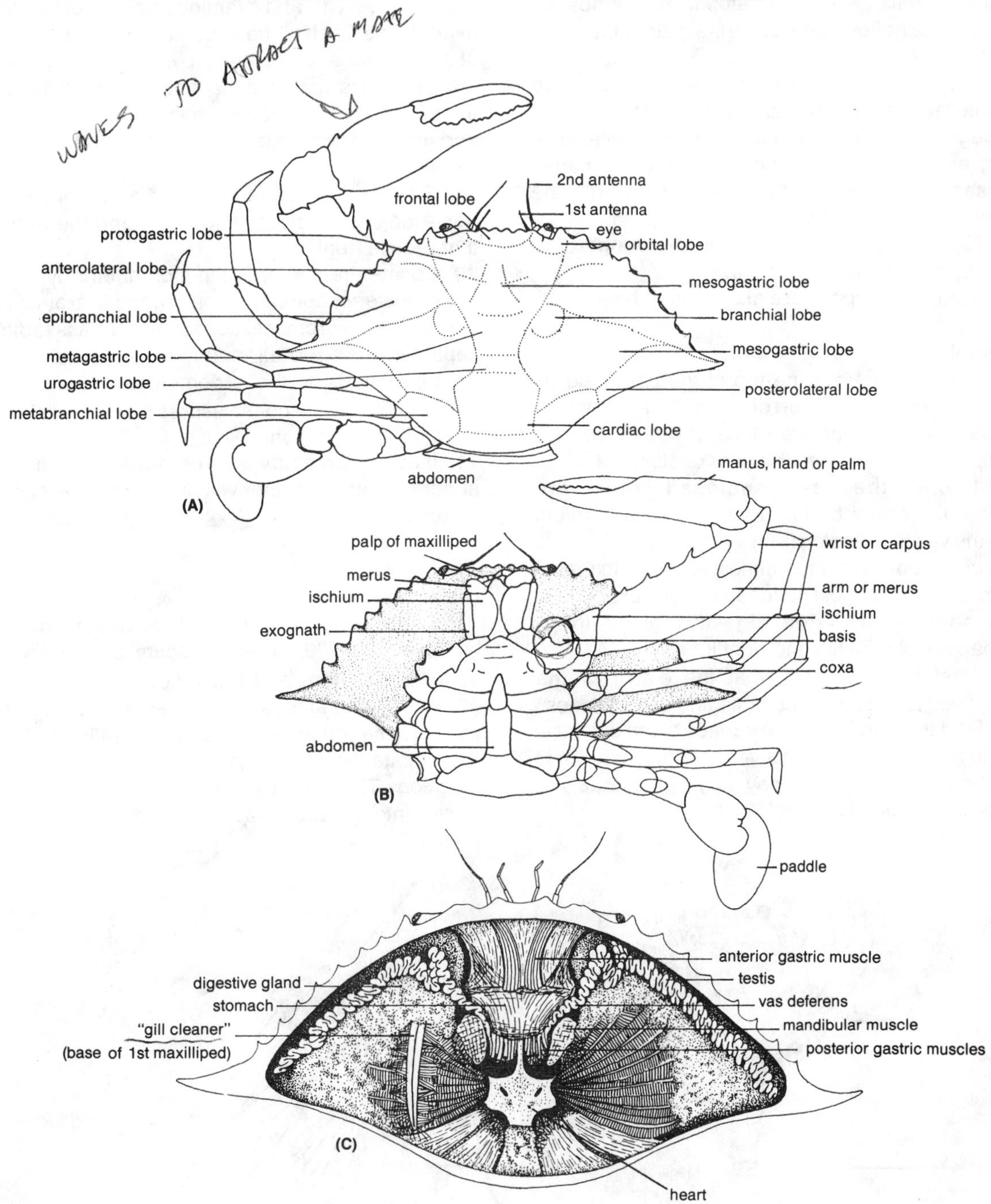

Figure 15.28. The blue crab *Callinectes.* **(A)** External dorsal view. **(B)** External ventral view. **(C)** Dorsal internal view.

and used in copulation. The first pair is larger and possesses an elongate, hollow spine (endopod). The second pair is similar to the first, but its spine is shortened and not hollow.

11. Observe that the spine of the second pleopod enters the first pleopod through a hole (anterior foramen) at its base. Spermatophores are pushed through the hollow spine of the first pair by the spine of the second.

12. Examine the pleopods of a female crab. In the female the first pleopods are absent, but the second through the fifth pairs are present and similar. The long endopod and exopod of these appendages possess many setae. These setae help hold the numerous embryos while they develop.

13. Examine a female specimen in **sponge**. The spongy material is the egg mass.

14. Note that uropods are absent in both sexes.

Internal Anatomy

1. Carefully remove the top of the carapace by making an incision a short distance inward from the dorsal border of the carapace and separate the underlying tissues from the exoskeleton.

2. Locate the heart enclosed within the pericardial cavity by the translucent pericardium and the vessels leading from the heart (Fig. 15.28C). The circulatory system is almost identical to that of the crayfish. The most noticeable organ exposed at this point is the large **digestive gland** (hepatopancreas) with left and right lobes.

3. Carefully determine what the extent of the digestive gland is without damaging other tissues.

4. Find the gonads on the surface of the digestive gland and trace their ducts ventrally just anterior to the heart. Ovarian size may vary considerably in different specimens.

5. Paired **antennary glands** that are excretory in function lie just lateral to the cardiac stomach; they usually are difficult to find.

6. Locate the eight pairs of gills on either side of the crab. The anterior four pairs may be obscured by the digestive gland and gonad (Fig. 15.28C).

7. Remove a gill and examine it using a dissecting microscope. Each gill has a central axis and consists of an anterior and posterior row of lamellae.

8. The gut is similar to that of the crayfish. Locate the esophagus, cardiac stomach, pyloric stomach, and intestine. Find the long, narrow midgut cecum extending from the pyloric stomach.

9. Remove the stomach region and identify the nerve ring around the esophagus and the nerves that radiate from it.

10. Locate the cerebral ganglion anteriorly and a large, posterior concentration of ganglia around the sternal artery. From this conspicuous mass radiate separate nerves to all mouthpart appendages, to the pereiopods, to the abdomen, and to a pair of circumesophageal commissures that connect with the cerebral ganglion.

11. If advanced study is to be done of blue crab anatomy, your instructor will provide the necessary directions.

Crab Larval Stages

1. Examine prepared slides of **zoea** and **megalops** larvae (Fig. 15.29). In each locate the carapace, abdomen, rostrum, and stalked compound eyes.

2. Find the biramous thoracic appendages in the zoea. These will develop into the maxillipeds.

3. Locate the chelipeds and walking legs in the megalops. This is the last stage before final metamorphosis into the adult crab.

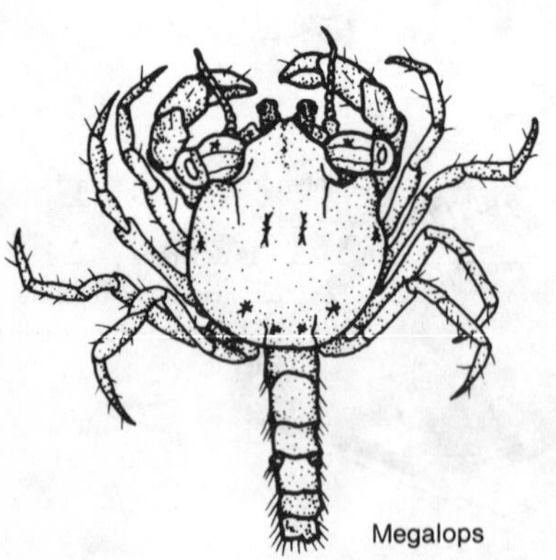

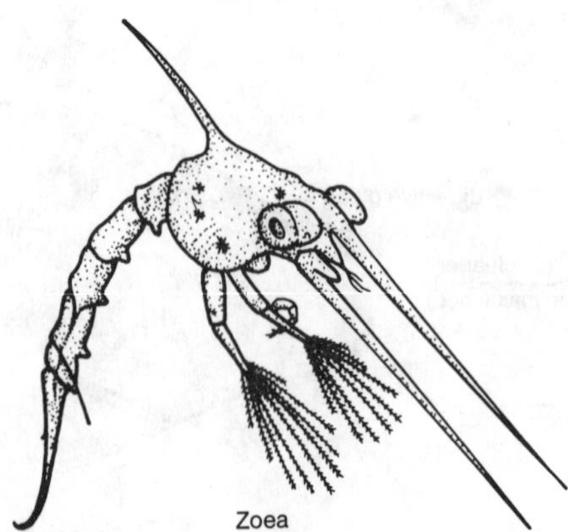

Figure 15.29. Zoea and megalops larvae of a crab.

Live Specimens. Observe the live specimens of crabs on display. How do they move about the aquarium? How do the various species react when threatened? Can you observe the bailing activity of the scaphognathites? As your instructor directs, place a small piece of shrimp to one side of the crab. How does the crab respond? Observe the elaborate movements of the mouthparts during feeding.

D. Subphylum Uniramia

Class Insecta

All insects belong to the class Insecta (in-SEC-ta; L., *insectum*, insect). These schizocoelomate protostomes are the dominant and most successful group of all invertebrates, especially in the invasion of terrestrial habitats. Insects are the only invertebrates capable of flight. There are probably not less than 750,000 described insect species; however, some recent estimates of the actual number of extant insect species may reach 10 million. The ecological and economical impacts insects have on local and worldwide populations of plants and animals are tremendous. The study of insects is called **entomology** (G., *entomon*, insect), often with emphasis in medical entomology and economic entomology.

The bodies of adults are divisible into three **tagmata**: an anterior **head**, middle **thorax**, and posterior **abdomen**. The head typically bears two antennae, five eyes (two compound and three ocelli), and mouthparts. Six legs and usually two pairs of wings in the adults are attached ventrolaterally and dorsolaterally to the thorax, respectively. The thorax is always composed of three segments.

Classification

Class Insecta is often divided into two large subclasses based on the presence or absence of wings. Insects lacking wings are believed to be the most primitive. Each subclass is further divided into orders based on features including morphology of mouthparts, modes of development, and wing venation. Familiarity with the principal characteristics of the orders is a necessity for the beginning student of insect biology. The more common orders are characterized below.

A. **Subclass Apterygota**. Wingless and most primitive of living insects.
   1. **Order Thysanura**. Silverfish and bristletails. Fast-moving insects with two or three styliform appendages on the abdomen; chewing mouthparts; destructive to clothing and books; metamorphosis lacking.
   2. **Order Collembola**. Springtails. Small, leaping insects with an abdominal organ for jumping; found in leaf litter; chewing mouthparts; metamorphosis lacking.

B. **Subclass Pterygota**. Primarily winged insects; the wingless condition, if present, has been secondarily acquired.
   1. **Order Ephemeroptera**. Mayflies. First pair of net-veined wings larger than second pair; two or three caudal appendages; vestigial chewing mouthparts of short-lived adults; metamorphosis gradual.
   2. **Order Odonata**. Dragonflies and damselflies. Long net-veined wings; chewing mouthparts; dragonflies are stout-bodied and stronger fliers than are damselflies; metamorphosis gradual.
   3. **Order Orthoptera**. Grasshoppers, katydids, crickets, roaches, and allies. The winged forms of this large order generally have the membranous hind wings folded beneath the leathery forewings (**tegmina**); femur of hind legs enlarged in many species; chewing mouthparts; metamorphosis gradual.
   4. **Order Isoptera.** Termites. Abdomen broadly joined to the thorax; bodies soft and often whitish; these colonial, social insects have winged and wingless individuals; wings are

equal in length; chewing mouthparts; metamorphosis gradual.

5. **Order Hemiptera**. True bugs. Basal part of forewings thickened and distal part membranous; the membranous sections overlap at rest; hind wings membranous; piercing-sucking mouthparts form a beak arising at the front of the head; metamorphosis gradual.
6. **Order Homoptera**. Aphids, cicadas, and leafhoppers. Winged and wingless individuals; forewings membranous unlike those of Hemiptera; piercing-sucking mouthparts form a beak arising from back of the head; metamorphosis gradual; many entomologists make this order a suborder of Hemiptera.
7. **Order Neuroptera**. Lacewings, ant lions, and dobsonflies. Both pairs of wings are veined and membranous; chewing mouthparts; metamorphosis complete; larva of ant lions commonly called doodlebug; larva of dobsonfly is a hellgrammite. The larvae of one family (Sisyridae) live in and eat freshwater sponges.
8. **Order Coleoptera**. Beetles. Anterior wings are thick and hard and called **elytra**; membranous hind wings when not in use are folded under the elytra; metamorphosis complete; chewing mouthparts; largest insect order, with over 300,000 species; larva commonly called a grub.
9. **Order Lepidoptera**. Butterflies and moths. Body and wings in the adult are covered with minute scales; coiled proboscis for sucking nectar from flowers; metamorphosis complete; larva commonly called a caterpillar.
10. **Order Diptera.** Flies and mosquitoes. Hindwings reduced to knoblike **halteres**; piercing-sucking and lapping mouthparts; metamorphosis complete. Larva commonly called a maggot.
11. **Order Hymenoptera.** Ants, bees, wasps. Winged and wingless species occur in this large order; few veins in the transparent wings; solitary or social species; mouthparts modified for chewing, lapping, or sucking; metamorphosis complete.
12. **Order Siphonaptera.** Fleas. Bodies small, wingless, and compressed; legs modified for jumping; adult with piercing-sucking mouthparts; metamorphosis complete.

Observational Procedure: *Romalea*

The grasshopper or locust (order Orthoptera) will be used to illustrate basic insect anatomy.

Obtain a large specimen such as the lubber grasshopper (*Romalea*), a common species found in the southern United States. Observe the body markings, coloration, and reduced wings that are characteristic of the species. Note that the grasshopper is segmented. What type of segmentation is represented?

As in all insects, the external, protective covering is a tough, nonliving **cuticle** or **exoskeleton** containing chitin. The hardened sclerotized plates of the cuticle are called **sclerites**. The major sclerites are typically separated by membranous areas called sutures that permit movement of body parts. A dorsal sclerite is the **tergum**, a ventral one the **sternum**, and a lateral one the **pleuron**. In areas (e.g., head) where individual segments are fused, the terga (or tergites) are fused as a solid hardened mass. The sclerites may also be divided into subplates.

Locate the three tagmata composing the grasshopper's body: (1) anterior head, (2) middle thorax, and (3) posterior abdomen (Fig. 15.30).

Head

1. The head is in form of a capsule (Figs. 15.30 and 15.31). The upper part is heavily sclerotized and the ventral area contains the mouth surrounded by several mouthparts used in feeding. Exact agreement by entomologists does not exist as to the number of segments in the insect head. Note the shape of the head and its position in relation to the long axis of the body. The mouthparts of the grasshopper are directed downward, a position called **hypognathous**.

2. The head capsule is divided by sutures into several regions. It is often difficult to discern where one region begins and the other ends. The dorsum of the head between and behind the eyes is the **vertex** (Fig. 15.31). The **frons** is the anteriofrontal region. Note on the frons the two arms or forks of the Y-shaped **epicranial suture** whose stem begins on the back of the head. The forks are called the **frontal suture** and the stem the **coronal suture**. The epicranial suture represents a line of weakness that splits the head capsule at molting.

3. Locate the **gena** or "cheek," the lateral lower part of the head posterior to the frons and below the eyes. The back of the head is the **occipital** that contains the **occipital foramen** (or foramen magnum).

4. Examine the anterior face of the head (Fig. 15.31). Locate the liplike sclerite, the **clypeus**, below the frons to which it is attached. Hanging down from the clypeus is the **labrum** or "upper lip."

5. Lift the clypeus with forceps to fully expose the bilobed labrum. Does the labrum articulate directly

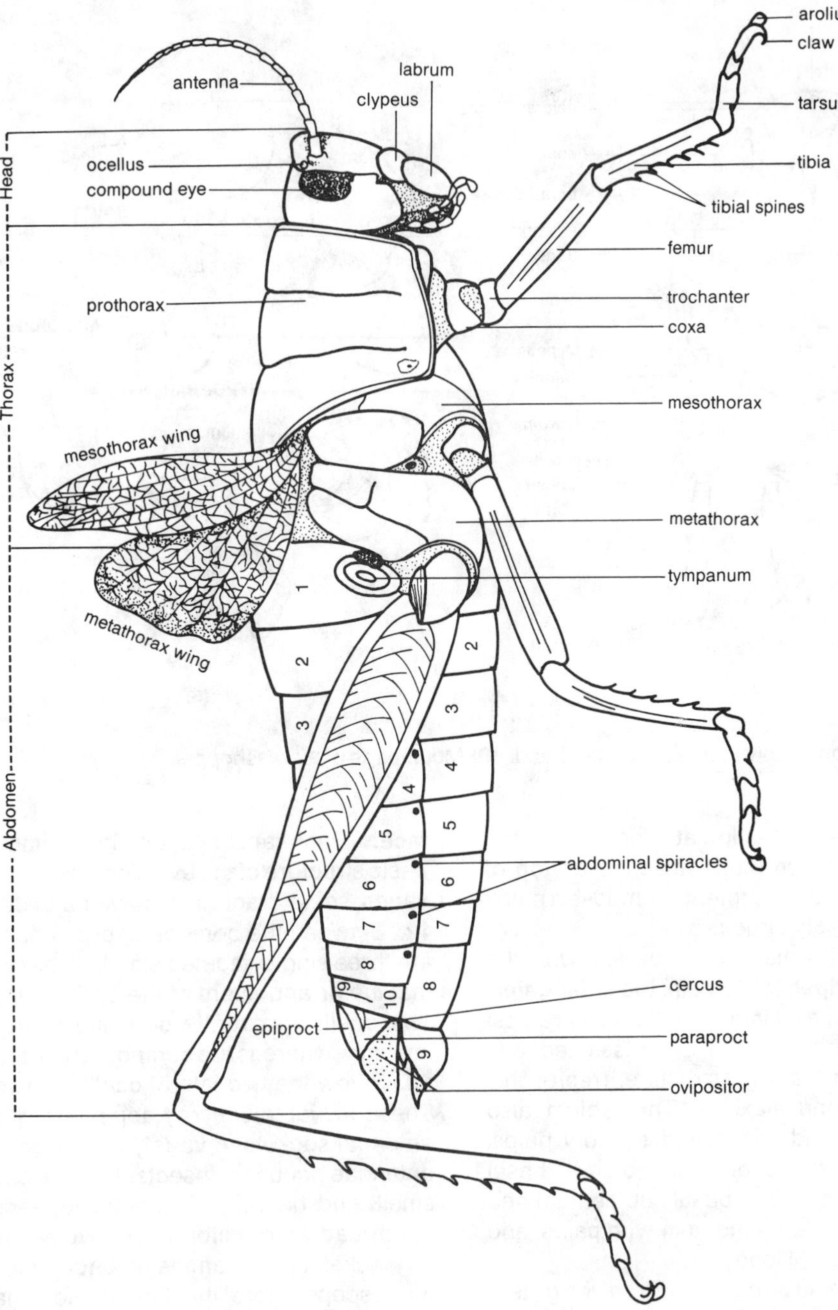

Figure 15.30. Lateral view of a female lubber grasshopper.

to the clypeus? A swollen area, the **epipharynx**, is prominent in many insects on the ventral or posterior side of the labrum. Does the grasshopper have an epipharynx?

6. Behind the labrum are paired, unsegmented **mandibles**. Lift up the labrum with your forceps to expose these paired, hard mouthparts with teeth used to cut, chew, and grind food. The mandibulate insects include other orthopterans, odonates, isopterans, and coleopterans, so-named because of the type of mandibles present.

7. Behind the mandibles are the paired maxillae.

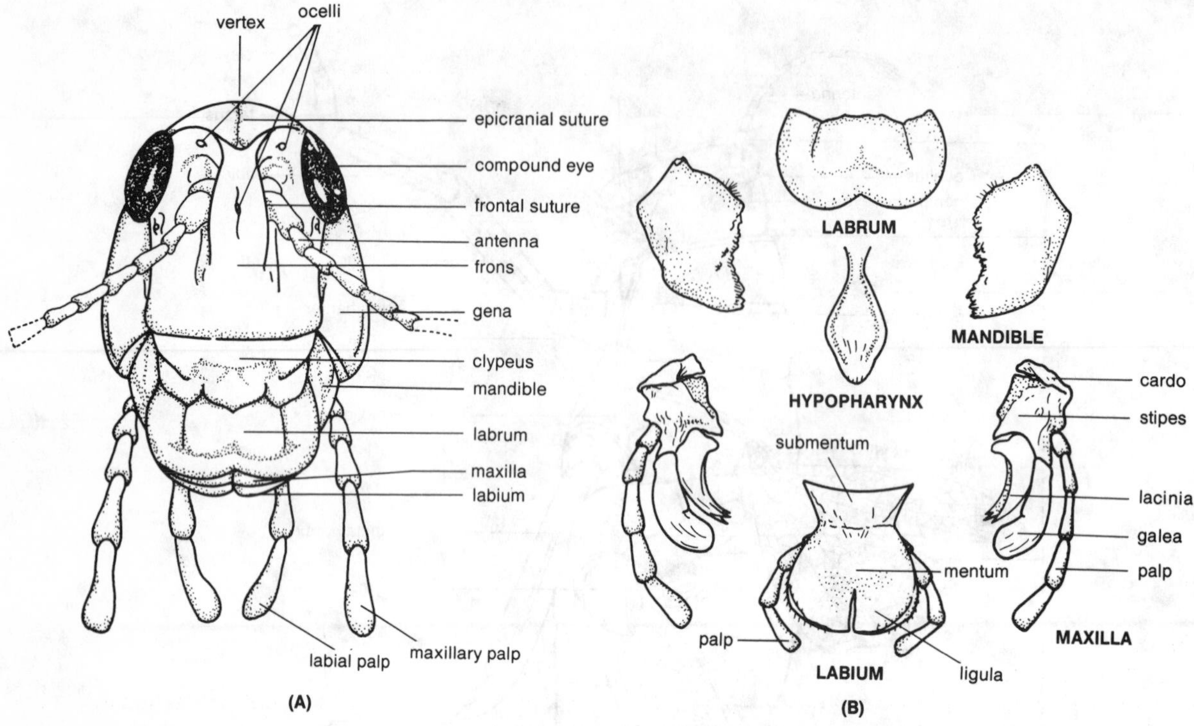

Figure 15.31. **(A)** Frontal view of a grasshopper head. **(B)** Mouthparts of a grasshopper.

These masticate and manipulate food. Note the antennalike, segmented **palps** attached on top of the maxillae. How many segments compose a palp? The palps are sensory structures.

8. Other parts of a maxilla are as follows: the basal **cardo**, the **stipes** with a palp, the outer **galea**, and the inner jawlike **lacinia**. These areas are best observed when the maxillae are dissected out. Although the **labium** is a single structure, it represents fused paired second maxillae. The labium also manipulates food and has paired sensory palps. Note the three regions on the labium: basal **postmentum** (divided into a basal submentum and distal mentum), middle **prementum** with palps, and distal **ligula** with apical lobes.

9. The labium should be dissected from the grasshopper and observed with the dissecting microscope.

10. Locate the median, unpaired tonguelike **hypopharynx** that protrudes in the preoral cavity. The hypopharynx is closely associated with the base of the labium. Your instructor may have you dissect out all of the mouthparts and examine them under the dissecting microscope.

11. Note the shape and position of the two large immovable **compound eyes**. Examine an eye under the dissecting microscope. Note the many hexagonal **facets**. Each facet is a lens to an **ommatidium**, the basic structure of the eye. The number of ommatidia varies; some dragonflies have up to 30,000.

12. Between the compound eyes and on the frons are three single-faceted simple eyes or **ocelli**. Note that the arrangement of the ocelli forms a triangle. The ocelli apparently perceive changes in light intensity, whereas the compound eyes form images.

13. Below the two lateral ocelli are two antennae. These movable sensory appendages articulate in antennal sockets. A variety of shapes and sizes of antennae occurs in insects. They function in touch, smell, and hearing. The grasshopper's antennae are threadlike or filiform. Remove an antenna from its socket and examine it under the dissecting microscope. Note the tiny sensory hairs on the antenna. Locate the basal **scape**, the middle **pedicel**, and the long **flagellum** consisting of many segments (Fig. 15.32).

Cervix or Neck. Connecting the head and thorax is a membranous region called the **cervix** or **neck**. It is not a separate body segment, but a contribution from the labial segment and the prothoracic segment. The cervix allows flexibility of movement between the head and thorax regions.

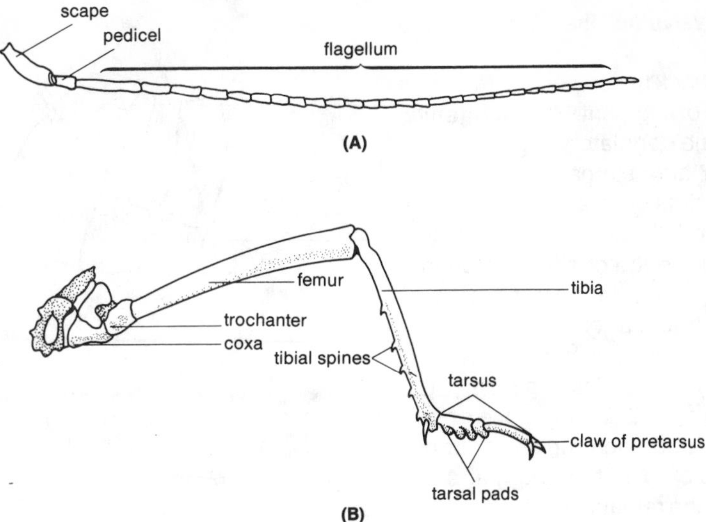

Figure 15.32. **(A)** Insect antenna. **(B)** Insect leg.

Thorax

1. All insects have three segments to the thorax: anterior **prothorax**, middle **mesothorax**, and posterior **metathorax** (Fig. 15.30). Each bears a pair of legs. The parchment-like forewings (or **tegmina**) are attached to the mesothorax and the hind wings are attached to the metathorax.

2. Observe the wing coloration and the folded shape of the hind wings. Flight is primarily by the hind wings. Note the many veins in the wings. What function might they perform? The venation pattern is important in insect taxonomy. Both wings are outgrowths of the body wall and lack internal muscles.

3. Examine the sclerites of the thorax. Each tergum or dorsal plate on the thorax is called a **notum**. The wings are attached to the lateral margins of the nota.

4. Note the greatly sclerotized sterum. The different specialized plates and sutures of the notum and sternum will not be discussed.

5. Each pleuron is divided by a suture forming two sclerites. The anterior sclerite is the **episternum** and the posterior one the **epimeron**. Observe these with the dissecting microscope.

6. Two pairs of **spiracles**, external openings to the tracheal respiratory system, occur on the thorax above the attachments of the second and third pairs of legs. You will observe that these locations are between the prothorax and mesothorax and between the mesothorax and metathorax, respectively.

7. Observe the legs. Each leg consists of six parts: **coxa**, **trochanter**, **femur**, **tibia**, **tarsus**, and **pretarsus** (Fig. 15.32). The short basal coxa articulates with the body.

8. Note the shape of the greatly enlarged femur of the last pair of legs. What adaptation might this enlarged part serve?

9. Note the two rows of spines on the tibia. Are the spines of both rows equal in size and number? Why are the spines located on the dorsal surface of the tibia instead of the ventral surface?

10. Count the segments of the tarsus. Note the tarsal pads or **pulvilli**. How many are there and what function might the pads serve? The short pretarsus also has an adhesive pad, the **arolium**, and a pair of **claws**.

Abdomen

1. Observe that the abdomen is distinctly segmented and lacks jointed appendages (Figs. 15.30 and 15.33). Count the abdominal segments. The first segment bears a **tympanum**, an oval membrane covering the organ of hearing. The last pair of segments is reduced and limited to a tergite, the **epiproct**, below which lies the **anus.**

2. Observe the tergite and anus under the dissecting microscope. Note the small spine or **cercus** projecting behind the tenth segment. The plate below the cercus is the **paraproct**. Although reduced in size, the cercus is one of the few appendages found on the abdomen. The other appendages on the abdomen are reproductive in function.

3. At the end of the abdomen in female grasshoppers is a pincer-shaped **ovipositor** for digging a hole in the ground for deposition of eggs. Observe the bladelike structures (**valvulae**) that compose the ovipositor (Fig. 15.33).

4. Male grasshoppers lack the ovipositor. Note that the expanded sternum of segment 9 (the **subgenital plate**) encloses the male copulatory organs.

5. Observe the terga and sternal plates of the abdominal segments. Note their arrangement. Are pleural sclerites present?

6. On the lower anterior edge of each abdominal tergite of segments 1 to 8 is a small oval spiracle that leads to an internal trachea of the respiratory system. Observe the spiracles with the dissecting microscope. Some may be situated in a pigmented area and be difficult to see. Counting those on the thorax and those on the abdomen, the grasshopper has ten pairs of spiracles. The first four pairs are used in inspiration and the remaining pairs function in expiration.

Internal Anatomy. Much of the internal anatomy of the grasshopper will be difficult to observe. Therefore, exercise care in dissection and observation.

1. With a pair of fine-pointed scissors, clip off the wings close to the body. Make a very shallow midsagittal cut through the dorsal body wall beginning at the head and ending at the posterior end. Pin down the two halves of the body wall to the dissection tray.

2. The large cavity exposed and containing the internal organs is the **hemocoel** (Fig. 15.34). This is not a true body cavity or coelom but part of the **open (lacunar)** circulatory system. The hemocoel is filled with blood or **hemolymph**. The **heart** is a elongated slender tube in the dorsal part of the abdomen. It extends anteriorly as an aorta. Although the heart is difficult to see, make an attempt to locate it.

3. Note the muscles along the body wall, especially those in the thorax associated with the legs and wings. The muscles of the grasshopper are in bands.

4. Note the whitish, delicate tubes extending from the body wall to the viscera. These are **tracheae**. Prepare a wet mount of a piece of a trachea and examine it under low and high powers of a compound microscope. Describe the appearance of the tracheae.

5. Components of the digestive system are the most conspicuous of the internal organs. Much of the hemocoel is filled with the digestive system (Fig. 15.34). Locate the tube-shaped **esophagus** leading from a short **pharynx** that connects to the mouth. The esophagus enlarges posteriorly as a thin-walled **crop** in the thorax. The crop leads to a narrow **proventriculus** (or **gizzard**) with internal gastric teeth for shredding food.

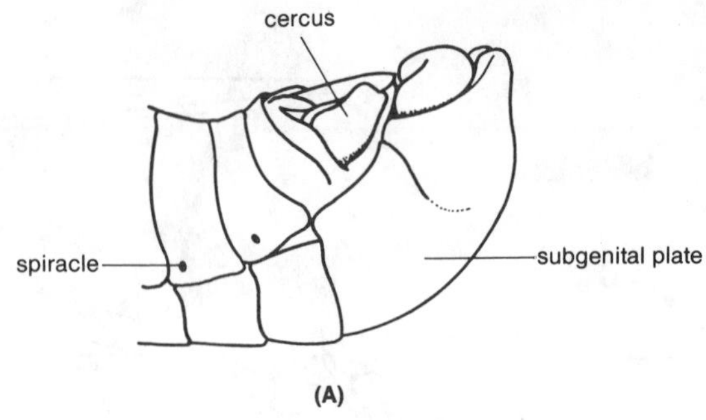

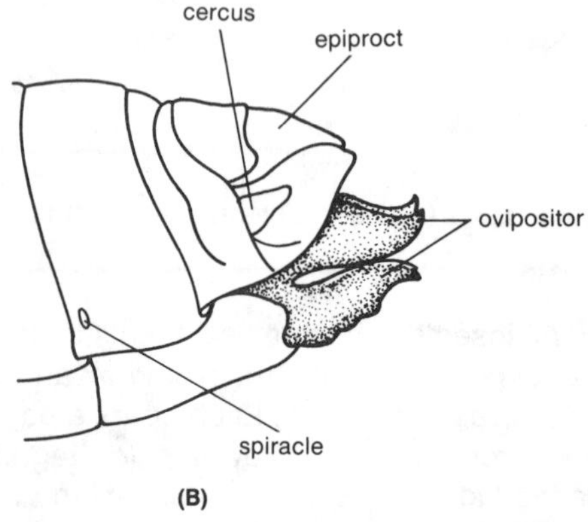

Figure 15.33. Posterior ends of **(A)** male and **(B)** female grasshoppers.

6. You may see salivary or labial glands below the crop, with a duct opening in the preoral cavity between the labium and hypopharynx. The components above compose the foregut (or stomodeum) of the grasshopper's digestive system.

7. Posteriorly the proventriculus leads to an elongated **stomach** or **ventriculus**. Much of the stomach is concealed by finger-shaped **gastric ceca**. These are involved in nutrient absorption. How many are present? The stomach and gastric ceca constitute the midgut (mesenteron).

8. Connecting to the stomach posteriorly is an **intestine**, with an enlarged anterior component and a slender posterior component, an enlarged **rectum**, and the **anus**. The intestine, rectum, and anus constitute the hindgut (proctodeum).

9. The many threadlike excretory organs called **Malphigian tubules** and other organs such as those originating from the reproductive system may conceal parts of the hindgut. Care must be exercised not to destroy these structures.

10. Grasshoppers are dioecious. Females possess the egg-laying ovipositor, whereas males lack this structure and thus have a rounded posterior end (Fig. 15.33). Females lay their eggs in the ground in late summer or early fall. In the following spring, the eggs hatch into immature grasshoppers (or nymphs). They undergo several molts before attaining the adult form.

11. Examine a male grasshopper and locate as many of the following parts as possible (Fig. 15.35). The two fused **testes** of the male lie dorsal to the intestine. A slender tube, the **vas deferens**, extends from each testis. The tube passes posterioventrally and becomes enlarged as a reservoir, the **seminal vesicle**, before uniting to form a common **ejaculatory duct** that enters the **penis**. Two **accessory glands** that secrete a fluid join the ejaculatory duct.

12. Examine a female grasshopper. Several tapering **ovarioles** or egg tubes where eggs are produced form an **ovary**. Extending from each ovary is an **oviduct**. The two oviducts join to form the **vagina**, an egg-holding chamber located under the intestine. Two **accessory glands** and a **seminal receptacle** (or **spermatheca**) enter the vagina separately. Sperm received from the male during copulation are stored in the receptacle.

13. Insects possess a highly developed nervous system (Fig. 15.34). It is similar to that of the crayfish and for the most part, the system lies below the digestive system. Attempt to observe the following components of the grasshopper's nervous system. The central nervous system consists of a **trilobed brain** located in the head above the esophagus. Extending off the brain is a pair of **circumesophageal connectives** that connect to a **subesophageal ganglion**. The ganglion innervates the mandibles, maxillae, and labium. Extending posteriorly in the insect is the paired **ventral nerve cord** with one pair of ganglia per segment.

Adaptability of Structure

Insects provide excellent opportunities for students to study specializations and adaptability of structures. Morphological modifications in insects can be seen in most structures, but especially in the mouthparts, legs, antennae, and wings. Selected examples of insects showing adaptations and specializations of these structures will be observed.

Mouthparts. Insect mouthparts are adapted for feeding on different types of food. As a result, different methods are employed for food procurement. Some of the major types are biting and chewing, chewing and sucking, lapping (sponging), piercing and sucking, and siphoning. The grasshopper and cockroach (order Orthoptera) are good examples of insects with biting and chewing mouthparts. Many zoologists believe the other specialized mouthparts observed in many insects have evolved from the biting-chewing type. As you study the following examples, it is imperative that you keep in mind the mouthparts of the grasshopper previously examined as a basis for making comparisons.

1. Chewing and sucking: honeybee (*Apis*). The mouthparts of the honeybee (order Hymenoptera) are used for biting and chewing wax and sucking or lapping nectar from flowers. Examine a preserved honeybee or prepared slide of the insect's mouthparts (Fig. 15.36A). Note the arrangement of hairs on

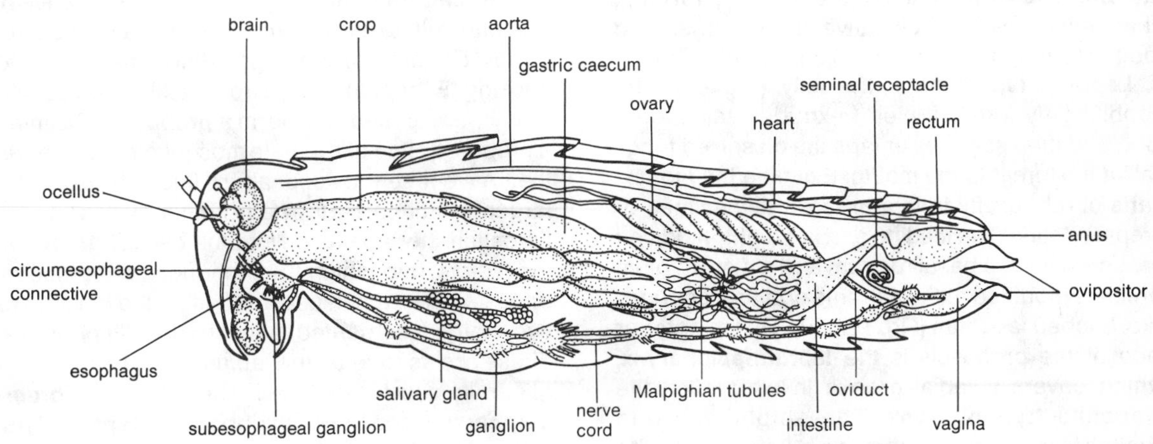

Figure 15.34. Internal anatomy of a female grasshopper.

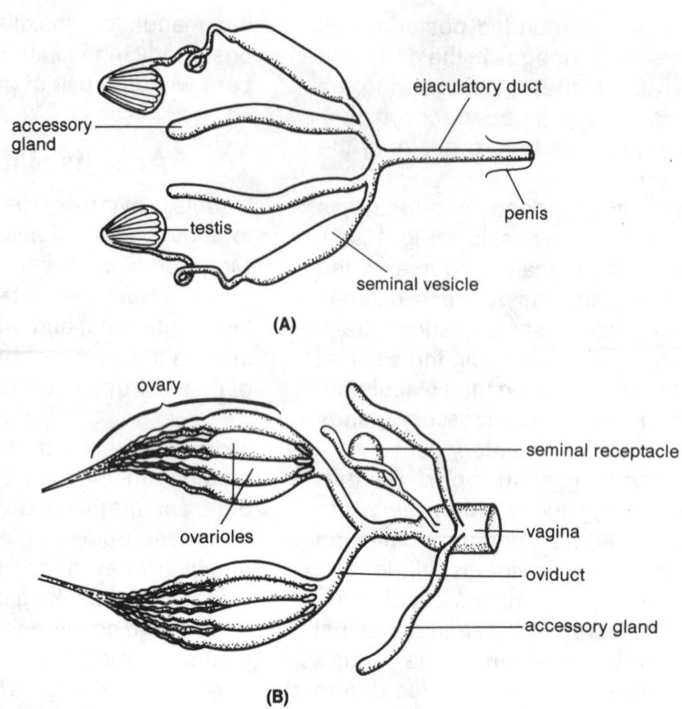

Figure 15.35. Reproductive systems of **(A)** male and **(B)** female insect.

the head and mouthparts. Below the clypeus is the narrow **labrum** and a smaller fleshy **epipharynx**. Note that the **mandibles** project laterally alongside the labrum. Are teeth similar to those observed on the grasshopper's mandibles present? Locate the medial, hairy **labium**. The specialized labium, also called the **glossa** or **alaglossa**, forms an extensible organ for probing into flowers. **Labial palps** and **maxillae** occur on each side of the tonguelike labium. These fit against the labium, forming channels through which saliva is discharged and food drawn up to the mouth.

2. Lapping (sponging): housefly (*Musca*). The nonbiting fly (order Diptera) extrudes saliva onto food and then sponges or laps the dissolved food-saliva mixture into the mouth. Examine the mouthparts of a housefly from a preserved specimen or prepared slide. The feeding part is a highly modified **proboscis** with a basal rostrum and a distal labium which is modified at the extreme end as a sponge-like, bilobed **labellum** (Fig. 15.36B). On the anterior face of the proboscis is the **labrum-epipharynx**, which covers a labial groove in which lies the bladelike **hypopharynx**. The epipharynx and hypopharynx interlock, forming a food tube to the esophagus. **Mandibles** are lacking as are most of the **maxillae** except for two maxillary palps. Unlike the housefly, other dipterans, such as horse-flies and deer flies, have sharp, bladelike mandibles to cut flesh and maxillae serve as long probing styles. These flies are blood feeders. In contrast to the mandibulate insects with mandibles designed for cutting, grinding, and chewing, the horsefly has **haustellate** mouthparts, in which the mandibles are stylelike for sucking or are completely absent.

3. Piercing and sucking: mosquito (*Anopheles*). The haustellate mouthparts of the female mosquito (order Diptera) are elongated for piercing and sucking. Except for the paired maxillary palps, the mouthparts collectively form a **proboscis**. Examine a prepared slide of a female mosquito's mouthparts that have been teased apart (Fig. 15.36C). The largest mouthpart is the hairy **labium**, with a flexible tip called the **labellum**. Note the central groove of the labium, into which the other needlelike mouth-parts lie when not in use. Alongside the labium are two hairy, segmented sensory **maxillary palps**. These are as long as the labium in *Anopheles*, but generally shorter in *Culex*. The nonhairy **labrum-epipharynx** and narrow ribbonlike **hypopharynx** form a tube for drawing blood into the mouth. The hypopharynx contains a small tube for ejecting

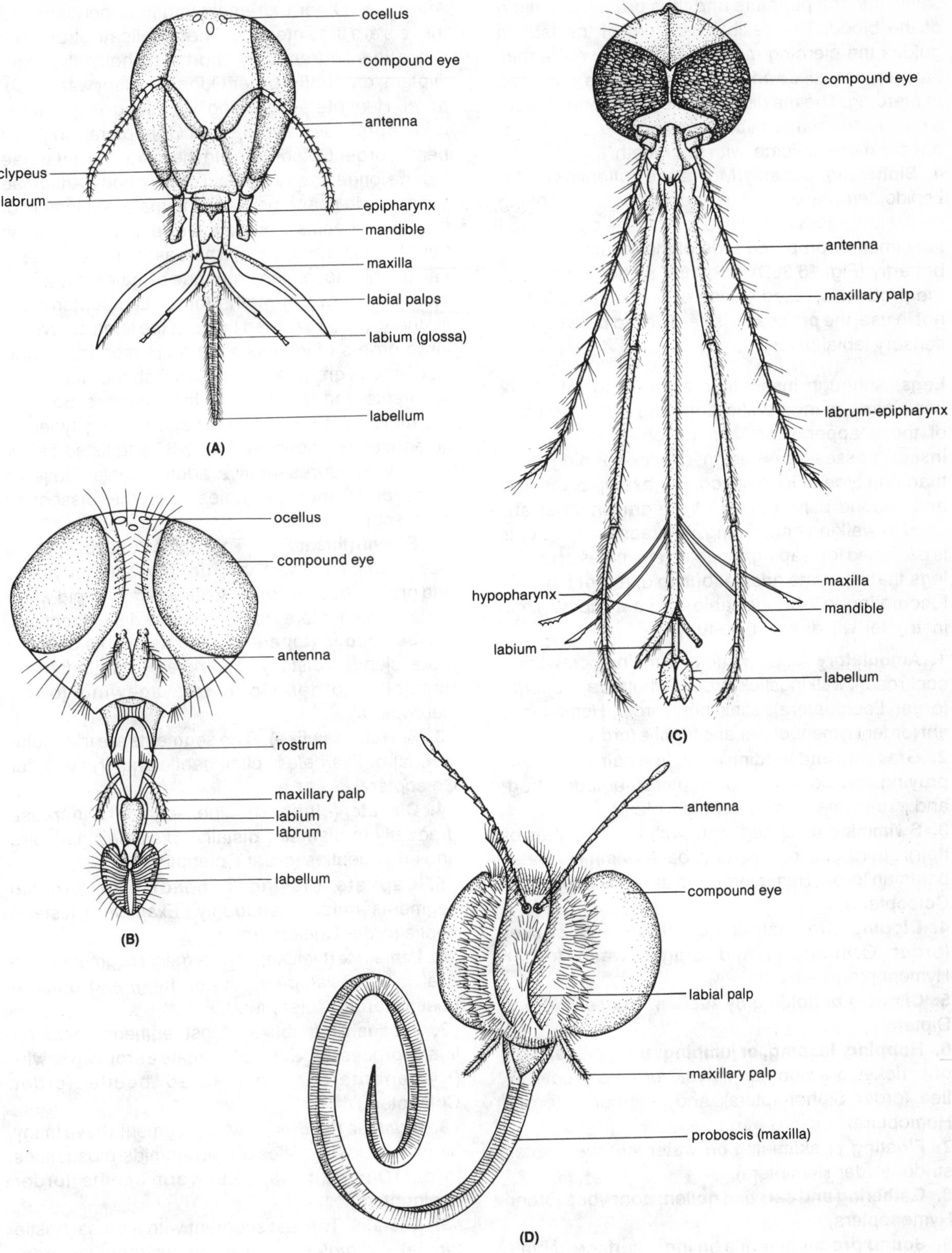

Figure 15.36. Types of insect mouth parts. **(A)** Honeybee. **(B)** Housefly. **(C)** Mosquito. **(D)** Butterfly.

saliva into the puncture and thus prevents clotting of the blood. The flexible labellum of the labium guides the piercing mouthparts. Observe the thin, flat paired **maxillae** and paired **mandibles** also used in piercing. The maxillae bear teeth on one edge of a narrow fin-shaped tip. The mandibles are similar but are more delicate, with finer teeth.

4. Siphoning: butterfly. Moths and butterflies (order Lepidoptera) have mouthparts designed for siphoning nectar and juices from flowers. Examine a preserved specimen or prepared slide of the mouthparts of a butterfly (Fig. 15.36D). The long siphoning tube is the **proboscis**, fused galeae of the maxillae. When not in use, the proboscis is held coiled between the sensory labial palps.

Legs. Although insect legs are composed of six basic parts, many modifications and specializations of these appendages can be observed. Certain insects possess legs designed for performing more than one type of locomotion. For example, the first and second pairs of legs of the grasshopper are used in walking or crawling, whereas the third pair is designed for leaping. Selected examples of insect legs that illustrate adaptations to different types of locomotion will be available for your study. Your instructor will direct your study.

1. Ambulatory (e.g., walking, running, crawling): cockroach, walkingstick (order Orthoptera), butterfly (order Lepidoptera), stink bug (order Hemiptera), ant (order Hymenoptera), and termite (order Isoptera)

2. Grasping and holding prey: first pair of legs of a praying mantis (order Orthoptera), ambush bug, and giant water bug (order Hemiptera)

3. Swimming (flattened parts with long brushes of hair): third pair of legs of a backswimmer, water boatman (order Hemiptera), and diving beetle (order Coleoptera)

4. Digging: first pair of legs of the mole cricket (order Orthoptera) and digger wasp (order Hymenoptera)

5. Clinging or holding by suction: housefly (order Diptera)

6. Hopping, leaping, or jumping: third pair of legs of a cricket, grasshopper, katydid (order Orthoptera), flea (order Siphonaptera), and leafhopper (order Homoptera)

7. Floating or skimming on water surface: water strider (order Hemiptera).

8. Gathering and carrying pollen: honeybee (order Hymenoptera).

9. Sound production: file on inner surface of hind femur of an acridid grasshopper (order Orthoptera).

Antennae. Other extremely varied appendages of insects are the antennae. These delicate structures range in size from very short (e.g., housefly, order Diptera; dragonfly, order Odonata; giant water bug, order Hemiptera) to being very long (e.g., camel cricket and true katydid, order Orthoptera; longhorn beetle, order Coleoptera). In the latter, the antennae may be longer than the insect's entire body. Antennae even vary in shape between the male and female of the same species. For example, female mosquitoes have beadlike antennae, whereas the male's antennae are plumose. The antennae of males in many of the giant silkworm moths (order Lepidoptera) are feathery and larger than those of the females. Within large orders of insects such as Hymenoptera and Coleoptera, great variations in the shape, number of segments, and size occur in the antennae. Some of the more common and perhaps striking types of antennae are shown in Fig. 15.37 and listed below along with representative adult insects. Observe these and other examples with the dissecting microscope.

1. Filiform (threadlike). The segments have a uniform size. *Examples*: silverfish (order Thysanura), dytiscid and ground beetles (order Coleoptera), vespid wasp (order Hymenoptera), mole cricket (order Orthoptera).

2. Setaceous (tapering). The segments become more slender distally. *Examples*: dragonfly and damselfly (order Odonata), lacewing (order Neuroptera).

3. Serrate (sawlike). The segments are triangular in shape. *Examples*: click beetle and firefly (order Coleoptera).

4. Clavate (clubbed). The segments increase gradually in diameter distally. *Examples*: ladybird and June beetles (order Coleoptera).

5. Capitate (having a head). The terminal segments enlarge suddenly. *Example*: histerid beetle (order Coleoptera).

6. Lamellate (leaflike). The terminal segments form oval lobes. *Examples*: June, dung, and unicorn beetles (order Coleoptera).

7. Pectinate (comblike). Most segments have long lateral processes. *Examples*: male European sawfly (Hymenoptera), fire-colored beetle (order Coleoptera).

8. Plumose (feathery). Most segments have many hairs. *Examples*: males of chironomids, mosquitoes (order Diptera); giant silkworm moths (order Lepidoptera).

9. Aristate. The last segment with a dorsal bristle (**arista**). *Examples*: housefly and syrphid fly (order Diptera).

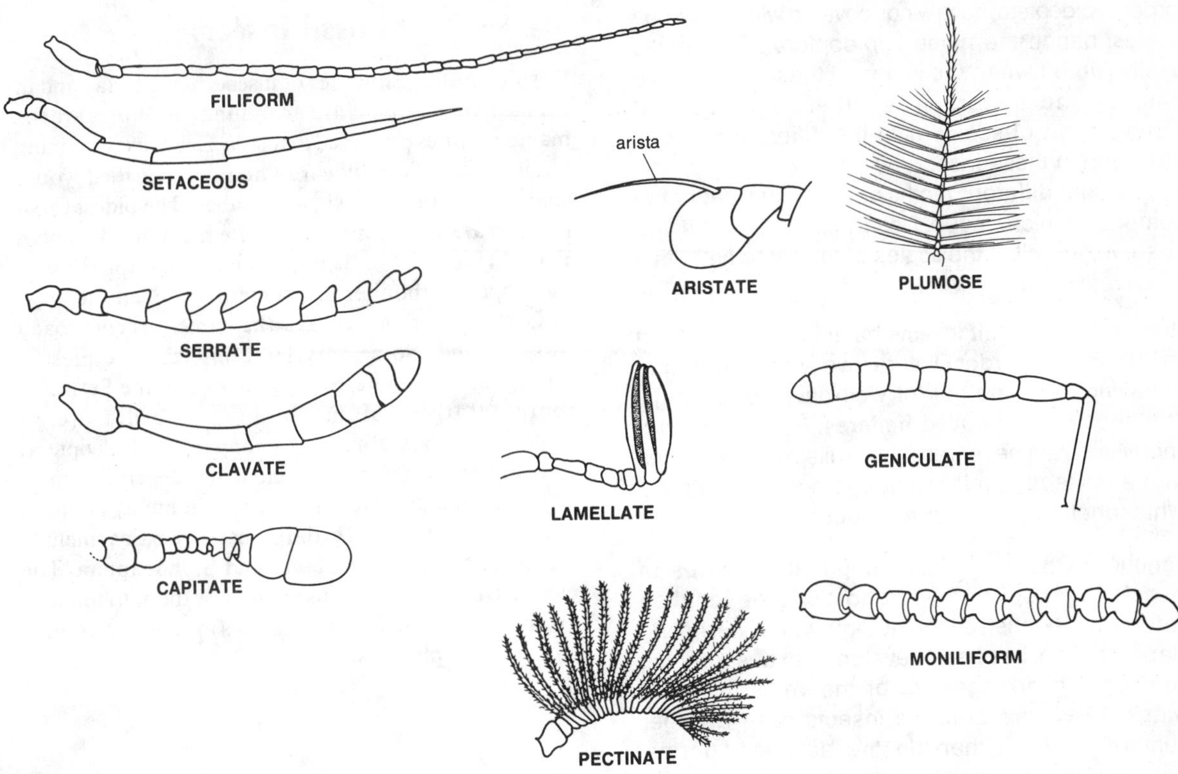

Figure 15.37. Types of insect antennae.

10. Geniculate (elbowed). The first segment is long and the following smaller segments are at an angle. *Examples*: ant (order Hymenoptera), stag and histerid beetles (order Coleoptera).

11. Moniliform (beadlike). The segments resemble a string of similar-shaped beads. *Examples*: wrinkled bark and tenebrionid beetles (order Coleoptera).

Wings. The presence of wings undoubtedly is a major feature that has contributed to the success of insects. Most insects have two pairs of wings. Insect wings are outgrowths of the body wall and structurally unlike those of bats and birds, which are modified forelimbs.

Considerable variation in the size, texture, shape, number, venation, and position held at rest exists in insect wings. Examples of these variations will be observed in selected representative insects.

Size. The more primitive winged insects have both pairs of wings equal or nearly equal in length. Observe a dragonfly or damselfly (order Odonata). Note the size of the wings in relation to the body length. Unlike many insects, mayflies (order Ephemeroptera) and dragonflies and damselflies cannot fold their wings over the abdomen when at rest. Examine the wing lengths of other insects from different orders.

Texture. Many insects, such as the dragonfly, have all four wings of the same membranous texture. Members of the orders Orthoptera and Coleoptera have the first pair of wings parchment-like or hard, respectively. The grasshopper's forewings are called **tegmina** because of their leathery texture. Examine other orthopterans, such as the field cricket and American cockroach. Note the texture of the forewings compared to that of the hindwings. The hard forewings of beetles are called **elytra**. Observe examples of beetles showing variations of the elytra; note how the membranous hindwings are folded under the forewings. True bugs of the order Hemiptera have the proximal part of the forewings harden and the distal part membranous. This half-wing condition (**hemelytron**) is reflected in the ordinal name, Hemiptera. Butterflies and moths (order Lepidoptera) have wings covered with numerous **scales**, hence the name Lepidoptera. The scales easily rub off when the wings of these insects are held by the fingers. The different types and arrangement of scales give the different patterns observed in these insects' wings. Examine several scales from different moths or butterflies under the compound microscope. Describe the patterns you observe. Are all of the scales of the same type on a given individual?

Number. Most adult insects have four wings. Adult members of the order Diptera (two wing) are unique in having only one pair of well-developed wings. The second pair, called **halteres**, is reduced and knoblike. Examine representative flies or mosquitoes with a dissecting microscope to see the halteres. What function might these reduced wings serve?

Venation. One of the most important features of insect wings used by the taxonomist is the venation pattern. An extensive terminology, which will not be elaborated on, has been developed to describe the number and arrangement of the veins in insect wings. The more primitive insects have a larger number of veins than do the advanced forms. Compare the venation of a dragonfly or damselfly (order Odonata) with that of a wasp or bee (order Hymenoptera). Examine the venation in the wings of insects representing different orders. Can you determine if a basic pattern exists in the venation? The veins are tracheae that extend into the wing. The areas between the veins are called **cells**.

Fossil Insects

Despite the large number of insects living today and in the past, insect fossils are not abundant. Wings are the main structures preserved in fossils. Deposits containing fossils of insects include iron nodules, petrified wood, coal, shale, volcanic ash, and amber. The oldest fossil remains are fragmentary and date from the **Devonian Period** (Fig. 15.38). The first pterygotes appeared in the Upper Carboniferous, sometimes called the "Age of Cockroaches" because of the numerous cockroach fossils found. Some very large insects were present during the Carboniferous. By the end of the Permian, representatives of many modern orders (e.g., Ephemeroptera, Odonata, Homoptera, and Coleoptera) were present. Orders represented in the early and mid-Mesozoic include Hemiptera, Lepidoptera, Diptera, and Hymenoptera. Tertiary fossil remains include insects of the orders Isoptera and Siphonaptera. The closest living relatives of insects may be the myriapodous arthropods. Both these and insects may have stemmed from onychophorans.

Observational Procedure

Examine representative fossils of insects. Note the similarities and differences to the present-day forms. Your instructor will give specific instructions to assist you in your observations.

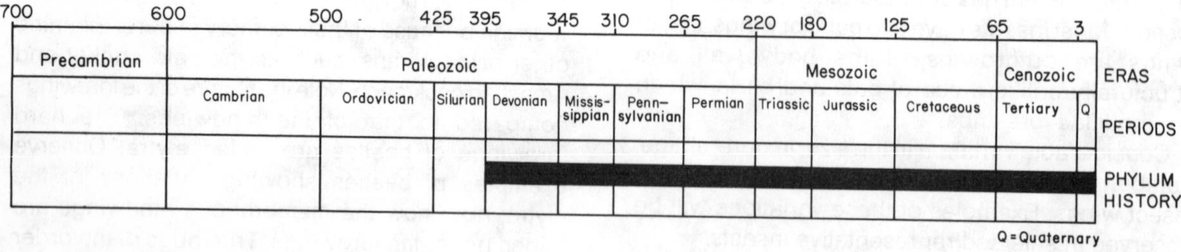

Figure 15.38. Geologic history of insects.

E. Myriapodous Arthropods

The myriapodous (G., *myrias*, number 10,000 + G., *podos*, foot) arthropods are wormlike, uniramous schizocoelomates that live in moist dark places such as beneath wood, rocks, leaf litter, and in unused parts of buildings. The 10,500 species are divided into four classes: Chilopoda (ki-LOP-o-da; G., *cheilos*, lip + G., *podos*, foot) the centipedes, Diplopoda (dip-LOP-o-da; G., *diplos*, double + G., *podos*, foot) the millipedes, Pauropoda (paur-OP-o-da; G., *pauros*, small + G., *podos* foot) the pauropods, and Symphyla (sim-PHY-la; G., *symphylos*, of the same race + G., *phylon*, a tribe) the symphylans.

Two tagmata, the head and trunk, compose the bodies of these arthropods. The head bears a pair of segmented antennae, a pair of mandibles, an upper lip (labrum), a tonguelike hypopharynx, and one (in millipedes and pauropods) or two (in centipedes and symphylans) pairs of maxillae. Some species of millipedes and centipedes have nonimage-forming ocelli or compound eyes on the head. Pauropods and symphylans lack eyes.

Most people are familiar with centipedes and millipedes, which comprise the vast majority of myriapods. However, pauropods and symphylans are seldom observed; the former are 2 mm or less in length, and the latter are no more than 8 mm in length. Emphasis will be placed on studying centipedes and millipedes in this exercise.

Classification

The four classes of myriapods are probably polyphyletic and may be related to insects. Pauropods and diplopods are considered closely related, whereas chilopods and symphylans form another group of similarity.

A. Class Chilopoda. Centipedes. Body flattened; trunk of many segments with one pair of legs per segment; first pair of trunk appendages (maxillipeds) are poisonous; two pairs of maxillae on head; genital opening at the posterior end; most centipedes are swift-moving, carnivorous predators. Examples are *Scolopendra* and *Scutigera*.

B. Class Symphyla. Symphylans. Body small and flattened, with trunk composed of 12 segments; one pair of legs per segment; two pairs of maxillae; genital opening on third trunk segment; plant feeders, often found in greenhouses. Examples are *Scutigerella* and *Scolopendrella*.

C. Class Diplopoda. Millipedes. Body usually cylindrical; trunk with diplosegments and two pairs of legs per segment; one pair of maxillae (gnathochilarium); genital opening anterior on trunk; most are slow-moving herbivores. Examples are *Julus* and *Spirobolus*.

D. Class Pauropoda. Pauropods. Small, soft-bodied arthropods with biramous antennae; one pair of maxillae (gnathochilarium); trunk of 11 segments; legs absent on segments 1 and 11; vegetarians and inhabitants of leaf litter and soil. Examples are *Pauropus* and *Allopauropus*.

Observational Procedure

1. Obtain a specimen each of a centipede and a millipede. Compare the shapes of their bodies, the number of segments, and number and arrangement of legs attached to the segments.

2. Note the less flexible and harder body of the millipede compared to that of the centipede. Calcium salts in the millipede's procuticle harden the sclerites, especially the tergites.

3. What type of segmentation occurs in the two animals? Are all segments the same size? Compare the tergites, pleurites, and sternites of the two animals. Are intertergites present? What adaptations might be correlated with a flattened body as in the centipede and the cylindrical body of the millipede?

4. Observe the specimens under the dissecting microscope. Locate the anterior **head** and the posterior **trunk** (Figs. 15.39 and 15.40). The greater part of the body of both animals is the trunk. The first four trunk segments of millipedes are often called collectively a **thorax**. The first thoracic segment is called the **collum** and is legless (Fig. 15.39). Segments 2 to 4 bear one pair of legs. Locate these segments. Note the flattened head of the centipede

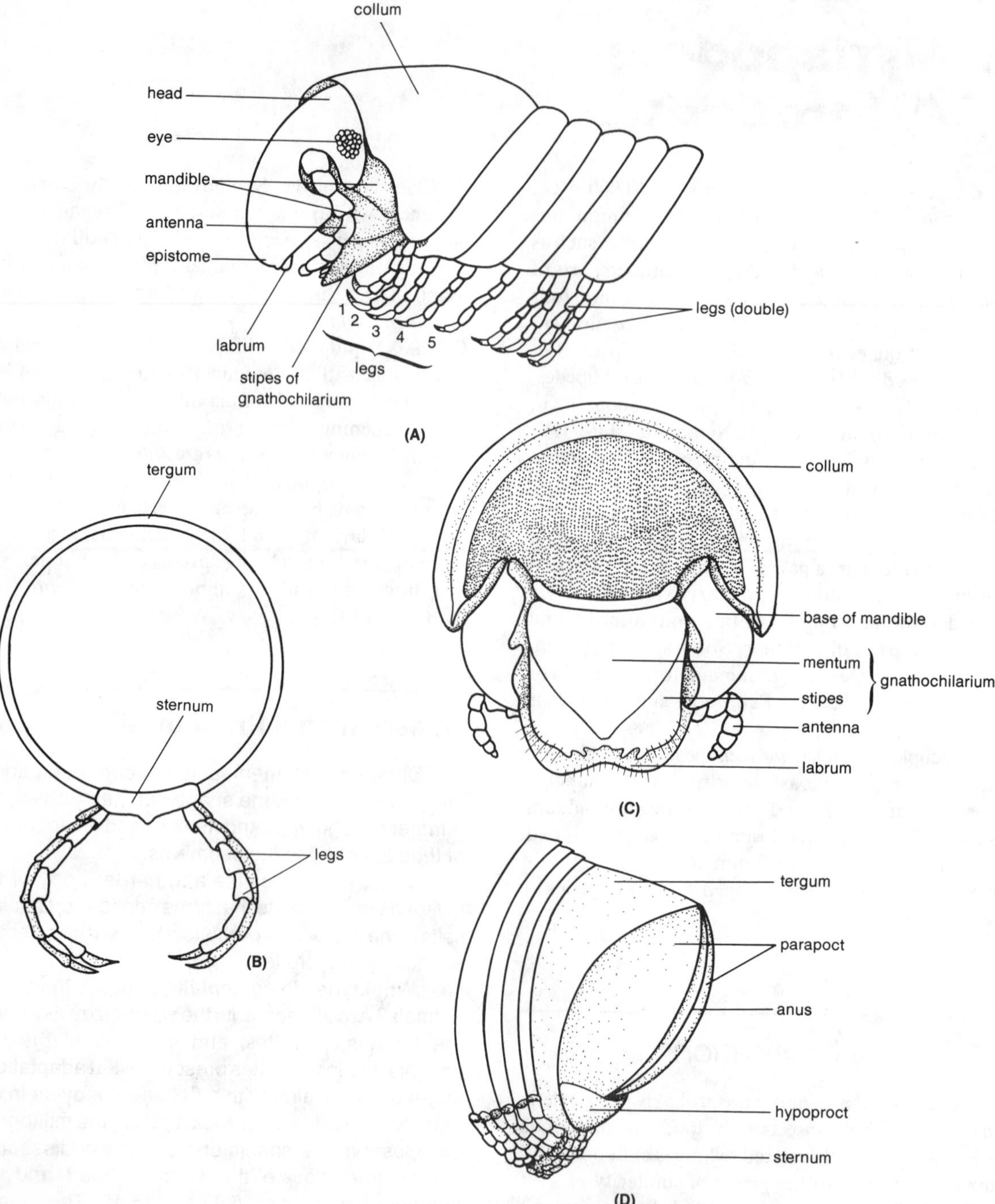

Figure 15.39. **(A)** Anterior lateral view of a millipede. **(B)** Cross section of a millipede. **(C)** Posterior view of a millipede's head. **(D)** Posterior lateral view of a millipede.

and the round head of the millipede. The individual segment composing the heads of both arthopods is not readily discernible.

5. Examine with the dissecting microscope the legs of both animals. Are all of the legs on each animal of the same size? Note the **diplocondition** in the millipede, which results from fusion of two originally separate segments.

6. Are claws or pads present similar to those on the grasshopper's legs? Are the number of leg

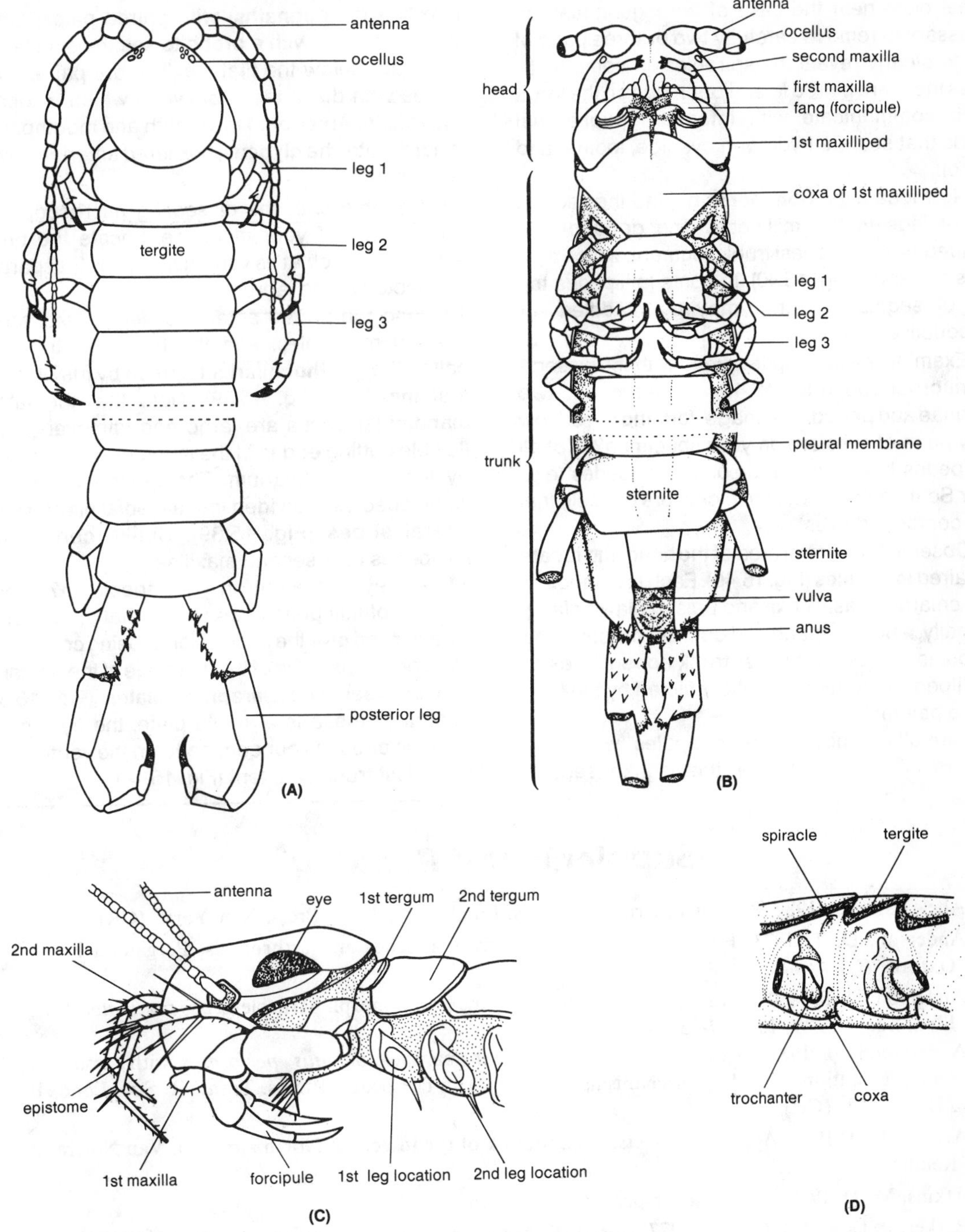

Figure 15.40. **(A)** Dorsal view of a centipede. **(B)** Ventral view of a centipede. **(C)** Anterior lateral view of a centipede. **(D)** Lateral view of a centipede's trunk, showing location of a spiracle.

segments for both animals the same? In centipedes the last pair of legs are longer than the others and are sensory or defensive in function.

7. Locate the **spiracles**. Look for the small openings in the pleural membrane dorsal to each leg (except the first and last three segments) of the centipede (Fig. 15.40). In the house centipede (*Scutigera coleoptrata*) each spiracle is centrally placed on a tergum and opens to a saclike tracheal lung. Millipedes have two spiracles per segment on the sternal plate near the base of the legs. It may be necessary to remove carefully two or three pairs of legs to clearly reveal the spiracles. In some millipedes there are paired lateral openings on the terga which communicate with internal repugnatorial glands that secrete hydrogen cyanide, iodine, and phenol.

8. Now locate the **gonopores** behind the second pair of legs in the millipede. The gonopore of centipedes is on the last trunk segment, which also bears the anus (Fig. 15.40). In some millipedes the legs of segment 7 or 8 may be modified for reproductive purposes.

9. Examine the centipede's head with the dissecting microscope (Fig. 15.40). Observe the two antennae and paired, nonimage-forming eyes. How many pairs of eyes are on your specimen? Not all centipedes have eyes, and some centipedes (e.g., order Scutigeromorpha) have clusters of ocelli that form compound eyes.

10. Observe the ventral area of the head and locate the paired **forcipules** (Fig. 15.40). Each is composed of an enlarged basal coxa and poison claw or fang; internally, a poison gland and duct are present. The forcipules represent the first trunk appendages or maxillipeds. A centipede's bite inflicted by the fang can be painful.

11. Carefully remove both forcipules with your scissors. This will expose the leglike second maxillae (**palpognaths**) with paired palps. Lift up these maxillae with a probe to see the smaller first maxillae. Below the first maxillae are paired mandibles with distal teeth, between which is located the mouth. Anterior to the mouth and mouthparts is a hard plate, the **clypeus**, bordered anteriorly by the **epistome**.

12. Observe with the dissecting microscope the round head of your millipede. Locate the broad epistome, which at its ventral margin is inseparably connected to the labrum.

13. Find the paired antennae, segmented, teeth-bearing mandibles, and the broad ventral plate called the **gnathochilarium** formed by fusion of the first maxillae (Fig. 15.39). Note that the lateral mandibular bases are large and immovable. The flexible cutting edges of the mandibles are covered by the gnathochilarium. The gnathochilarium in some species is divided into a medial **mentum** and lateral **stipes** (Fig. 15.39). Unlike centipedes, millipedes lack second maxillae.

14. Are eyes present on your specimen? Some species of millipedes have ocelli that are grouped together and give the impression of being compound.

15. Locate the anus of the millipede at the posterior end between two **paraproct plates** (Fig. 15.39). Ventral to these is a single plate, the **hypoproct.** Find the anus of your centipede on the ventral area of the last trunk segment (Fig. 15.40).

Supplemental Reading*

Abele, G. (ed.). 1982. Biology of Crustacea. Vols. 1 and 2. Academic Press, New York. (CR)

Anderson, D. T. 1973. Embryology and phylogeny in annelids and arthropods. Pergamon Press, Oxford. (G)

Anderson, D. T. 1980. Cirral activity and feeding in the lepadomorph barnacle *Lepas pectinata*. Proc. Linn. Soc. N.S.W. 104:147-159. (CR)

Anderson, D. T. 1981. Cirral activity and feeding in the barnacle *Balanus perforatus*, with comments on the evolution of feeding mechanisms in thoracican cirripedes. Philos. Trans. R. Soc. Lond. B 291:411-449. (CR)

Arnett, R. H. 1985. American insects: a handbook of the insects of North America. Van Nostrand Reinhold, New York. (I)

Atkins, M. D. 1978. Insects in perspective. Macmillan, New York. (I)

Barker, P. L., and R. Gibson. 1977. Observations on the feeding mechanism, structure of the gut, and digestive physiology of the European lobster *Homarus gammarus*. J. Exp. Mar. Biol. Ecol. 26:297-324. (CR)

Bergström, J. 1973. Organization, life, and systematics of trilobites. Fossils Strata 2:1-69. (T)

Bergström, J. 1984. Legs in the trilobite *Rhenops* from the Lower Devonian Hunsrück Slate. Lethaia 17:69-72. (T)

*The supplemental literature for Arthopoda is coded to indicate the following groups: G, General; T, Trilobites; CH, Chelicerates; CR, Crustaceans; I, Insects; M, Myriapods.

Bland, R. G., and H. E. Jaques. 1978. How to know the insects. 3rd ed. Wm. C. Brown, Dubuque, IA. (I)

Bliss, D. E. (ed.). 1982-1985. The biology of Crustacea. Vols. 1-10. Academic Press, New York. (CR)

Blower, J. G. (ed.). 1974. Myriapoda. Symposia of the Zoological Society of London, 32. Academic Press, New York. (M)

Blum, M. S. (ed.). 1985. Fundamentals of insect physiology. Wiley-Interscience, New York. (I)

Borror, D. J., and R. E. White. 1974. A field guide to the insects of America north of Mexico. Houghton Mifflin, Boston. (I)

Borror, D. J., D. M. DeLong, and C. A. Triplehorn. 1981. An introduction to insect biology and diversity. W. B. Saunders, Philadelphia. (I).

Boudreaux, H. B. 1979. Arthropod phylogeny with special reference to insects. Wiley, New York. (I)

Briggs, D. E. G. 1983. Affinities and early evolution of Crustacea: the evidence of the Cambrian fossils. Crustacean Issues:1-22. (CR)

Brownell, P. H. 1984. Prey detection by the sand scorpion. Sci. Am. 251(6):86-97. (CH)

Budd, T. W., J. C. Lewis, and M. L. Tracy. 1978. The filter feeding apparatus in crayfish. Can. J. Zool. 56:695-707. (CR)

Burgess, J. W. 1976. Social spiders. Sci. Am. 234:101-106. (CH)

Calman, W. T. 1909. Crustacea. *In*: E. R. Lankester (ed.). A Treatise on zoology. Vol. 7. A. & C. Black, London. (CR)

Camatini, M. (ed.). 1979. Myriapod biology. Academic Press, New York. (M)

Cameron, J. N. 1985. Molting in the blue crab. Sci. Am. 252:102-109. (CR)

Chapman, R. F. 1982. The insects: structure and function. 3rd ed. Harvard University Press, Cambridge, MA. (I)

Cisne, J. L. 1973. Life history of an Ordovician trilobite *Triarthrus eatoni*. Ecology 54:135-142. (T)

Cisne, J. L. 1974. Trilobites and the origin of arthropods. Science 186:13-18. (T)

Cisne, J. L., G. O. Chandlee, B. D. Rabe, and J. A. Cohen. 1980. Geographic variation and episodic evolution in a trilobite. Science 209:925-927. (T)

Clarke, K. U. 1973. The biology of the Arthropoda. American Elsevier, New York. (G)

Clarkson, E. N. K. 1975. The evolution of eyes in trilobites. Fossils Strata 4:7-32. (T)

Clarkson, E. N. K. 1979. The visual system of trilobites. Paleontology 22:1-22. (T)

Cloudsley-Thompson, J. L. 1968. Spiders, scorpions, centipedes, and mites. Pergamon Press, Elmsford, NY. (CH, M)

Connell, J. H. 1961. The influence of interspecific competition and other factors on the distribution of the barnacle *Chthamalus stellatus*. Ecology 42:710-723. (CR)

Crompton, J. 1987. The spider. (Reprint of 1950 ed.). Nick Lyons, New York. (CH)

Denno, R. F., and H. Dingle. 1981. Insect life history patterns. Springer-Verlag, New York. (I)

Downer, R. G. H. (ed.). 1981. Energy metabolism in insects. Plenum Press, New York. (I)

Ehrlich, P. R., and A. H. Ehrlich. 1961. How to know the butterflies. Wm. C. Brown, Dubuque, IA. (I)

Elzinger, R. J. 1981. Fundamentals of entomology. 2nd ed. Prentice-Hall, Englewood Cliffs, NJ. (I)

Enright, J. T. 1977. Diurnal vertical migration: adaptive significance and timing. 1. Selective advantage: a metabolic model. Limnol. Oceanogr. 22:856-872. (CR)

Enright, J. T. 1977. Diurnal vertical migration: adaptive significance and timing. 2. Test of the model: details of timing. Limnol. Oceanogr. 22:873-886. (CR)

Evans, H. E. 1984. Insect biology. Addison-Wesley, Reading, MA. (I)

Foelix, R. F. 1982. Biology of spiders. Harvard University Press, Cambridge, MA. (CH)

Fortey, R. A., and S. F. Morris. 1978. Discovery of nauplius-like trilobite larvae. Paleontology 21:823-833. (T)

Gillot, C. 1980. Entomology. Plenum Press, New York. (I)

Govind, C. K. P. and J. Pearce. 1986. Differential reflex activity determines claw and closer muscle activity in developing lobsters. Science 233:354-356. (CR)

Gupta, A. P. 1979. Arthropod phylogeny. Van Nostrand Reinhold, New York. (G)

Hadley, N. F. 1986. The arthropod cuticle. Sci. Am. 255(1):104-112. (G)

Hamner, W. M., P. P. Hamner, S. W. Strand , and R. W. Gilmer. 1983. Behavior of Antarctic krill, *Euphausia superba*: chemoreception, feeding, schooling, and molting. Science 220:433-435. (CR)

Hessler, R. R., and W. A. Newman. 1975. A trilobite origin for the Crustacea. Fossils Strata 4:437-459. (T, CR)

Hoffman, R. L., and J. A. Payne. 1969. Diplopods as carnivores. Ecology 50:1096-1098. (M)

Kaston, B. J. 1978. How to know the spiders. 3rd ed. Wm. C. Brown, Dubuque, IA. (CH)

Kerfoot, W. C. (ed.). 1980. Evolution and ecology of zooplankton communities. University Press of New England, Hanover, NH. (CR)

Kettle, D. S. 1984. Medical and veterinary entomology. Wiley, New York. (I)

Kim, K. C. (ed.). 1985. Coevolution of parasitic arthropods and mammals. Wiley, New York. (CH, I)

King, P. E. 1973. Pycnogonids. St. Martin's Press. New York. (CH)

Labandeira, C. C., B. S. Beal, and F. M. Hueber. 1988. Early insect diversification: evidence from a Lower Devonian bristletail from Québec. Science 242:913-916.

Levi-Setti, R. 1975. Trilobites: a photographic atlas. University of Chicago Press, Chicago. (T)

Lewis J. G. E. 1981. The biology of centipedes. Cambridge University Press, Cambridge. (M)

Lynch, M. 1980. The evolution of cladoceran life histories. Rev. Biol. 55:23-42. (CR)

Manton, S. M. 1977. The Arthropoda: habits, functional morphology and evolution. Clarendon Press, Oxford. (G)

Matthews, R. W., and J. R. Matthews. 1978. Insect behavior. Wiley, New York. (I)

Mauchline, J., and L. R. Fischer. 1969. The biology of the euphausiides. Adv. Mar. Biol. 7:1-454. (CR)

McCafferty, W. P. 1981. Aquatic entomology. Jones & Bartlett, Boston. (I)

McDaniel, B. 1979. How to know the mites and ticks. Wm. C. Brown, Dubuque, IA. (CH)

McLaughlin, P. A. 1980. Comparative morphology of recent Crustacea. W. H. Freeman, San Francisco. (CR)

Moore, R. C. 1959. Treatise on Invertebrate Paleontology. Part O. Arthropoda l. Geol. Soc. Amer., Univ. Kansas Press, Lawrence, KS.

Novak, V. J. A. 1975. Insect hormones. 2nd ed. Wiley, New York. (I)

Paffenhöfer, G. A., J. R. Strickler, and M. Alcaraz. 1982. Suspension feeding by herbivorous calanoid copepods: a cinematographic study. Mar. Biol. 67:193-199. (CR)

Polis, G. A., and S. J. McCormick. 1987. Intraguild predation and competition among desert scorpions. Ecology 68:332-343. (CH)

Richards, O. W., and R. G. Davies. 1978. Imm's outline of entomology. 6th ed. Chapman & Hall, London. (I)

Rosmoser, W. S. 1981. The science of entomology. 2nd ed. Macmillan, New York. (I)

Ross, H. H., J. R. P. Ross, and C. A. Ross. 1982. A textbook of entomology. 4th ed. Wiley, New York. (I)

Schmitt, W. L. 1968. Crustaceans. University of Michigan Press, Ann Arbor, MI. (CR)

Schram, F. R. 1978. Arthropods: a convergent phenomenon. Fieldiana Geol. 39:61-108. (G)

Schram, F. R. 1986. Crustacea. Oxford University Press, New York. (CR)

Shaw, S. R. 1969. Optics of arthropod compound eyes. Science 165:88-90. (G)

Shear, W. A. E. (ed.). 1986. Spiders: webs, behavior and evolution. Stanford University Press, Stanford, CA. (CH)

Stürmer, W., and J. Bergström. 1973. New discoveries on trilobites by x-rays. Palaeontol. Z. 47:104-141. (T)

Tiegs, O. W., and S. M. Manton. 1958. The evolution of the Arthropoda. Biol. Rev. Camb. Philos. Soc. 33:255-337. (G)

Watson, P J. 1986. Transmission of female sex pheromone thwarthed by males in the spider *Linyphia litigiosa* (Linyphiidae). Science 233:219-221. (CH)

Weygoldt, P. 1969. The biology of pseudoscorpions. Harvard University Press, Cambridge, MA. (CH)

Whittington, H. B. 1975. Trilobites with appendages from the Middle Cambrian, Burgess Shale. Fossils Strata 4:97-136. (T)

OTHER PROTOSTOME PHYLA

Six small, schizocoelomate, protostome phyla are grouped here as a matter of convenience: Echiura, Pentastomida. Pogonophora, Priapulida, Sipuncula, and Tardigrada. The Pentastomida may be arthropods and closely related to members of the class Branchiura. Priapulida is believed by some workers to be a pseudocoelomate phylum. Of the remaining four phyla, only Pogonophora exhibits segmentation and may be related to the annelids. The specific relationship of the other phyla to the protostome line is uncertain.

EXERCISE 16

Phylum Sipuncula (Sipunculida)

Sipuncula (sigh-PUN-ku-la; L., *sipunculus*, little siphon) is a small phylum of about 320 species of vase-shaped, wormlike, marine organisms. They are commonly called peanut worms because their contracted bodies resemble a peanut. When extended, the **trunk** is usually vaselike or plump (Fig. 16.1). The anterior part of the body or the neck region, called the **introvert,** constitutes at least one-half to two-thirds the length of

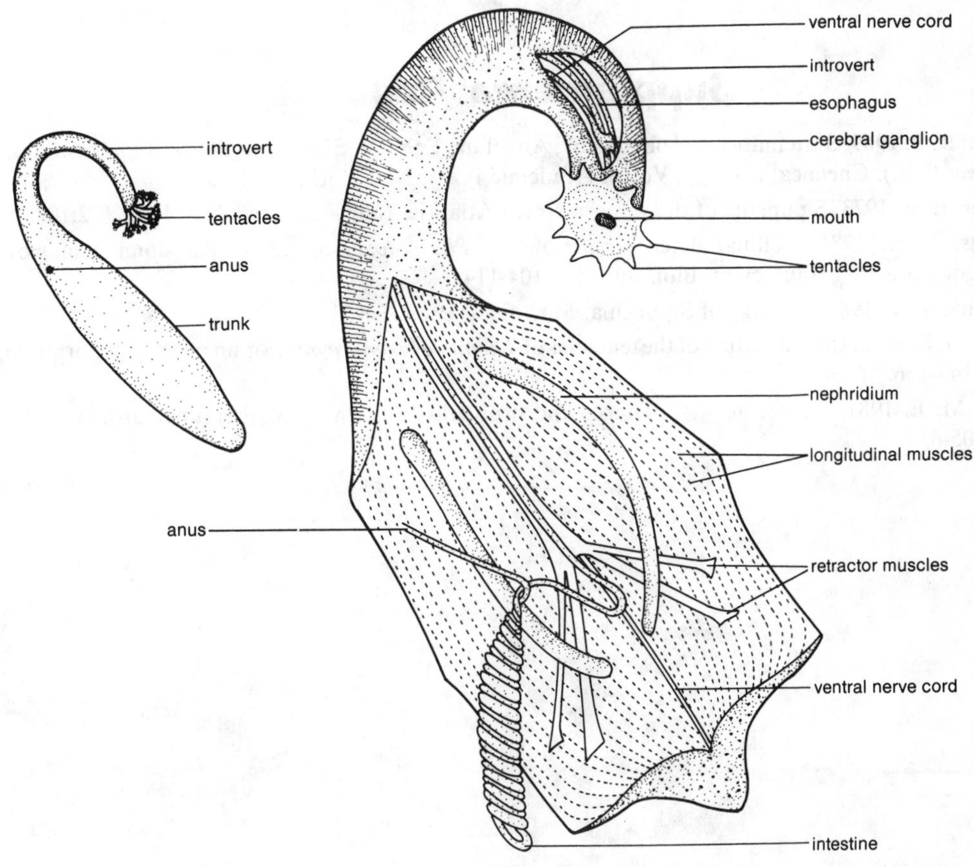

Figure 16.1. External and internal anatomies of the sipunculan, *Phascolosoma agassizii.*

the worm, and is capable of being rapidly everted and retracted. Segmentation and setae are lacking. At the end of the introvert is the mouth, surrounded by modified, lobate, ciliated tentacles. The tentacles are hollow, but do not communicate with the body cavity. Most sipunculans live a sedentary life in burrows or tubes. Internally, sipunculids have a spacious coelom and an elongated, U-shaped, digestive system (Fig. 16.1). The mouth is located at the anterior part of the body, from which the gut is directed posteriorly only to be recurved, and opened at the anus, which is located near the junction of the introvert and trunk.

Classification

Four families are recognized. Examples include *Golfingia*, *Phascolosoma*, and *Sipunculus*.

Observational Procedure

Examine the external anatomy of several peanut worms (Fig. 16.1). Locate the anterior and posterior ends, trunk, introvert, mouth with tentacles, and anus. Note the variations in size and shape of tentacles among species available for study.

If specimens are available for dissection, make a shallow, midsagittal, dorsal incision with scissors through the body wall slightly to one side of the anus. Extend the cut forward to the tip of the introvert and backward as far as possible. Pin the body wall open and locate the internal organs (Fig. 16.1). Observe the introvert under a dissecting microscope to see the many minute **spines** and **sensory papillae**. Examine the inner side of the tentacles surrounding the mouth and find the groove that conveys food to the mouth.

Supplemental Reading

Clark, R. B. 1969. Systematics and phylogeny; Annelida, Echiura, Sipuncula. *In*: M. Florkin and B. T. Scheer (eds.). Chemical zoology. Vol. 4. Academic Press, New York, pp. 1-68.

Cutler, E. B. 1973. Sipuncula of the western North Atlantic. Bull. Am. Mus. Nat. Hist. 152:103-204.

Dybas, L. K. 1981. Cellular defense reactions of *Phascolosoma agassizii*, a sipunculan worm; phagocytosis by granulocytes. Biol. Bull. 161:104-114.

Murina, V. V. 1984. Ecology of Sipuncula. Mar. Ecol. Prog. Ser. 17:1-7.

Pilger, J. F. 1982. Ultrastructure of the tentacles of *Themiste lageniformis* (Sipuncula). Zoomorphology 100:143-156.

Rice, M. E. 1981. Larvae adrift: patterns and problems in life history of sipunculans. Am. Zool. 21:605-619.

EXERCISE **17**

Phylum Tardigrada

Phylum Tardigrada (TAR-di-grad-a; L., *tardi*, slow + L., *grada*, walk) comprises a small group of bilaterally symmetrical, arthropod-like animals. About 400 species of tardigrades, or water bears as they are commonly known, are recognized. These minute (50 to 1000μm) protostomes possess short cylindrical bodies with four pairs of short, stubby, nonjointed legs extending ventrolaterally as if they were attached to a fat little bear (Fig. 17.1). Each leg has two or more toes, claws, or a combination of both at the distal end for grasping the substrate. Many semiterrestrial species have eyespots anteriorly. The bearlike image in tardigrades is reinforced when one observes the slow pawing motion that they make with their legs when moving.

Tardigrades are principally freshwater, but marine species have been found in the littoral area and at considerable depths (5000 m). Semiterrestrial forms often are found in the thin water films in soils or covering mosses, lichens, and liverworts. In both freshwater and marine habitats, they may be found on surfaces of aquatic plants and in interstitial spaces among sedimentary particles. Many species have a cosmopolitan distribution.

Classification

Three classes are recognized (Heterotardigrada, Mesotardigrada, and Eutardigrada) based on the presence or absence of cephalic appendages, Malpighian tubules, morphology of solid supports in the pharynx, and types and number of claws and toes. Class Mesotardigrada consists of a single species collected from a Japanese hot spring. Examples include *Macrobiotus* and *Echiniscus*.

Observational Procedure

Observe the movements of a live tardigrade under a compound microscope. Note the four pairs of legs, each of which possesses a set of claws. Describe the locomotory activity of your specimen. How fast does the animal move? Does the specimen stop to feed? How is feeding accomplished? Using live or preserved specimens, attempt to locate the following internal structures: (1) muscular sucking pharynx, (2) salivary glands, (3) esophagus, (4) midgut, (5) Malpighian tubules, (6) rectum, (7) anus, and (8) unpaired gonad (Figs. 17.1 and 17.2). Eggs may be present in the ovaries of females. Locate several of the body muscle fibers. Trace a muscle from its attachment to the body wall at one point to another. What is the net result of the contraction of this single muscle fiber?

If time and specimens are available for experimental work, follow the directions given by your instructor to initiate and revive animals from the **anabiotic** resting state.

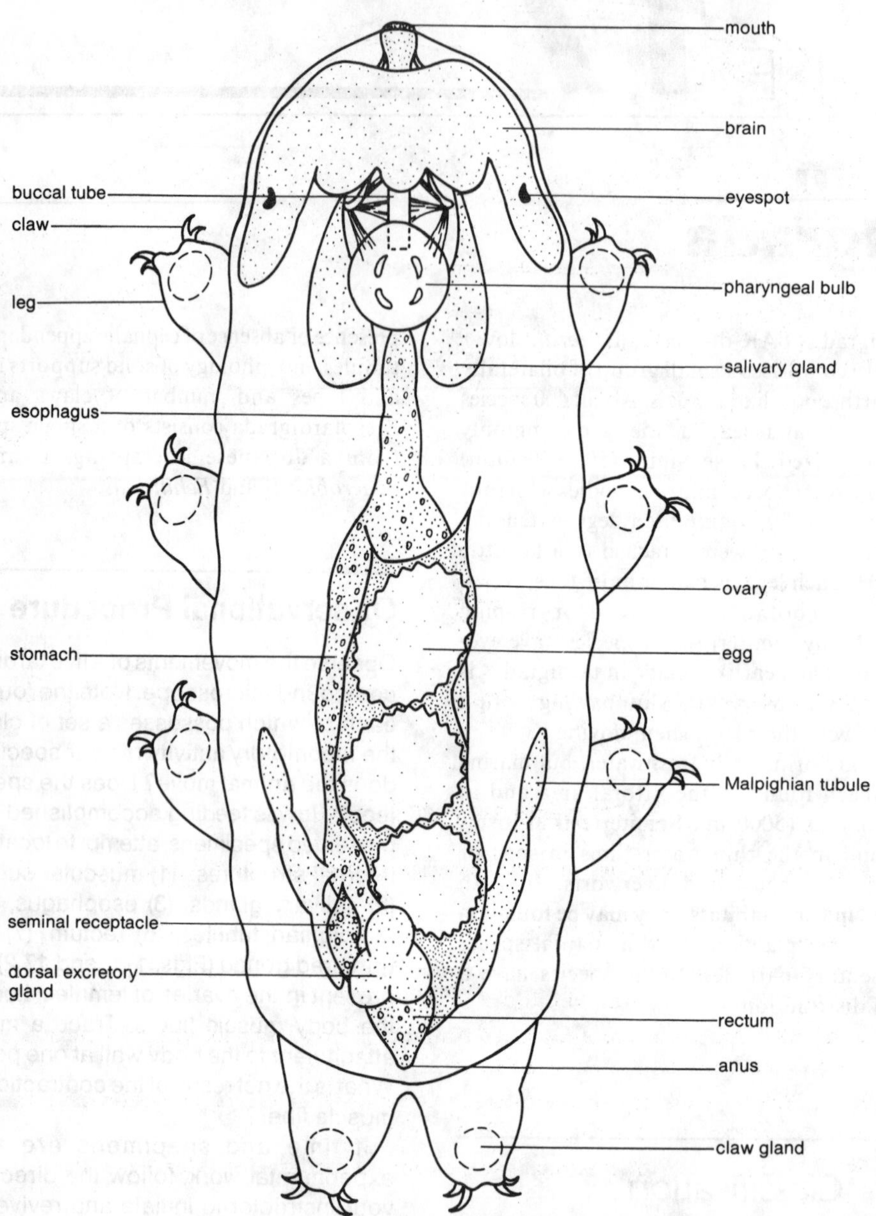

Figure 17.1. Dorsal internal view of a female tardigrade of the genus *Macrobiotus*. (After Pennak.)

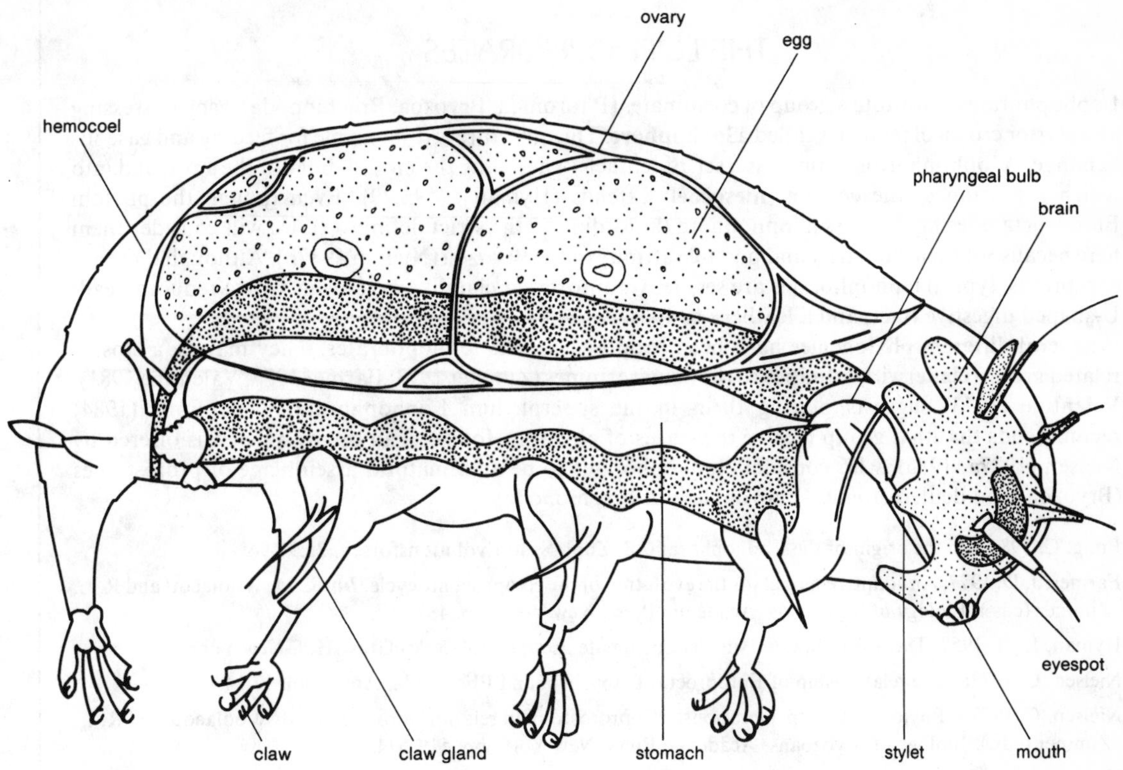

Figure 17.2. Lateral internal view of a female marine tardigrade of the genus *Styraconyx*. (After Kristensen and Higgins.)

Supplemental Reading

Crowe., J. H. 1975. The physiology of cryptobiosis in tardigrades. Mem. Ital. Idrobiol. 32 (Suppl.): 37-59.

Crowe., J. H., and A. F. Cooper, Jr. 1971. Cryptobiosis. Sci. Am. 225 (12):30-36.

Crowe., J. H., and R. P. Higgins. 1967. The revival of *Macrobiotus areoatus* Murry (Tardigrada from the cryptobiotic state. Trans. Am. Microsc. Soc. 86:286-294.

Everitt, D. A. 1981. An ecological study of an Antarctic freshwater pool with particular reference to Tardigrada and Rotifera. Hydrobiologia 83:225-237.

Higgins, R. P. 1959. Life history of *Macrobiotus islandicus* Richters with notes on other tardigrades from Colorado. Trans. Am. Microsc. Soc. 78:137-154.

Higgins, R. P. (ed.). 1975. International symposium on tardigrades. Mem. 1st. Ital Idrobiol. 32 (Suppl.): 1-469

Nelson, D. R. (ed.). 1982. Proceedings of the 3rd international symposium on the Tardigrada, East Tennessee state University Press. Johnson City, TN.

Pollock, L. W. 1970. Distribution and dynamics of intersitial Tradigrada at woods Hole, Massachusetts, USA. Ophelia 7:145-166.

Schuster, T. R., and A. A. Grigarick. 1965. Tradigrada from western North America with emphasis on the fauna of California. Univ. Calif. Publ. Zool. 76:1-67.

THE LOPHOPHORATES

Lophophorates constitute a group of coelomates (Phoronida, Bryozoa, Brachiopoda), each possessing an anterior crown of tentacles called a **lophophore**. This structure is responsible for feeding and gaseous exhange. A lophophore is defined as a set of tentacles encircling the mouth, but not the anus, and into which a portion of the coelom (**mesocoel**) extends (Hyman 1959:229). Even though the phylum Entoprocta does not possess a lophophore (according to the strict definition), we will consider them here because of their similarity and possible affinity to the Bryozoa (Nielsen 1977b). Although there are exceptions, typical lophophorates are sessile, vermiform, or clamlike organisms with a reduced head, U-shaped digestive tract, and a leathery-to-hard case.

Several different phylogenies have been proposed for the lophophorates. They may be a closely related group (**clade**) with affinities to the deuterostomes (Zimmer 1973, Farmer 1977, Valentine 1981). Valentine (1973) suggests placing them in the superphylum Lophophorata, while Emig (1984) recommends that that group receive the status of phylum. However, an opposing view is offered by Nielsen (1977a, 1985), who considers lophophorates to be an unnatural assemblage of protostomes (Bryozoa) and deuterostomes (Phoronida and Brachiopoda).

Emig, C. 1984. On the origin of the Lophophorata. Z. Zool. Syst. Evolutionsforsch. 22:91-94.

Farmer, J. D. 1977. An adaptive model for the evolution of the ectoproct life cycle. *In*: R. M. Woolacott and R. L. Zimmer (eds.). Biology of bryozoans. Academic Press, New York, pp. 487-517.

Hyman, L. H. 1959. The invertebrates: smaller coelomate groups. Vol. 5. McGraw-Hill, New York.

Nielsen, C. 1977a. The relationship of Entoprocta, Ectoprocta and Phoronida. Am. Zool. 17:149-150.

Nielsen, C. 1977b. Phylogenetic considerations: the protostomian relationships. *In*: R. M. Woolacott and R. L. Zimmer (eds.). Biology of bryozoans. Academic Press, New York, pp. 519-534.

Nielsen, C. 1985. Animal phylogeny in the light of the trochaea theory. Biol. J. Linn. Soc. 25:243-299.

Valentine, J. W. 1973. Coelomate superphyla. Syst. Zool. 22:97-102.

Valentine, J. W. 1981. The lophophorate condition. Lophophorates, notes for a short course. University of Tennessee, Knoxville TN, pp. 190-204.

Zimmer, R. L. 1973. Morphological and developmental affinities of the lophophorates. *In*: G. P. Larwood (ed.). Living and fossil Bryozoa. Academic Press, New York, pp. 593-599.

EXERCISE 18

Phylum Phoronida

Phylum Phoronida (fo-RON-i-da; G., *Phoronis*, from Greek mythology) contains only about ten species of nonsegmented, tube-dwelling, marine worms. Ranging in length from a few millimeters to over 30 cm, these sedentary animals secrete a chitinous, parchment-like tube to which sand, small stones, shells, and other debris may be attached (Fig. 18.1). Although usually not common, they may be found carpeting shallow to moderately deep (about 400 m) tropical and temperate waters (Emig 1979, 1982; Johnson 1959). Phoronids usually live as solitary individuals buried in soft sediments (mud or sand), or as aggregates (Fig. 18.2) inhabiting burrows among rocks and mollusk shells, or attached to pilings and other organisms. Some species penetrate calcareous shells or rocks. Phoronids move freely in their tubes, but they do not leave them. The horseshoe-shaped **lophophore** is located at the anterior end of the animal. Normally, this structure is extended from the tube for feeding, but upon disturbance it is quickly retracted. The gut is elongate, U-shaped, and extends from the lophophore to the posterior end and back toward the anterior end. Phoronids have a complex larval stage known as an **actinotroch** (Fig. 18.3), sometimes interpreted as a modified **trochophore** larva, which it is not (Zimmer 1973).

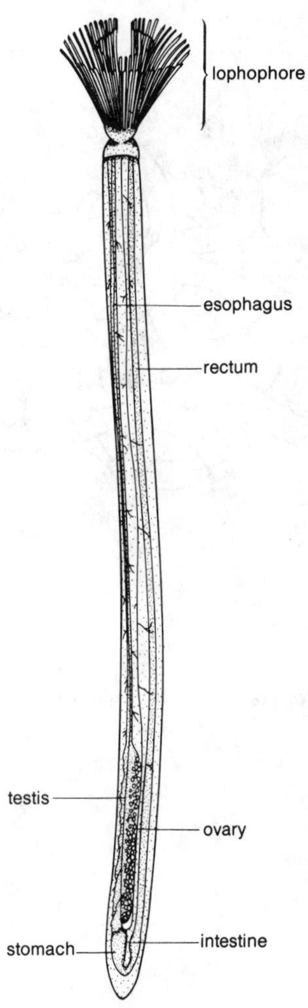

Figure 18.1. Generalized phoronid.

Classification

No taxonomic division higher than genus (*Phoronis* and *Phoronopsis*) is recognized.

Observational Procedure

Observe a phoronid specimen under a dissecting microscope and identify the structures discussed above and shown in Fig. 18.1. Note the lophophore at the anterior end with its tentacles surrounding the mouth. A small flap of tissue, the **epistome**, may be seen above the mouth. The anus is external to the lophophore and may be located by carefully rotating the animal about its longitudinal axis. The paired **nephridiopores**, which are located on either side of the anus, will probably not be seen without considerable effort. Given adequate lighting and a sufficiently transparent specimen, you may be able to trace the gut from mouth to anus.

Figure 18.2. Aggregate of the phoronid, *Phoronis hippocrepis*.

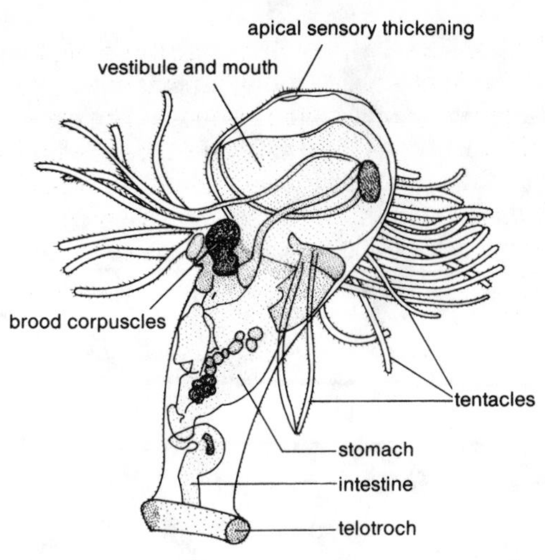

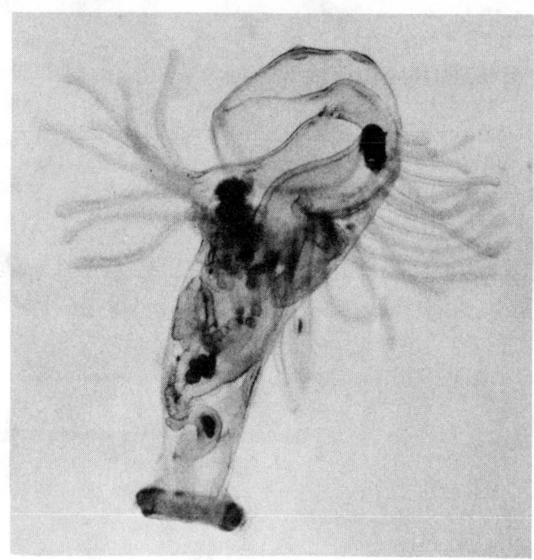

Figure 18.3. Photomicrograph of an actinotroch larva of the phoronid, *Phoronis muelleri*, with a line drawing showing interpretation. (Photomicrograph courtesy of K. Herrmann.)

Supplemental Reading

Abele, L. G., T. Gilmour, and S. Gilchrist. 1983. Size and shape in the phylum Phoronida. J. Zool. (Lond.) 200:317-323.

Emig, C. C. 1979. A synopsis of British and other phoronids. Academic Press, London.

Emig, C. C. 1982. The biology of Phoronida. Adv. Mar. Biol. 19:1-89.

Johnson, R. C. 1959. Spatial distribution of *Phoronopsis viridis* Hilton. Science 129:1221.

Marsden, R. C. 1957. Regeneration in *Phoronis vancouverensis*. J. Morphol. 101:307-324.

Nielsen, C. 1977. The relationship of Entoprocta, Ectoprocta, and Phoronida. Am. Zool. 17:149-150.

Strathmann, R. 1973. Function of lateral cilia in suspension feeding of lophophorates (Brachiopoda, Phoronida, Ectoprocta). Mar. Biol. 23:129-136.

Zimmer, R. L. 1973. Morphological and developmental affinities of the lophophorates. *In*: G. P. Larwood (ed.). Living and fossil Bryozoa. Academic Press, New York, pp. 593-599.

Zimmer, R. L. 1978. The comparative structure of the preoral hood coelom in Phoronida and the fate of this cavity after metamorphosis. *In*: F.-S. Chia and M. E. Rice (eds.). Settlement and metamorphosis of marine invertebrate larvae. Elsevier, New York, pp. 23-40.

EXERCISE 19

Phylum Bryozoa (Ectoprocta)

Phylum Bryozoa (BRI-o-ZO-a; G., *bryo*, moss + G., *zoa*, animals) comprises a group of sessile, predominantly colonial lophophorates, some of which superficially resemble hydroids. Many of the 4000 living species of bryozoans are found in the marine intertidal zone, but specimens have been recovered from depths of over 8000 m. There are a few common freshwater genera. The common name "moss animal" is fitting, because bryozoans grow in water as colonies, sometimes forming mosslike blankets of growth over the object to which they are attached. Two alternative names are used for this phylum. **Ectoprocta** emphasizes that the anus is located outside the lophophore in contrast to phylum Entoprocta, (Exercise 20) while **Polyzoa** emphasizes the colonial aspect of members of the phylum.

Although the individual organism, or **zooid**, is microscopic in size (usually less than 1.0 mm), the colony or **zoarium** may attain many centimeters in length and contain thousands of zooids. Zooids consist of two basic parts: a protective exoskeleton called the **zoecium** (zooecium) and the softer, internal viscera (Figs. 19.1 and 19.2). Zoecia may be gelatinous, chitinous, or calcareous and are secreted by the epidermis of the body wall. The body wall and the zoecium together are referred to as the **cystid**, while the rest of the animal, including the lophophore, digestive apparatus, and associated viscera, constitutes the **polypide**. In forms with calcareous zoecia, an orifice permitting eversion and retraction of the lophophore may be covered by a lid or **operculum**.

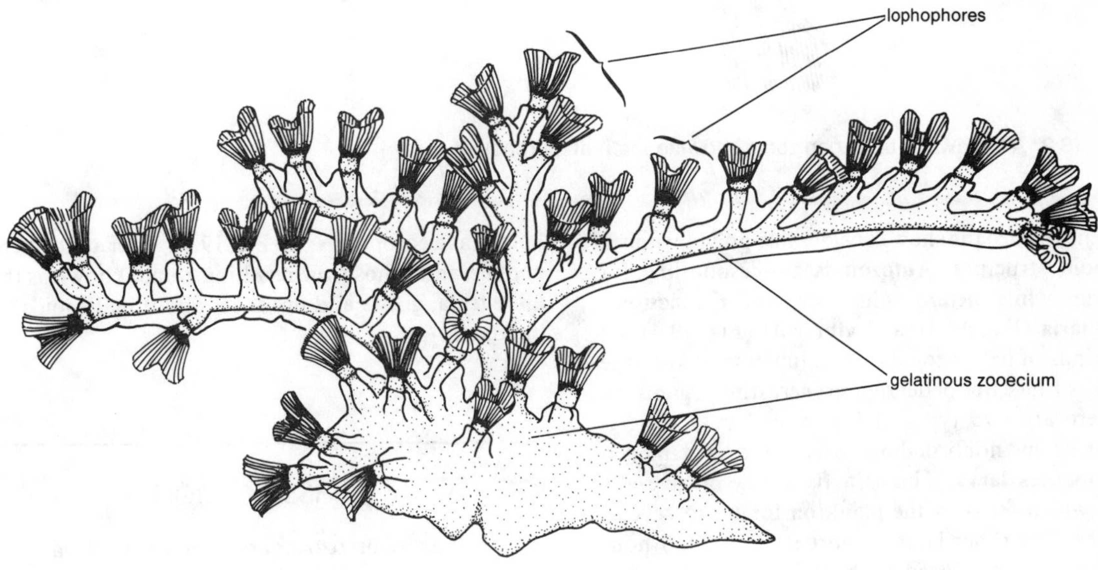

Figure 19.1. Freshwater bryozoan colony, *Hyalinella*.

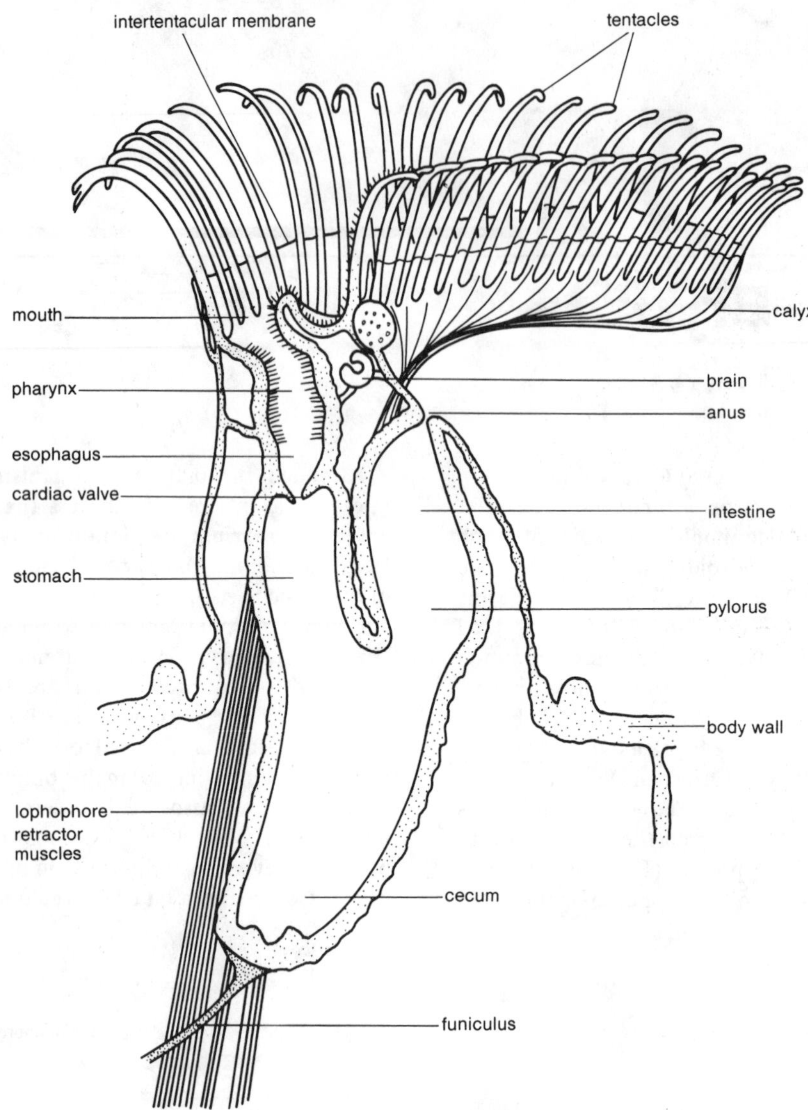

Figure 19.2. A freshwater bryozoan zooid showing internal morphology.

Some bryozoans show considerable polymorphism in zooid structure. **Autozooids** are responsible for feeding, while **heterozooids** have other functions. **Avicularia** (Fig. 19.3). and **vibracula** (Fig. 19.4) are two kinds of heterozooids which function to keep the colony surface free of debris and encrusting organisms.

There are two types of bryozoan larvae. One is produced by nonbrooding species and is named a **cyphonautes larva**. This is a flattened, bivalve-like larva which feeds in the plankton for up to several months. The other larva is more spherical and non-feeding; it is produced by brooding species and is known as a **coronate larva** (Fig. 19.5). After attachment and metamorphosis the resulting zooid, known as the **ancestrula**, grows and produces additional zooids by budding.

Classification

Three classes of bryozoans are recognized: Phylactolaemata, Stenolaemata, and Gymnolaemata.

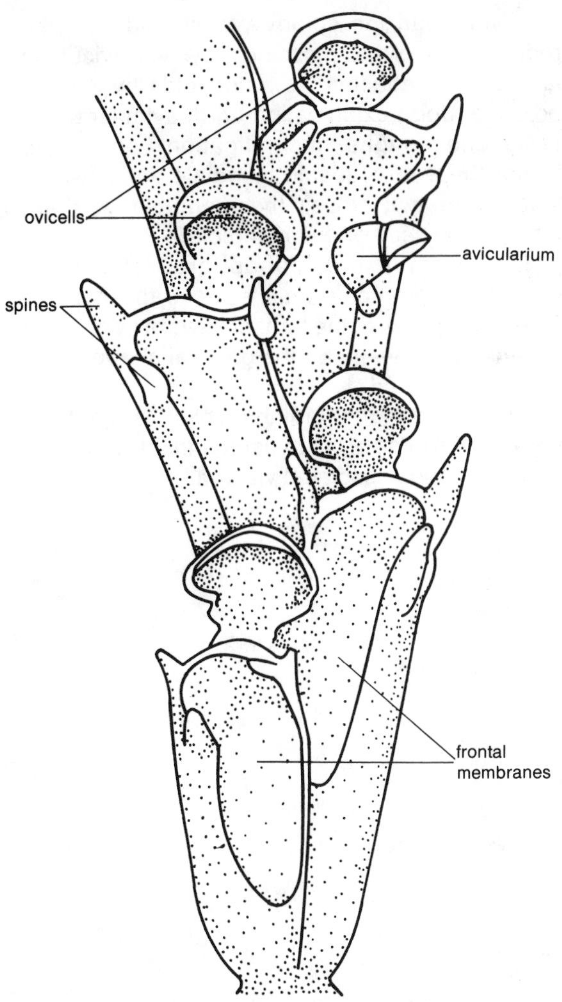

Figure 19.3. Portion of a colony of *Bugula*.

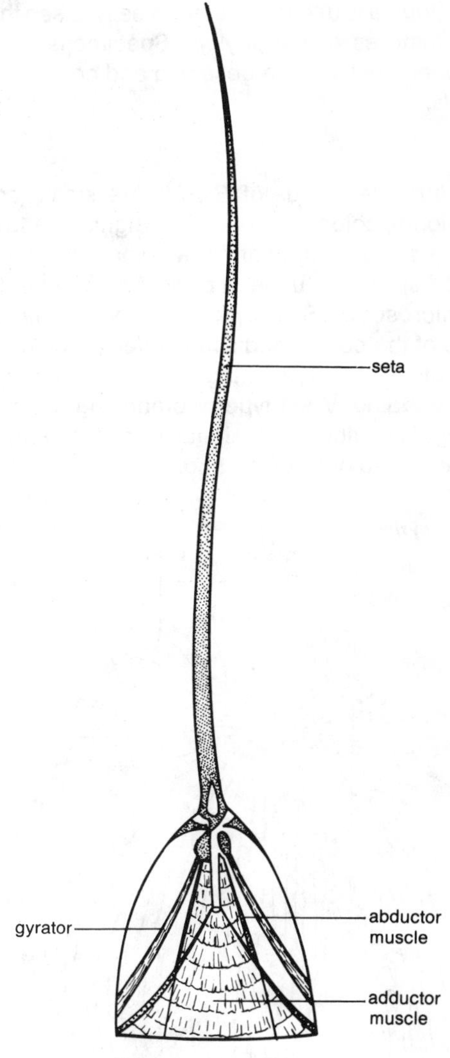

Figure 19.4. Vibraculum.

1. Class Phylactolaemata. One order, Plumatellida, comprising about 50 species of freshwater bryozoans usually with a horseshoe-shaped lophophore; lacking a calcified zoecium; no polymorphism; statoblasts present; examples include *Fredericella*, *Pectinatella*, and *Plumatella*, and an unusual form, *Cristatella*, in which the entire colony is slowly motile, gliding on a muscular sole.

2. Class Stenolaemata. Marine, with a tubular, calcified zoecium; one extant order (Cyclostomata) and several orders that became extinct during the Paleozoic; example, *Crisia*.

3. Class Gymnolaemata. Mostly marine bryozoans with a circular lophophore and polymorphic zooids; two orders: Ctenostomata (noncalcified zoecium and stoloniferous to somewhat flattened growth forms; nonoperculate zooids; marine, except for two freshwater genera, *Paludicella* and *Pottsiella*) and Cheilostomata (a large group of widely distributed marine bryozoans having a boxlike, variably calcified zoecium, with an operculum: polymorphic zooids with avicularia and/or vibracula present); examples include those with plantlike growth forms (*Bugula*) and encrusting types (*Electra*, *Membranipora*, *Schizoporella*).

Observational Procedure

We begin our observations of bryozoans with a prepared slide of the common marine cheilostome,

Bugula. *Bugula* is used because it is easy to see the general features of the phylum. Specimens from other genera will then be observed and compared to *Bugula*.

Bugula

1. Although the zooids of *Bugula* are small (ca. 500 μm long), colonies may attain heights of a few centimeters (Fig. 19.3). Examine a prepared slide or preserved specimen under a dissecting or a compound microscope. Start by focusing on a zooid at the base of the colony and scan the entire colony, taking note of the branching pattern formed by asexual budding. What type of branching pattern does *Bugula* exhibit? Occasionally, one may find a specimen with a developing bud.

2. As you scan the colony you should note the predominance of the autozooid. However, avicularia may be seen attached to autozooids by a short peduncle. Upon examination of an avicularium in profile under higher magnification you should observe the **rostrum** (beak), mandible, and powerful musculature that closes the **mandible**. A sharp tooth may be seen on the mandible.

3. Now observe the lophophores of several autozooids. How many tentacles does an autozooid have? If the bases of the tentacles were connected by a line, what would be the general shape of the figure thus created?

4. Try to locate the mouth, which is centrally located at the base of the tentacles. Inside the zooid lies the gut and associated viscera.

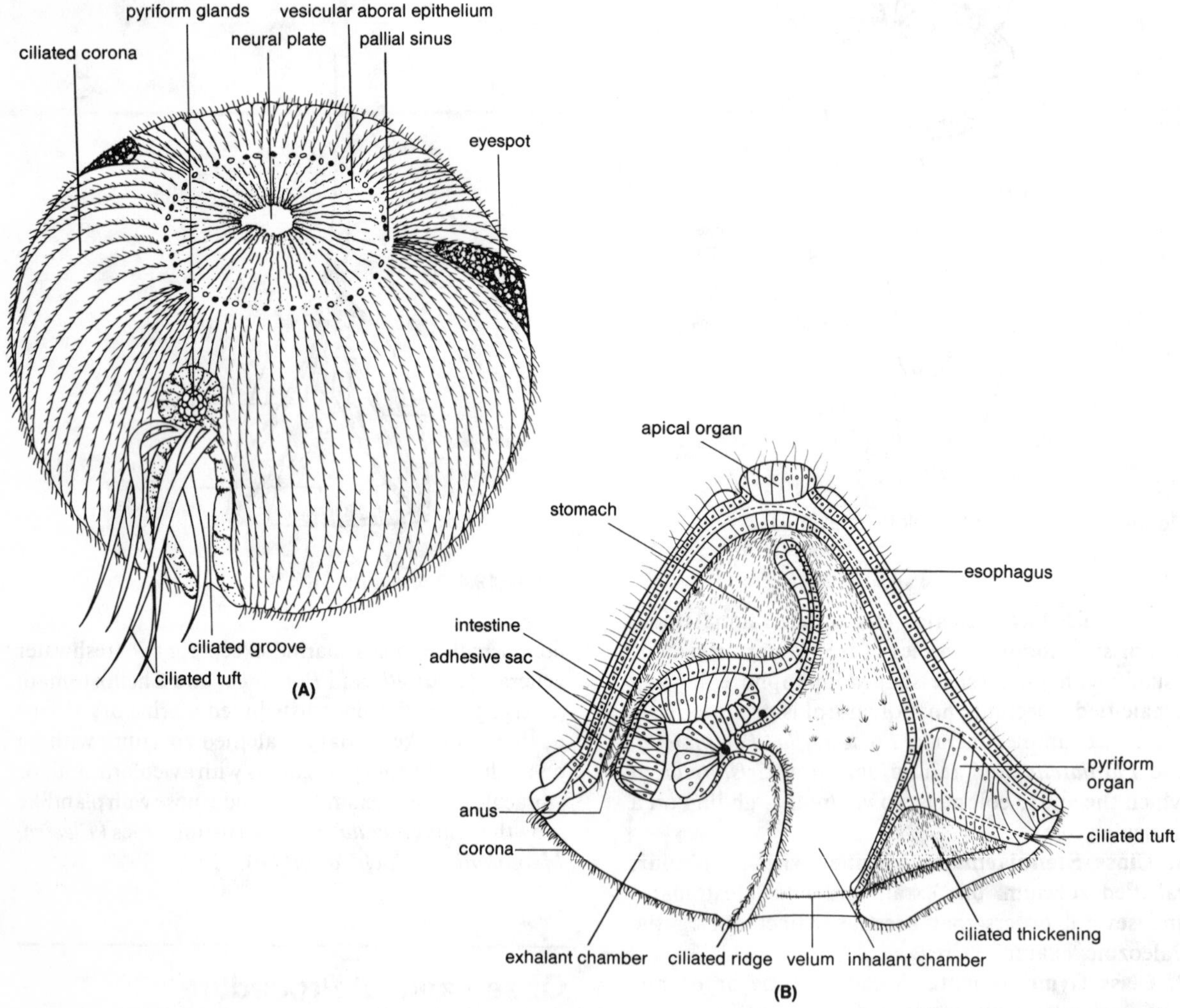

Figure 19.5. Generalized bryozoan larval types. **(A)** Coronate. **(B)** Cyphonautes. (After Woolacott and Zimmer, and Nielsen, respectively.)

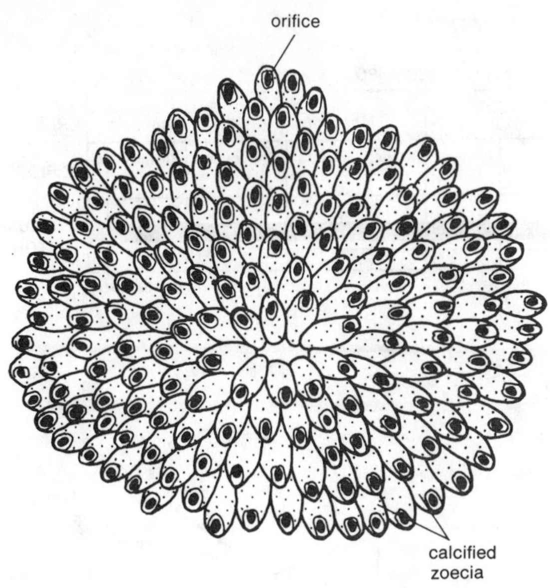

Figure 19.6. Encrusting marine cheilostomate bryozoans.

5. Scan the colony and find an individual in which you can trace the entire digestive tract. Beginning with the mouth, locate the esophagus, cardiac valve, stomach with cecum, pyloric valve, rectum, and anus. Note in particular that the anus empties outside the lophophore. As you trace the gut, you may be able to determine what food this specimen of *Bugula* ate.

6. If the specimen were part of a fully grown colony, it may have **ovicells**. These globular or hemispherical structures will be situated at the distal (upper) end of some of the zoecia. Inside you may be able to see a developing embryo.

7. Not all of the autozooids are still functional. At the base of the colony several almost empty zoecia may be seen, and in some of these you may see a small, darkstained structure. This is the **brown-body**; it consists of the cellular remnants of a degenerated zooid.

8. Compare *Bugula* and a cnidarian hydroid such as *Obelia*, noting similarities and differences in each.

Encrusting Marine Bryozoans. Observe a specimen of an encrusting bryozoan (Fig. 19.6). Note how different the growth form is from that of *Bugula*. The autozooids are found in calcareous, boxlike zoecia; the entire colony resembles a brick wall. At one end of the zooid an orifice can be seen which may or may not have a lidlike operculum. Ovicells, avicularia, vibracula, and spines may be present, depending on the species.

Freshwater Bryozoans. In contrast to the two marine gymnolaemates just studied, we will now observe a freshwater phylactolaemate such as *Pectinatella* and/or *Plumatella*. Is the shape of their lophophore different from that of *Bugula*? How would you describe its shape? Running from the lophophore to the base of the animal you may see the lophophore retractor muscle; it should not be confused with the funiculus. Note the numerous tentacles in this species. How many are there? Trace the gut of your specimen, which is similar to that of *Bugula*. The funiculus if present, may be seen attached to the base of the stomach and connected to the body wall. Scattered along the length of the **funiculus** may be several **statoblasts** (internal buds) in various stages of development (Fig. 19.7).

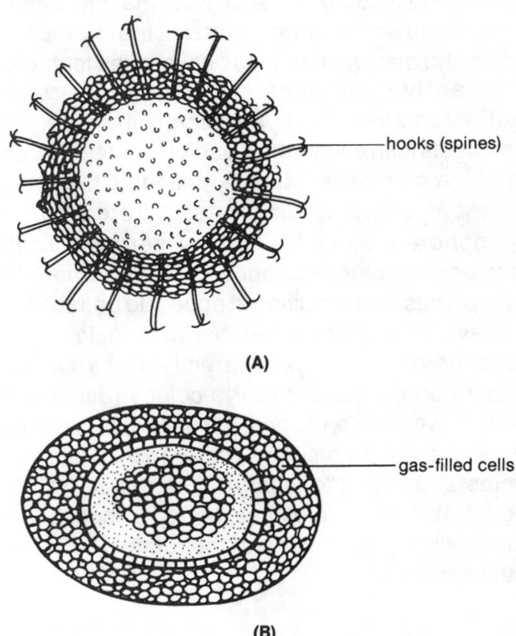

Figure 19.7. Phylactolaemate statoblasts. **(A)** Spinoblast. **(B)** Floatoblast. (From Barnes after Allman.)

Live Specimens

1. Observe the manner of rapid retraction and gradual extension of the lophophore. Describe how the lophophore is retracted and extended. (N.B.: This is a good example of a muscular/fluid hydraulic mechanism.)

2. Note the movement of the cilia on the tentacles in expanded individuals. One also may observe flicking of the whole tentacles. This process, which varies among species, is a mode of transporting food particles toward the mouth and rejecting unwanted materials.

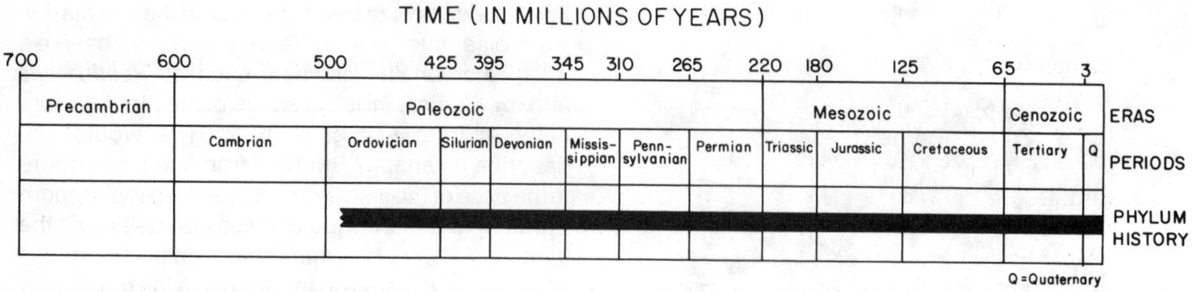

TIME (IN MILLIONS OF YEARS)

700	600	500	425	395	345	310	265	220	180	125	65	3	
Precambrian	Paleozoic							Mesozoic			Cenozoic		ERAS
	Cambrian	Ordovician	Silurian	Devonian	Missis-sippian	Penn-sylvanian	Permian	Triassic	Jurassic	Cretaceous	Tertiary	Q	PERIODS
													PHYLUM HISTORY

Q = Quaternary

Figure 19.8. Geologic history of the phylum Bryozoa.

3. Using a Pasteur pipette, gently add a few drops of water containing some carmine red (or other suspended particles) near the lophophore and observe the feeding currents. Note that the feeding currents run down into the center of the lophophore crown and then out the sides between the tentacles.

4. In active colonies of some encrusting cheilostomates, such as *Membranipora*, the autozooids lean away from a particular area, thereby leaving a blank site called a **chimney**. Chimneys represent areas where water, filtered by the lophophores, exits the colony surface. These transient structures cannot be seen when the lophophores are retracted but appear to be important in preventing interference between neighboring lophophores. Chimneys apparently have a constant location on the surface of the colony (Banta et al. 1974). To verify this, observe the pattern of chimneys across the colony's surface before you disturb the zooids and after the lophophores reemerge. You should also be able to see periodic ingestion of particles by the pharynx and the rotating food cord in the stomach.

Fossil Forms

First appearing in the early **Ordovician** nearly 500 million years ago (McLeod 1978), bryozoans have a rich geologic history (Fig. 19.8) with more than 16,000 described species. Unfortunately, a formidable nomenclature has been developed within the literature on fossil bryozoans. We will not take time to present the details of this discipline. Identification and detailed study are done from longitudinal and thin tangential sections which have been prepared from fossils (Moore et al. 1952).

Observe the fossil specimens available for study using a hand lens or dissecting microscope. Students should be able to observe the orifice of each individual zooid. One interesting genus,

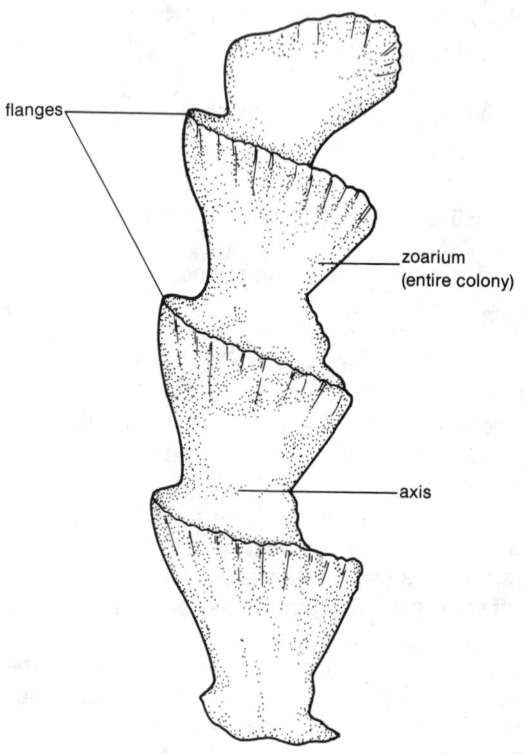

Figure 19.9. The fossil bryozoan *Archimedes*.

Archimedes, which occurred from the **Mississippian** through **Permian Periods**, resembles a water-screw and was named after the inventor of that device (Fig. 19.9). Examine several fossil bryozoans for the presence of numerous small bumps scattered fairly uniformly over the surface. Many fossil stenolaemates have these bumps. They are known as **monticules** and are believed to be the **Paleozoic** equivalent of the chimneys discussed above (Banta et al. 1974).

Supplemental Reading

Banta, W. C. 1977. Body wall morphology of the sertellid cheilostome bryozoan, *Reteporellina evelinae*. Am. Zool. 17:74-91.

Banta, W. C., F. K. McKinney, and R. L. Zimmer. 1974. Bryozoan monticules: excurrent water outlets. Science 185:783-784.

Bushnell, J. H., and K. S. Rao. 1974. Dormant or quiescent stages and structures among the Ectoprocta: physical and chemical factors affecting viability and germination of statoblasts. Trans. Am. Microsc. Soc. 93:524-543.

Buss, L. W. 1981. Group living, competition, and the evolution of cooperation in a sessile invertebrate. Science 213:1012-1014.

Harvell, C. D. 1984a. Predator-induced defense in a marine bryozoan. Science 224:1357-1359.

Harvell, C. D. 1984b. Why nudibranchs are partial predators: intracolonial variation in bryozoan palatability. Ecology 65:716-724.

Larwood, G. P. (ed.). 1973. Living and fossil Bryozoa: recent advances in research. Academic Press, New York.

Mayr, E. 1968. Bryozoa versus Ectoprocta. Syst. Zool. 17:213-216.

McLeod, J. D. 1978. The oldest bryozoans: new evidence from the early Ordovician. Science 200:771-773.

Moore, R. C., C. G. Lalicker, and A. G. Fischer. 1952. Invertebrate fossils. McGraw-Hill, New York.

Osman, R. W., and J. A. Haugsness. 1981. Mutualism among sessile invertebrates: a mediator of competition and predation. Science 211:846-848.

Ross, J. R. P. 1977. Microarchitecture of body wall of extant cyclostome ectoprocts. Am. Zool. 17:93-105.

Ryland, J. S. 1970. Bryozoans. Hutchinson University Library, London.

Ryland, J. S., and A. R. D. Stebbing. 1971. Settlement and orientated growth in epiphytic and epizooic bryozoans. *In*: D. J. Crisp (ed.). Fourth European marine biology symposium. Cambridge University Press, Cambridge, pp. 283-300.

Seed, R., and R. J. O'Connor. 1981. Community organization in marine algal epifaunas. Annu. Rev. Ecol. Syst. 12:49-74.

Sutherland J. P. 1978. Functional roles of *Schizoporella* and *Styela* in the fouling community at Beaufort, North Carolina. Ecology 59:257-264.

Woolacott, R. M., and R. L. Zimmer (eds.). 1977. Biology of Bryozoans. Academic Press, New York.

Yoshioka, P. M. 1982. Role of planktonic and benthic factors in the population dynamics of the bryozoan *Membranipora membranacea*. Ecology 63:457-468.

EXERCISE 20

Phylum Entoprocta

Entoprocta (EN-to-PROK-ta; G., *ento*, within + G., *proct*, anus) is a little known phylum of mostly sessile invertebrates numbering about 100 species. Although considered by some to be pseudocoelomates, their relationship to that group is uncertain. Currently, many invertebrate zoologists place entoprocts as a separate phylum related to the Bryozoa or a subphylum of Bryozoa (see Nielsen 1971, 1977 and Farmer 1977 for an in-depth discussion).

Except for a single freshwater genus (*Urnatella*, Fig. 20.1) entoprocts are marine, growing in short tufts or mats on mollusk shells, pilings, rocks, and on living animals such as polychaetes and sponges. The small, somewhat transparent zooids are composed of two main parts: a slender **stalk** and an enlarged body or **calyx** with a crown of tentacles (Figs. 20.2 and 20.3). The stalk elevates the calyx a few millimeters above the substrate and may be smooth, spined, or have swellings that resemble beads on a string. The swellings are muscular joints that permit bending movements in the animal. The viscera are completely enclosed in the calyx, which typically is vase- to boat-shaped. Enclosed by the crown of tentacles is a flattened to slightly concave region termed the vestibule. The mouth is at the anterior end of the vestibule and the anus at the posterior end. The term *entoproct* is derived from the fact that the anus is located inside the crown of tentacles.

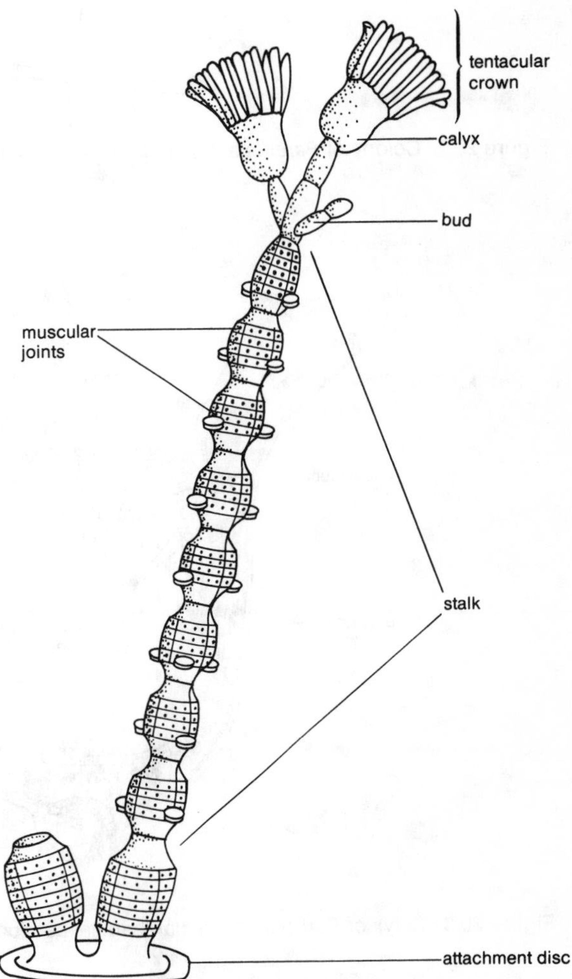

Figure 20.1. Colony of *Urnatella gracilis*. (After Leidy, from Pennak.)

Classification

Three families occur in the phylum: Loxosomatidae (solitary entoprocts, e.g., *Loxosoma* and *Loxosomella*);

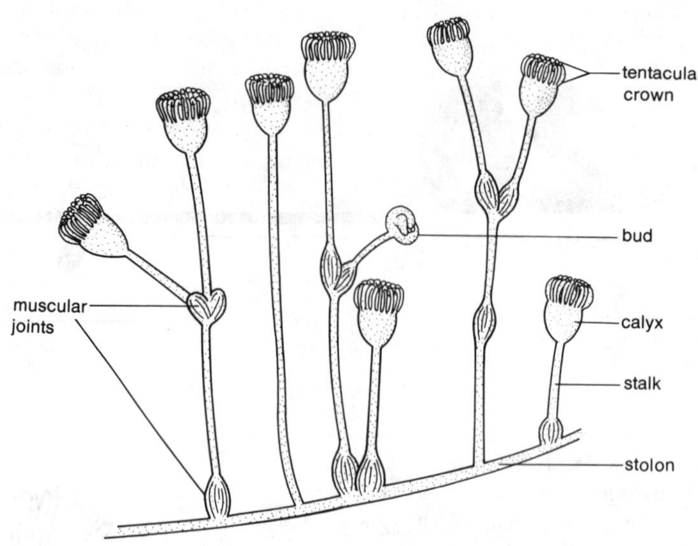

Figure 20.2. Colony of *Barentsia*. (After Robertson, from Hyman.)

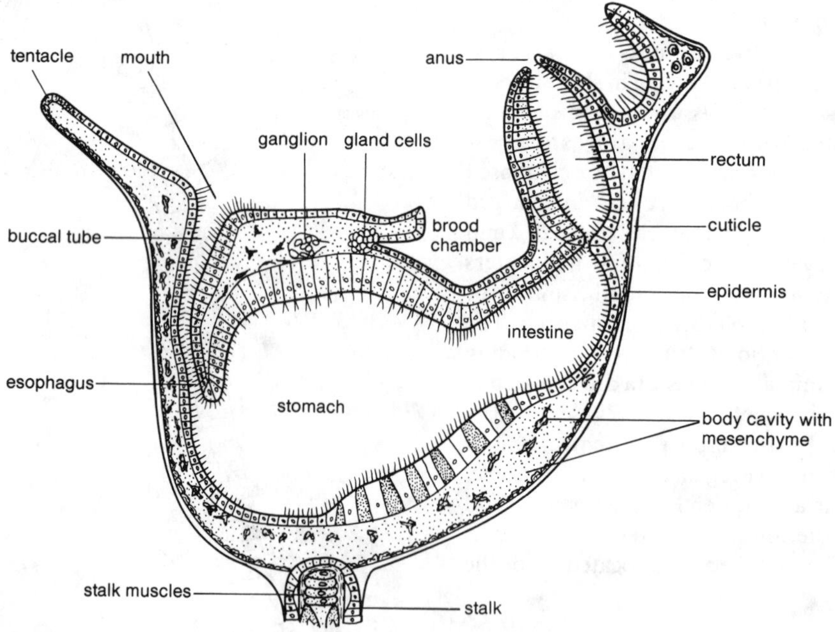

Figure 20.3. Calyx of *Pedicellina* (median sagittal section). (After Becker, from Hyman.)

Pedicellinidae (colonial marine and brackish-water forms such as *Pedicellina* , *Myosoma*, and *Barentsia*); Urnatellidae (with one freshwater genus, *Urnatella*).

Observational Procedure

Examine a whole-mount slide of an entoproct and identify the calyx, with its crown of tentacles, and stalk (Figs. 20.1 to 20.3). Internally the body is filled with a gelatinous material containing mesenchymatous cells, a U-shaped ciliated digestive tract, and reproductive and excretory systems. There is a cuticular covering of variable thickness over the entire animal, except for the tentacles and vestibule.

Compare the entoproct's external and internal structure to those of a hydrozoan (e.g., *Campanularia*) and a bryozoan (e.g., *Bugula*). Take particular note of the (1) striking differences in overall body architecture and reproductive structures, (2) lack of tentacular ciliation in hydrozoans, and (3) lack of nematocysts in entoprocts and bryozoans.

Supplemental Reading

Farmer, J. D. 1977. An adaptive model for the evolution of the Ectoproct life cycle. *In*: R. M. Woolacott and R. L. Zimmer (eds.). Biology of Bryozoans. Academic Press, New York, pp. 487-517.

Leidy, J. 1884. *Urnatella gracilis,* a freshwater polyzoan. J. Acad. Nat. Sci. (Phila.) 9:5-16.

Nielsen, C. 1971. Entoproct life cycles and the entoproct/ectoproct relationship. Ophelia 9:209-341.

Nielsen, C. 1977. The relationship of Entoprocta, Ectoprocta, and Phoronida. Am. Zool. 17:149-150.

Nielsen, C., and J. Rostgaard. 1976. Structure and function of an entoproct tentacle with a discussion of ciliary feeding types. Ophelia 15:115-140.

Tracy, B. H., and D. H. Hazelwood. 1983. The phoretic association of *Urnatella gracilis* (Entoprocta: Urnatellidae) and *Nanocladius downesi* (Diptera: Chironomidae) on *Corydalus cornutus* (Megaloptera: Corydalidae). Freshwater Invertebr. Biol. 2:186-191.

Weise, J. G. 1961. The ecology of *Urnatella gracilis* Leidy: phylum Endoprocta. Limnol. Oceanogr. 6:228-230.

EXERCISE 21

Phylum Brachiopoda

Brachiopoda (brak-e-OP-o-da; G., *brachio*, arm + G., *pod*, foot) is a small group (ca. 400 species) of lophophorates found in marine intertidal areas to depths up to 7600 m. The animals are <10 cm long and are enclosed in two shells. At first glance brachiopods resemble bivalve mollusks. However, they are different in two major ways: (1) Brachiopods possess a lophophore and (2) The brachiopod shell is very different from that of bivalves. Shells of bivalves are arranged laterally, consisting of right and left valves, the plane of symmetry passing between the valves. Valve orientation in brachiopods is such that the two valves assume a dorsoventral position and are normally bilaterally symmetrical; the plane of symmetry passes through the valves (Fig. 21.1). Furthermore, brachiopod valves are often dissimilar in size and shape (Fig. 21.2). In most brachiopods the larger **pedicle valve** is ventral and the smaller **brachial valve** dorsal; in the epifaunal inarticulate brachiopods, this size relationship is reversed. Another feature of most modern brachiopod shells is the posterior extension of the ventral valve to form a **beak** that bears a **foramen** for exit of an attachment **stalk** or **pedicle**. This makes the shell resemble a Roman oil lamp and it is for this reason that brachiopods are called lampshells (Fig. 21.2). The two valves may articulate by teeth and sockets at the posterior margin; if teeth and sockets are absent, a complex musculature permits sliding and rotation of the valves.

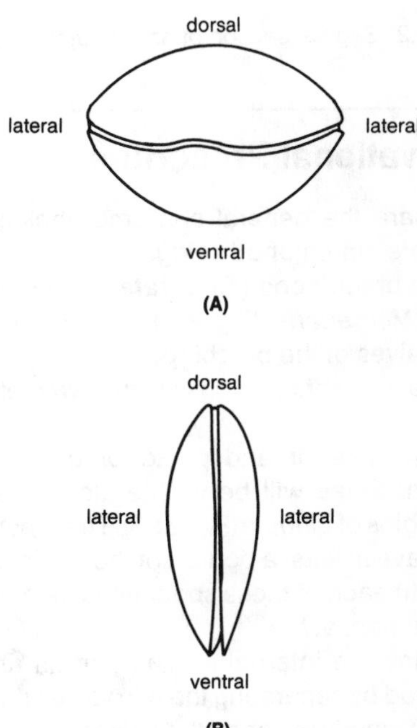

Figure 21.1. Comparison of the shell morpology of **(A)** a brachiopod and **(B)** a bivalve mollusk. Both are shown in an anterior view.

Classification

The phylum is divided into two classes: Inarticulata and Articulata. The latter possesses articulating teeth and sockets at the posterior margin of the shell, whereas the former does not.

1. Inarticulata. Brachiopods with a large pedicle that serves as an anchor in sediments or attaches to hard substrates, or with their ventral valve cemented firmly to the substrate; complete digestive tract; most shells of $Ca_3(PO_4)_2$; examples include *Crania, Discinisca, Glottidia*, and *Lingula*, a genus in existence since the **Silurian Period**.

2. Articulata. Brachiopods possessing a foramen in the beak of the ventral valve through which the pedicle passes, attaching the animal to hard substrates; digestive tract incomplete, ending as a blind intestine; shell of $CaCO_3$ (calcite); examples include *Lagueus, Terebratalia*, and *Terebratulina*.

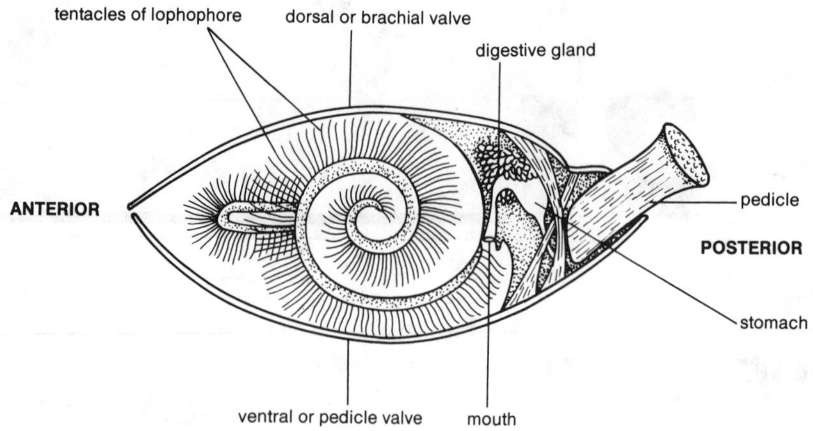

Figure 21.2. Sagittal section of an articulate brachiopod, showing the internal anatomy.

Observational Procedure

1. Compare the general shell morphology of an inarticulate brachiopod (*Lingula* or *Glottidia*), an articulate brachiopod (*Terebratalia*), and a bivalve mollusk (*Mercenaria*) (Fig. 21.1). Identify dorsal and ventral valves of the brachiopod.
2. Contrast that orientation with the symmetry in the bivalve.
3. Identify anterior and posterior ends of each specimen. **Setae** will be visible along the lateral shell margins of *Lingula* (Fig. 21.3). The brachiopods should have at least a portion of the pedicle visible. How might each of these specimens be oriented in its natural habitat?
4. Examine the internal anatomy of an articulate brachiopod by separating the two valves (Fig. 21.4). To do this, gently pry apart the valves at the anterior end until the hinge breaks.During this procedure, one or more sets of muscles may be pulled off their attachments. Note where they were attached as you pull the valves apart. Sometimes it may be necessary to cut the adductor and/or diductor muscles using a fine dissection blade in order to separate the valves completely. Take care not to destroy the lophophore and other internal organs.
5. Observe the looped lateral arms and central coiled portion of the lophophore.
6. The digestive gland should be visible at the posterior end of the shell (Fig. 21.2). It is composed of many spherical structures called **acini** (L., *acinus*, a berry) and resemble bunches of grapes.
7. The gonads are often best seen in the posterior parts of the branches of the coelom that extend into the mantle as structures called mantle canals.

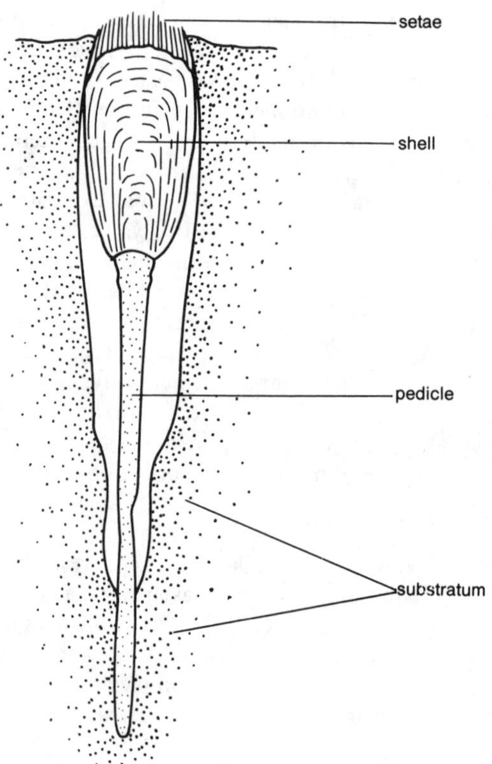

Figure 21.3. In situ view of the inarticulate brachiopod, *Lingula*.

Your instructor will provide additional directions if live brachiopods are to be examined (e.g., a study of feeding currents).

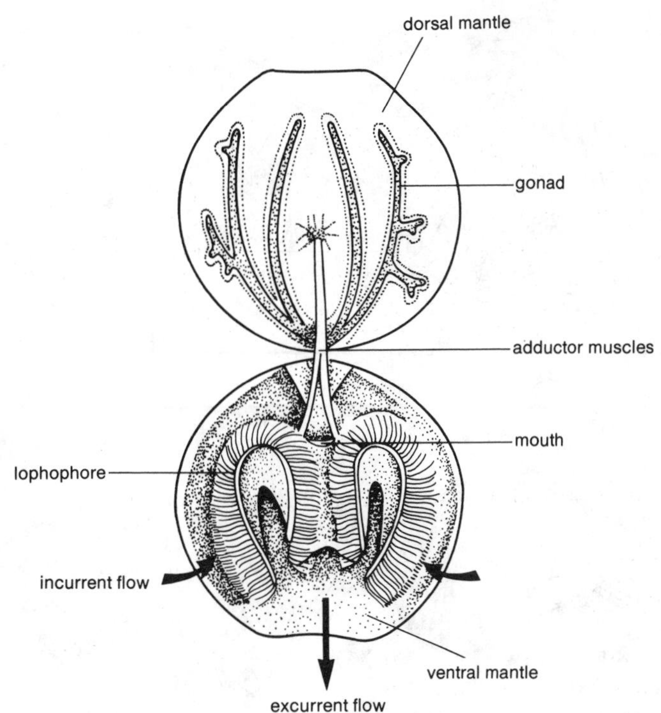

Figure 21.4. Interior of the articulate brachiopod, *Lagueus californianus*. Arrows indicate the course of water currents.

Fossil Forms

In comparative terms, fossil brachiopods (ca. 40,000 species) far outnumber present-day forms. They are found in **Cambrian** deposits, but achieved their greatest diversity during the **Ordovician** and **Devonian Periods** and their greatest specialization in the **Permian Period** (Fig. 21.5). Some fossil specimens were much larger than the current species, measuring more than 35 cm wide. As in other animal groups that possess a rich fossil history, a complex nomenclature has been developed by paleontologists.

Observe specimens of fossil brachiopods comparing their general shell morphology with specimens of living species (Fig. 21.6). Note the dorsal and ventral valves, beak with foramen, presence of any external ornamentation, and/or growth lines. Occasionally, specimens are found with the shell cavity exposed. Such specimens should be examined for the presence of hinge dentition, lophophore skeletal supports, and scars indicating muscle attachments.

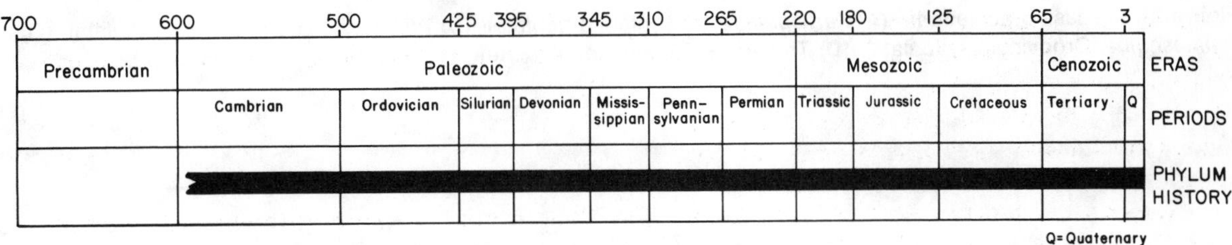

Figure 21.5. Geologic history of the phylum Brachiopoda.

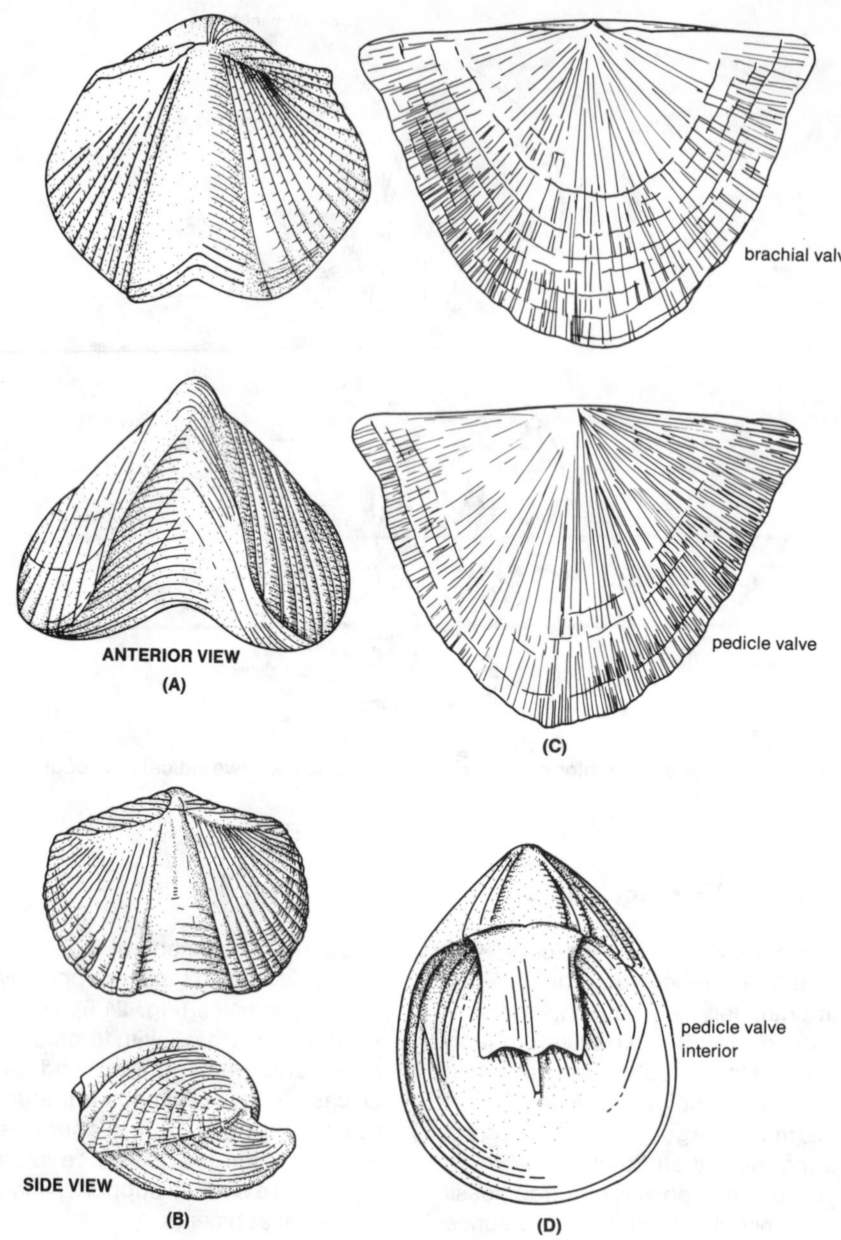

Figure 21.6. Fossil brachiopods. **(A)** *Paraspirifera* (Devonian, articulate). **(B)** *Platyrachella* (Devonian, articulate). **(C)** *Rafinesquina* (Ordovician, articulate). **(D)** *Trimerella* (Silurian, inarticulate).

Supplemental Reading

Broadhead, T. W. 1981. Lophophorates, notes for a short course. Studies in Geology 5. University of Tennessee, Department of Geological Sciences, Knoxville, TN.

Chuang, S. H. 1956. The ciliary feeding mechanism of *Lingula unguis* (L.) (Brachiopoda). Proc. Zool. Soc. Lond. 127:167-189.

Cowen, R. 1971. The food of articulate brachiopods—a discussion. J. Paleontol. 45:137-139.

Craig, G. Y. 1952. A comparative study of the ecology and paleoecology of *Lingula*. Trans. Edinb. Geol. Soc.15:110-120.

Hammen, C. S. 1977. Brachiopod metabolism and enzymes. Am. Zool. 17:141-147.

Jones, G. F., and J. L. Barnard. 1963. The distribution and abundance of the inarticulate brachiopod *Glottida albida* (Hinds) on the mainland shelf of southern California. Pac. Nat. 4 (2):27-52.

McCammon, H. M. 1969. The food of articulate brachiopods. J. Paleontol. 43:976-985.

McCammon, H. M., and W. A. Reynolds. 1976. Experimental evidence for direct nutrient assimilation by the lophophore of articulate brachiopods. Mar. Biol. 34:41-51.

Paine, R. T. 1963. Ecology of the brachiopod *Glottidia pyramidata*. Ecol. Monogr. 33:255-280.

Richardson, J. R. 1986. Brachiopods. Sci. Am. 255 (3):100-106.

Rudwick, M. J. 1970. Living and fossil brachiopods. Hutchinson University Library, London.

Smith, S. A., C. W. Thayer, and C. E. Brett. 1985. Predation in the Paleozoic: gastropod-like drillholes in Devonian brachiopods. Science 230:1033-1035.

Thayer, C. W. 1985. Brachiopods versus mussels: competition, predation, and palatability. Science 228:1527-1528.

Thayer, C., and M. Steele-Petrovic. 1975. Burrowing of the lingulid brachiopod *Glottidia pyramidata*: its ecological and paleoecologic significance. Lethaia 8:209-221.

Wright, A. D. 1979. Brachiopods radiation. *In*: M. R. House (ed.). The origin of major invertebrate groups. Academic Press, New York, pp. 235-252.

DEUTEROSTOME PHYLA

All deuterostomes are enterocoelomates with radial, indeterminate cleavage. If we ignore for the moment the putative relationship of certain lophophorates (see Nielsen 1977, 1985), only four phyla make up the deuterostome line: Echinodermata, Hemichordata, Chaetognatha, and Chordata. These four phyla are very heterogeneous in structure, function, and development. Echinodermata is a large phylum exhibiting the greatest diversity among the deuterostomes, including forms familiar to people visiting the seashore (e.g., sea stars, brittle stars, sand dollars, sea urchins, and sea cucumbers). Hemichordata and Chaetognatha are small phyla of marine worms, together containing fewer than 200 species. Most members of the phylum Chordata are vertebrates and possess a **vertebral column.**

Nielsen, C. 1977. The relationship of Entoprocta, Ectoprocta and Phoronida. Am. Zool. 17:149-150.

Nielsen, C. 1985. Animal phylogeny in the light of the trochaea theory. Biol. J. Linn. Soc. 25:243-299.

EXERCISE 22

Phylum Echinodermata

Members of the phylum Echinodermata (e-KI-no-DER-ma-ta; G., *echinos*, a hedgehog + G., *derma*, skin) are among the most familiar invertebrates observed by people visiting the seashore. Echinoderms get their name from the fact that most have **spines** and **tubercles** which project from the surface of the animal. The endoskeleton is composed of calcareous **ossicles** or **plates** that vary in shape and their connection with each other. The skeleton is covered with a thin ciliated epidermis or is embedded in a rather thick, leathery body wall.

There are more than 7000 living species of echinoderms, which include sea lilies, sea stars (or starfishes), brittle stars, sand dollars, sea urchins, and sea cucumbers. Approximately 13,000 fossil species are known. Echinoderms are distributed widely, being especially abundant in the Indo-Pacific areas. This strictly marine phylum lacks parasitic and colonial members. None are segmented.

A major characteristic of these largely bottom dwellers is the basic **pentamerous**, radial symmetry of the adults. This means that an echinoderm's body is arranged typically into five (or multiples thereof) similar and equal parts that radiate from a central axis. The radiating areas, which bear the **tube feet** (or **podia**) of the unique **water-vascular** (or **ambulacral**) system, are called **arms** (or **ambulacra, rays,** or **radii**). Between two ambulacra is an **interambulacral** (or **interradial**) area. The oral surface contains the mouth and the aboral surface is away from the mouth.

The water-vascular system is the most distinguishing feature of echinoderms. It is basically an organ system of canals or tubes derived from the spacious, ciliated **enterocoelom**. The water-vascular system serves in locomotion, food gathering, respiration, chemosensation, and excretion. It will be studied in more detail in your observations on the representative groups.

Unlike most adult echinoderms, the microscopic larvae are bilateral in symmetry. This is why the echinoderms are placed in the Bilateria and not in the Radiata. The echinoderms are the major invertebrate phylum of **enterocoelous deuterostomes**.

Classification

The phylum Echinodermata is divided currently into four subphyla: Homalozoa, Crinozoa, Asterozoa, Echinozoa.

I. Subphylum Homalozoa. Carpoids, all extinct. Four classes of Paleozoic echinoderms lacking radial symmetry or the pentamerous condition.

II. Subphylum Crinozoa. Radially symmetrical forms with a cup-shaped body (theca) to which are attached five, ten, or more long, hollow arms used to gather food. Contains both stalked and unstalked forms. The oral surface is directed upward. Several extinct classes and one class, Crinoidea, with living members.

- **A. Class Crinoidea**. Sea lilies and feather stars (or comatulids). The oral surface, bearing the mouth and anus, faces upward. Crown or calyx contains all the visceral organs. Tube feet lack suckers. No madreporite, spines, or pedicellariae. Examples are *Florometra* and *Cenocrinus* (sea lilies); *Neometra* and *Antedon* (feather stars).

III. Subphylum Asterozoa. Radially symmetrical, unattached, flattened, star-shaped echinoderms. Mouth is in a central position and on the underside of the body. Some workers designate the following three classes as subclasses under the class Stelleroidea.

A. **Class Somasteroidea**. Paleozoic sea stars; all extinct.

B. **Class Asteroidea**. Sea stars. Arms hollow, not demarcated sharply from the central disc, aboral madreporite, pedicellariae in many species, and hepatic ceca extend into the arms. Open ambulacral grooves, containing tube feet with or without suckers, on the oral side of the arms. Examples are *Asterias*, *Pisaster*, *Astropecten* , *Luidia*, and *Solaster*.

C. **Class Ophiuroidea**. Brittle stars (or serpent stars) and basket stars. Elongated and movable, solid arms that are demarcated sharply from the central disc. Oral madreporite, no pedicellariae, arms lack hepatic ceca, and the ambulacral grooves are closed. Examples are *Ophioderma*, *Ophiothrix* (brittle stars), and *Gorgonocephalus* (basket star).

IV. **Subphylum Echinozoa**. Discoid, globoid, or cylindrical echinoderms lacking arms. Burrowing and surface-dwelling forms. Several extinct classes. The two classes with living members are described below.

A. **Class Echinoidea**. Sea urchins, heart urchins, and sand dollars. Discoid and globoid body shapes, no arms, madreporite on the aboral surface, and closed ambulacral grooves. Most have a masticatory apparatus called Aristotle's lantern. Skeleton (or test) is rigid with movable long and short spines. Examples are *Strongylocentrotus*, *Arbacia* (sea urchin), *Echinocardium*, *Moira* (heart urchins), and *Mellita* and *Clypeaster* (sand dollars).

B. **Class Holothuroidea**. Sea cucumbers. Body cylindrical and elongated on an oral-aboral axis. No arms. Madreporite internal, modified tube feet form oral tentacles, and endoskeleton of microscopic, separate ossicles embedded in a usually leathery body wall. No spines and no pedicellariae. Ambulacra represented by closed canals. Examples are *Cucumaria*, *Thyone*, and *Leptosynapta*.

A. Class Asteroidea

The Asteroidea (AS-ter-OID-e-a; G., *aster*, star + G., *eidos*, form) is perhaps the best known group of echinoderms, containing the sea stars or starfishes. The 1600 species of sea stars are distributed widely throughout coastal waters of the world. The northwest Pacific waters, especially from Puget Sound to the Aleutians, have the largest concentration of shallow-water asteroid species. Most sea stars are more active at night than during the day.

Some are voracious carnivores feeding on oysters, barnacles, clams, and other marine animals; others are detritus feeders, mud swallowers, or opportunists. The digestive system is well developed. The ciliated enterocoelom is filled with fluid containing free-moving phagocytic amoebocytes called **coelomocytes**. These cells, as well as the tube feet of the water-vascular system and **dermal branchiae (papulae)**, aid in the removal of the nitrogenous wastes (NH_4^+). Gaseous exchange also occurs through the tube feet and dermal branchiae. The skeleton is made up of calcareous **ossicles** bound together by connective tissue and muscle fibers. Outer circular and inner longitudinal fibers lie beneath the dermis. The tube feet contain well-developed longitudinal muscles.

The common North Atlantic sea star, *Asterias forbesi*, will be studied as our representative asteroid. Other genera, such as *Pisaster*, may be available for observation, and the following account will apply to either sea star. *Asterias forbesi* occurs commonly from Cape Cod southward, being found rarely to Maine. In life the five rays are stout and cylindrical and the body coloration is variable (greenish black seen commonly); the madreporite is bright orange. *A. forbesi* moves along the substrate from the interidal zone to about 30 fathoms.

Observational Procedure

External Anatomy. Obtain both preserved and dried specimens of the sea star. For the dried specimen, one that has been opened and cleaned is preferred. Place the preserved sea star in a dissection pan and cover with water. The radial symmetry is very apparent (Fig. 22.1). The five rays radiate from the **central disc** as in brittle stars (Fig. 22.12).

Although adult sea stars and other echinoderms are radial, they retain vestiges of bilateral symmetry. To appreciate this, a brief consideration will be made of Carpenter's method of designating the radii and interradii. The system can be applied to other adult echinoderms; however, we will use the sea star for instructive purposes.

With the dried (or preserved) specimen, locate the calcareous, button-shaped **madreporite** on the aboral surface (Fig. 22.1). Draw an imaginary line from the madreporite to the opposite ray. This ray is designated A and the remaining rays (B, C, D, and E) are lettered counterclockwise from ray A when viewing the sea star from the aboral surface. The **interradii** are designated by the letters of the rays that enclose them (i.e., AB, BC, CD, DE, and EA). If the animal were cut beginning at the madreporite and extending through the center of ray A, the two halves would be mirror images of one another. Therefore, the animal would conform to bilateral symmetry. If the animal were cut at right angles to the plane of symmetry, the body consists of three rays (the **trivium**), and the two opposite rays, C and D, comprise the **bivium**. The madreporite lies on the center of the bivium and in the CD interradius.

Take the preserved sea star and place its oral surface (i.e., the surface with the centrally located mouth) next to the bottom of the dissecting pan. Observe the aboral surface (Fig. 22.2)

Aboral Surface. The surface is spinose with short, calcareous spines covering the area throughout. There is a thin, ciliated epidermis covering the sea star's body, including the spines. Observe the structural details of the outer surface with a dissecting microscope. Around the bases of the spines are fingerlike, fleshy, grayish sacs. These are the **dermal branchiae** (or **dermal papulae**). They are ciliated, hollow projections continuous with the body cavity, serving as respiratory and excretory organs (Fig. 22.3A). Around the spines and throughout the sea star's surface are the minute, pincerlike **pedicellariae**, which keep the body surface free of debris and small organisms. Pedicellariae are modified spines, covered with epidermis, and unique to the echinoderms. The pedicellariae appear as tiny, whitish specks in your specimen, especially on the dried one. Pedicellariae may be stalked or unstalked. Gently scrape the aboral surface of the

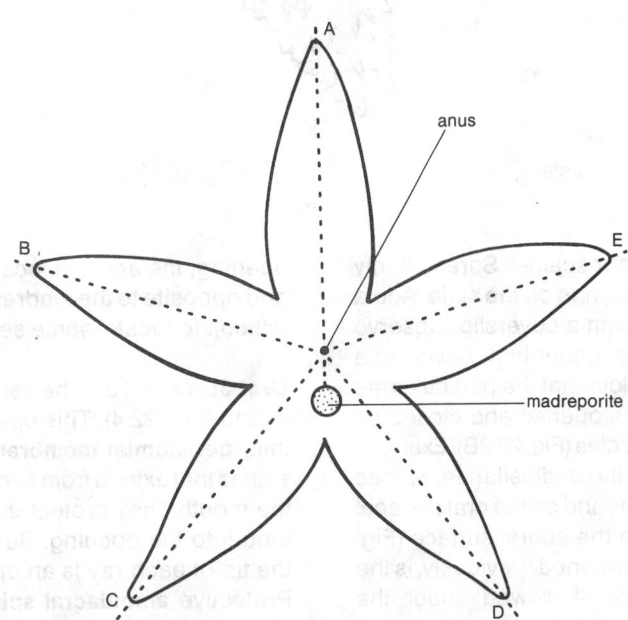

Figure 22.1. Aboral view of a sea star showing Carpenter's system of designating the radii and interradii. See the text for a discussion of the system. (From Meglitsch 1972.)

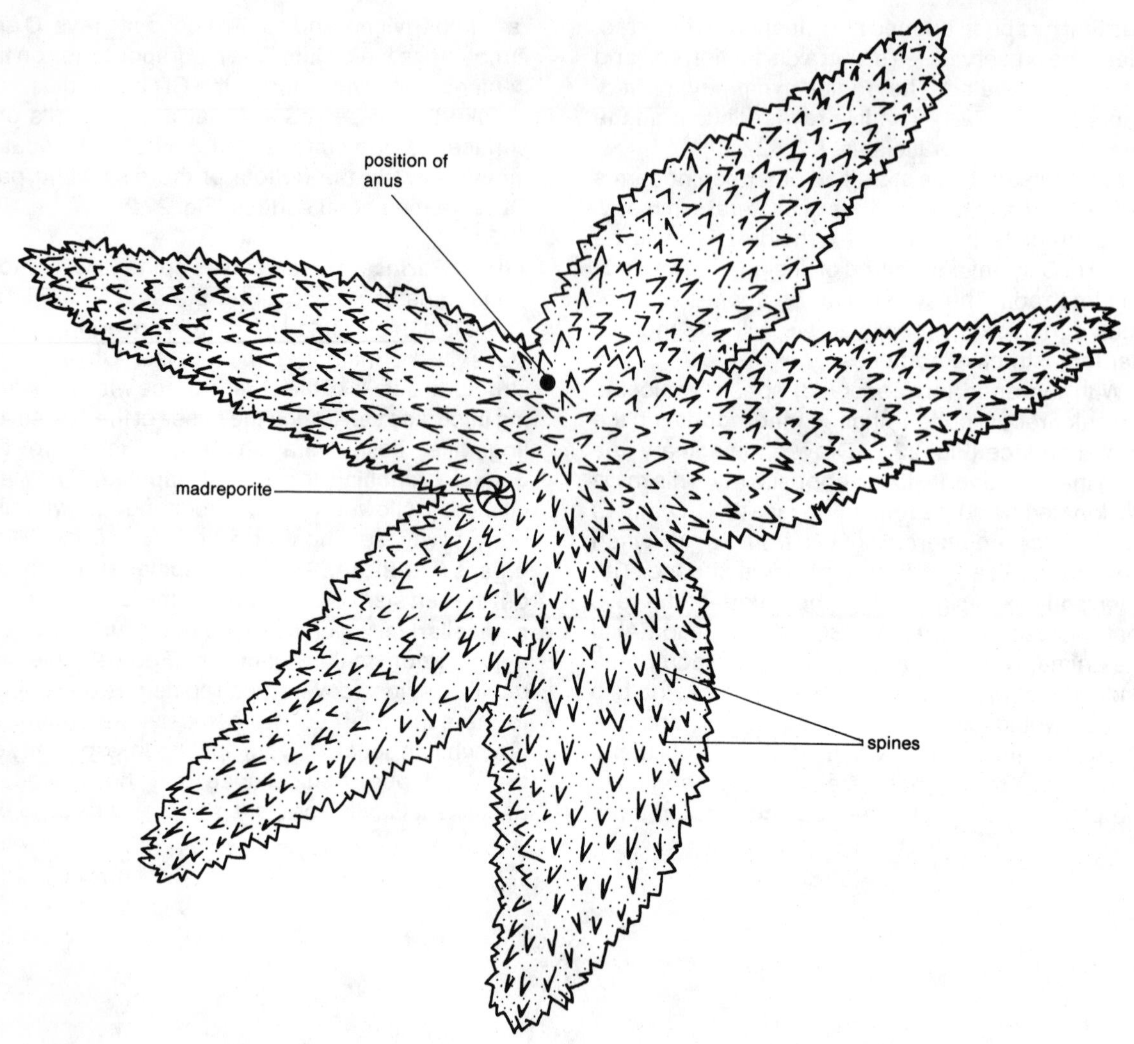

Figure 22.2. Aboral view of a sea star.

preserved sea star with your scalpel. Spread thinly a small amount of the scrapings on the slide. Add a drop of water and cover with a coverslip. Observe under low power and then under high power of a compound microscope. Note that the pedicellariae are pincerlike, and each is opened and closed by adductor and abductor muscles (Fig. 22.3B). Examine a prepared slide showing the pedicellariae. Spines and pedicellariae are also found on the oral surface.

Two openings occur on the aboral surface (Fig. 22.2). The madreporite, mentioned previously, is the hard, button-shaped plate. If viewed under the dissecting microscope, the madreporite's grooves can be seen. The madreporite is part of the water-vascular system to be studied later. The second opening, the anus, is located in the BC interradius and opposite to the madreporite. It is very small and difficult to locate; some sea stars lack the anus.

Oral Surface. Turn the sea star over and locate the mouth (Fig. 22.4). This opening is surrounded by a thin, **peristomial membrane**. Note the paired **oral spines** that extend from each interradius and around the mouth; they protect the mouth and help move food into the opening. Running from the mouth to the tip of each ray is an open, **ambulacral groove**. Protective **ambulacral spines** cover each groove. These and the oral spines are movable. In each ambulacral groove are the fleshy **tube feet** (or **podia**), elements of the water-vascular system.

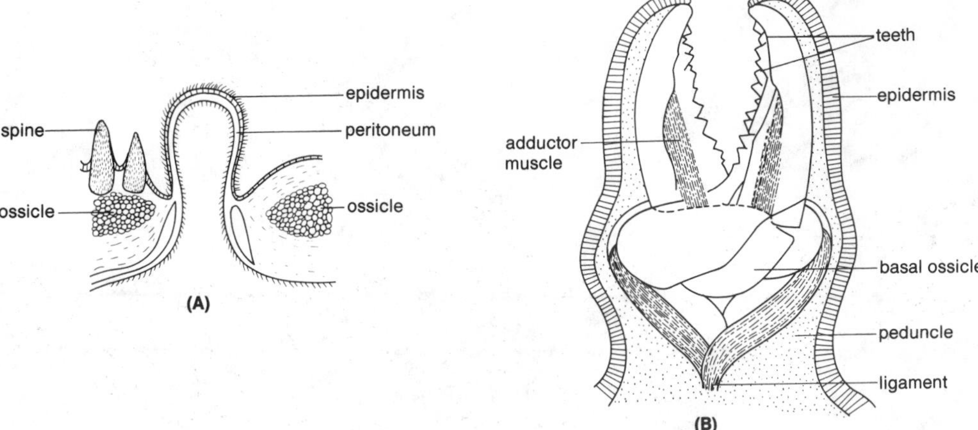

Figure 22.3. **(A)** Section through the body wall of a sea star showing a dermal branchia. (From Cuenot, *Traité de Zoologie*, Vol. XI. Grassé, Ed.) **(B)** Pedicellaria of a sea star.

These highly contractile tubes terminate in sucker discs. Not all sea stars have suckers on the tube feet. The combined efforts of many tube feet provide attachment to the substrate and are the means by which locomotion is accomplished. The tube feet are also used in feeding. At the distal end and on the oral surface of each arm is an **eyespot** formed by numerous pigment-cup ocelli. These are sensitive to light. Near the eyespot is a small modified tube foot (**tentacle**), which lacks a sucker.

Internal Anatomy. To study the internal anatomy of the sea star, you will remove the specimen's aboral covering. Place the preserved sea star with the oral surface down. With scissors, cut off the tips of each arm. Insert the scissors inside the tip of one arm and cut along the edges (i.e., dorsolateral margins) of each arm. Do not cut too deeply. Make sure that the body wall at each interradius is cut fully. Cut around the madreporite to leave it attached to the lower body. Carefully lift the aboral surface off the oral. As you remove the aboral surface, detach with your probe or scalpel all fleshy material adhering by mesenteries to the inner surface of the aboral covering. If the dissection is done properly, you should have all fleshy materials intact and in the lower half of the sea star's body. The exposed body cavity is the spacious enterocoelom.

Digestive System. Refer to Figs. 22.5 and 22.6 as you read the remainder of this paragraph. Locate the short **esophagus**, which leads internally from the mouth. The esophagus opens into the large, muscular **cardiac** portion of the stomach, which everts through the mouth when the animal feeds. After digestive enzymes have acted on the food, the stomach and partially digested contents are retracted into the body by **retractor muscles** attached to the ambulacral ridge (Fig. 22.7). Aborally, the cardiac stomach leads into the smaller, thin-walled **pyloric** stomach. This pentagonal stomach receives ducts from five pairs of **pyloric ceca** (or **digestive glands** or **hepatic ceca**). Each arm contains two of these digestive glands, which produce enzymes for digestion. The pyloric stomach leads into a very short, inconspicuous **intestine**. A branched sac, the **rectal ceca** (or **intestinal ceca**), of unknown function, lies on top of the pyloric stomach, but has a connection with the intestine. Beyond the opening of the rectal ceca, the short **rectum** leads to the **anus**.

Skeletal System. To study the skeleton of the sea star's arm, remove the pyloric ceca in one ray. Note the distinct **ambulacral ridge** composed of middle and lateral rows of ambulacral ossicles (Figs. 22.7 and 22.8). Examine the ridge on your dried specimen. The large, ovoid ossicle at the end of the ridge in the central disc area is the **odontophore**. Locate the two **retractor muscles**, which are whitish, threadlike strands attached to the ambulacral ridge and to the cardiac stomach. Note that the lateral ambulacral ossicles form the sides of the ambulacral groove on the oral surface, and the fleshy tube feet of the water-vascular system extend through these ossicles. The lower ends of the lateral ossicles rest on a series of **adambulacral ossicles** (Fig. 22.8). The

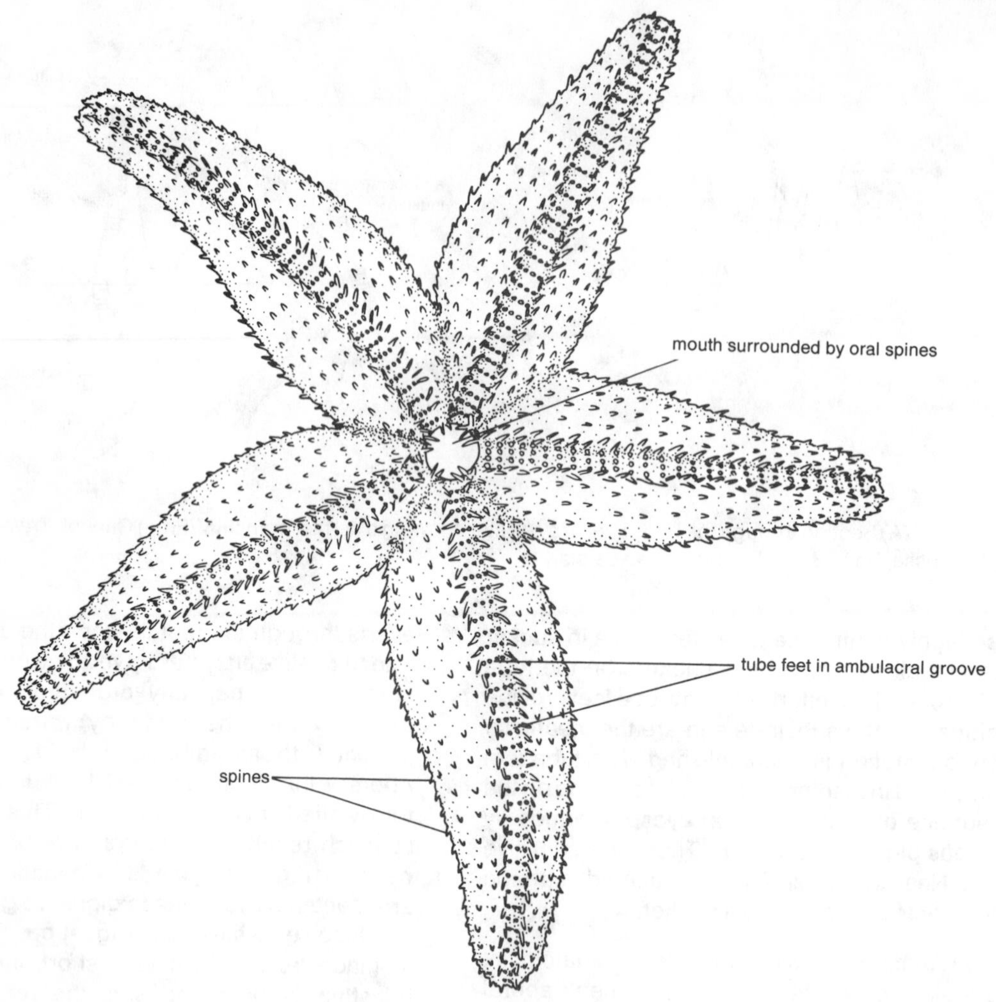

Figure 22.4. Oral view of a sea star.

latter bear the long, slender, movable **ambulacral spines** observed along the groove on the oral surface. The **inferomarginal ossicles** compose the bottom, lateral row of strutlike ossicles. Lateral to these, and forming the upper lateral row along the length of the arm, are the **superomarginal ossicles**. Make a cross section of a piece of the arm and observe it under the dissecting microscope to see more clearly the above-mentioned ossicles and their relationship to each other. If you have a cross section of an arm from a dried specimen, or prepared slide, examine these under a dissecting microscope. Study the inner face of the aboral covering and note its ossicles.

Reproductive System. *Asterias* is dioecious. There are two **gonads** located at the base of each arm (Fig. 22.7). These resemble a cluster of grapes. During the reproductive season the gonads nearly fill each arm. At other times, they are small and occupy a small space. Two gonads are connected by a single gonoduct, which terminates as a single gonopore or gonopore cluster located between the base of the arms in each interradius. Examine a prepared slide of sea star development and locate the large, unfertilized egg, showing the nucleus and nucleolus (Fig. 22.9).

Fertilization occurs after the gametes are shed in the water. The zygote undergoes cleavage and

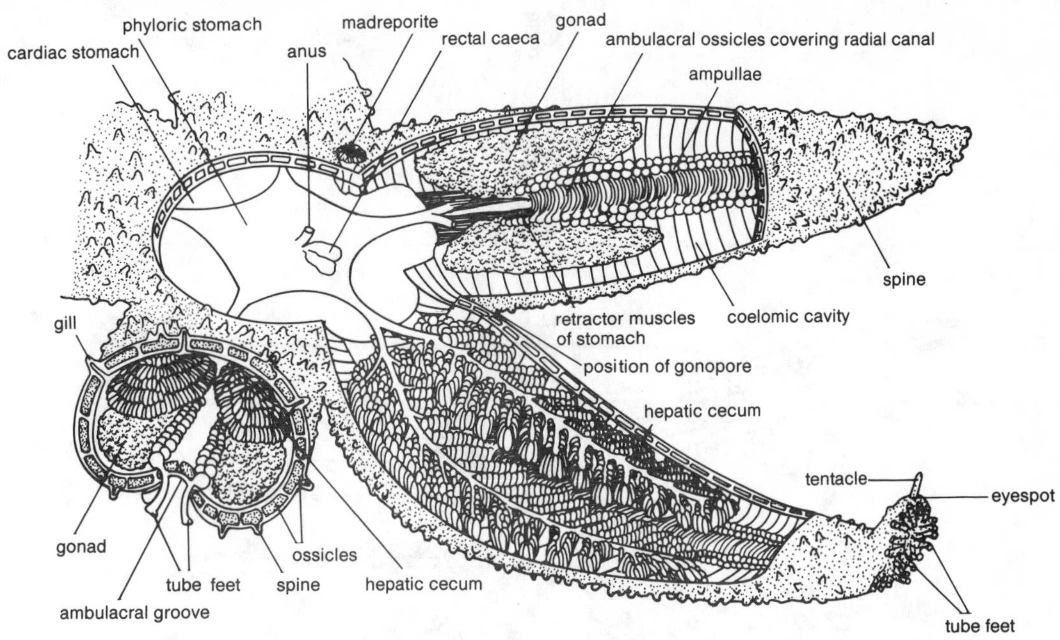

Figure 22.5. Internal aboral view of a sea star, showing the digestive system.

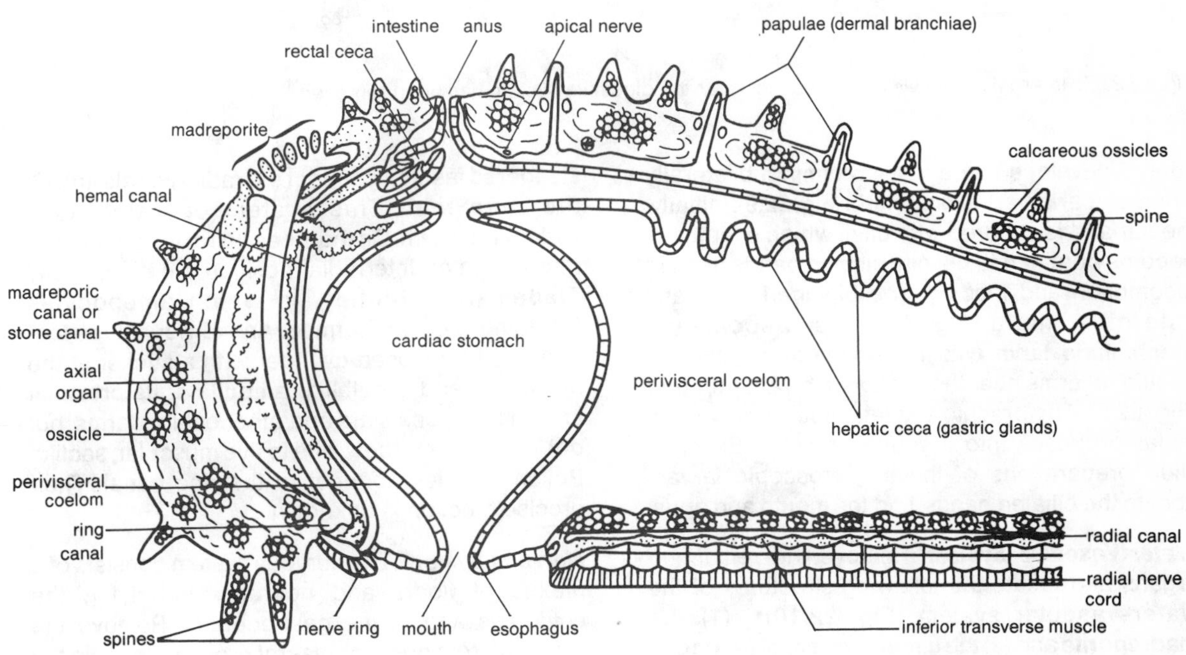

Figure 22.6. Diagrammatic vertical section through a ray and the central disc of *Asterias*. (From Chadwick.)

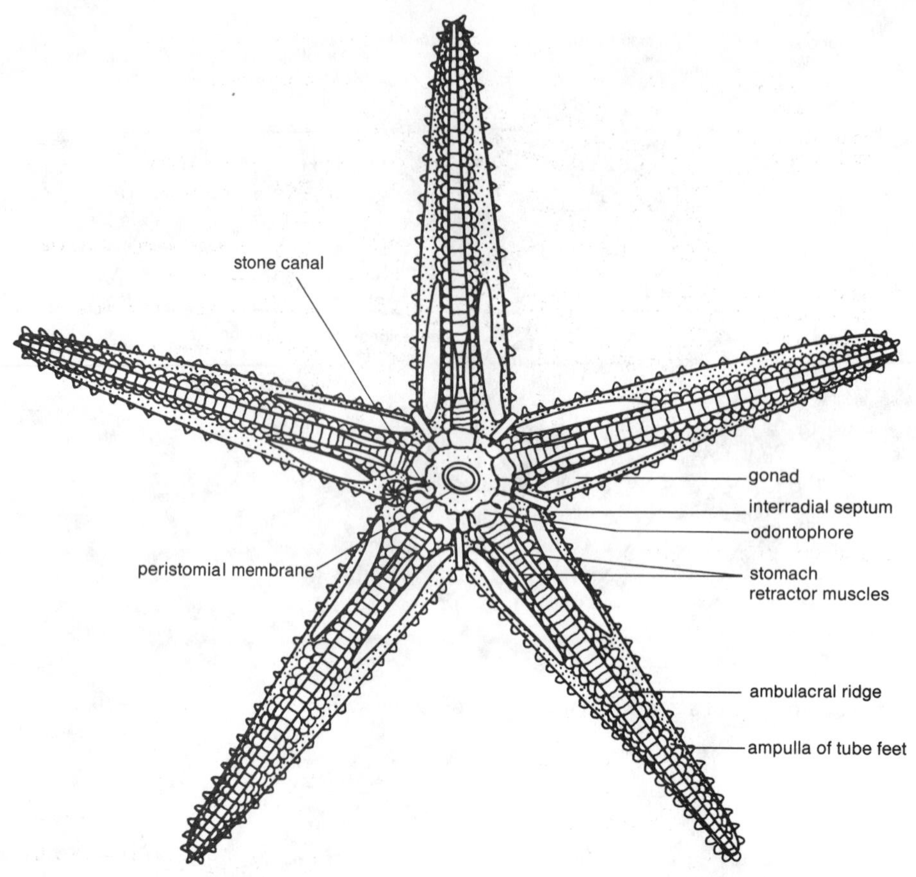

Figure 22.7. Internal aboral view of a sea star after the digestive system has been removed.

soon it develops into a free-swimming, bilaterally symmetrical larva, the **bipinnaria** (Fig. 22.9). Initially, the surface is covered with cilia, which eventually become confined to definite ciliated bands used in locomotion and feeding. The bipinnaria in many asteroids, including *Asterias*, becomes a **brachiolaria larva** with the appearance of three additional arms near the sucker at the anterior end (Fig. 22.9). The brachiolaria later settles down and metamorphoses into a young sea star. Examine slide preparations of these microscopic larvae. Locate the ciliated bands; find the mouth and anus.

Water-Vascular System. Remove the remaining organs and locate the following structures of the water-vascular system (Fig. 22.10): (1) the **madreporite** and (2) the **circumoral canal** (or **ring** or **water canal**) are connected by (3) the calcified **stone canal**; (4) five **radial canals**, one in each arm on the oral side, lead off the **circumoral canal**; (5) short **transverse** (or **lateral**) **canals** arranged in a staggered fashion connect the radial canals and (6) the **tube feet**. This system operates as a hydraulic system. On the inner wall of the circumoral ring and in an interradial position are nine, fleshy **Tiedemann's bodies**, in which amebocytes (coelomocytes) are formed. An individual tube foot consists of a closed cylinder with a sucker at the distal end and a bulblike **ampulla** at its proximal end. In many sea stars and other echinoderms, but not in *Asterias forbesi*, one to five muscular, saclike **Polian vesicles** connect to the ring canal. Their precise function is unknown.

Nervous System. The nervous system consists of a plexus of fibers and neurons underlying the epidermis and in the gastrodermis. Remove the tube feet from one ambulacral groove. Observe the minute yellowish or brown cord that runs along the bottom of the groove along the entire length of the arm. This is the **radial nerve cord** (Fig. 22.8). The above constitutes the **ectoneural** portion of the

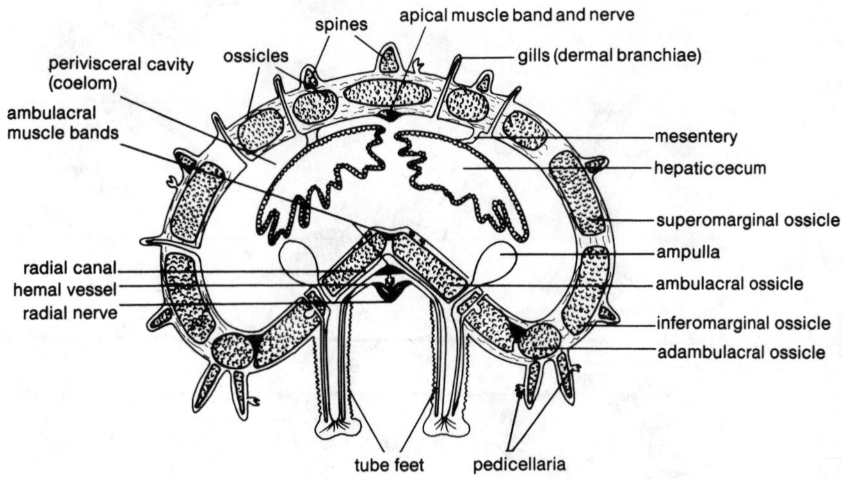

Figure 22.8. Cross section of a sea star's arm.

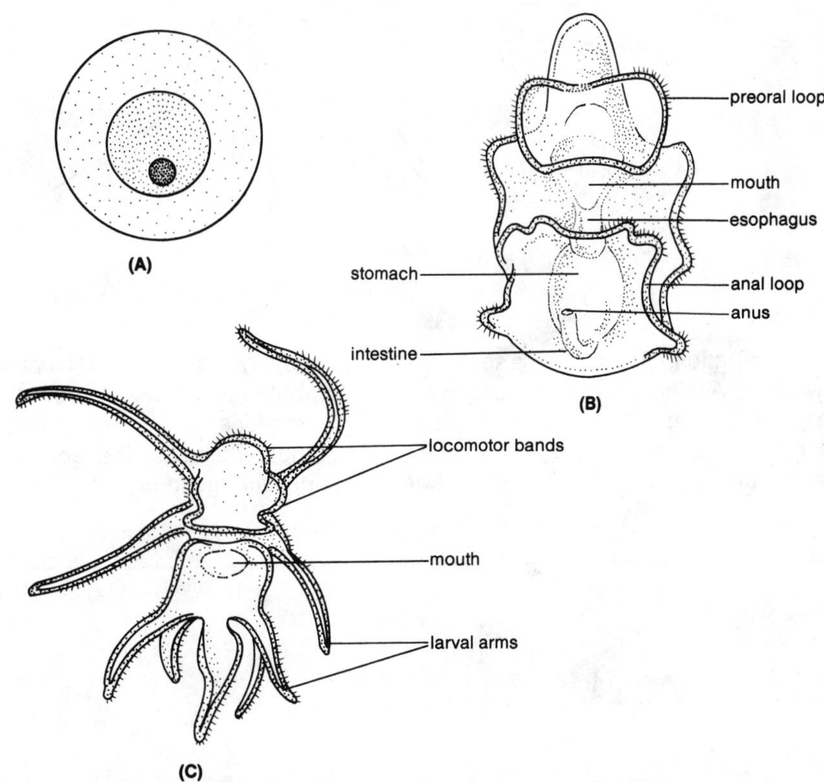

Figure 22.9. (**A**) Unfertilized egg, (**B**) bipinnaria larva, and (**C**) brachiolaria larva of a sea star.

nervous system. The **endoneural** portion may be seen in the oral walls of the cardiac and pyloric portions of the stomach and in the median folds of the hepatic ceca. An apical portion of the nervous system occurs medially on the inner surface of the aboral body wall. This single cord is fused with the other four cords in the center of the disc. The apical portion is difficult to find.

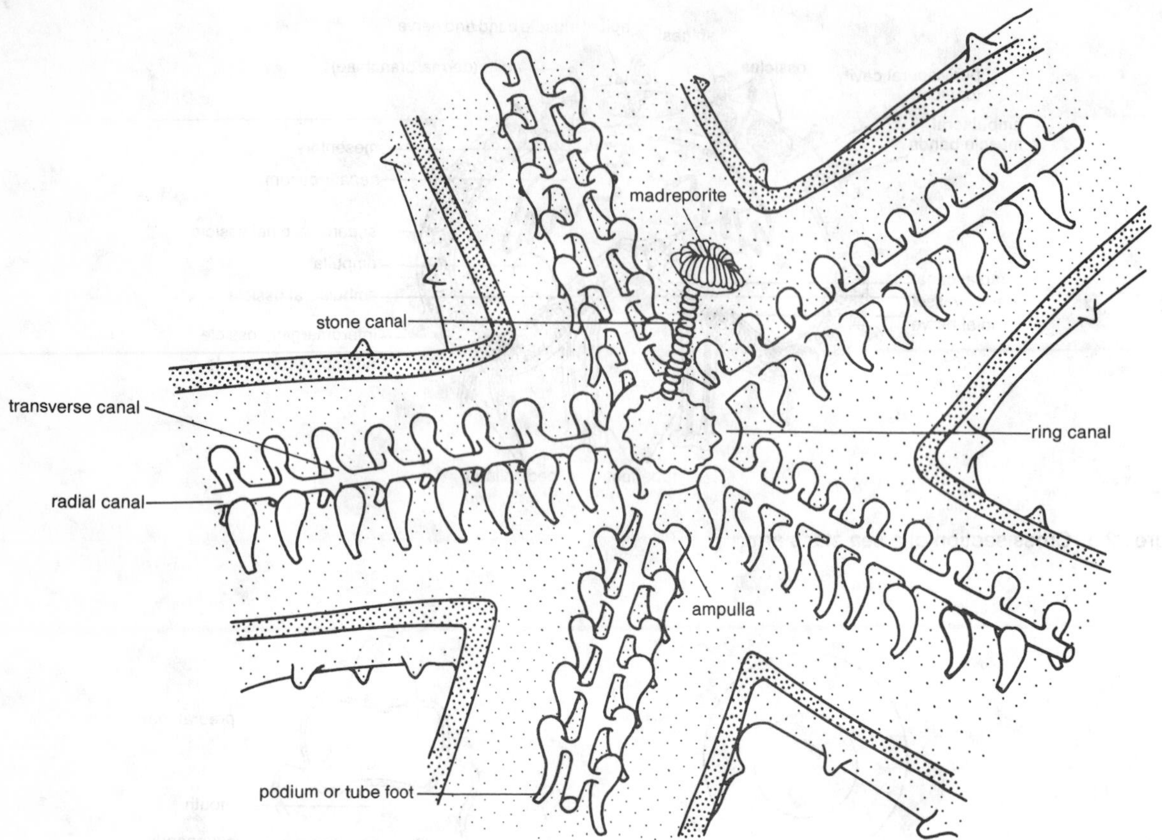

Figure 22.10. Water-vascular system of a sea star.

Microscopic Examination. Obtain a cross section of a sea star arm and with the aid of Fig. 22.8, locate the following structures: epidermis, dermis, peritoneum, the various ossicles, enterocoelom, dermal branchiae, pyloric ceca, tube feet, transverse canal, radial canal, radial nerve, and ambulacral groove. If your slide was made from a young sea star, you will not see the gonads or spines. Correlate the observed structures in the slide with those on the dissected specimen.

B. Class Ophiuroidea

Ophiuroids (O-fee-u-roids; G., *ophis*, serpent + G., *oura*, tail + G., *eidos*, form) are called brittle stars because of their tendency to fragment when disturbed, or serpent stars because their long, slender arms wriggle in a snakelike motion. Included in this class are the largest echinoderms, the basket stars, which have highly branched arms whose ends tend to coil. *Gorgonocephalus* is a basket star that is often available for classroom observation (Fig. 22.11).

Because of the basic star-shaped body and pentamerous condition, ophiuroids resemble the asteroids. As you study these organisms you will see distinct differences between the two groups. Many zoologists consider ophiuroids to be the most successful of echinoderms

Figure 22.11. *Gorgonocephalus*, a basket star.

because they are the largest group, with 2000 living species. Ophiuroids often occur in dense aggregations on the seafloor. Certainly, the brittle stars are the most mobile members of the phylum. There are burrowing ophiuroids and a few can swim. The only commensal echinoderms are ophiuroids. For our laboratory study, *Ophioderma* or some similar brittle star will be used.

Observational Procedure

External Anatomy. Obtain both dried and preserved individuals. The central disc is composed of calcareous plates or **shields** arranged in a distinct fashion (Fig. 22.12). Study the pattern in your specimen. The small disc is flattened and rather smooth, lacking the conspicuous and numerous spines observed in the sea star. The disc of some species is spiny, and there is variation in the pattern of the shields.

Observe the oral surface containing the mouth (Fig. 22.13). Use the dissecting microscope to better observe the plates and other external features. The plates on the oral surface of the central disc consist of the large, interradial **oral shields** (or **buccal plates**), two **aboral** (or **angular**) **plates** lying on either side of the oral shield, and the inward-pointing jaws, which extend toward the mouth (Fig. 22.13). The jaws bear teeth at their apices and **oral papillae** along their sides.

One of the oral shields is modified as the madreporite of the water-vascular system. The madreporite's oral position differs from that observed in the sea stars, where it is located aborally. Examine the arms of your specimen and note the lack of ambulacral grooves. Ophiuroids also lack dermal branchiae and pedicellariae.

The arms are very flexible because they are composed internally of a series of disc-shaped, calcareous **vertebrae** connected by large, intervertebral muscles (Fig. 22.14). The vertebrae represent the ambulacral ossicles that have become located internally. Four longitudinal rows of plates surround the vertebrae externally: one dorsal (or aboral) arm plate, two lateral arm plates, and one ventral (or oral) arm plate (Fig. 22.14). Along the outer margins of the lateral arm plates are holes through which the tube feet extend. Note the

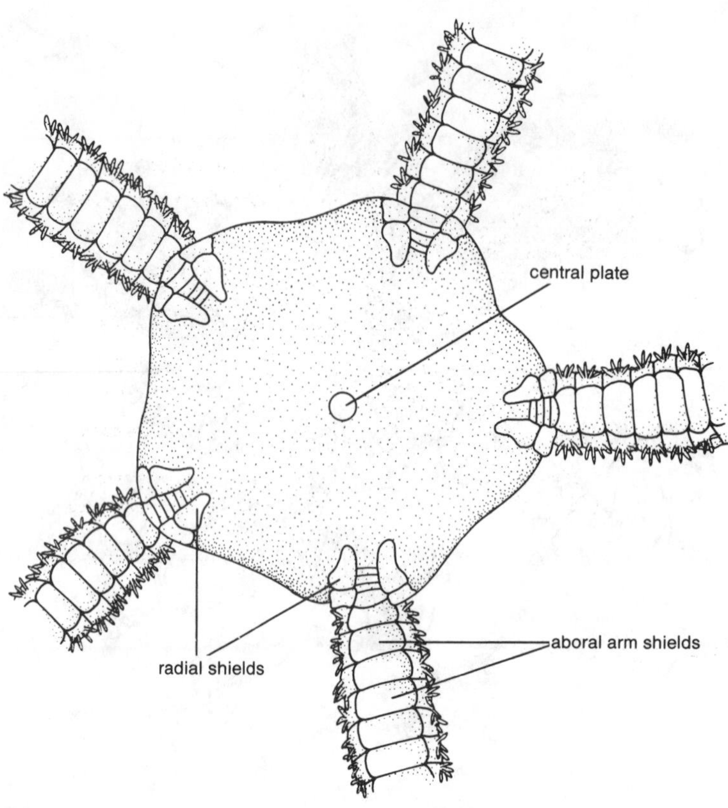

Figure 22.12. External aboral surface of a brittle star.

vertical row of spines protruding between the lateral arm plates (Fig. 22.13). How does their position compare to that of the podia? Unlike the condition in most asteroids, ophiuroid tube feet lack suckers.

Break off an arm to form a cross section and locate the four muscles, vertebrae, and the four plates described above (Fig. 22.14).

Locate the small, external openings on the oral surface within the central disc and alongside the arm margins. These are **bursal slits** (Fig. 22.13). There are five pairs of them and each slit leads inward to a **bursa**, which is an internal sac formed by an invagination of the disc's oral surface. Gaseous exchange occurs across the bursae as seawater is pulled in and expelled. In *Ophioderma* and other brittle stars, the gametes are stored temporarily in these slits. In some species (not *Ophioderma*) the bursae serve as brood chambers for the developing brittle stars as well as in respiration. In addition to gas exchange and reproduction, the bursae may be the principal sites for nitrogenous waste removal.

Internal Anatomy. Using a preserved specimen, cut away the surface of the aboral disc. The ophiuroid's enterocoelom is much reduced. Locate the large, saclike **stomach** with ten marginal pouches (Fig. 22.15). Alternating in position with these pouches are the **gonads** and **bursae**. The stomach is connected to the mouth by a short esophagus. An intestine, hepatic ceca, and anus are lacking. Unlike asteroids, the digestive system in ophiuroids is confined to the central disc. Ophiuroids are scavengers, filter feeders, or suspension feeders. The tube feet are used to capture food.

The water-vascular and nervous systems are similar to those of asteroids, with some minor differences. These systems will not be studied in your dissections.

Although most ophiuroids are dioecious, hermaphroditic species are not uncommon. Ophiuroid reproduction is similar to that of asteroids. The bilaterally symmetrical larva, when present (many brooding ophiuroids are known), is an **ophiopluteus**, characterized by eight arms (Fig.

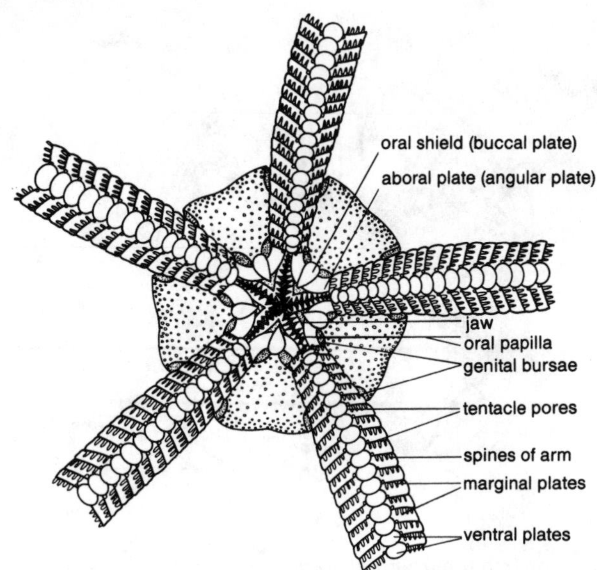

Figure 22.13. External oral view of *Ophioderma*.

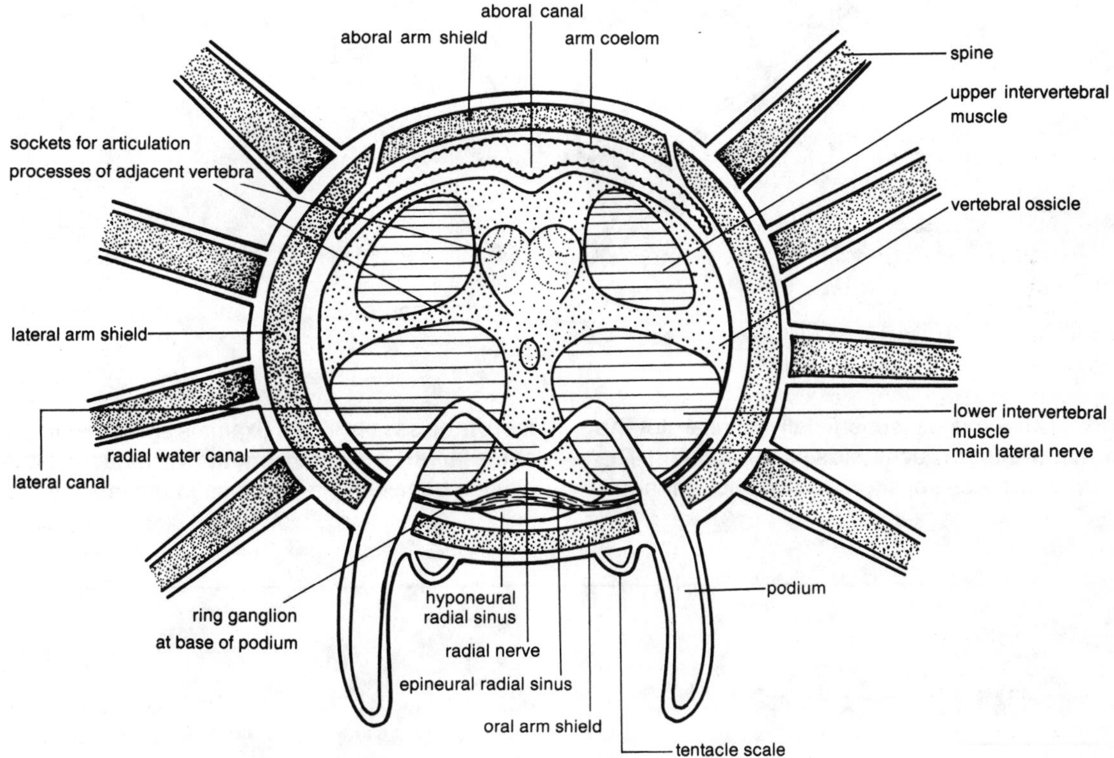

Figure 22.14. Cross section of a generalized ophiuroid arm. (After Barnes.)

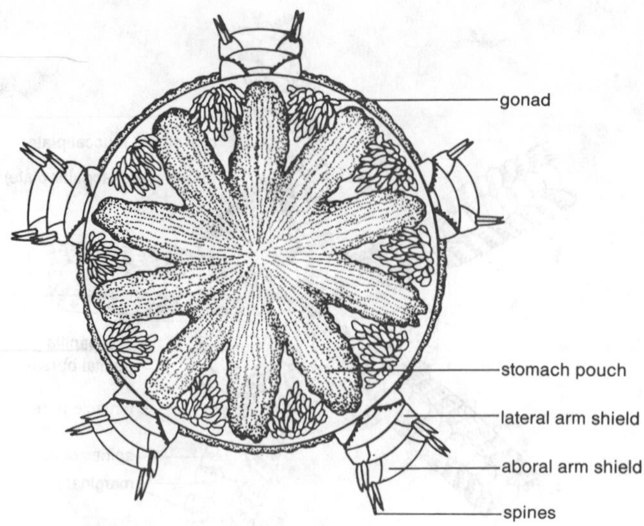

Figure 22.15. Internal anatomy of an ophiuroid, aboral view.

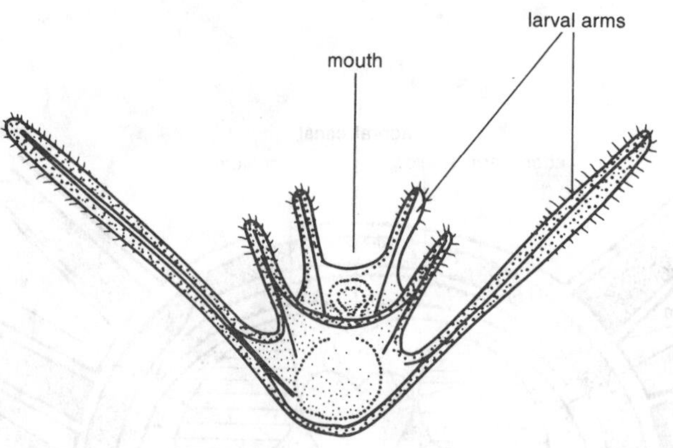

Figure 22.16. Ophiopluteus of a brittle star.

22.16). The larva is superficially similar to the **echinopluteus** of echinoids (Fig. 22.20). The ophiopluteus develops gradually into an adult; there is no sessile stage. Examine slide preparations of the larva. Look for the mouth and anus. Are there ciliated bands similar to those found in asteroids?

C. Class Echinoidea

Sea urchins, sand dollars, and heart urchins (spatangoids) belong to the class Echinoidea (ECH-i-NOI-de-a; G., *echinos*, hedgehog + G., *eidos*, form). There are approximately 900 living species of these free-moving echinoderms found in the littoral and benthic zones of the oceans. The body is globular to disclike. Movable

spines cover their bodies; those of "regular" sea urchins are longer than those of the "irregular" sand dollars and heart urchins. The spines function in protection, burrowing, keeping the body surface clean, and righting the body. The gonads of some "regular" sea urchins are used for food by some people.

Echinoids have a rigid endoskeleton called the **test**. The test and spines are covered with a ciliated epidermis. The ossicles of the test are flattened and sutured together firmly so that in most echinoids the individual ossicles are immobile, unlike the condition in most echinoderms. A body musculature is lacking on the inner surface of the test. The interlocking ossicles radiate in rows from the aboral apex to the mouth, which is located in the center of the oral surface. The ambulacra and interambulacra are formed by ossicles arranged in ten alternating double columns which converge at the two poles. The ambulacral plates have paired holes or pores through which the tube feet extend; holes are lacking on the interambulacral ossicles. Echinoids lack projecting, free-moving arms similar to those of asteroids and ophiuroids.

Although the basic symmetry is radial, as observed in sea urchins, some echinoids have a secondary, bilateral symmetry. Because of this difference in symmetry, echinoids are divided into two groups: (1) the "regular" echinoids (sea urchins), with radial symmetry, and (2) the "irregular" echinoids (heart urchins and sand dollars), with bilateral symmetry.

Observational Procedure

Sea Urchin. Obtain a preserved complete specimen of a sea urchin, such as *Arbacia* or *Strongylocentrotus*, and a specimen that has been cleaned and the spines removed. Note the globular or ovoid shape; the oral surface is flatter than the aboral one.

Cleaned Specimen. Locate on the oral surface the five **ambulacra** composed of rows of double ossicles and the five double rows that constitute the **interambulacra** (Fig. 22.17). The ambulacra have pores for the tube feet, whereas the interambulacra lack the holes. The anus is at or near the center of the aboral surface; a membrane (the **periproct**) with embedded, minute ossicles bears the anus. Five

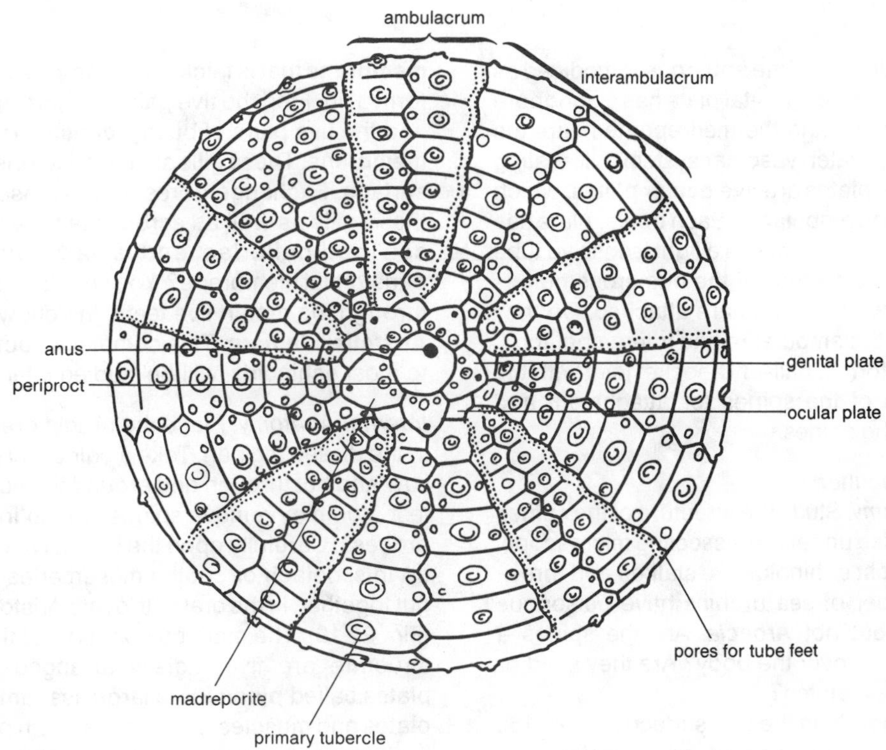

Figure 22.17. Test of *Arbacia*, aboral view.

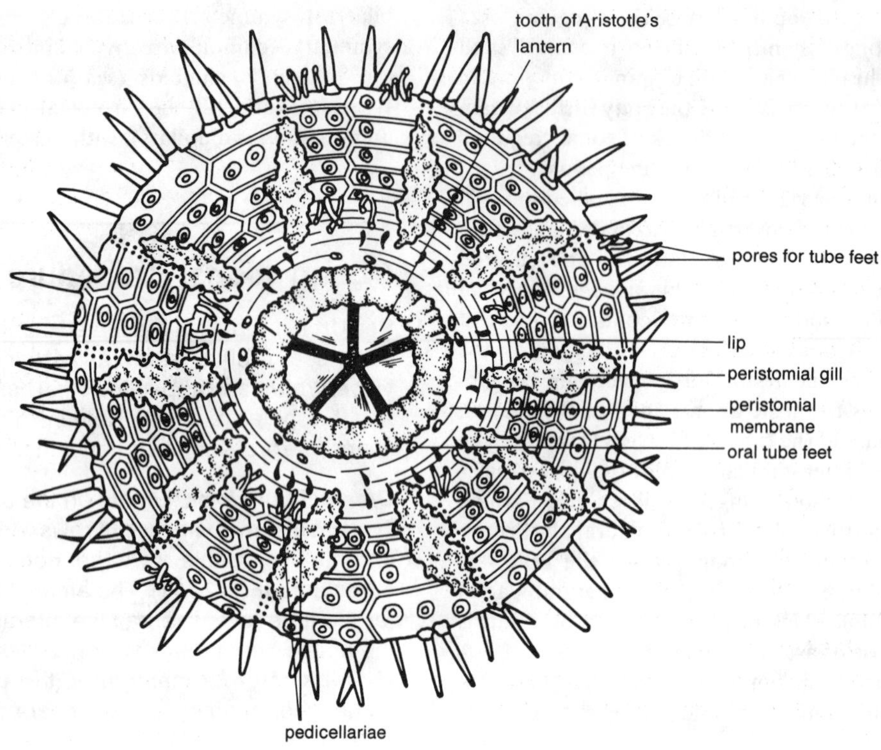

Figure 22.18. Oral view of a sea urchin.

genital plates surround the anus; one is modified as the madreporite. Each genital plate has a gonopore for exit of gametes and the madreporite bears the openings of the water-vascular system. Alternating with the genital plates are five **ocular plates**, which coincide with the ambulacra. Each ocular plate has a pore through which passes a light-sensitive tube foot. Observe the pattern of the numerous bumps (**tubercles**) to which spines are attached. Do they occur on both the ambulacra and interambulacra? The tubercles form a ball-and-socket joint with the concave ends of the spines for attachment and movement of the spines.

Preserved Specimen

External Anatomy. Study the structure of the **spines** and **pedicellariae** under the dissecting microscope. The latter in most echinoids are stalked and three-jawed. A number of sea urchins have poisonous pedicellariae, but not *Arbacia.* Are the spines a uniform length all over the body? Are they solid or hollow? Are they jointed?

Locate the **mouth** on the oral surface (Fig. 22.18). The opening is surrounded by a **peristomial membrane** that is thickened along the inner edge to form a lip. Find the five pairs of short, **buccal podia** and the five pairs of bushy, **peristomial gills** in the membrane. These gills are evaginations of the body surface. Exchange of respiratory gases occurs in these gills as well as across the tube feet. Do you see pedicellariae and small spines in the peristomial membrane? You should see protruding through the mouth five teeth, the chewing parts of **Aristotle's lantern.** This complex structure, unique to most echinoids, will be studied later.

Internal Anatomy. Place the urchin oral side down in the dissecting pan. Take a pair of scissors or fine saw and cut through and around the equator of the test. You may remove spines to help in the cutting process. Carefully open the two halves, and as you do this, detach or cut the mesenteries holding the gut together. In the oral half locate Aristotle's lantern (Fig. 22.19). The main components of this complex structure are five radially arranged calcareous plates called **pyramids**, a large assortment of other plates and muscles, and five oral teeth projecting at the tip of each pyramid. The lantern can be protruded or retracted through the mouth. Sea urchins use their sharp teeth to scrape, tear, and pull food materials. Most sea urchins are herbivores, feeding primarily on algae growing on rocks and similar surfaces.

Surrounding the lantern, the **ring canal** with its associated **Polian vesicles** may be seen. The vesicles are opposite each interambulacrum. The **stone canal** and **axial organ** ascend aborally off the ring canal. The former terminates at the madreporite on the aboral surface. The axial organ is part of the hemal system. This system, found also in other echinoderms, is composed of a complex of channels that are difficult to follow in preserved specimens; the system will not be studied. The exact function of the axial organ is unknown. The **radial canals** leave the ring canal. At that juncture you may have difficulty locating the canals, but they, along with ampullae of the tube feet, may be observed along the inner surface of the test. Note the lack of body muscles attached to the inner test wall. What would account for the lack of this musculature?

The digestive system consists of the oral **mouth**, a **buccal** cavity, and a **pharynx** (the interior of Aristotle's lantern), which leads to a long **esophagus** (Fig. 22.19). The esophagus leads to a **small intestine** that curves around the inside of the test in a counterclockwise direction as viewed from the aboral surface. The small intestine leads to a **large intestine** that makes an almost complete circuit in the opposite direction. Some workers call the two circuits of the intestine the stomach. The large intestine leads to the short **rectum**, which opens at the **anus**. Running along the inner surface of the oral part of the intestine is the ciliated **siphon** connecting the esophagus at one end and the intestine at the other end. The precise function of the siphon is unknown. It may remove water from the intestinal contents, thus concentrating enzymes and food during the initial stages of digestion. Digestion and absorption occur evidently in the intestine.

Five **gonads** follow the interambulacral areas in the inner aboral surface. Each gonad opens to the outside by a genital pore, observed when the external anatomy was studied. All echinoids are dioecious. Fertilization is external in the seawater and a bilateral larva, the **echinopluteus**, may form. The larva metamorphoses into a juvenile urchin without settling down. Study slides of the echinopluteus (Fig. 22.20). How does this larva compare to the ophiopluteus of brittle stars?

Sand Dollar. Obtain a preserved and cleaned sand dollar such as *Mellita*. Note the disc-shaped body with a covering of small, numerous spines. Unlike "regular" sea urchins, the anus of sand dollars and other "irregular" echinoids has moved from the aboral center to the posterior edge of the body or to the posterior **lunule** (Fig. 22.21). Lunules are the elongated notches or holes on the sand dollar and their numbers vary among different species. At least in some forms, the lunules aid the sand dollar in burrowing in the sand, although spines are the principal structures for that purpose. Lunules are absent in the sea biscuit (*Clypeaster rosaceus*) of Florida and the West Indies (Fig. 22.22).

On the clean specimen, examine the aboral surface under the dissecting microscope and observe the five **petaloids** (the ambulacra), which resemble a flower. Find the holes that contained the tube feet. These aboral podia function in gas exchange. Peristomial gills like those of sea urchins do not occur in the "irregular" echinoids. Can you see the tube feet on the preserved sand dollar?

At each interradius and near the base of the petaloids are the **gonopores**, each surrounded by a **genital plate**. One of the genital plates functions as the madreporite. At the base of each petaloid there is an **ocular plate** with a hole that bore a light-sensitive tube foot in life.

Locate the **mouth** on the flat oral surface. Sand dollars possess a modified Aristotle's lantern which cannot be protracted through the mouth. The internal anatomy of the sand dollar is similar to that of sea urchins. Because of the difficulty in dissection, the sand dollar's internal anatomy will not be studied.

Heart Urchin. Heart urchins (spatangoids) are "irregular" echinoids that are similar to sand dollars (Fig. 22.23). The body is egg- or heart-shaped and on the aboral surface there are **petaloid ambulacra** containing tube feet for gaseous exchange. The mouth and peristome are located toward the anterior end of the flat, aboral surface, and the anus is at the posterior or more pointed end. There are three anterior ambulacra and two posterior ambulacra on the aboral surface. These ambulacra contain pores through which the tube feet protrude. Small, dense spines cover the body surface; the spines are similar to those of sea urchins and sand dollars. Heart urchins lack Aristotle's lantern. Both sand dollars and heart urchins feed on minute organic materials.

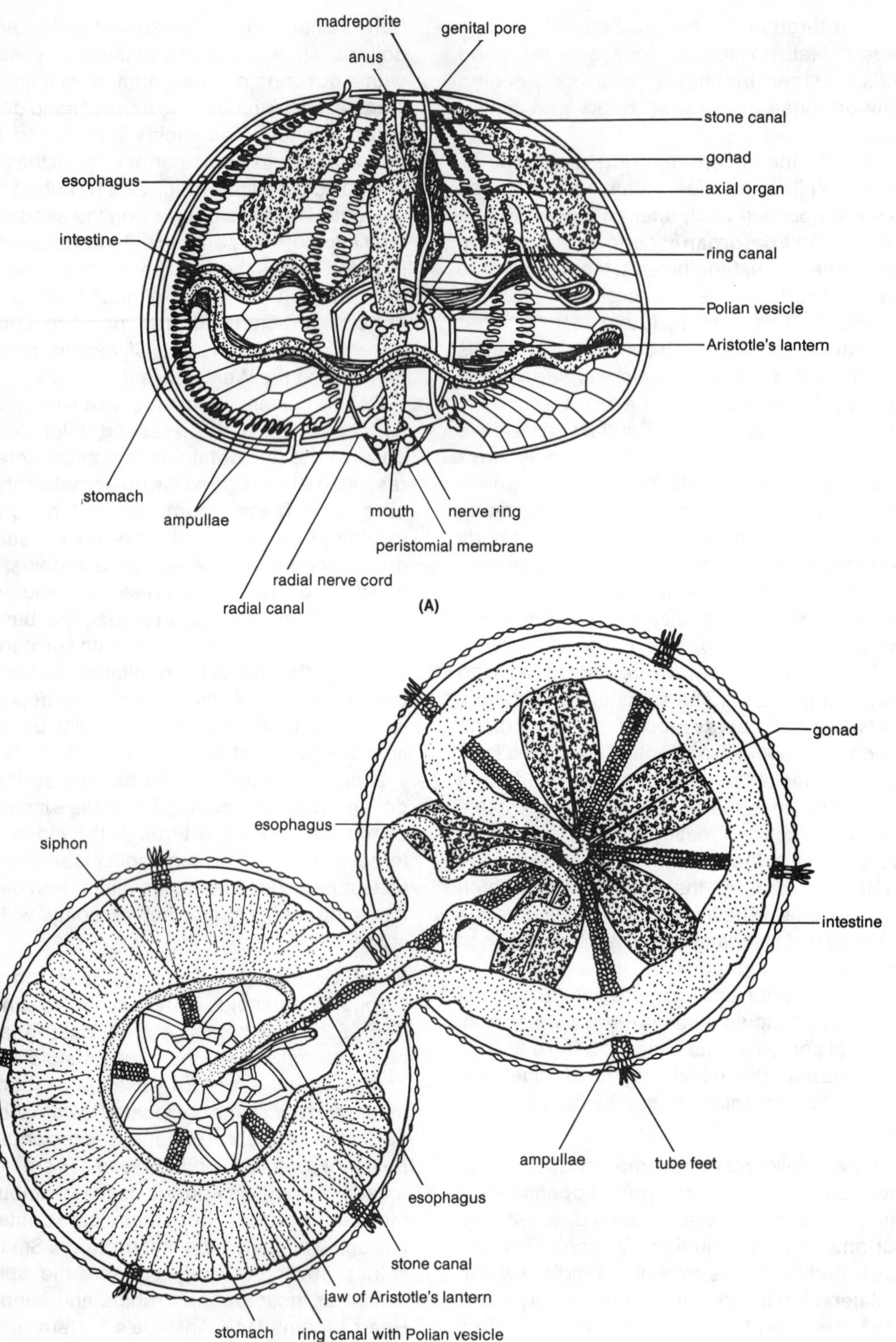
madreporite
genital pore
anus
stone canal
gonad
esophagus
axial organ
intestine
ring canal
Polian vesicle
Aristotle's lantern
stomach
ampullae
mouth
nerve ring
peristomial membrane
radial nerve cord
radial canal
(A)
gonad
esophagus
siphon
intestine
ampullae
tube feet
esophagus
stone canal
jaw of Aristotle's lantern
stomach
ring canal with Polian vesicle
(B)

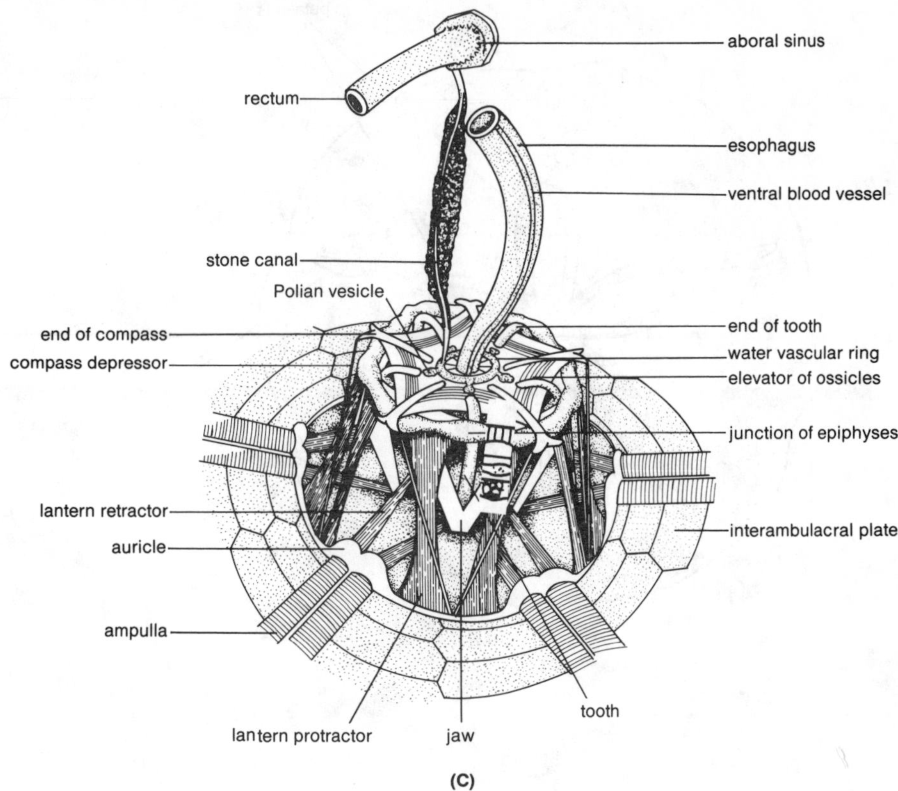

Figure 22.19. Internal anatomy of a sea urchin. (**A**) Lateral view. (**B**) Oral-aboral view. (**C**) Aristotle's lantern.

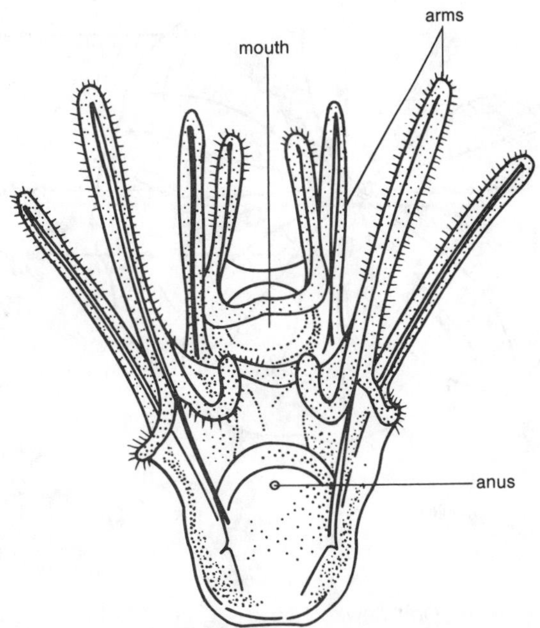

Figure 22.20. Early echinopluteus of an echinoid.

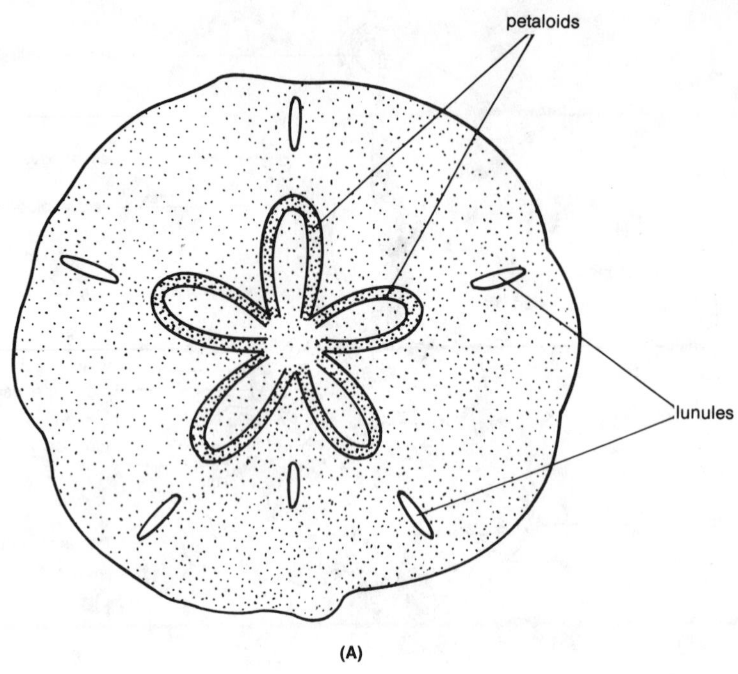

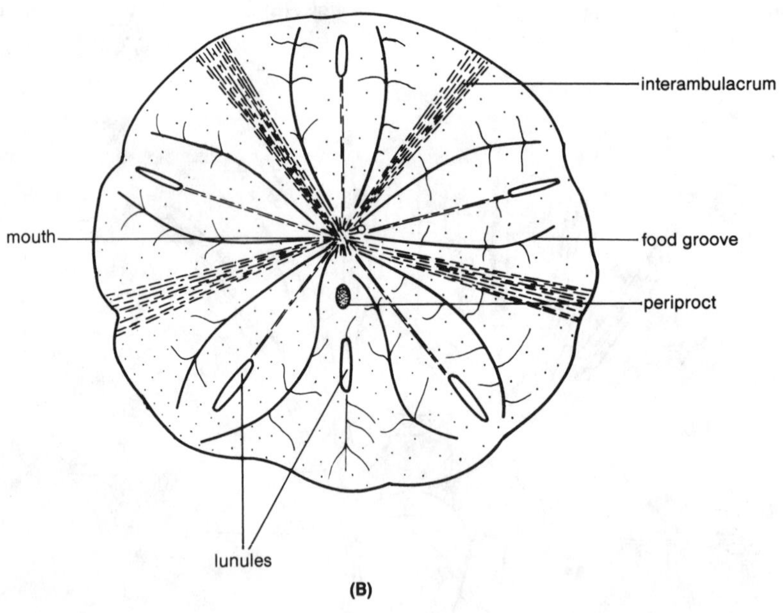

Figure 22.21. Sand dollar. **(A)** Aboral view. **(B)** Oral view.

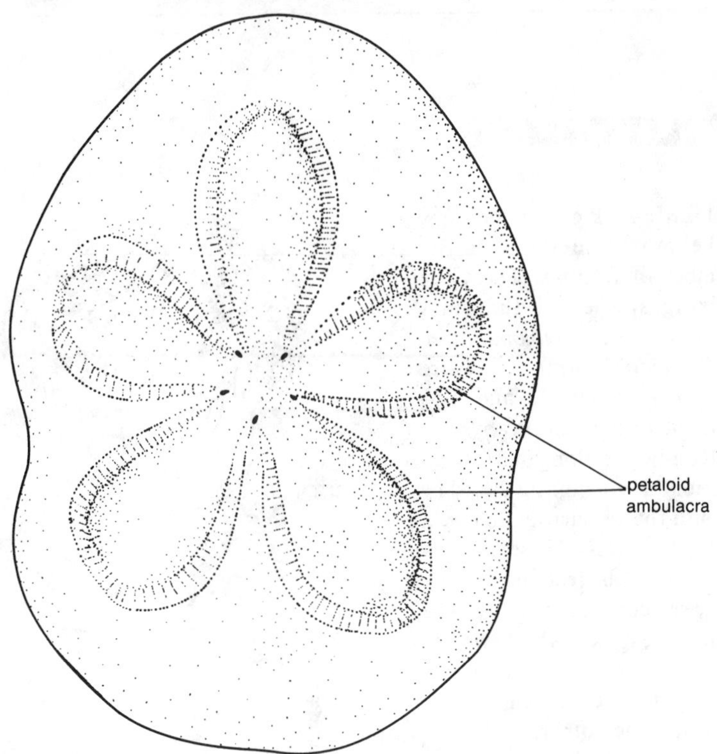

Figure 22.22. Sea biscuit (*Clypeaster*).

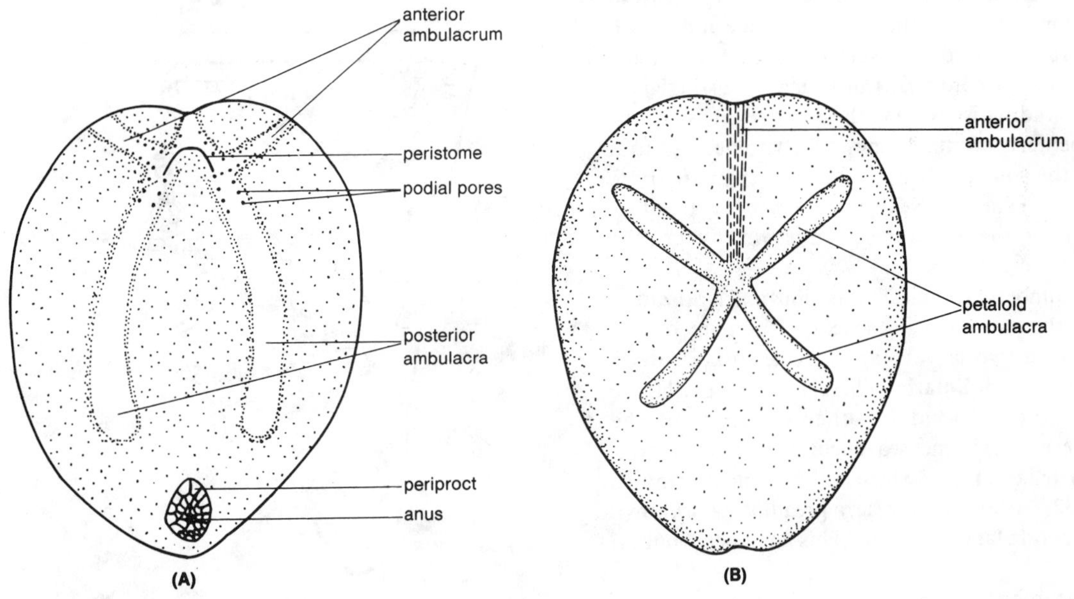

Figure 22.23. Heart urchin. **(A)** Oral view. **(B)** Aboral view.

D. Class Holothuroidea

Sea cucumbers are rather bizarre-looking echinoderms. Indeed, they are unlike other members of the Echinodermata in a number of features. There are approximately 1100 living species of the class Holothuroidea (HOL-o-thu-ROI-de-a; G., *holothourion*, a kind of zoophyte + G., *eidos*, form). They are common in the littoral areas of the sea and also constitute an important fauna of the marine abyss.

The body of a sea cucumber is elongated on an oral-aboral (anteroposterior) axis (Fig. 22.24). The oral end contains the **mouth** that is surrounded by a number of food-gathering **tentacles**. These mucus-covered structures are modified tube feet and usually can be retracted within the sea cucumber's body. The opposite end bears the **anus**. The spineless body wall of most sea cucumbers is leathery and contains an embedded endoskeleton of microscopic, calcareous ossicles that are not fused with each other.

Tube feet are scattered over the body surface (e.g., *Thyone*), grouped in five ambulacral areas (*Cucumaria*), or totally lacking (*Leptosynapta*). In those forms with five distinct ambulacral areas, the tube feet are arranged in three ambulacra on the ventral surface and in two ambulacra on the dorsal surface. Ray A is on the midventral line according to Carpenter's (1884) designation of the radii, discussed in the exercise on asteroids. The tube feet on the somewhat flattened, ventral surface (the **sole**) are better developed than are those on the dorsal surface. Sea cucumbers move about by muscular contraction of their body, aided by the tube feet when present.

Sea cucumbers are typically dioecious and fertilization occurs in the water. During development many species pass through two larval stages: first, an **auricularia** and second, a **doliolaria**. These microscopic larvae resemble those found in other members of the Echinodermata. Some sea cucumbers have a non-feeding **vitellaria larva**, which is also found in crinoids (Fig. 22.32). After these larvae, a preadult stage known as a **pentactula larva** develops. This larva bears buccal tentacles.

Sea cucumbers are noted for their great powers of regeneration. When disturbed or under stress, the animal is unique among echinoderms in being able to **eviscerate** the internal organs through the anal opening. This is perhaps a means of distracting predators. Regeneration of the lost organs then occurs. *Holothuria* and a few others can eject a sticky mass of **Cuvierian tubules**, which entangle the predator as the sea cucumber crawls to safety.

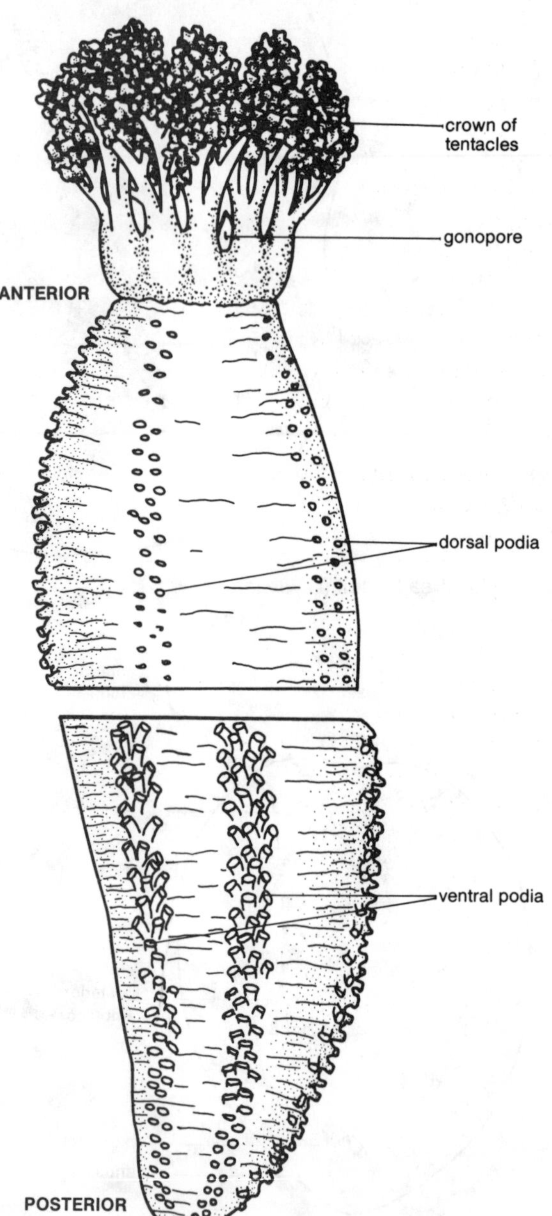

Figure 22.24. External view of a sea cucumber.

Observational Procedure

Obtain a specimen of *Cucumaria*, *Thyone*, or similar species. Study the external features, paying attention to the locations of the mouth and anus, arrangement of the tube feet, and the nature of the body shape and covering. Examine slide preparations of the microscopic **ossicles** of the body wall (Fig. 22.25).

Beginning at the posterior end, make a ventral incision through the body wall and extend this incision through the anterior end. Open the body wall laterally and pin it to the dissecting pan. Cover the specimen with water. As you study the internal anatomy, refer to Fig. 22.26.

The enterocoelom in sea cucumbers is spacious. In life the body cavity is filled with coelomic fluid containing various types of amoebocytes. The peritoneal cilia produce a current that moves the coelomic fluid throughout the enterocoelom.

Digestive System. This system consists of the **mouth**, expanded **pharynx**, short **esophagus**, muscular **stomach**, and long **intestine**, which leads to the muscular **rectum** (Fig. 22.26). In *Cucumaria*, but not in *Thyone*, the esophagus is absent and the stomach is poorly developed. Note the muscles that attach the rectum to the inner surface of the body wall. A pair of highly branched **respiratory trees** extends off the rectum (Fig. 22.26). The muscular rectum pumps water into the trees and expels it back to the sea. Exchange of respiratory gases occurs within the trees. They are unique to sea cucumbers; however, the burrowing forms, such as *Leptosynapta*, lack them and exchange of respiratory gases occurs across the body surface.

Water-Vascular System. Surrounding the pharynx is a hard structure called the **calcareous ring**, to which **retractor muscles** are attached posteriorly. These muscles retract the entire oral region inside the body. The calcareous-ring complex may be homologous with the Aristotle's lantern of echinoids. Just behind the ring is the **ring canal** at the base of the pharynx. The dorsal madreporite is connected to the ring canal by the stone canal. The internal position of the madreporite is unlike that of other echinoderms. One to several (often two) saclike **Polian vesicles** arise from the ring canal. Follow these where they join the canal (Fig. 22.26). Some specimens may have small vesicles, whereas others may have greatly expanded vesicles. The Polian vesicles may maintain pressure within the water-vascular system. Five **radial canals** off the ring canal run anteriorly to the oral tentacles and then descend to the aboral end of the body. The canals follow the large **longitudinal muscles** that are attached to the inner surface of the body wall. These lie under the five ambulacral areas. The canals lead into numerous side canals which connect with the tube feet in the ambulacra. The longitudinal muscles also join the retractor muscles at the anterior end. In addition to these muscles, the inner body wall is lined with **circular muscles**. The circular and longitudinal muscles allow the sea cucumber to expand and contract its body.

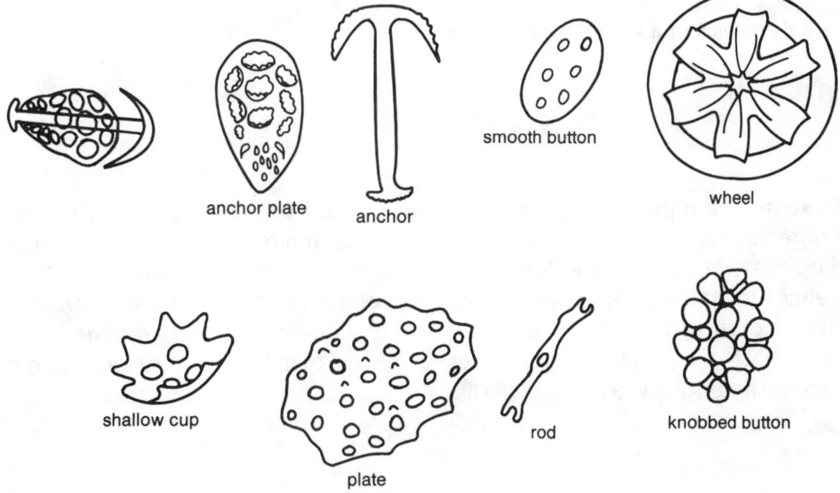

Figure 22.25. Ossicles of several types of sea cucumbers.

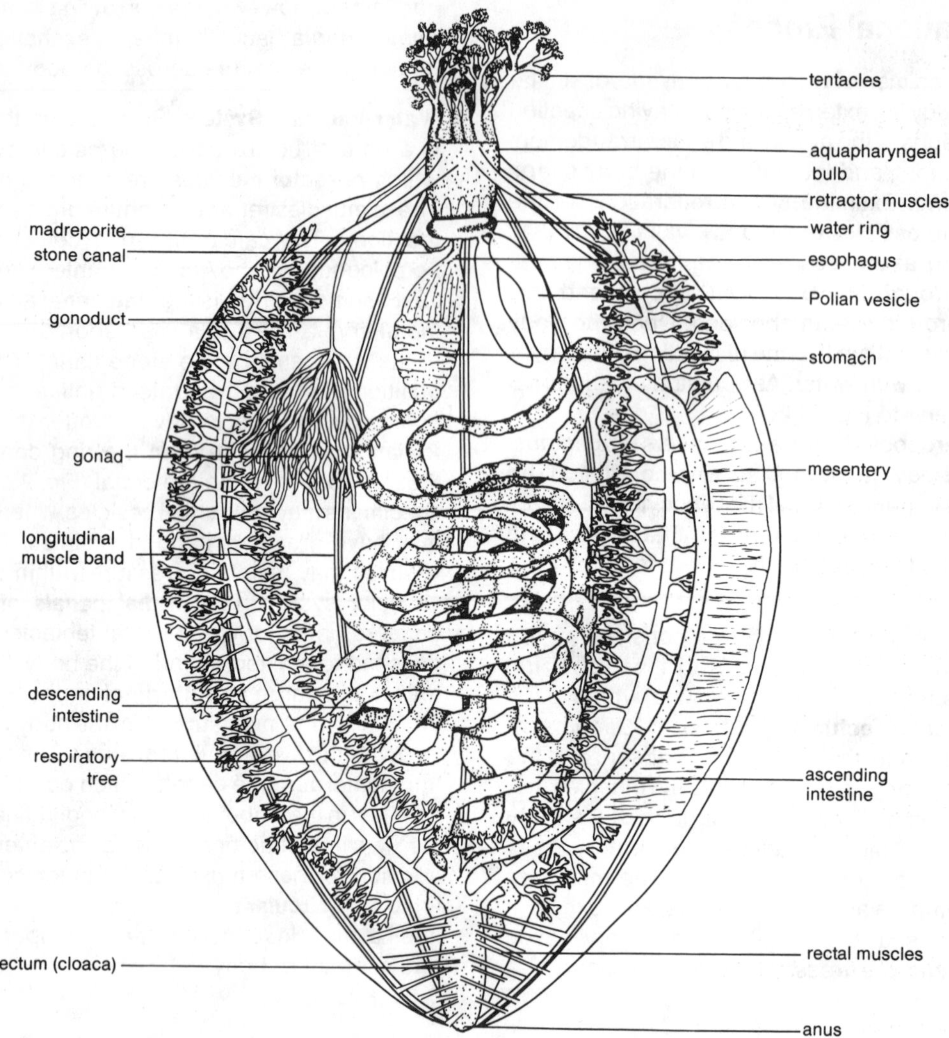

Figure 22.26. Internal anatomy of a sea cucumber (*Thyone*).

Reproductive System. The reproductive system is simple in these dioecious animals (Fig. 22.26). The moplike **gonad** consists of a tuft of fine filaments. The gonad increases in size with the approach of sexual maturity. Holothurians are the only echinoderms with a single gonad. A gonoduct carries eggs or sperm anteriorly to a genital papilla located between two dorsal tentacles. Microscopic examination of crushed gonadal filaments will reveal the sex of a specimen. The round head and long tail of the sperm can be seen when stained with methylene blue and examined under oil immersion. Eggs of the female are large and ovoid.

E. Class Crinoidea

The class Crinoidea (cri-NOID-e-a; G., *krion*, a lily + G., *eidos*, form) contains sea lilies (about 100 living species) and feather stars or comatulids (KO-mat-u-lids; L., *comatus*, hairy) (about 600 living species). They have a worldwide distribution in warm and frigid waters and live at depths ranging from a few meters to over 5000 m. Crinoids are gregarious and often form huge local populations.

The living species represent a mere remnant of a once-thriving group in the geologic past. The crinoid fossil record dates back to the **Cambrian**, and over 5000 fossil species have been described from the **Paleozoic Era** (Fig. 22.27).

Most crinoids living today are the stemless comatulids, whereas in the past the majority of species were attached permanently to the seafloor by a **stem** (or **stalk**). Comatulids are attached to a substrate by a stem in their early development, but break away and move above on their jointed **cirri**, somewhat in a walking fashion, or swim by lashing their arms (Fig. 22.28). Sea lilies are similar in structure to the comatulids, but have a stalk for attachment (Fig. 22.29).

Observational Procedure

Obtain a fossil or living crinoid provided with a stem, and a stemless comatulid such as *Antedon*. As you read the following description, refer to Figs. 22.28 and 22.29. The body consists of the cup-shaped **calyx (theca)**, housing the internal organs; the **arms**, constituting a food-gathering apparatus; and the attachment device (**stem** and/or **cirri**), used for clinging to the substratum. The combination of the calyx and arms is called the **crown**. Feather stars lack the stalk but have cirri for grasping the substrate.

The cup-shaped calyx is composed of external calcareous plates and is attached by its aboral surface. The membranous covering of the calyx is the **tegmen**; it is the oral surface of the organism (Fig. 22.30). Embedded in the tegmen are minute calcareous plates. Also in the tegmen are tiny ciliated canals that open into the enterocoelom. These openings may function as a madreporite. The tegmen is divided radially into five ambulacral and five interambulacral areas (Fig. 22.30). The **mouth** is situated at or near the center of the tegmen and the **anus** occurs in one of the interambulacral areas. The anus is on an elevated **anal cone** that directs feces away from the mouth. Unlike other echinoderms, crinoids have the oral surface directed upward.

Extending from the periphery of the crown are the five arms. Each arm, along with its ambulacral groove, may branch once or several times, forming ten or more arms. Note the rows of spherical yellow-brown dots along the sides of each ambulacral

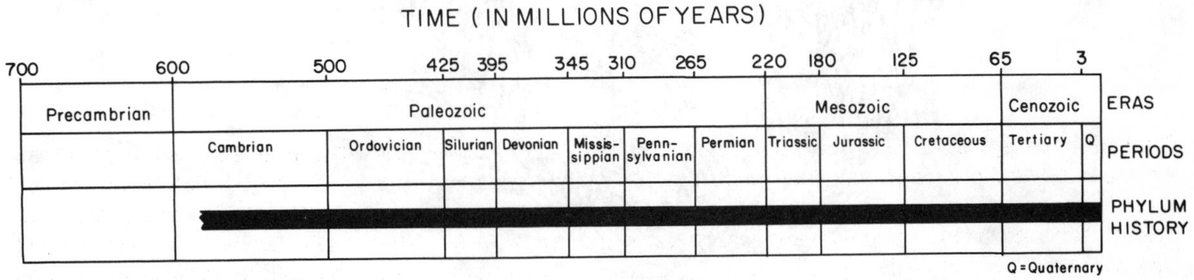

TIME (IN MILLIONS OF YEARS)

700–600	600–500	500–425	425–395	395–345	345–310	310–265	265–220	220–180	180–125	125–65	65–3	3	
Precambrian	Paleozoic							Mesozoic			Cenozoic		ERAS
	Cambrian	Ordovician	Silurian	Devonian	Mississippian	Pennsylvanian	Permian	Triassic	Jurassic	Cretaceous	Tertiary	Q	PERIODS
													PHYLUM HISTORY

Q = Quaternary

Figure 22.27. Geologic history of the phylum Echinodermata.

groove (Fig. 22.31). These bodies are **saccules** and may serve as waste deposit areas. To see the saccules, observe them with a dissecting microscope. The margins of the ambulacral grooves bear movable plates called **lappets**. These can close over the ambulacral groove. Within the grooves are the tentacle-shaped tube feet bearing mucus-secreting papillae used for food capture. Ampullae are lacking on the tube feet. Cilia within the ambulacral groove propel the food to the mouth. Crinoids are suspension feeders.

Extending laterally off each arm are numerous jointed **pinnules** (Fig. 22.30). The ambulacral grooves of the pinnules are continuous with the main arm grooves; both podia and lappets occur in the pinnules. Gametes develop from the germinal epithelium in those pinnules located along the proximal half of the arm length. Both the arms and pinnules are composed of ossicles of the endoskeleton.

The stem is of stacked disc-shaped ossicles, each perforated by a central axial canal for passage of components to certain organs. The discs are known as **columnals**. Because of the internal skeletal ossicles, the stalk has a jointed appearance. View these structures on a fossil crinoid (Fig. 22.29).

Crinoids are dioecious. The gametes rupture through the pinnule walls. The fertilized eggs remain attached to the pinnules as in *Antedon* or they are shed in the seawater. The free-swimming larva is a nonfeeding **vitellaria**, similar to that of sea cucumbers (Fig. 22.32). The larva settles to the seafloor and eventually develops into an adult.

Because of the difficulty involved in the dissection, the internal anatomy of the crinoid will not be studied.

Figure 22.28. External view of a feather star (comatulid).

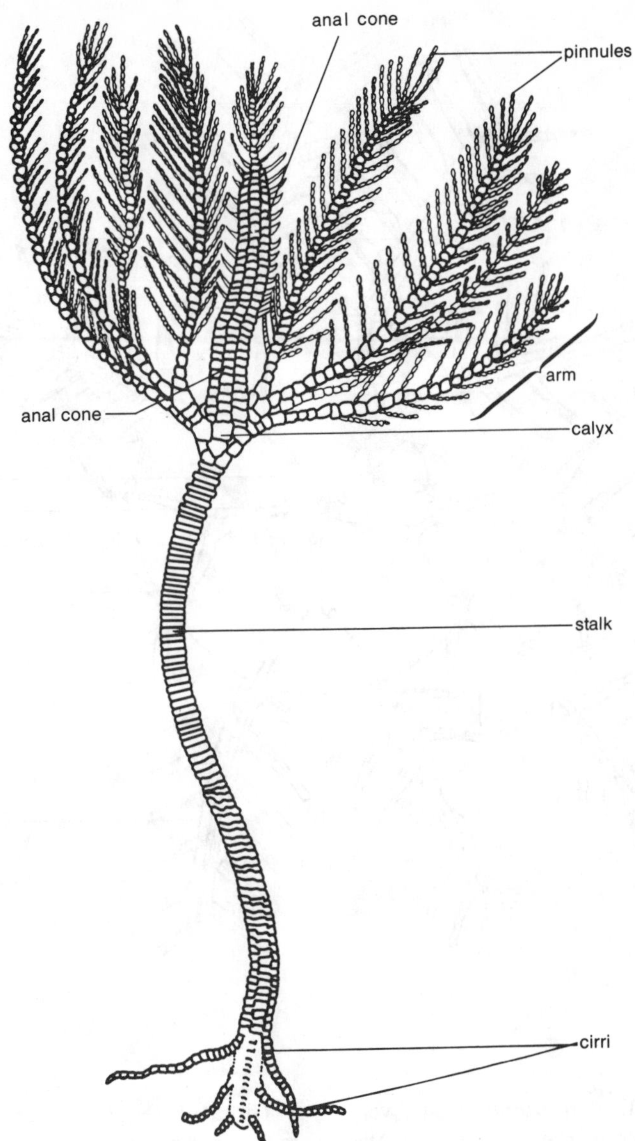

Figure 22.29. External view of a sea lily.

F. Fossil Echinoderms

Because of their hard, calcareous endoskeleton, the fossil record of echinoderms is rich. This ancient group, according to many zoologists, dates to the **Lower Cambrian Period** (Fig. 22.27). At least 16 extinct classes from the **Paleozoic** are recognized. Despite this rich fossil record, the exact relationship of

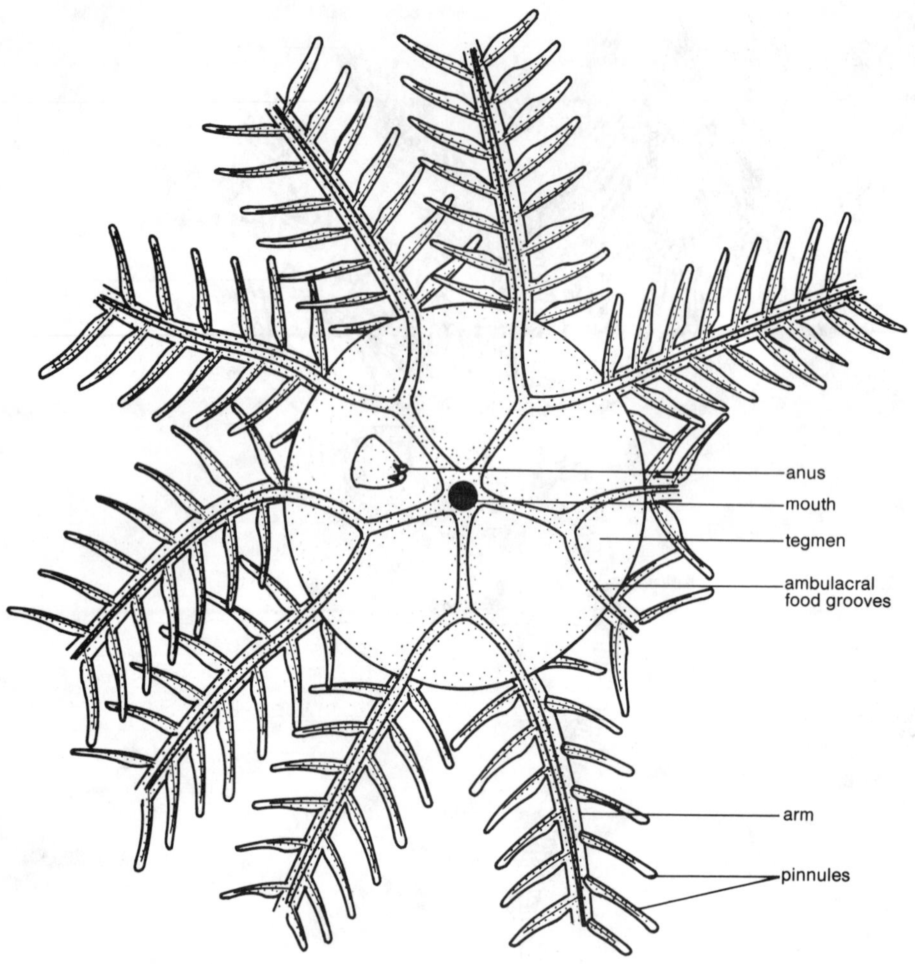

Figure 22.30. Oral surface of a crinoid.

the echinoderms to the other invertebrates and vertebrates is unknown. Some workers believe that the echinoderms may be ancestral to the lophophorates or to the hemichordates. In the past, the echinoderms were incorrectly thought to be related closely to the Cnidaria because of the radial symmetry common to both groups.

Most of the fossil species were stalked, sessile, and had both the oral surface and the ambulacral groove directed upward. Many possessed extensions of the groove in a pinnulelike projection called a **brachiole** (Fig. 22.33). These fossils were similar to modern sea lilies. The ambulacral system of tube feet was probably first used for feeding rather than for locomotion. Later, the podia served for locomotion, and several independent lines toward bilateral symmetry ensued.

Three of the oldest groups (Cambrian Period) recognized currently are Helicoplacoidea, Carpoidea, and Blastozoa. The Helicoplacoidea was asymmetrical or bilateral, but not radial. The spindle-shaped bodies of helicoplacoids had a spiral ambulacrum. *Helicoplacus curtisi* is an example (Fig. 22.34).

The Carpoidea was a diverse group. Currently, four classes are recognized. Carpoids were asymmetrical and may have had a crown of tentacles similar to those of lophophorates. Some had at least one brachiole. *Gryocystis* and *Dendrocystites* are examples (Fig. 22.33).

The Blastozoa contained either stalked or unstalked animals. This group includes some of the oldest of the fossil groups and their bodies resembled those of modern crinoids. The Cystoidea and Blastoidea were

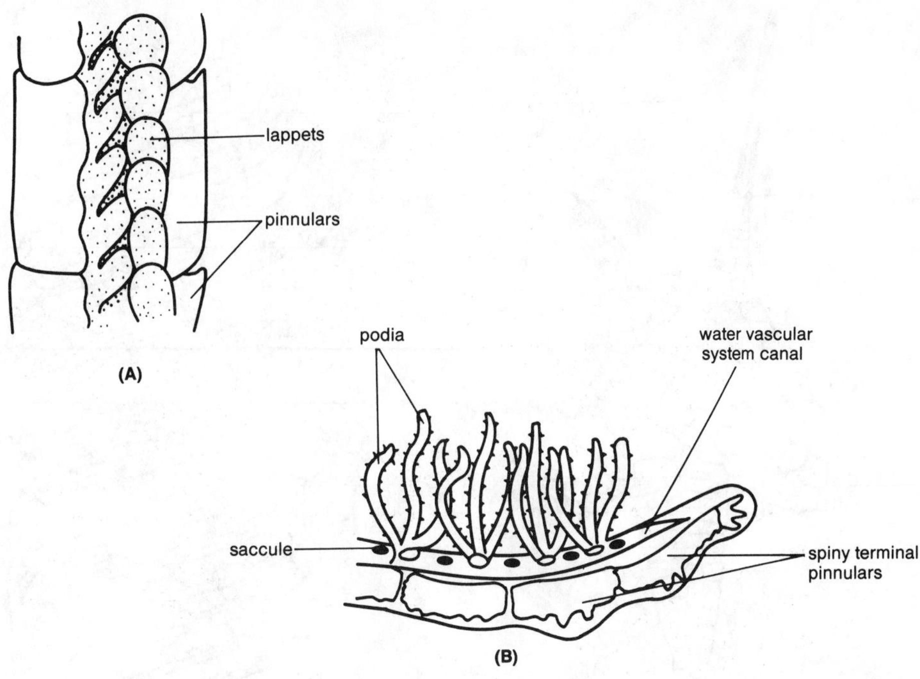

Figure 22.31. **(A)** Part of a crinoid pinnule, showing lappets and ambulacral groove. **(B)** Section through the tip of a crinoid pinnule, showing the podia and saccules.

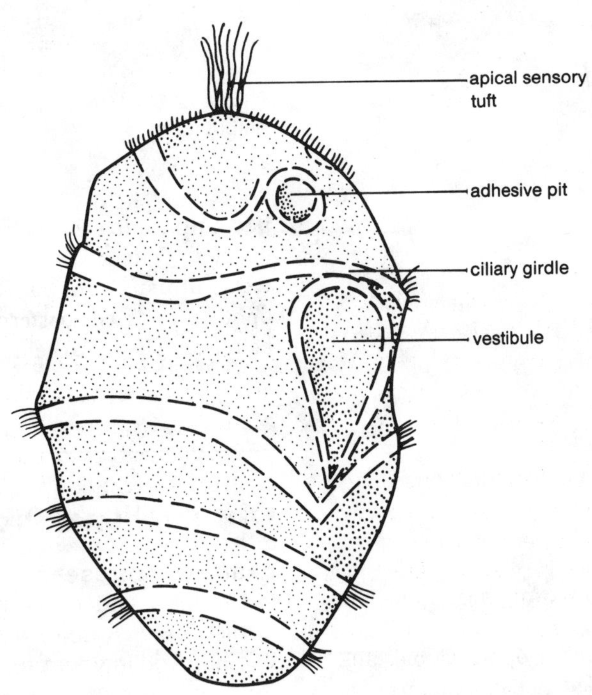

Figure 22.32. Crinoid vitellaria larva.

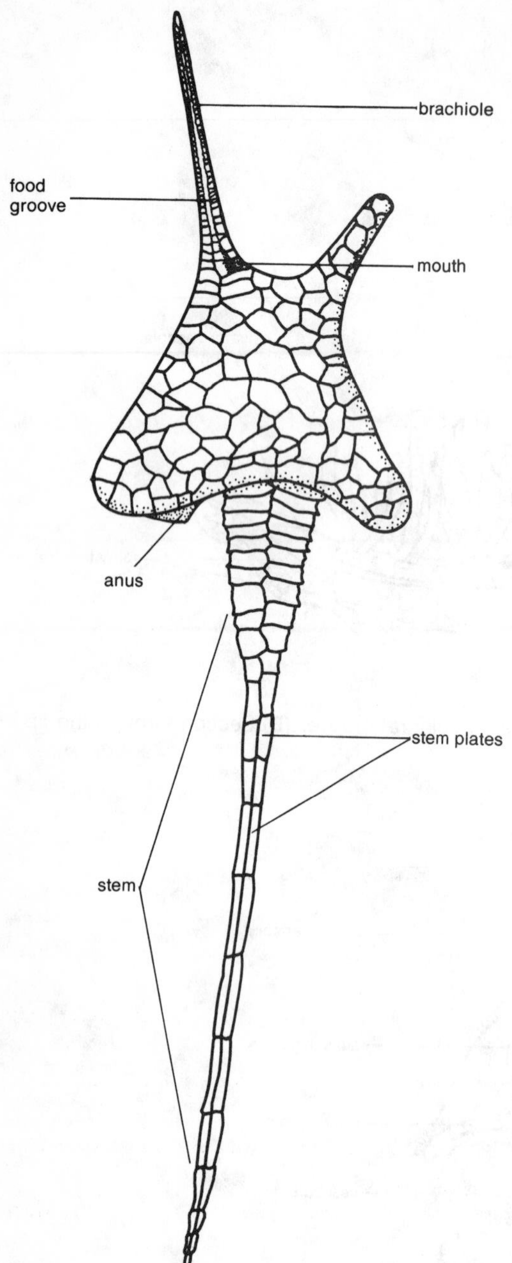

Figure 22.33. *Dendrocystites*, a carpoid with one brachiole.

sessile, often stalked with an oval theca, and had brachioles (Fig. 22.35).

Two later fossil groups were the ophiocistioids and edrioasteroids. Some resembled modern sea stars with

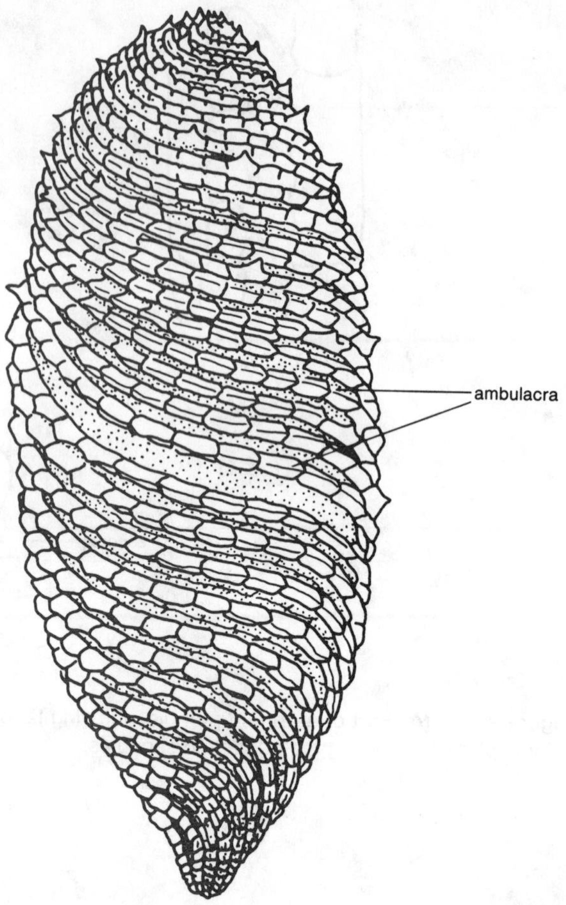

Figure 22.34. *Helicoplacus*, showing the spiral ambulacrum.

podia projecting through openings in the ambulacral ossicles. Ophiocistioids resembled echinoids. They were unusual in having huge tube feet. The ophiocistioids, edrioasteroids, and helicoplacoids may be related closely to echinoids and holothuroids.

Observational Procedure

Examine representative fossils and note the similarities and differences to the present-day forms. Your instructor will give specific instructions to guide you in your observations.

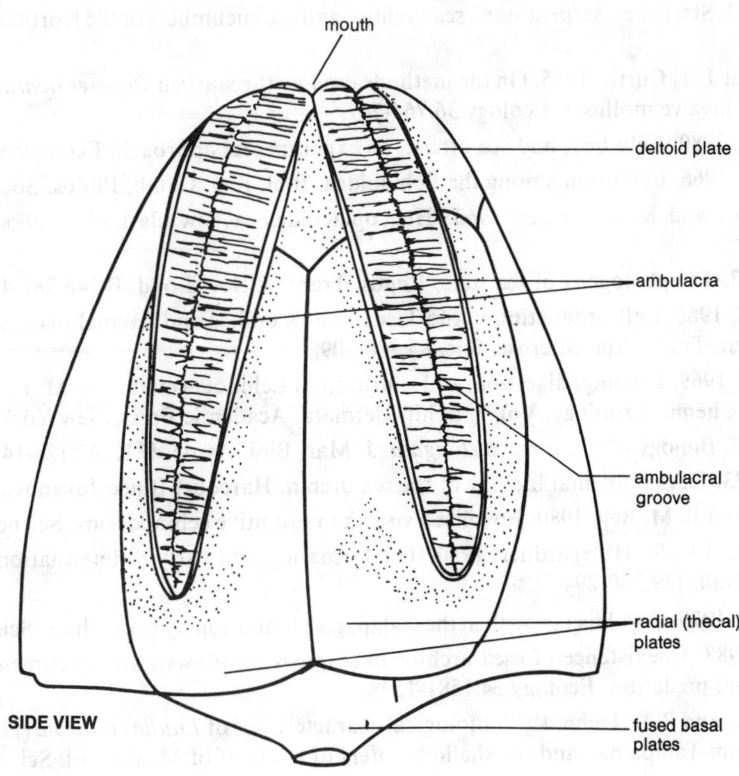

Figure 22.35. *Pentremites*, a blastoid showing ambulacra.

Supplemental Reading

Anderson, J. M. 1965. Studies on visceral regeneration in sea-stars. III. Regeneration of the cardiac stomach in *Asterias forbesi* (Desor). Biol. Bull. 129:454-470.

Angerer, R. C., and E. H. Davidson. 1984. Molecular indices of cell lineage specification in sea urchin embryos. Science 226:1153-1160.

Ausich, W. I., and D. J. Bottjer. 1982. Tiering in suspension-feeding communities on soft substrate throughout the Phanerozoic. Science 216:173-174.

Binyon, J. 1964. On the mode of functioning of the water vascular system of *Asterias rubens*. J. Mar. Biol. Assoc. U.K. 44:577-588.

Binyon, J. 1972. Physiology of echinoderms. Pergamon Press, Elmsford, New York.

Broom, D. M. 1975. Aggregation behaviour of the brittle-star *Ophiothrix fragilis*. J. Mar. Biol. Assoc. U.K. 55:191-197.

Burke, R. D. 1980. Podial sensory receptors and the induction of metamorphosis in echinoids. J. Exp. Mar. Biol. Ecol. 47:223-234.

Burnett, A. L. 1960. The mechanism employed by the starfish *Asterias forbesi* to gain access to the interior of the bivalve *Venus mercenaria*. Ecology 41:583-584.

Christensen, A. M. 1970. Feeding biology of the sea star *Astropecten irregularis*. Ophelia 8:1-134.

Clark, A. M. 1962. Starfishes and their relations. British Museum, London.

Coe, W. R. 1972. Starfishes, serpent stars, sea urchins, and sea cucumbers of the Northeast. Dover, New York.

Cottam, G., and J. T. Curtis. 1955. On the methods used by the starfish *Pisaster ochraceus* in opening three types of bivalve molluscs. Ecology 36:764-767.

Duggins, D. O. 1980. Kelp beds and sea otters: an experimental approach. Ecology 61:447-453.

Durham, J. W. 1966. Evolution among the Echinoidea. Biol. Rev. Camb. Philos. Soc. 41:368-391.

Durham, J. W., and K. E. Caster. 1963. Helicoplacoidea, a new class of echinoderms. Science 140:820-822.

Fell, H. B. 1963. The phylogeny of sea stars. Philos. Trans. R.Soc. Lond. B 246:381-485.

Ferguson, J. C. 1966. Cell production in the Tiedemann's bodies and haemal organs of the starfish *Asterias forbesi*. Trans. Am. Microsc. Soc. 85:200-209.

Ferguson, J. C. 1969. Feeding, digestion, and nutrition in Echinodermata. *In*: M. Florkin and B. T. Scheer (eds.). Chemical zoology. Vol. 3. Echinodermata. Academic Press, New York, pp. 71-100.

Fish, J. D. 1967. Biology of *Cucumaria elongata*. J. Mar. Biol. Assoc. U.K. 47:129-143.

Giudice, G. 1973. Developmental biology of the sea urchin. Harcourt Brace Jovanovich, New York.

Haugh, B. N., and B. M. Bell. 1980. Fossilized viscera in primitive echinoderms. Science 209:653-657.

Houk, M. S., and R. T. Hinegardner. 1980. The formation and early differentiation of sea urchin gonads. Biol. Bull. 159:280-294.

Howarth, R. W. 1979. Coral reef growth in the Galapagos: limitation by sea urchins. Science 203:47-50.

Keller, B. D. 1983. Coexistence of esea urchins in seagrass meadows: an experimental analysis of competition and predation. Ecology 64:1581-1598.

Lawrence, J. M., and P. K. Dehn. 1979. Biological characteristics of *Luidia clathrata* (Echinodermata: Asteroidea) from Tampa bay and the shallow waters of the Gulf of Mexico. Fl. Sci. 42:9-13.

Lawrence, J. M. 1987. Functional biology of Echinoderms. Johns Hopkins Univ. Press, Baltimore, MD.

Lovtrup, S. 1975. Validity of the Protostomia-Deuterostomia theory. Syst. Zool. 24:96-108.

Macurda. D. B., Jr. 1978. These reef animals blossom at night. Smithsonian 9:86-88.

McClintock, J. B. 1983. Escape response of *Argopecten irradiand* (Mollusca; Bivalvai) to *Luidia clathrata* and *Echinaster* sp. (Echinodermata: Asteroidea). Fl. Sci. 46:95-100.

Nichols, D. 1960. The histology and activities of the tube-feet of *Antedon bifida*. Q. J. Microsc. Sci. 101:105-117.

Nichols, D. 1969. Echinodermata. 4th ed. Hutchinson University Library, London.

Pentreath, R. J. 1970. Feeding mechanism and the functional morphology of podia and spines in some New Zealand ophiuroids. J. Zool. 161:395-429.

Seilacher, a. 1979. Constructional morphology of sand dollars. Paleobiology 5:191-221.

Swan, E. F. 1961. Seasonal evisceration in the sea cucumber *Parastichopus californicus*. Science 133:1078-1079.

Vedas, R. L. 1977. Preferential feeding: an optimization strategy in sea urchins. Ecol. Monogr. 47:337-371.

Watts, S. A., R. E. Scheibling, A. G. Marsh, and J. B. McClintock. 1983. Induction of aberrant ray numbers in *Echinaster* sp. (Echinodermata: Asteroidea) by high salinity. Fl. Sci. 46:125-128.

Wilkie, I. C. 1978. Arm autonomy in brittle stars. J. Zool. 186:311-330.

EXERCISE 23

Phylum Hemichordata

Hemichordata (he-me-CHOR-da-ta; G., *hemi*, half + G., *chorda*, string) contains about 85 species of enterocoelous, burrowing or sessile marine animals. Their bodies are divisible into an anterior **proboscis**, middle **collar**, and a long posterior **trunk** (Fig. 23.1). **Pharyngeal gill slits** for gas exchange occur in most hemichordates. Other features include a complete digestive system, open circulatory system, nervous system combining invertebrate and vertebrate features, and separate sexes. Two classes of this widely distributed phylum are represented by acorn or tongue worms and pterobranchs.

Pterobranchia (ter-o-BRAN-chi-a; G., *pteron*, wing + G., *branchia*, gills) consists of fewer than 15, small (1 to 5 mm) species. Most of these uncommon, deep-water invertebrates live in secreted tubes forming colonies or aggregations. Extending dorsally from the collar region are one or more arms bearing several food-gathering, hollow, ciliated tentacles. Because of the rarity, unavailability, and small size of pterobranchs, they will not be studied in this exercise.

There are approximately 70 species of acorn worms in the class Enteropneusta (en-TER-op-NEUS-ta; G., *enteron*, an intestine + G., *pnein*, to breathe). These slow-moving worms range mostly from 10 to 45 cm in length. They are found in shallow waters of intertidal and littoral zones beneath shells and rocks, in seaweed, or burrows dug in mud or sand. The burrows are generally U-shaped and mucus-lined.

Enteropneusts are either suspension feeders or deposit feeders. Those of the former type trap minute food particles in mucus on the proboscis and move them to the mouth by ciliary action. Deposit feeders extract organic materials from vast amounts of consumed sand and mud. Undigested remains are deposited as coiled castings at one end of the burrow.

Classification

1. **Class Enteropneusta.** Acorn or tongue worms; 70 species. Solitary inhabitants of shallow waters; numerous gill slits present. Examples are *Saccoglossus* and *Balanoglossus*.
2. **Class Pterobranchia**. Pterobranchs; 15 species. Most are colonial or live in aggregations. These largely deep-water inhabitants live in secreted tubes and have one or more pairs of tentacular arms arising from the dorsal area of the collar. One pair of gills present or lacking. Pterobranchs may be more closely related to phoronids (phylum Phoronida) and brachiopods (phylum Brachiopoda) than to enteropneusts. Graptolites, an enigmatic group of fossils, may be pterobranchs (Armstrong et al. 1984). Examples are *Rhabdopleura* and *Cephalodiscus*.

Observational Procedure

Place in a small dish containing water an acorn worm such as *Saccoglossus kowalevskii* of the North American and European Atlantic coasts.

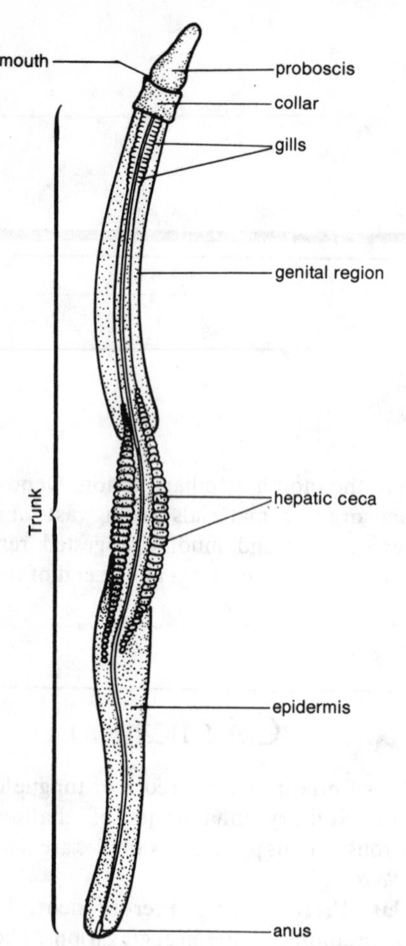

Figure 23.1. External view of a hemichordate.

Examine it under the dissecting microscope. Note the vermiform, soft body that is covered by a ciliated epidermis and richly supplied with mucous glands. Measure the length of your specimen. Locate the three body regions: proboscis, collar, and trunk (Fig. 23.1). Each region corresponds to the body divisions of a typical deuterostome (e.g., **protosome**, **mesosome**, and **metasome**). Internally, each region has its own coelomic cavity: a single protocoel of the proboscis, paired mesocoels of the collar, and paired metacoels of the trunk.

Examine the conical-shaped proboscis used for burrowing in the sand. Locate the narrow **proboscis stalk** that is attached to the proboscis and inner surface of the dorsal side of the bandlike collar (Fig. 23.2). A small proboscis pore on the stalk allows water to enter and leave the protocoel during burrowing. Arising from the dorsal area of the collar and extending into the proboscis is a narrow **buccal diverticulum** formerly thought to be a **notochord** (Fig. 23.2).

On the anterior margin of the ventral side of the collar and below the proboscis stalk is the mouth (Fig. 23.2). Plankton and other food materials making contact with the proboscis are caught in mucus and carried to the mouth by ciliary action. From the mouth food goes to a short buccal cavity of the collar and then into the ventral portion of the pharynx. Posterior to the pharynx is a muscular esophagus that leads to a long intestine where digestion and absorption occur. Undigested materials are eliminated by the terminal anus. Attempt to locate the anus in your specimen.

Most of the animal's body consists of the trunk, which has a midventral and middorsal **longitudinal ridge** (Fig. 23.1). The trunk can be divided into three regions: (1) an anterior **branchial** region, containing numerous U-shaped gill slits; (2) a middle **genital** region, containing internal gonads; and (3) a posterior **abdominal** region, containing intestine, paired and lateral hepatic cecal pouches of the intestine, and anus. Locate the **gill pores** in paired rows along the dorsolateral area of the branchial trunk region. Each pore leads to an internal branchial sac that communicates with the slit.

The **open circulatory system** of acorn worms consists of a middorsal contractile vessel that transports colorless blood anteriorly and a midventral contractile vessel that carries blood posteriorly (Fig. 23.2). A pulsating heart vesicle occurs in the proboscis. Near the vesicle and at the base of the proboscis is a network of vessels, the **glomerulus**, which may have excretory functions.

A **nerve plexus** beneath the surface epithelium is locally concentrated in longitudinal cords. The proboscis and trunk contain middorsal and midventral cords that are connected to a nerve ring at the junction of the collar and trunk (Fig. 23.2). The collar contains only a hollow, middorsal cord that may be homologous to the dorsal tubular nerve cord of chordates.

Hemichordates are dioecious. Gametes from the gonads in the trunk region escape via a pore into the water. Fertilization is external and development is either direct or indirect. Many species produce a free-swimming, ciliated **tornaria larva** that is similar to the bipinnaria larva of sea stars (Echinodermata: Asteroidea). *Saccoglossus kowalevskii* lacks a larva; the eggs hatch as young worms.

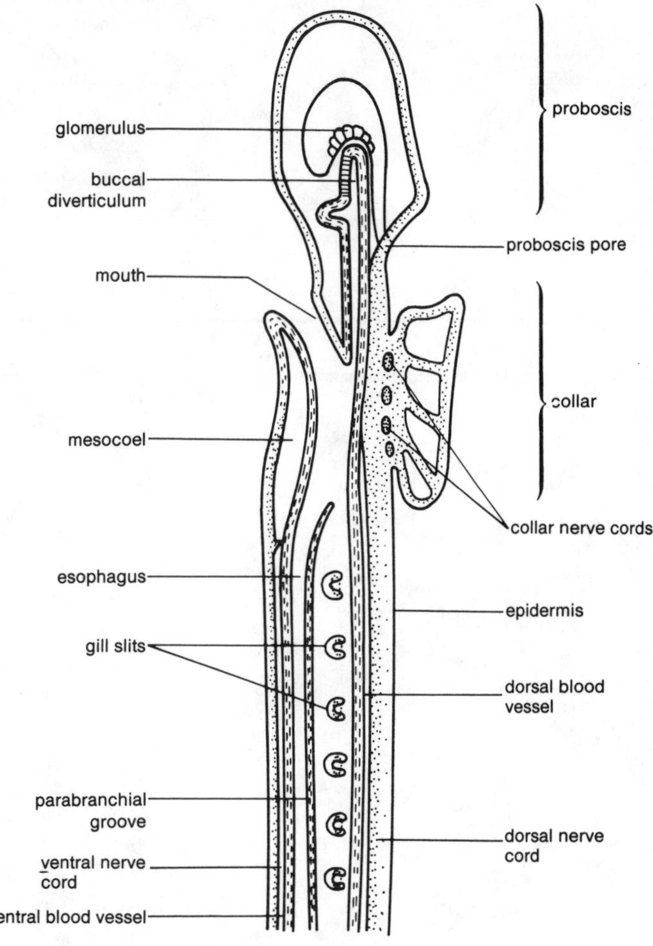

Figure 23.2. Internal sagittal view of a hemichordate.

Supplemental Reading

Armstrong, W. G., P. N. Dilly, and A. Urbanek. 1984. Collagen in the pterobranch coenecium and the problem of graptolite affinities. Lethaia 17:145-152.

Barrington, E. 1965. The biology of Hemichordata and Protochordata. W.H. Freeman, San Francisco.

Hadfield, M. G. 1975. Hemichordata. *In*: A. C. Geise, and J. S. Pearse (eds.). Reproduction of marine invertebrates. Vol. II. Academic Press, New York, pp. 185-240.

Knight-Jones, W. E. 1952. On the nervous system of *Saccoglossus cambrensis*. Philos. Trans. R. Soc. Lond. B 236:315-354.

Neilsen, C. 1985. Animal phylogeny in the light of the trochaea theory. Biol. Linn. Soc. 25:243-299.

Stebbing, A. R. D., and P. N. Dilly. 1972. Some observations on living *Rhabdopleura compacta*. J. Mar. Biol. Assoc. U.K. 52:443-448.

EXERCISE 24

Phylum Chaetognatha

Chaetognatha (ke-TO-na-tha; G., *chaeto*, bristle + *gnathos*, jaw) is a small phylum of about 70 species of unsegmented, bilaterally symmetrical deuterostomes. Chaetognaths are commonly called arrowworms because they resemble short arrows or darts. They range in length from a few millimeters to more than 10 cm, most are less than 4 cm. The arrowworm body consists of three regions: (1) a rounded head armed with grasping spines, (2) an elongate trunk with one to two pairs of lateral fins, and (3) a postanal tail with spatulate tail fin (Fig. 24.1). The epidermis is covered by a thin transparent cuticle. Sensory hair-fans, important in detecting vibrations produced by potential prey (e.g., copepods), are arranged in a pattern of several rows. In the neck is the **ciliary loop**, consisting of two concentric rings of ciliated epidermal cells, whose function is unknown. Internally, arrowworms possess several coelomic compartments. A single, small coelomic head compartment is separated from larger, paired trunk compartments by a septum in the neck. At the junction of the trunk and tail, coelomic compartments are separated by the trunk-tail septum.

Arrowworms are exclusively marine. Found in all oceans, they are especially common in tropical and subtropical waters. Most arrowworms are planktonic, except for species of the genus *Spadella*, which inhabit the benthos of shallow waters.

Classification

A single class, Sagittoida, consists of two orders based on the presence or absence of ventral transverse muscles.

Observational Procedure

Observe several specimens of an arrowworm, such as *Sagitta*, using a dissecting microscope. Identify the three body regions: **head**, **trunk**, and **tail** (Fig. 24.1). In the head region, locate the large, chitinous spines used to seize prey. Anterior to the spines are two rows of small teeth that aid in prey capture. The head and spines may be covered by a protective fold of the body wall called the **hood**; the hood may be folded back. Locate the vestibule on the ventral side of the head. In this small chamber prey are held before being swallowed. Occasionally, specimens are found with prey still present in the vestibule. Note the eyes on the dorsal side of the head.

While illuminating your specimens from above, find the two pairs of lateral fins and the tail fin, which internally have raylike supports. The fins are delicate structures and are often distorted or lost in preserved specimens. Some species (e.g., *Spadella*) have one pair of lateral fins. The sensory hair-fans are difficult to see, but sometimes may be seen in stained specimens premanently mounted on slides. Note the longitudinal muscle bands that run the length of the animal.

The gut is a simple straight tube. At the anterior end is a muscular pharynx that leads to the long, straight intestine. An anus is located on the ventral side in the region of the trunk-tail septum. Find the paired ovaries in the posterior part of the trunk. Carefully manipulate your specimen so that you can see the female **gonopores** just before the level of the trunk-tail septum. In the tail, locate the paired testes and seminal vesicles. The latter are embedded laterally in the body wall.

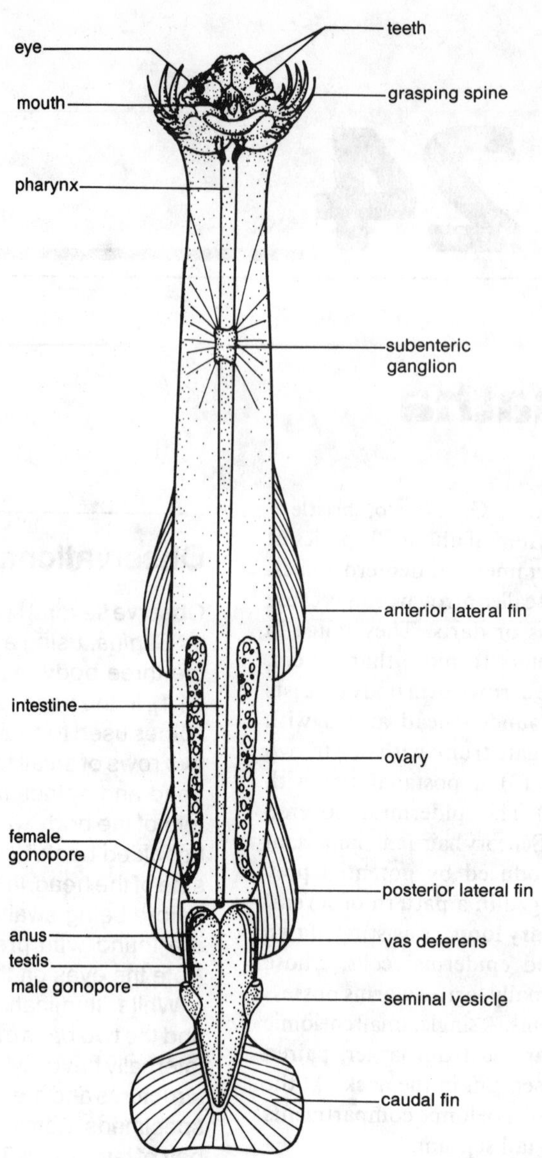

Figure 24.1. General anatomy of the chaetognath, *Sagitta*.

Supplemental Reading

Alvariño, A. 1965. Chaetognaths. Oceanogr. Mar. Biol. Annu. Rev. 3:115-194.

Feigenbaum, D. 1978. Hair-fan patterns in the Chaetognatha. Can. J. Zool. 56:536-546.

Feigenbaum, D. 1982. Feeding by the chaetognath, *Sagitta elegans*, at low temperatures in Vineyard Sound, Massachusetts. Limnol. Oceanogr. 27:699-706.

Feigenbaum, D., and M. R. Reeve. 1977. Prey detection in the Chaetognatha: response to a vibrating probe and experimental determination of attack distance in large aquaria. Limnol. Oceanogr. 22:1052-1058.

Ghirardelli, E. 1968. Some aspects of the biology of the chaetognaths. Adv. Mar. Biol. 6:271-375.

Newbury, T. K. 1972. Vibration perception by chaetognaths. Nature (Lond.) 236:459-460.

Parry, D. A. 1944. Structure and function of the gut in *Spadella* and *Sagitta*. J. Mar. Biol. Assoc. U.K. 26:16-36.

Reeve, M. R. 1970. Complete cycle of development of a pelagic chaetognath in culture. Nature (Lond.) 227:381.

Welsch, U., and V. Storch. 1982. Fine structure of the coelomic epithelium of *Sagitta elegans*. Zoomorphology 100:217-222.

PHYLUM CHORDATA

Nearly 50,000 extant species comprise the phylum Chordata, but most of them are members of the subphylum Vertebrata (VER-ta-BRA-ta; L., *vertebratus*, jointed or articulated). Two additional subphyla, Urochordata (U-ro-kor-DA-ta; G., *uro*, tail + G., *chordata*, string) and Cephalochordata (SEF-a-lo-kor-DA-ta; G., *cephala*, head), are of some interest to both invertebrate and vertebrate zoologists as members may be similar to the ancestral vertebrates. All chordates possess paired **pharyngeal clefts**, a dorsal **tubular nerve cord**, a supportive **notochord**, and a **postanal tail** during some part of their lives.

Urochordates, commonly called tunicates, are nonsegmented. Only their larvae have a nerve cord and notochord. Adults of the urochordate class Ascidiacea (sea squirts) are sessile, while those of the classes Larvacea and Thaliacea (salps) are planktonic. Cephalochordates, commonly known as lancelets, possess an elongate, fishlike, segmented body. The nerve cord and notochord are present through the entire lives.

Since many textbooks of vertebrate zoology discuss the chordates, we will not treat them in this manual. Your instructor will provide directions if the two nonvertebrate subphyla are to be studied.

GENERAL REFERENCES

There is no single vehicle where invertebrate zoologists publish their work. Below are some of the important comprehensive pieces of literature on invertebrates. After each exercise, we have listed additional literature that deals specifically with the phylum under consideration (Supplemental Reading). For the most part, the literature cited in the Supplemental Reading sections avoids duplication of the literature cited by major textbooks of invertebrate zoology.

Alexander, R. McN. 1979. The invertebrates. Cambridge University Press, Cambridge.

Anderson, J. N., A. D. A. Rayner, and D. W. H. Walton. 1984. Invertebrate-microbial interactions. Cambridge University Press, New York.

Ax, P. 1987. The phylogenetic system. John Wiley & Sons, New York.

Barnes, R. D. 1986. Invertebrate zoology. 5th ed. W. B. Saunders, New York.

Barnes, R. S. K. (ed.). 1984. A synoptic classification of living organisms. Sinauer Associates, Sunderland, MA.

Barnes, R. S. K., P. Calow, and P. J. W. Olive. 1988. The invertebrates: a new synthesis. Blackwell Scientific, Boston.

Barrington, E. J. W. 1979. Invertebrate structure and function. 2nd ed. Wiley, New York.

Barth, R. H., and R. E. Broshears. 1982. The invertebrate world. W.B. Saunders, New York.

Beklwemishev, W. N. 1969. Principles of comparative anatomy of invertebrates. Vol. 1. Promorphology. Vol. 2. Organology. University of Chicago Press, Chicago.

Bereiter-Hahn, J., A. G. Matoltsy, and K. S. Richards (eds.). 1984. Biology of the integument. Vol. 1 Invertebrates. Springer-Verlag, New York.

Boardman, R. S., A. H. Chechmam, and A. J. Rowell. 1986. Fossil invertebrates. Blackwell Scientific, Oxford.

Brown, F. A., Jr. (ed.). 1950. Selected invertebrate types. Wiley, New York.

Cable, R. M. 1977. An illustrated laboratory manual of parasitology. 5th ed. Burgess, Minneapolis, MN.

Calow, P. 1981. Invertebrate biology. Halsted Press, New York.

Casanova, R., and R. P. Ratkevich. 1981. An illustrated guide to fossil collecting. Naturegraph, Happy Camp, CA.

Cheng, T. C. 1986. General parasitology. 2nd ed. Academic Press, New York.

Cloud, P., and M. F. Glaessner. 1982. The Ediacarian period and system: Metazoa inherit the Earth. Science 218:783-792.

Conway Morris, S., J.D. George, R. Gibson, and H. M. Platt (eds.). 1985. The original and relationships of the lower invertebrates. Systematics Association special volume 28. Clarendon Press, Oxford.

Cox, F. E. G. 1982. Modern parasitology. Blackwell Scientific Oxford.

Crawford, C. S. 1981. Biology of desert invertebrates. Springer-Verlag, New York.

Dales, R. P. 1981. Practical invertebrate zoology. 2nd ed. Wiley, New York.

Delly, J. G. 1985. Narcosis and preservation of freshwater animals. Am. Lab. (April):31-40.

Diehl, F. A., J. B. Feeley, and D. G. Gibson. 1971. Experiments using marine animals. Aquarium Systems, Inc. Eastlake, OH.

Edmondson, W. T. (ed.). 1959. Fresh-water biology. 2nd ed. Wiley, New York.

Field, K. G., G. J. Olsen, D. J. Lane, S. J. Giovannoni, M. T. Ghislin, E. C. Raff, N. R. Pace, and R. A. Raff. 1988. Molecular phylogeny of the animal kingdom. Science 239:748-753.

Freeman, W. H., and B. Bracegirdle. 1982. An atlas of invertebrate structure. Heinemann, London.

Giese, A. C., and J. S. Pearse, 1974-1979. Reproduction of marine invertebrates. Vols. 1-5. Academic Press, New York.

Glaessner, M. F. 1984. The dawn of animal life. Cambridge University Press, New York.

Gosner, K. L. 1971. Guide to identification of marine and estuarine invertebrates. Wiley, New York.

Hickman, C. P. 1973. Biology of the invertebrates. C. V. Mosby, St. Louis, MO.

House, M. R. (ed.). 1979. The origin of major invertebrate groups. Systematics Association special Vol. 12. Academic Press. New York.

Hyman, L. H. 1940-1967. The invertebrates. Vols. I-VI. McGraw-Hill, New York.

Kaestner, A. 1967-1970. Invertebrate zoology. Vols. I-III. Wiley, New York.

Kerfoot, W. C. 1980. Evolution and ecology of zooplankton communities. University Press of New England, Hanover, NH.

Laverack, M. S., and J. Dando. 1987. Lecture notes on invertebrate zoology. 3rd ed. Blackwell Scientific, Oxford.

Lehmann. U., and G. Hillmer. 1983. Fossil invertebrates. Cambridge University Press, Cambridge.

Lincoln, R. J., and J. G. Sheals. 1979. Invertebrate animals. Collection and preservation. Cambridge University Press, Cambridge.

Lutz, P. E. 1985. Invertebrate zoology. Addision-Wesley, Reading , MA.

MacGinitie, G. E., and N. MacGinitie. 1968. Natural history of marine animals. 2nd ed. McGraw-Hill, New York.

MacInnis, A. J., and M. Voge. 1970. Experiments and techniques in parasitology. W.H.Freeman, San Francisco.

Marquardt, W. C., and R. S. Demaree, Jr. 1985. Parasitology. Macmillan, New York.

Marshall, A. J., and W. D. Williams (eds.). 1972. Textbook of zoology. Invertebrates . American Elsevier, New York.

Meglitsch, P. A. 1972. Invertebrate zoology. 2nd ed. Oxford University Press, Oxford.

Meyer, M. C., and O. W. Olsen. 1980. Essentials of parasitology. Wm. C. Brown, Dubuque, IA.

Moore, R. C., et al. (eds.). 1954-continued. Treatise on invertebrate paleontology. 32 parts, Geological Society of America and University of Kansas Press, Lawrence, KS.

Murray, M. 1967. Hunting for fossils. Collier Books, Macmillan, New York.

Noble, E. R., and G. A. Noble. 1982. Parasitology: the biology of animal parasites. 5th ed. Lea & Febiger, Philadelphia.

Parker, S. P. 1982. Synopsis and classification of living organisms. Vols. 1 and 2. McGraw-Hill, New York.

Pearse, V., J. Pearse, M. Buschsbaum, and R. Buschsbaum. 1987. Living invertebrates. Boxwood Press, Pacific Groove, CA.

Pechenik, J. A. 1985. Biology of the invertebrates. Prindle, Weber, & Schmidt, Boston.

Pennak, R. W. 1978. Fresh-water invertebrates of the United States. 2nd ed. Wiley, New York.

Pierce, S. K., T. K. Maugel, and L. Reid. 1987. Illustrated invertebrate anatomy. Oxford University Press. New York.

Ricketts, E. F., J. Calvin, J. W. Hedgpeth, and D. W. Phillips. 1986. Between Pacific tides. 5th ed. Stanford University Press, Stanford, CA.

Russell-Hunter, W.D. 1979. A life of invertebrates. Macmillan, New York.

Schmidt, G. S., and L. S. Roberts. 1989. Foundations of parasitology. 4th ed. C. V. Mosby, St. Louis, MO.

Shaw, A. C., S. K. Lazell, and G. N. Foster. 1974. Photomicrographs of invertebrates. Fletcher and Sons, Norwich, England.

Sherman, I. W., and V. G. Sherman. 1976. The invertebrates: function and form. Macmillan, New York.

Tasch, P. 1980. Paleobiology of the invertebrates. Wiley, New York.

Tomlinson, J. T. 1976. Invertebrate behavior. San Francisco State University, San Francisco.

Trager, W. 1986. Living together. Plenum Press, New York.

Valentine, J. W. (ed.). 1985. Phanerozoic diversity patterns. Princeton University Press, Princeton, NJ.

Vermeij, G. J. 1978. Biogeography and adaptation. Harvard University Press, Cambridge, MA.

Vernberg, F. J., and W. B. Vernberg (eds.). 1981. Functional adaptations of marine organisms. Academic Press, New York.

Vogel, S. 1981. Life in moving fluids. Prindle, Weber, & Schmidt, Boston.

Welsh, J.H., R. I. Smith, and A. E. Kammer. 1968. Laboratory exercises in invertebrate physiology. Burgess, Minneapolis, MN.

Whittington, H. B. 1985. The Burgess Shale. Yale University Press, New Haven, CT.